Remote Sensing and GIS GPS based Resource Management

Dr. T.S. Chouhan
Department of Geography
University of Jaipur
Jaipur - 342 004 (Raj.)

SCIENTIFIC PUBLISHERS

Published by:

SCIENTIFIC PUBLISHERS (INDIA)
5A, New Pali Road, P.O. Box 91
Jodhpur 342 001, India
E-mail: info@scientificpub.com
Website: www.scientificpub.com

Print: 2026

ISBN: 978-93-86102-86-7
eISBN: 978-93-87869-87-5

Printed in India

PREFACE

India as a developing country, largely depend on the utilisation of natural resources for their immediate advancement. Advancement involves a change in the existing use of resources to a higher level or more economical methods of production or the more complete use of resources that are at present partially or completely neglected. Every region within our country has its own array of natural resources, suited to different forms of use and offering different opportunities of production. These resources set the pattern for primary production and to a degree, secondary production. We need to make decisions which natural resources can be easily and effectively developed, which are most likely to contribute our needs in both short and long term on a sustained basis and where and when other resources can be most gainfully employed. Efforts have been made in this book to prepare a GIS and GPS based database of resources of Rajasthan, being a prerequisite for sustainable developmental planning. Important among these resources are soil, surface and ground water, vegetation, minerals, climatic conditions, land-use and human resources. In addition to these resources a few other physical parameters are landforms, slope and physical infrastructures. The integration of resource data in the area of geology, geomorphology, soil, agriculture, forestry, hydro-geology, land-use, land-cover, environment, ecology, etc. can assist in identification of homogeneous land units having unique combination of characteristics and evaluated for site specific and suitable land-use practices in respect of scientific land utilization. Geographic Information System (GIS) today is an indispensable tool in the planning for tomorrow's requirements. Basic fundamentals and advance analysis in Geographical Information system have been discussed in length in this present book.The Global Positioning System is a burgeoning technology, which provides unequalled accuracy and flexibility of positioning having tremendous amount of applications in GIS data collection, surveying, and mapping. Basic fundamentals, components and operational methodology of the GPS system have been discussed in this volume. Application of GPS and its future in developing regions have also been covered inside the book.

The Research book addresses one fundamental developmental problem of Rajasthan i.e., the problem of easy access to a wide variety of data relating to natural resources, socio-economic factors and physical infrastructure in the state. These data are necessary to understand the extent of development in different parts of the state so that improvements can be planned on a priority basis. Developmental strategies can then be outlined and programmes implemented in specific regions. Process adopted for building the Rajasthan Resources Assessment Database has been discussed in book. An attempt has also been made towards generation of a 3D Digital Terrain Model using GIS software for the state of Rajasthan in the research book. Geological features, climatological characteristics, forest and flora and fauna of the Rajasthan have also been discussed in detail in this book.

Mineral resources are the backbone for the industrial and socio-economic development of Rajasthan state as it ranks seconds after Jharkhand in mineral deposits in India. The research book detailed information about the mineral resources were collected, analysed and then represented and also briefs about the non-conventional sources of energy in the book. The importance of livestock in agriculture is of great significance in state. The entire field operation from ploughing to the harvesting of crops is carried out by the animals mainly bullock and camel. Animal husbandry occupies an important role in the state economy. Since more than half of the area is arid or semi-arid and is not suitable for traditional crop farming.

A number of Remote Sensing and GIS, GPS based rural and urban case studies have been successfully accomplished. These studies have been focused on the burning issues in the context of Rajasthan. This book will be very useful for anyone, researchers, teachers, inventor, administrators, social workers or politicians with having interest in the state's resources. Remote Sensing and GIS, GPS data user community could also get the benefit from the Rajasthan state resource database.

T.S. Chouhan

ACKNOWLEDGEMENT

My sincere thanks to the Officers of State and Center Government Departments, CARE India, UNICEF, Jaipur, ICHAP, Jaipur, Malviya National Institute of Technology, Jaipur, Central Ground Water Board, Jaipur, State Ground Water Department, Jaipur, Central Arid Zone Research Institute, Jodhpur, Non Governmental Organisations working in Bikaner, Udaipur and Jaipur district for providing their support and literature/ data at various stages of preparation of book.

The author are appreciative of the sincere efforts by Mr. Tanay Sharma and his production team, including the reviewers, editors and the graphics designers, of Scientific Publishers (India), Jodhpur. The author like to extend their thanks to all the team for their commendable efforts, as a result of which the book has been brought out speedily in an excellent published form.

T.S. Chouhan

LIST OF FIGURES

Fig. No.	Description	Page
1.1.	Location Map (Source: Census of India)	2
1.2.	Administrative Tehsil Map (Source: Census of India)	3
1.3.	Rajasthan Image Map (Source: NRSA)	4
2.1.	Integrated Elements of GIS	16
2.2.	The component parts of GIS	16
2.3.	Major Hardware Components of a Geographical Information System	17
2.4.	Classical three-tier architecture of a GIS software system.	18
2.5.	The Main Software Components of a GIS	18
2.6.	Data Collection and Input	19
2.7.	The Components of the Geographic Data Base	19
2.8.	Data Output	20
2.9.	Data Transformation	20
2.10.	GIS as a Management Tool	21
2.11.	Overlay of Thematic Maps in GIS	24
2.12.	Various Functions of GIS	27
2.13.	Spatial and temporal characteristics of commonly used remote sensing system and their sensors	33
2.14.	Applications of GIS	40
3.1.	Triangulation Method	42
3.2.	Components of GPS	43
3.3.	Control Segment	44
3.4.	Space Segment	44
3.5.	Concept of Geopositioning	45
3.6.	Concept of Differential Geopositioning	46
3.7.	Position Dilution of Precision (PDOP).	50
4.1.	Fundamentals of e-governance	57
4.2.	Impacts of e-governance	60
4.3.	Fundamentals of Geo-informatics	62
4.4.	Flow Diagram of Land Management System (LMS)	65
4.5.	SuiCorridor Sites using Geo-informatics	67

Fig. No.	Description	Page
6.1.	Landforms	80
6.2.	Physiography and Drainage	81
6.3.	Relative Relief	82
6.4.	Digital Terrain Model of Rajasthan	83
6.5.	X-Sectional Profile of Rajasthan	84
6.6.	Geology	87
6.7.	Climatic Zones	112
6.8.	Mean Annual Rainfall	113
6.9.	Mean seasonal Rainfall (Source: IMD)	120
6.10.	Mean Annual Water surplus and Deficit (Source: IMD)	121
6.11.	Annual Average Temperature (Source: IMD)	122
6.12.	Mean Annual Potential Evapotranspiration (Source: IMD and Dr. T.S. Chouhan, Agricultural Geography)	125
6.13.	Forest Type (Source: Forest Survey of India and Forest Department of Rajasthan)	130
7.1.	Soils	140
7.2.	Landuse Pattern	153
7.3.	Depth to Water Level	157
7.4.	Water Level Fluctuation	158
7.5.	Groundwater Potential Zones	160
7.6.	Gross and Net Irrigation Areas	162
8.1.	Mining Area Locations Map	172
8.2.	Mineral Types Map	173
9.1.	Jurisdiction changes (Source: Census of India)	196
9.2.	Population Distribution (Source Census of India)	198
9.3.	Decadal Growth Rate (Source : Census of India)	201
9.4.	Sex Ratio (Source: Census of India)	204
9.5.	Population Age 0 to 6 years (Source: Census of India)	206
9.6.	Literacy (Source: Census of India)	209
9.7.	Livestock Population (Source: Board of Revenue, Ajmer)	213
9.8.	Cattle, Buffaloes, Sheep and Goat Population (Source: Board of Revenue, Ajmer)	214
9.9.	Pack Animal and Camel distribution (Source: Board of Revenue, Ajmer)	215
10.1.	Transport Network Map (Source: RSRTC)	218
10.2.	Post and Telegraph	222
11.1.1.	Study Area Location map	226
11.1.2.	Village Map	227

Fig. No.	Description	Page
11.1.3.	Depth to water.	238
11.1.4.	Slope	245
11.1.5.	Erosion	246
11.1.6.	Geo units	247
11.1.7.	Geo divisions	248
11.1.8.	Land type	249
11.1.9.	Soils	251
11.1.10.	Level of economic development	262
11.2.1.	Project Area Location Map (Source: Census of India)	266
11.2.2.	Study village Location Map (Source: Census of India)	267
11.2.3.	Transport Network Map (Source : Census of India)	271
11.2.4.	Groundwater Potential Map (Source: GWD and CGWD)	274
11.2.5.	Groundwater Portability Map (Source: GWD and CGWD)	275
11.2.6.	Activity Location Map (Source: Census of India)	277
11.2.7.	Increased in storage capacity of Talai's (Source: TALAI's)	280
11.2.8.	Cost v/s Direct Benefit (Source: Talai)	281
11.2.9.	Increase in General Command Area (Source: CAD)	282
11.2.10.	Cost v/s Benefit (Canal Works) (Source: CAD)	283
11.2.11.	Project Impact on Ground water Scenario (Source: CAD, GWD &CGWD)	284
11.3.1.	Administrative setup of the districts (Source: GWD and CGWD)	289
11.3.2.	Geology (Source: GWD and CGWD)	291
11.3.3.	Groundwater Potential zones (Source: GWD and CGWD)	292
11.3.4.	Groundwater Yield (Source: GWD and CGWD)	298
11.3.5.	Groundwater depth (Source: GWD and CGWD)	299
11.3.6.	Groundwater Fluctuation (Source: GWD and CGWD)	300
11.3.7.	Irrigation Sources (Source: GWD and CGWD)	304
11.3.8-8A.	Hydrogeological and Water Level Approach (Source: GWD and CGWD)	306
11.3.9.	Groundwater Potability Approach (Source: GWD and CGWD)	307
11.3.10.	Proposed Area for Groundwater Harvesting Works (Source: GWD & CGWD)	308
11.3.11.	Proposed Area for Groundwater Harvesting Works (Source: GWD and CGWD)	309
11.3.12.	Proposed area for Groundwater Harvesting Works (Source: GWD and CGWD)	310
11.3.13.	Proposed Area for Groundwater harvesting Works (Source: GWD and CGWD)	311
11.3.14.	Proposed area for Groundwater Availability (Source: GWD and CGWD)	312
11.3.15.	Proposed area for groundwater availability (Source: GWD and CGWD)	313

Fig. No.	Description	Page
11.3.16.	Proposed area for groundwater availability (Source: GWD and CGWD)	314
11.3.17.	Proposed area for groundwater availability (Source: GWD and CGWD)	316
11.4.1.	Vulnerability Index by Methods of Difference (Source: Table)	324
11.4.2.	Vulnerability Index by Methods of Standard Deviation (Source: Table)	326
11.4.3(A).	District map of Rajasthan (Criteria-2 first 30 parameters) (Source: Table)	327
11.4.3(B).	District Map of Rajasthan (Criteria -2, Remaining Parameters) (Source : Table)	328
11.4.4(A).	Vulnerability Index specific to Demographic Arabibles by Method of Different (Source Table)	329
11.4.4(B)	Vulnerability Index specific to Demographic Arabibles by Method of Standard Deviation (Source Table)	330
11.4.5(A).	Vulnerability Index specific to Economic Development Variables by Method of Different (Source Table)	331
11.4.5(B).	Vulnerability Index specific to Economic Development Variables by Method of Standard Deviation (Source Table)	331
11.4.6(A).	Vulnerability Index specific to Health Variables by Method of Different (Source Table)	333
11.4.6(B).	Vulnerability Index specific to Health Variables by Method of Standard Deviation (Source Table)	333
11.4.7(A).	Vulnerability Index specific to Social Development Variables by Method of Difference (Source Table)	335
11.4.7(B).	Vulnerability Index specific to Social Development Variables by Method of Standard Deviation (Source Table)	335
11.4.8(A).	Vulnerability Index specific to Socio-Cultural & Other Variables by Method of Difference (Source Table)	337
11.4.8(B).	Vulnerability Index specific to Socio-Cultural & Other Variables by Method of Standard Deviation (Source Table)	337
11.4.9.	Vulnerability Index specific to Presence of Rural Traditional Sex Workers (Source Table)	339
11.4.10.	Vulnerability Index specific to Length of National High-way passing through a District (Source Table)	341
11.4.11.	Vulnerability Index specific to Reported Cases of AIDS	342
11.4.12.	Vulnerability Index specific to Number if Inmates in Jail	343
11.4.13.	Vulnerability Index specific to Reported Cases of TB	344
11.5.1.	Rajasthan Mining Area Location Map	
11.5.2.	Rajasthan Mines Number Map	351
11.5.3.	Rajasthan Mineral Type Map	354
11.5.4.	Rajasthan Mine Workers Map	354
11.5.5.	Rajasthan Revenue Generation Map	357

Fig. No.	Description	Page
11.5.6.	Social Map of HRBGs and Mines in Rajasthan	358
11.5.7.	Rajasthan Composite Map	361
11.6.1.	Location Map of Study Blocks	369
11.6.2.	Location Map of Study Villages	369
11.6.3.	Rainfall Distribution	371
11.6.4.	Water Balance	372
11.6.5.	Pre and Post Monsoon Water Level Depth – Kotra Block	374
11.6.6.	Pre and Post Monsoon Water Level Depth – Girwa Block	374
11.6.7.	Pre and Post Monsoon Water Leve l Depth – Salumbar Block	375
11.6.8.	Water Level Fluctuation – Kotra Block	375
11.6.9.	Water Level Fluctuation – Girwa Block	376
11.6.10.	Water Level Fluctuation – Salumbar Block	376
11.6.11.	TDS Distribution	378
11.6.12.	Nitrate Distribution	379
11.6.13.	Total Hardness Distribution	380
11.6.14.	Chloride Distribution	380
11.6.15.	pH Distribution	381
11.6.16.	Population Distribution and Irrigation Methods	383
11.6.17.	Cropping Pattern – Rabi and Kharif Seasons	384
11.6.18.	Cost Benefit Analysis - Study Joyra Naka	387
11.6.19.	Cost Benefit Analysis - Joyra Naka	389
11.6.20.	Cost Benefit Analysis - Study Site Hingawat Nal	390
11.6.21.	Cost Benefit Analysis - Hingawat Nal	392
11.6.22.	Cost Benefit Analysis - Study Site Hama Talav	393
11.6.23.	Cost Benefit Analysis - Hama Talav	395
11.6.24.	Cost Benefit Analysis - Study Site Khedia	396
11.6.25.	Cost Benefit Analysis - Khedia (Semliya)	398
11.6.26.	Cost Benefit Analysis - Study Site Sirwal	399
11.6.27.	Cost Benefit Analysis - Sirwal (Dang)	401
11.6.28.	Cost Benfit Analysis - Patoo Khera	402
11.6.29.	Cost Benefit Analysis - Charnot Land (Patu Khera)	403
11.6.30.		405
11.6.31.		406
11.6.32.	Comparative Graph of Optimum Direct Benefits	408
11.7.1.	Location of Study Area	415
11.7.2.	Ward map of Sunel town	416
11.7.3.	Water supply coverage	417

Fig. No.	Description	Page
11.7.4.	PSP/ WELL location map	419
11.7.5.	WS Distribution network map	420
11.7.6.	WS Problematic area map	420
11.7.7.	Drainage Network map	423
11.7.8.	Drainage problematic map	424
11.7.9.	Spot Level Map	425
11.7.10.	Contour map	426
11.7.11.	Garbage disposal Location map	428
11.7.12.	Solid Waste Management Points.	429

LIST OF TABLES

Table No.	Description	Page
1.1.	Source and Information	12
1.2.	Topo Sheet Scale and numbers for Rajasthan	13
2.1.	GIS Supporting Spatial Statistics	22
2.2.	Classification of geographic data for data collection purposes	28
3.1.	Accuracy in GPS	47
3.2.	GPS Frequencies	47
4.1.	Application areas for e-governance using Geo-informatics	63
6.1.	Mean Annual Rainfall	117
6.2.	Mean Rainy Days	118
6.3.	District wise Area of Forest According to Legal Status of Rajasthan (as on 31.3.1998) (In Sq. Km.)	128
6.4.	Forest Type in Rajasthan	131
6.5.	Number of Floral Species	135
6.6.	Floral Species	135
6.7.	List of Rare and Endangered Plants	137
7.1.	Physiography and Soil Association	141
7.2.	Soil Fertility Status	142
7.3.	Land Use Classification and Utilisation	149
7.4.	Land Use Pattern	150
7.5.	Change in Water Level	156
7.6	Irrigation Facilities in Rajasthan	161
8.1.	Category-Wise Recoverable Reserves of Minerals in Rajasthan (million tones)	186
8.2.	Mineral Production in Rajasthan (In Million tones)	187
9.1.	Details of Districts created after 1991 Census	197
9.2.	Rural-Urban Composition of Population in India and Rajasthan	199
9.3.	Percentage decadal variation of population from 1901 to 2001 by Rural-Urban Composition in Rajasthan	200
9.4a.	First five and last five districts of Rajasthan as per their rural decadal growth rate.	202

Table No.	Description	Page
9.4b.	First five and last five districts of Rajasthan as per their Urban decadal growth rate.	202
9.5.	Sex ratio in India and Rajasthan 1901-2001	203
9.6.	Sex ratio for Rural-Urban areas at state level for 1901-2001 (Rajasthan)	203
9.7	Highest and Lowest Sex Ratios in rural and urban areas of Rajasthan	205
9.8	Population in the age-group 0-6 in Rural and Urban areas by sex in 1991 and 2001	207
9.9	Literacy rates by sex at state level for rural and urban areas in 1991 and 2001	208
10.1.	Road Length in Rajasthan	219
10.2	Road Transport Route Nationalized	220
10.3	Villages Connected with Roads	221
10.4.	Post, Telegraph and Telephone Offices	224
11.1.1.	Number of Rainy days in Osian	230
11.1.2.	Rainfall Percentages	230
11.1.3.	Mean annual and extreme rainfall in Osian and its variability	231
11.1.4.	Mean Percentage Probabilities	231
11.1.5.	Percentage frequencies of weekly rainfall data (in mm) of Osian station.	232
11.1.6.	Highest Probability of Measurable Rainfall	233
11.1.7.	Depth to water below ground level (bgl)	239
11.1.8.	Water yielding capacities of different lithological formations.	241
11.1.9.	Main Feature of Occupation in Osian Tehsil	254
11.1.10.	Occupational set up in Osian tehsil	255
11.1.11.	Land utilization pattern (Area in Hectare)	255
11.1.12.	Occupational Structure in Study Villages (1991)	256
11.1.13.	Occupational structure in study villages	257
11.1.14.	Percentage of schedule caste and schedule tribes	258
11.1.15.	Literacy in Study Villages	258
11.1.16.	The Amenities available in Study Villages (Year 1991)	259
11.1.17.	Agrodemographic indicators affecting economic development – Structural indicators	260
11.1.18.	Agro-demographic indicators affecting economic development – Occupational indicators	261
11.1.19.	Agrodemographic indicators affecting economic development – Productivity indicators	261
11.1.20.	Levels of Economic Development	262
11.2.1.	Mean Annual Rainfall	272
11.2.2.	Mean Annual Evapo-transpiration in Study Area.	272

Table No.	Description	Page
11.2.3.	Type of Soil	272
11.2.4.	Depth to Ground Water	273
11.2.5.	Groundwater Depth Fluctuation	273
11.2.6.	Groundwater Yield	273
11.2.7.	Village Distribution	276
11.2.8.	Distribution of activities	276
11.3.1.	Tehsil-wise population	294
11.3.2.	Tehsil-wise population	295
11.3.3.	Tehsil-wise population	295
11.3.4.	Tehsil-wise population	296
11.3.5.	District wise Rainfall	296
11.3.6.	District wise Evapo-transpiration	296
11.3.7.	Groundwater Extraction (District wise)	301
11.4.1.	Ranking of Districts on the Basis of Vulnerability Index by Method of Difference	324
11.4.2.	Ranking of Districts on the Basis of Vulnerability Index by Method of Standard Deviation	325
11.4.3.	Ranking of Districts on The Basis of Vulnerability Index Specific to Demographic Variables.	328
11.4.4.	Ranking of Districts on the Basis of Vulnerability Index Specific to Economic Development Variables.	330
11.4.5.	Ranking of Districts on the Basis of Vulnerability Index Specific to Health Variables.	332
11.4.6.	Ranking of Districts on the Basis of Vulnerability Index specific to Social Development Variables.	334
11.4.7.	Ranking of the Districts on the Basis of Vulnerability Index Specific to Socio-Cultural and Other Variables	336
11.4.9.	Ranking of Districts on the Basis of Vulnerability Index Specific to Presence of Rural Traditional Sex-Workers.	339
11.4.10.	Ranking of Districts on the Basis of Vulnerability Index Specific to the Length of National Highway in the District	340
11.4.11.	Ranking of the Districts on the Basis of Vulnerable Index Specific to Reported Cases of Aids	341
11.4.12.	Ranking of the Districts on the basis of Vulnerable Index Specific to Number of Inmates in Jails	343
11.4.13.	Ranking of the Districts on the Basis of Vulnerable Index Specific to Reported Cases of TB	344
11.5.1.	Mine Numbers	350
11.5.2.	Mineral Type	351

Table No.	Description	Page
11.5.3.	Mineral Type	353
11.5.4.	Mine Workers	356
11.5.5.	Priority Rating of Districts	366
11.6.1.	Study Villages	368
11.6.2.	Rainfall Distribution (Year 2000)	372
11.6.3.	Mean Potential Evapo-Transpiration of Udaipur	373
11.6.4.	Water Quality Standards	377
11.6.5.	Population Distribution	382
11.6.6.	Rabi Season	382
11.6.7.	Kharif (Rainy) Season	382
11.7.1.	Data Collection from secondary sources	411
11.7.2.	Data Collection from Primary sources	412
11.7.3.	Proportion of population served with piped water supply and other alternative sources	421
11.7.4.	Ward-wise Gap in water demand and supply in Sunel town	421
11.7.5.	Simulation of residual water pressure at tail end of the network pipeline	422
11.7.6.	Carrying capacity of the drainage lines in each ward	427

CONTENTS

1	Introduction	1-14
2.	Geographical Information System	15-40
3.	Global Positioning System	41-55
4.	Blazing Technologies	56-68
5	Formulation of GIS for Resource Assessment	69-78
6.	Physio- Climo and Biotic Resources	79-138
7.	Land, Water and Irrigation Resources	139-169
8	Mineral, Power and Industrial Development	170-194
9	Human and Livestock Resources	195-216
10	Transportation and Communication	217-224
11.		
	Case Study 11.1. Case Studies	225-263
	Case Study 11.2. Impact Assessment Study of Rainwater Harvesting/ Recharging Structures in Bikaner District	264-285
	Case Study 11.3. Assessment of Potential Recharge Areas in Four Districts of Rajasthan	286-316
	Case Study 11.4. Mapping of High Risk Behaviour Population Groups in Rajasthan	317-345
	Case Study 11.5. GIS based Mapping of Mines, Mineworkers and High Risk Behaviour Groups in Rajasthan	346-366
	Case Study 11.6. Benefit Assessment of Rain Water Harvesting Structures In Three Blocks of Udaipur District	367-409
	Case Study 11.7. GIS Based Mapping of Water Supply System, Drainage and Solid Waste Management System of Sunel Town of Jhalawar District	410-430
12.	Sustainable Development Plan	431 441
	Bibliography	442-446

Chapter - 1

INTRODUCTION

Preparation of resource inventory, its analysis, spatial distribution and utilization in Rajasthan state is a prerequisite for sustainable developmental planning. Important among these resources are soil, surface and ground water, vegetation, minerals, climatic conditions, land-use and anthropogenic resources. In addition to these resources a few other parameters are landforms, slope, physical and institutional infrastructures. The integration of resource data of geology, geomorphology, soil, agriculture, forestry, hydro-geology, land-use, land-cover, environment, ecology, etc. can lead to identification of homogeneous land units having unique combination of characteristics and evaluated for site specific and suitable land-use practices in respect of scientific land utilization. India as a developing country and particularly Rajasthan state, largely depend on the utilisation of natural resources for their immediate advancement. Advancement involves a change in the existing use of resources to a higher level or more economical methods of production or the more complete use of resources that are at present partially or completely neglected. Every region within our country has its own array of natural resources, suited to different forms of use and offering different opportunities of production. These resources set the pattern for primary production and to a degree, secondary production. We need to make decisions which natural resources can be easily and effectively developed, which are most likely to contribute our needs in both short and long term on a sustained basis and where and when other resources can be most gainfully employed.

1.1 STUDY AREA

Rajasthan, located in the northwestern region of India, consisted of largest geographical area among the all the states of India covering about one-tenth geographical area of the country after the formation of new state as Chattisgarh from Madhya Pradesh (Figure 1.1). The state of Rajasthan has geographical coverage of 3,42,239 sq. Km. administratively divided in 32 districts, sub-divided into 241 tehsil, 237 Panchayat Samities and 41,353 villages as shown in (Figure 1.2). Rajasthan has a total number of 22 towns of which 216 falls in the category of Urban Agglomerations/towns. As per 2001 census, the population of the state is 56.47 Millions which is about 5.5% of the total population of India. Population density and sex ratio in Rajasthan as per the census 2001 is 165 persons per sq. km. and 890 female per 1000 male, which is about 50% and 1.22% less than the nation as a whole respectively.

Rajasthan is broadly, divided into two climatic and geographic regions by the Aravalli range. To its northwest lies the ill-watered sandy tract of the Thar Desert, and to its east lies the elevated Malwa Plateau. In the northeast, the districts of Bharatpur and Bundi lie on a large alluvial plain drained by the Chambal and Banganga rivers.

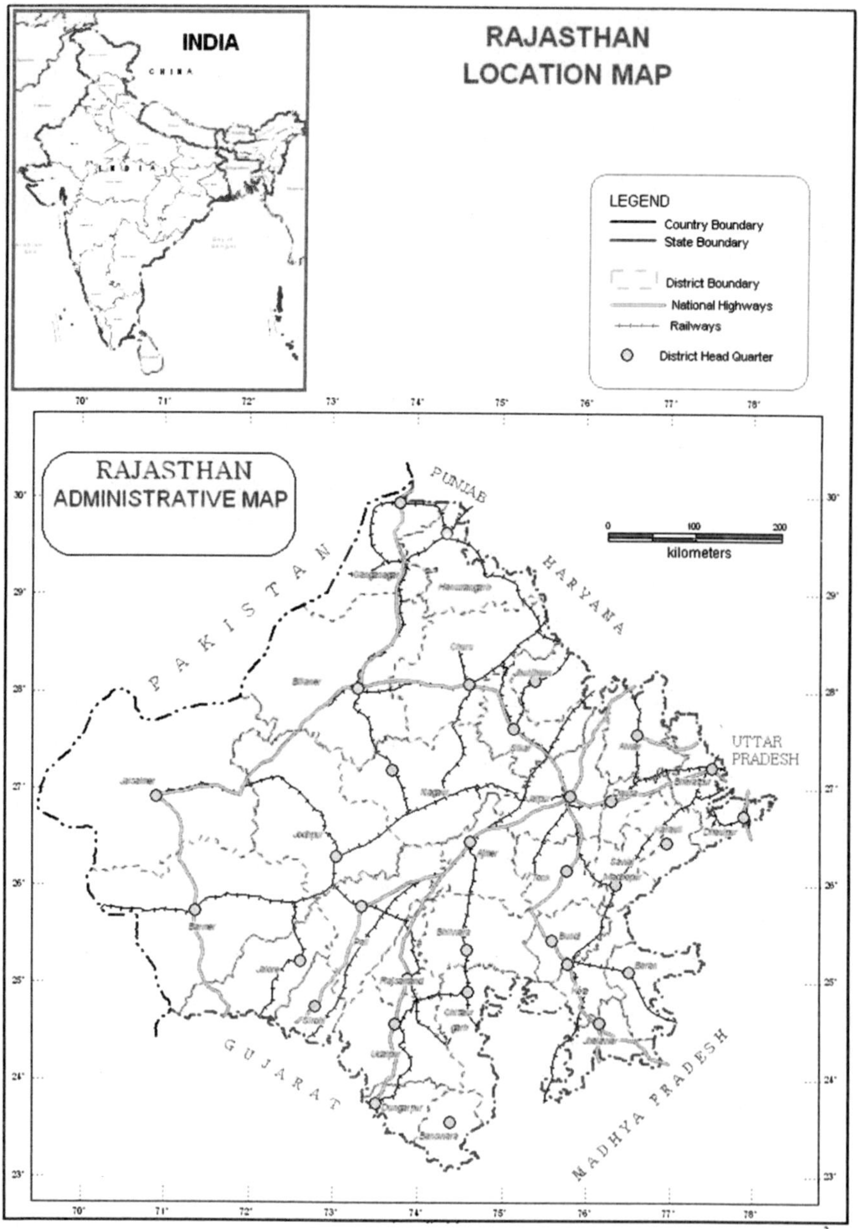

Figure 1.1. Location Map (Source: Census of India)

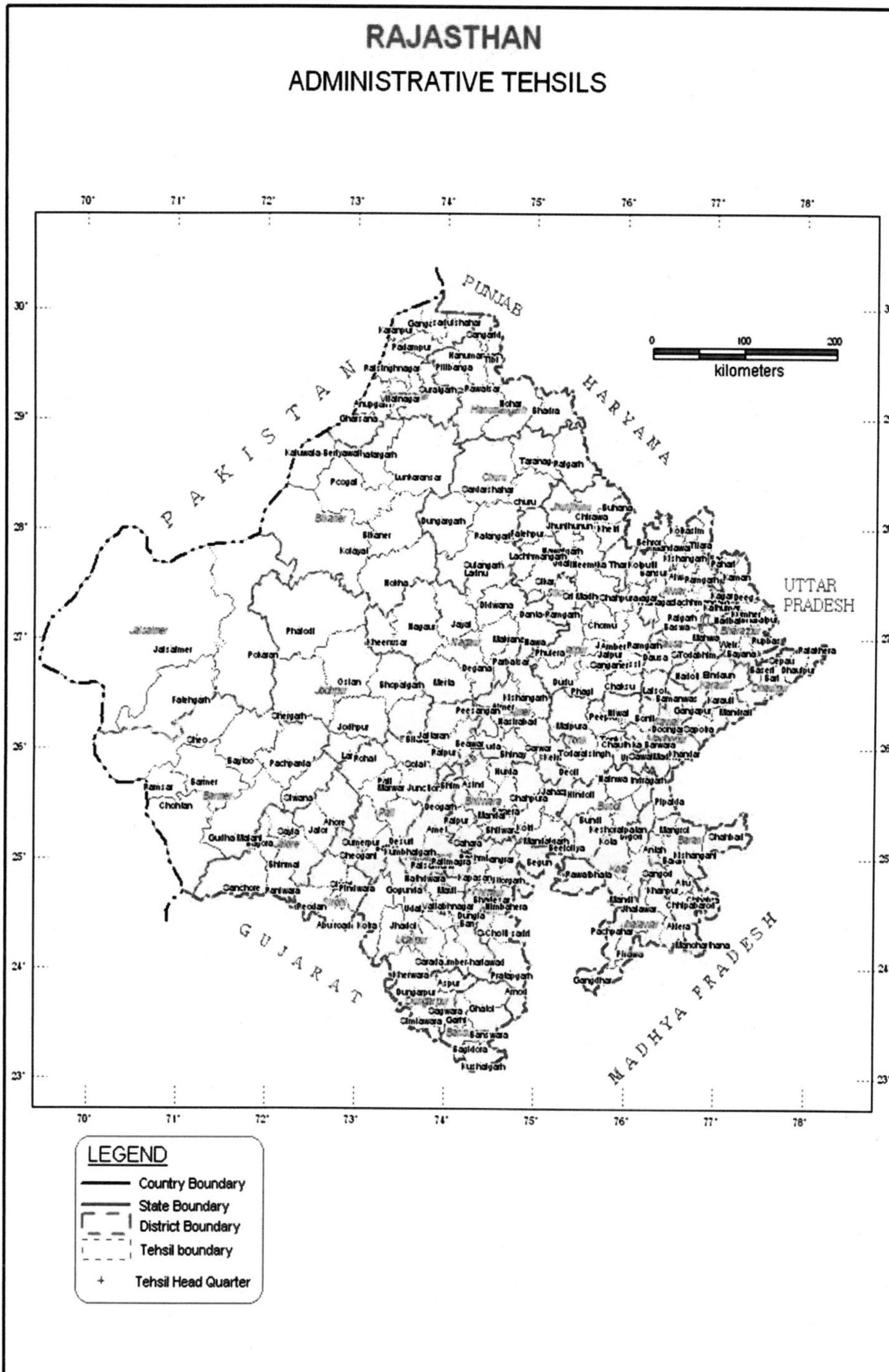

Figure 1.2. Administrative Tehsil Map (Source: Census of India)

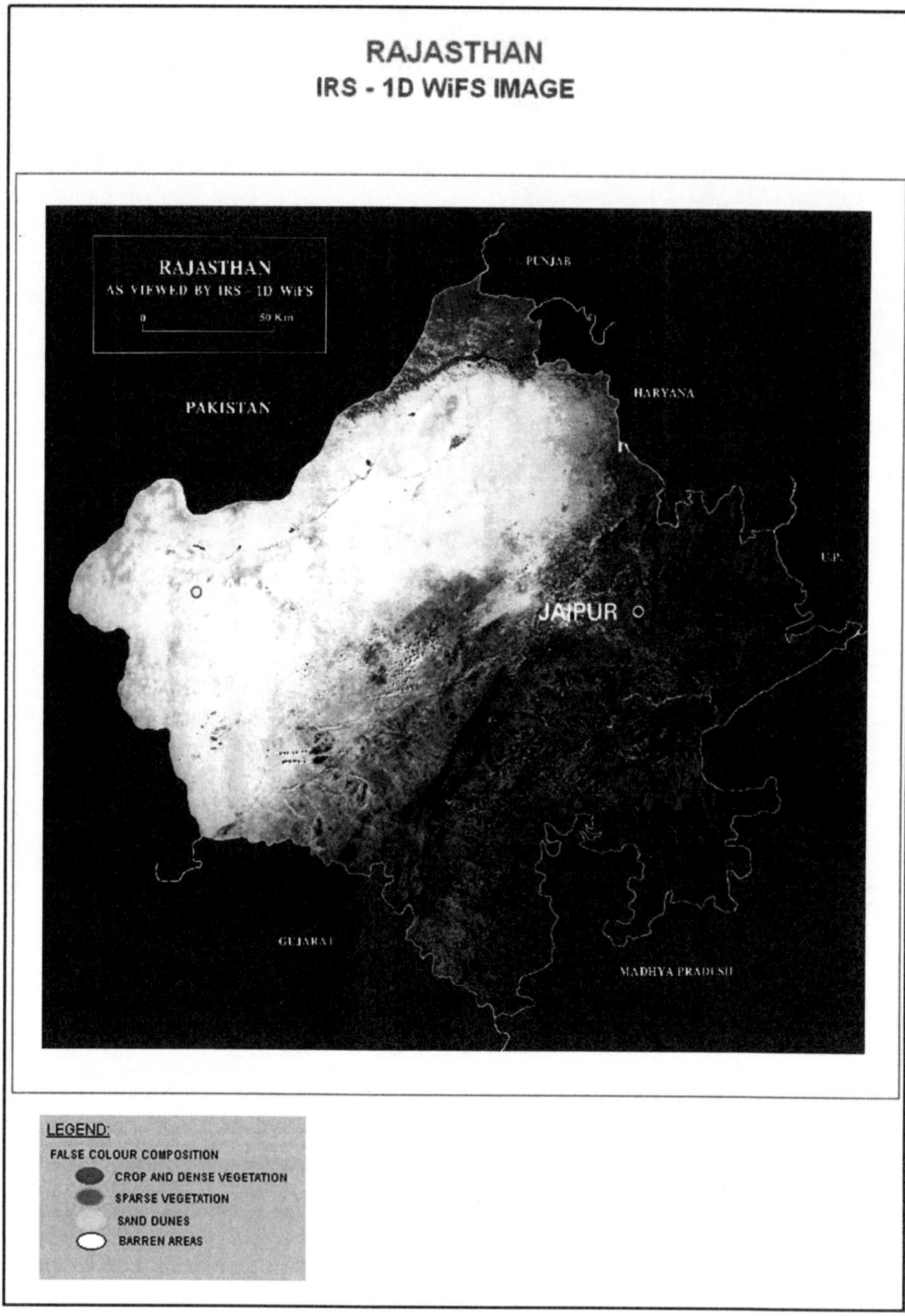

Figure 1.3. Rajasthan Image Map (Source: NRSA)

The northwestern desert tract gradually improves from an arid desert in the far west to a comparatively habitable and fertile tract towards the northeast. Cultivation in the desert region is poor and precarious, though some tracts have better soils and are more productive. The main towns of the region like Jodhpur and Jaisalmer are nevertheless well built. The Luni River, which rises in the Aravalli hills, runs from Ajmer to the Rann of Kutch, draining a large part of the arid tract, and substantial villages are found along its banks.

These features could be visualized in the IRS-1D image of the Rajasthan State as shown in Figure 1.3. The arid parts of Rajasthan are a near rainless desert. In some areas, rainfall scarcely averages more than 120 mm. Rain clouds from the oceans empty most of their water on the high ranges in Kathiawar in Gujarat or on the slopes of the Aravali range. Mount Abu, which is at an exceptional elevation for the region, about 1,220 m, receives 1,525-1,780 mm annually. Irrigation is limited by the scarcity of water in the west of the desert, and has traditionally been restricted to deep wells and rainwater harvesting systems. Rajasthan is rich as far as mineral deposits are concern. Rajasthan occupies a place of pride in non-metallic minerals of India, engaged in mining and marketing primarily of Rock phosphate; SMS grade Limestone, and Gypsum and Selenite and Green Marble. It is not only a market leader in mining and selling of Rock phosphate and natural Gypsum across the country, but also a global pioneer in technology both in open-cast mining and mineral beneficiation of Carbonate Rock phosphate. Depositions of natural gas and oil have been identified recently in the areas of Barmer district making Rajasthan as a focus state in terms of mines and geological settings.

1.2 GEOGRAPHICAL INFORMATION SYSTEM

A Geographic information system (GIS) is a composite of computer-based tools and methods for the integration of spatial data from many different sources and for the analysis, modeling and visual display of this data. It is well suited to collating data from such diverse sources as the census of India, state government departments, other research institutions and topographic maps. Data from large-scale village cadastral maps known as Khasara map (used for keeping land records by land departments) and remotely sensed data from satellites can be stored in the same GIS. The Geographical Information System is a very powerful tool of handling and analysis of spatial data into useful information and has direct utility where large volumes of data are generated, that may be physical, socio-economic, demographic or cultural. With the help of computer, the data are acquired, processed and stored in an organised manner so that the same can be retrieved, manipulated, analysed and formatted in GIS. It can be updated and also presented in the form of maps within no time. It is, therefore, GIS is an excellent tool for the management of extensively large and spatial data with all the advantages of the computer environment: speed, precision, consistency and absence of computation errors. GIS has been widely used in the fields of hydrology, geology, and hydrogeology, especially in developed countries. In the last decade it has been used in developing countries in the field of forestry, natural resource, disaster management and agriculture. Socio-economic applications of GIS are, however, still limited as compared to the developing countries.

The GIS holds immense potential in the field of regional and micro-level spatial planning. For example, it is often necessary to demarcate specific areas, such as tribal areas, hill areas, areas where industry may not be located, and so on, on the basis of guidelines developed by the government. Very often these cannot be done correctly manually because of the vast areas to be considered or the complex nature of the guidelines. Sometimes it is necessary to select suitable sites for the location of large physical infrastructure for the development of an area, such as a factory, hospital, embankment or bridge. GIS is eminently suitable for this. It is also of interest to programme administrators who would need to monitor changes regularly and base future programmes on the results achieved in the past.

This does not imply that GIS will only be of use to those in the business of planning for physical infrastructure and services, communication or area development. It is, in fact, extremely useful in market research and advertising of products. For instance, advertising funds prudently used if advertising is aimed at the population in the area most likely to buy the product advertised. For this, a database containing population characteristics of the potential market areas would have to be maintained. Many large retail companies need to keep a close and continuous watch over a wide market. This could reduce wastage by pointing to geographical areas where commodities move fast and where they move sluggishly. For those in the service industries, information on where best to locate services and how to get there is most useful. Establishing a GIS would enable such companies to monitor in detail the flow of commodities between every supply and sale point. The flow and control of goods can be judiciously regulated on the basis of information generated from the GIS. In all the examples above, one thing is common; people need to make decision based on spatial information.

1.3 OBJECTIVES

The planners can make good decisions only if they are based on reliable resource data or facts of existing resources. But in order to be able to show patterns, trends and exceptional situations, even the most reliable data of the resources must be supported by in–depth analysis and insight into ground conditions. To understand the data, it is necessary first of all to get access to them. This is particularly true in the field of spatial and development planning, where the source of data is usually fragmentary and dispersed. The main purpose of the Resource Assessment Study for Rajasthan was to first collect, store and manipulate data; and then to analyze and graphically present this data. This was done with a view to making information available to a wide public and to specialized users who need to make a variety of decisions relating to sustainable development planning.

The specific objectives of the Rajasthan Resource Assessment Study are listed below:

- To gather demographic, social and infrastructure data available on Rajasthan and to keep them under one platform. This would enable administrators, planners, researchers and other groups with an interest in the state to easily and quickly access the data. This involved making an inventory of data from various sources and creating the Rajasthan Resource Assessment database.

- To present this data in a manner that can provide meaningful insights into the present position of infrastructure and can point to areas where greater development effort should be made. To enable this, it was decided to use computer-based procedures to present output of analyzed data in graphic and mapped from wherever possible.
- To create the Rajasthan Resource database in such a manner that it would be possible to update and expand it later with a view to making it a permanent source of information on the region. Thus a reference system in the form of a data dictionary was maintained to keep a record of all data that was added to the database.
- To keep a record of the operational problems those are encountered during execution of the study. Information regarding operational costs, time schedules, equipment, software, and sources of data, quality and resolution of data.
- To carryout GIS based case studies relevant to the Rajasthan State and addressing the burning issues in various parts of the state.
- On the basis of the findings of the case studies draw conclusions and suggestions regarding policy related to data and information system.

GIS has been used in India recently by government agencies for the mapping of wasteland and of subsoil water, mainly using remotely sensed data. This study differs as it uses predominantly map, census and departmental data and is aimed at demonstrating the use of a geographical information system for resources assessment for the purpose sustainable development planning. It is to illustrate the value of GIS as a tool for socio-economic and physical planning with the testing of GIS tools and methods for making spatial decisions in a state like Rajasthan, the dissemination of knowledge of GIS and the creation of a database for the study area.

1.4 HYPOTHESIS

All regions of the state have some information about their own natural resources although few are nearly complete but in many cases the available technical information is not in a form that can be jointly assessed in a comparative as it have been obtained at different levels and scales according to different concepts. This is mainly on account of the fact that the data on different resources have been collected by different organisations. This is not a satisfactory situation hence; there should be some mechanism by which this information is brought together on a uniform pattern so that it could be considered for policy making. Hence, present research focuses following hypothesis:

To make decisions wisely and to establish an order of priority for state efforts, a thorough knowledge of our natural resources, their native and geographical distribution, magnitude, limitations, present stage of utilization and scope for further better use is essential. Unfortunately, so far, there is no comprehensive account of these resources, their interaction, vulnerability, and future possibilities for many states in India and particularly in Rajasthan. It is in this context that resource inventories would be a useful tool with great practical applications. The usefulness of the present study could be understood by the following facts:

- *Resource data on single platform:* Data exists with each state government departments for its own use. These data are often not shared with other department and many times studies are duplicated. There have been no efforts till now, to put them together on single platform. There are many good reasons for putting departmental data together in a single platform:
  - This will make it possible to get a composite picture of a unit area (e.g. block or district) with respect to a number of variables. These, in case of Rajasthan Resource Assessment Study, include natural and demographic resources such as geology drainage, water bodies, demography, education, health, communication, postal services etc.
  - It would becomes possible to see if there is any link between the status of one resource variable and another.
  - It is necessary to subjectively check the extent and quality of data. This is important because this data is the 'raw material' for government decisions in the planning process. It would also be a worthwhile exercise in pointing out the areas where more data need to be collected.
- *Linking resource data to maps*: Although the resource database can easily be stored in a tabular matrix, which involves much less labour and skill than the constructions of a digital map but it will be worthwhile to link them to places on a map due to following reasons:
  - Rajasthan covers a phenomenal 3,42,239 sq. km large size surface area made up of 32 revenue districts and 237 community development blocks. The numbers of settlements are no less than 222 towns and 41353 villages according to 2001 census. Besides these there are hundreds of kilometers of roads, railway lines, rivers and streams. Along with there are the many man made features, like aerodromes, post offices, hospitals, schools, bank branches, telephone exchanges and more. Storing this immense amount of data would require many volumes of statistics, which would render the data unmanageable. Also, if such data like this is presented in tables, few facts could be lost in rows and columns, which actually exist spatially.
- *GIS based computer environment*: Some government departments maintain resource data and maps, which are prepared manually, but the whole exercise of maintaining a database need to be done in GIS based computer environment due to the following factors:
  - *Time for upgradation:* The 1991 Census atlas of Rajasthan was published (after manual preparation) in 1998 and the last published District Census Handbook for Rajasthan is of Year 1991. Although the census is completed for year 2001 and provisional figures are published but the district handbook is still not available even in the year 2004. With this

speed one can imagine how much time it will take to publish the Census atlas for year 2001. By that time much of the population data represented will become out of date.

- The computer environment ensures speed of analysis and output without any hazard of errors, which creep in during computations of distances and areas from maps of varying scales. It also permits easy execution of complex techniques such as defining buffer zones and overlaying of different sets of features.
- Furthermore, in the case of maps, changing map scale (or size) and updating of information becomes a very quick and easy process, saving valuable time.
- The manually prepared map is largely dependent on the cartographer's artistic ability, an element that should not be underestimated as the appearance of the end product largely depends on it. Computer maps represent lines, colors and symbols consistently and thus even those with no artistic skill can produce useful maps.
- In the case of non-spatial data too, the compute enables analysis and updating of data speedily. The possibility of storing very large volumes of data makes the computer an ideal tool for development of a database.

- *GIS based Mapping*: The maps prepared in geographical information system are much advantageous tools than convention mapping in optimizing the decision-making procedure. This can be understood in following points:
  - Output from a GIS, especially if it happens to be in the form of a map, is very distinct from conventional mapping. GIS concerns the processes of collecting, compiling, correcting and editing data. But its greatest strengths are in integrating, modifying, analyzing and displaying the data. This repackaging of data leads to the conversion of data or fact to information that a planner may require and understand.
  - In India, where GIS is in its infancy stage, it may very often be the case that users are not aware of the potential of their applications. Administrators could apply GIS tools to their routine work, such as in the physical monitoring and evaluation of development programmes. Allocation of funds to different areas on the basis of guidelines, or the planning for relocation of services, or the establishment of physical infrastructure are only a few examples where GIS can be used to support decisions.
  - GIS has features that make it particularly suited to micro-level planning. One great advantage is that in GIS the data remains unaltered irrespective of the degree of aggregation required to reduce maps to fit into a certain size of paper or reduce the map scale.
  - In contrast, in manual work, information is lost when data is represented in a map of a smaller scale. This, which is often overlooked, is perhaps the most useful aspect of GIS and makes it a perfect tool for the 'bottom-up'

approach in planning. It also means that the optimal use of a GIS can be made when data is recorded at the largest possible scale and in the smallest indivisible unit. For instance, if some one has to make a comparative study of the districts in Rajasthan on the basis of length of roads surfaced every year. He or she possibly does a competent exercise using statistical methods and presenting results in manual maps. If, on the other hand, one has to do a block level study to see if the level of road development had anything to do with other characteristics of the blocks, such as, slope, elevation, nearness to towns, and the like, then a GIS would be most useful. In fact, without it, results would not fully reflect the variations within the state with respect to road construction and maintenance.

- The output of information from a GIS, after processing the contents of the cartographic database, can be in a number of different forms. It may generate reports in the form of tables, maps, graphs, summary figures or even text. This flexibility makes the output particularly useful for understanding many different kinds of data and especially data that is recorded from a map.
- In a poor country like India the establishment of a GIS can be justified in terms of improvement in services, reduction of wastage due to uniformed decisions and increased efficiency of resource utilization.

Finally, it appear that if technology is 'the totality of the means employed to provide objects necessary for human sustenance and comfort' (Webster's), then it seems rather appropriate that contemporary computer technology be used for the purpose of making decisions relating to the sustainable development of a large section of humanity.

1.5 METHODOLOGY

At this time it is very clear that there are many serious lacunae of data in the state and it would be useful to create a database for the sub-units (district-wise) of Rajasthan. It is also known that much of the monitoring and evaluation of development programmes is on the basis of financial figures relating to allocation of funds and their expenditure. But there is much less concern for the physical assets that are created. A data-base without a spatial component is of limited value. The resource database to be developed therefore, needed to reflect the physical infrastructure that is available in the state, since infrastructure is closely associated with development in any area. To achieve the objectives of the resource assessment study following methodology have been adopted.

1.5.1 Data Collection

Since the study required deeper understanding of natural as well as demographic resources of Rajasthan data were collected and used from both secondary as well as primary sources to generate geographical information system for further analysis. The details of the data collected from different sources based on the checklist developed are discussed as following:

1.5.1.1 *Secondary Sources*

In Rajasthan, the source of data are numerous, sometimes inaccessible and quite frequently only available where they are generated. Each state government department collects its own data, usually from the block level upwards, through its own personnel. Thus each department at its district headquarter maintains a record (theoretically) of district level data, although the block level data are not usually maintained here. Data can be unreliable, incomplete or difficult to obtain from such sources. Those working in the information sciences feel the problem of data acquisition most keenly. To obtain block level data, one must visit the concerned officials in the departments in the district. There is, however, a district statistical officer (DSO) in each district, whose function is to maintain all data of the district. Some kept records of revenue and agricultural data, aggregated for the whole district, but certainly not for health, education, transport, banking and other sectors. A similar situation exists at the state capital. There is a directorate of statistics and a directorate of evaluation, but from neither of these two departmental offices can one get data pertaining to infra structures in the districts, except for wells, irrigation, crops and other items relating to agriculture. In other words, data for agriculture are available with the department of statistics, but if needed the kind of data that were used in this project, then all the concerned ministries and department would have to be visited, as was done in this case. Even on dong so, sub-district level data were not always available.

The 'vertical' linkage from block to the state functioned for the purpose of data collection with varying degrees of efficacy within each district, but the 'horizontal' linkage between departments, either at the district level through the DSOs or at the state headquarters with the directorate of statistics, has virtually snapped. The implication in today' context is that although much is said at the national level about devolution of the planning process and district level planning, there is little realization among the policy makers that effective district level planning can only take place if there is a repository of district level data. It is necessary, therefore, to strengthen data collection at the district itself. The DSO can maintain a database for a broad cross-section of variables of the sub-units of the district. RAJCOMP, a state government department aimed at providing computer and IT facilities at the district level and the creation of a database for rural development for the use of the district administration. But the data generated do not, in fact, reflect the creation or lack of physical infrastructure in a district. This is because these databases are non-spatial in character.

Thus, for the study purpose existing literature already available with these departments was collected for review and making GIS based analysis. The type of information retrieved from different sources based on the checklist is reflected in the Table 1.1.

1.5.1.2 *Primary Sources*

Collection of primary data was considered necessary for the study to understand the operation and maintenance of a Geographical Information System (GIS). Some case studies involving primary data collection were carried out. Focused group discussions with Panchayat, Government departments, community members along with field survey methods were conducted to fulfill the objectives. Field survey was primarily conducted to triangulate the data collected from secondary sources.

Table 1.1. Source and Information

Source	Information/ Data
A. Spatial Data	
• Survey of India • Resource Atlas of Rajasthan • Watershed atlas of Rajasthan • Groundwater Atlas of Rajasthan • Census Atlas of Rajasthan (1991) • Irrigation Department • Public Health Engineering Department • Research Studies/ papers • Case studies • Project reports • Internet	Following information in layers have been retrieved from the data maps: • National and International Boundary • State boundary • District Boundary • Tehsil/block boundary • Forest and hills boundary • Drainage and Water bodies • Elevation/ spot levels/ contours • Geology • Geomorphology • Soil type • Climatological characteristics • Land-use and land cover • Irrigation • Agriculture • Transportation network o All type of roads o Railways o Airways • Communication network
B. Non-spatial Data	
• Census of India Publications (2001) • Statistical Abstract – 2001-02 • Gazetteers • State government publications • State department's annual reports • Research Studies/ papers • Case studies • Project reports • Internet	• Human resources • Population and settlements • Schedule caste and tribes • Language and religions • Urbanization • Health • Education • Livestock resources • Mineral resources • Power bases • Industrial development

1.5.2 Process of Data Collection

Before initiation of study a preliminary meeting with study guide was organised to discuss the objectives, scope of work, methodology and timeline of the study. Format of reporting and its submission was also discussed and agreed. The preliminary information and related components of the study was collected from the department and university's library at Jaipur. Based on the information, checklist for data collection was designed and finalized in consultation with study guide. As state and central government departments are the prime source of information/data, a formal meeting with each department head and concerned persons was organised at their respective offices to discuss the study design and solicit their support in data collection. Information/ data already available with them was initially collected and reviewed. Literature available with these departments on the concerned issue was also reviewed and relevant data retrieved from them. Some nongovernmental organisations and research institutions working in these areas were also visited and relevant information were collected and updated.

1.5.2.1 *Spatial Data*

All available maps/data (Topo-sheets of 1:250,000 and 1:50,000 scales) were collected from the Survey of India (SOI). Number of Topo sheets and coverage area along with the scale are shown in Table 1.2.

Table 1.2. Topo Sheet Scale and numbers for Rajasthan

Map Scale (R. F.)	Area of one Topo sheet	Number of sheets covering Rajasthan (partial and full)
1:10,00,000	4° Longitude x 4° latitude	5
1:250,000	1° Longitude x 1° latitude	44
1:50,000	15° Longitude x 15° latitude.	574

Note: one Topo sheet of R. F.. 1:50,000 covers about 707 sq km and 1:250,000 about 708 x 16 = 11328q km approximately.

1.5.2.2 *Non-Spatial Data*

The main heads under which attribute data have been collected for the locations digitized are: population, education, health, roads, railways, telecommunication, banking, electricity, postal services and tribal areas. These are all collected from secondary sources. Published data were collected, whenever available (for example; for population, railway). Most of the remaining data are, however, gathered from the departments of government of Rajasthan and occasionally, GOI. There are a number of sample survey reports of Rajasthan available for diverse areas of interest. They also contain useful data. However, for the purpose at hand, these did not prove useful because the database did not cover the entire state and was usually collected for small sample units. No discussion on demographic and socioeconomic data in the India can be complete without reference to the Census of India publications. The Census of India is the most comprehensive source of such data in the country. Its great strength lies in the variety of data that it updates every ten years. For GIS users one major limitation, however, is that the volumes containing data for the subunits of the district are published many years after data is collected. In addition, there are also some other problems regarding the reporting of data that can easily be corrected. The obvious example would be the changed format of the economic tables, which make historical comparisons of data a problem.

1.5.2.3 *Access to Maps*

In this country, the Survey of India (SOI) under Department of Science and Technology is the prime map-making organization. Maps are supplied to the public through a number of SOI map sales offices. In procuring SOI maps numerous problems like acquisition of maps of scale 1:250,000 and larger, inadequate stocks and 'presently not available' to 'restricted' are generally faced. For acquiring restricted area maps, a clearance from the Ministry of Defense is required. Even when given, it can take up to a year from applying for a map to receiving it. This naturally causes a tremendous loss of time. But the long-term ramification of this unspoken discouragement of the public use of maps has been that researchers areas. These include topographic maps of areas that fall within a fifty-kilometer belt of the international land and coastal boundaries of the country and those maps that are gridded. Another negatives result has been that are remarkably small number of the literate population have ever used a map for purpose at all. Using GIS techniques it is clear that the SOI, for the sake of convenience, actually restricts maps of areas well beyond the 50 km band along the international boundary. Perhaps there is need for rationalizing man on these rules regarding distribution of maps. Today, a number of reasons may be given for reconsidering the restrictive practices relating to map distribution. First, these maps are sold openly in a number of countries in the world. Second, when there is free movement of people without passports or visas between two nations (for example, India and Nepal), there seems to be little justification in withholding maps of the area. Third, one most realize that access o information cannot be defined only as that made available through the press, radio or television. Map information is also important in our lives. It is, of course, conceivable that some features on maps need not be recorded for security reasons, but entire sheets need not be restricted. Lastly, if village and district level planning is ever to become a reality, large-scale maps (Topo-sheets) must become easily available.

■■■■

Chapter - 2

GEOGRAPHICAL INFORMATION SYSTEM

Geographic Information System (GIS) today is an indispensable tool in the planning for tomorrow's requirements, such as rail network, dam site location, urban development etc. Until computers were available, geographic data were drawn on a piece of paper, as points, lines and areas. With the advent of computers, accurate and rapid analyses of diverse geographic data sets are perceived as a requirement for more effective planning. The information about the resources, thus collected from the real world, converted into digital form and is integrated and analysed in a GIS to make quick decisions as well as their implementation on the real world. Work on GIS began in late 1950s, but first GIS software came only in late 1970s from the lab of the ESRI. Canada was the pioneer in the development of GIS as a result of innovations dating back to early 1960s. Evolution of GIS has transformed and revolutionized the ways in which planners, engineers, managers etc. conduct the database management and analysis.

2.1 BASICS OF GIS

Early definitions of GIS treat it as a spatial database, but the recent trend is to consider it in following several ways (ESRI, 1996).

1. Information technology which stores, analyses and displays spatial and non-spatial data.
2. Decision support system involving the integration of spatially referenced data in a problem solving environment.
3. Tool for management, analysis and manipulation of spatially referenced data information in a problem solving environment.

From above definitions, it is clear that the common factor between all is the notion of spatial data processing. In this context, spatial data may be defined as objects that have physical dimensions on the Earth. "Every object present on the Earth can be geo-referenced", is the fundamental key of associating any database to GIS. Here, term 'database' is a collection of information about things and their relationship to each other, and 'geo-referencing' refers to the location of a layer or coverage in space defined by the co-ordinate referencing system.

2.2 COMPONENTS OF GIS

GIS refers to three integrated elements; (i) Geography which is the real world, (ii) Information which is data, their meaning and uses, and (iii) Systems which is computer technology and its support infrastructure (Figure 2.1 and Figure 2.2).

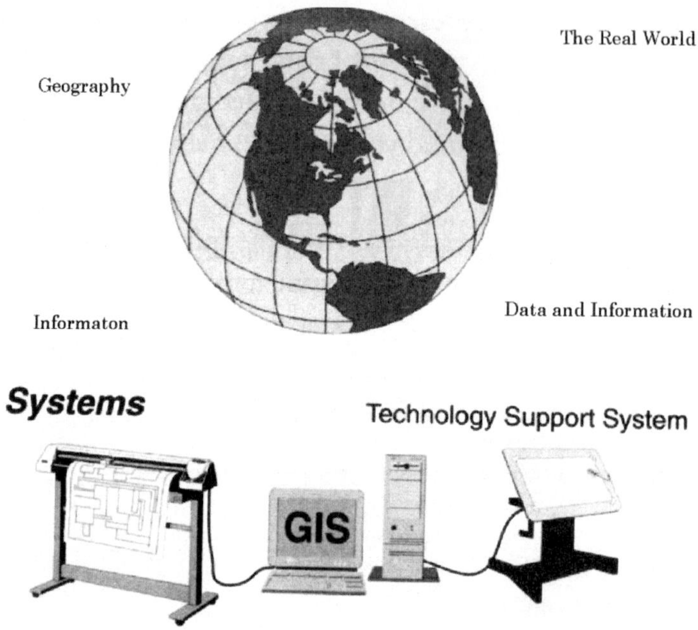

Figure 2.1: Integrated Elements of GIS

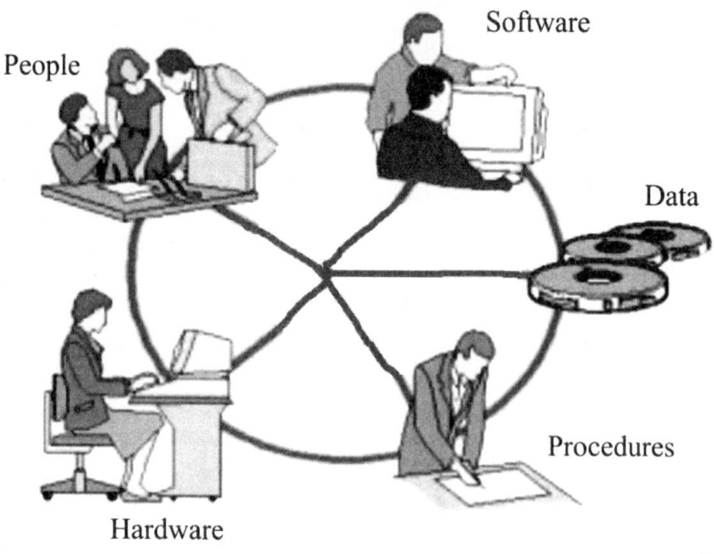

Figure 2.2: The component parts of GIS

The technology of GIS has developed so fast over the past one and a half decade that it is now accepted as an essential tool for the effective use of geographic information. GIS offers the capability of collection, compilation, storage retrieval, analysing, manipulation, displaying an integrating environmental, economic and social data in a single system.

2.2.1 Computer Hardware

The general hardware components of a Geographical Information System are presented in Figure 2.3. The computer has a hard disk drive for storing data and programs, but extra storage can be provided via a network or by digital tape cassettes, optical CD –ROMs, and other devices. A digitizer or a scanner is used to convert maps and documents into digital form so that they can be used by the computer programs. A plotter or a printer or any other kind of display device is used to present the results of the data processing. Inter-computer communication is provided by local and global electronic networks using special data lines with optical fibers or over ordinary telephone lines by using known as a 'modem'. The user controls the computer and the peripherals (a general term for plotters, printers, digitizers, and other apparatus liked to the computers) via the computers screen and keyboard, aided by a 'mouse' or pointing device.

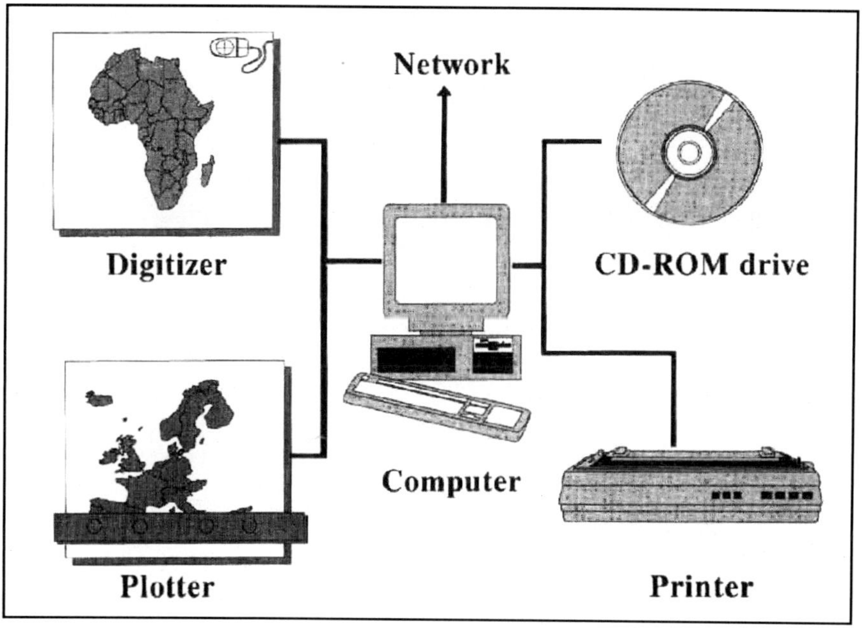

Figure 2.3: Major Hardware Components of a Geographical Information System

2.2.2 GIS Software

The software for a geographical information system may be split into following functional groups (Figure 2.5) with classical three tier system (Figure 2.4).

a. Data input and verification
b. Data storage and database management
c. Data output and presentation
d. Interaction with the user.

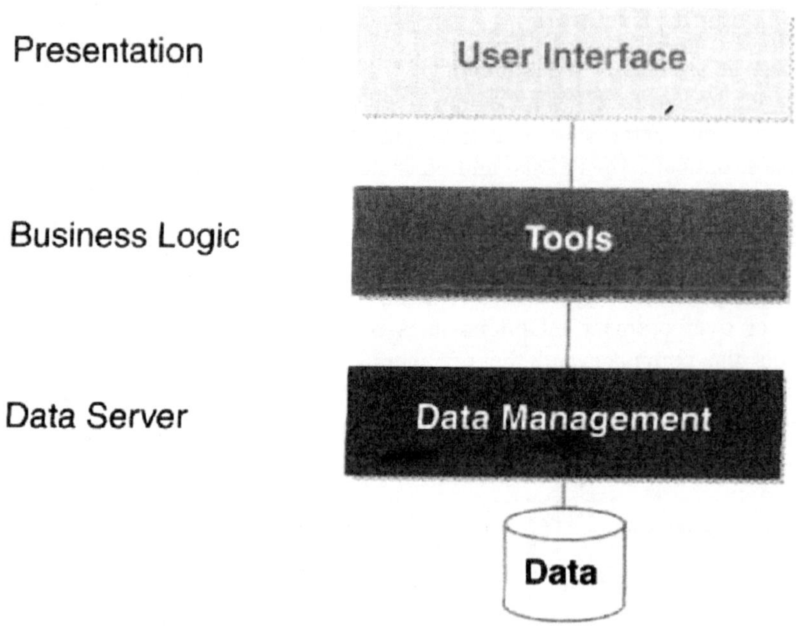

Figure 2.4: Classical three-tier architecture of a GIS software system.

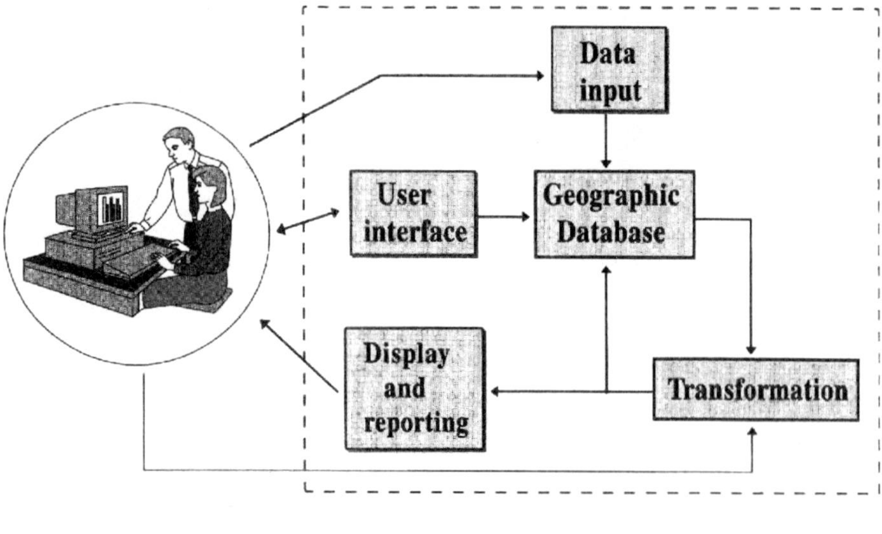

Figure 2.5 : The Main Software Components of a GIS

2.2.2.1 *Data Input*

It covers all aspects of capturing spatial data from existing maps, field observations and sensors (including aerial photography, satellites, and recording instruments) and converting them to a standard digital form (Figure 2.6). Many tools are available including the interactive computer screen and mouse, the digitizer, word processors and spreadsheet programs, scanners (in satellites or airplanes for direct recording of data or for converting maps and photographic images), and devices necessary for reading data already written on magnetic media such as tapes or CD-ROMs.

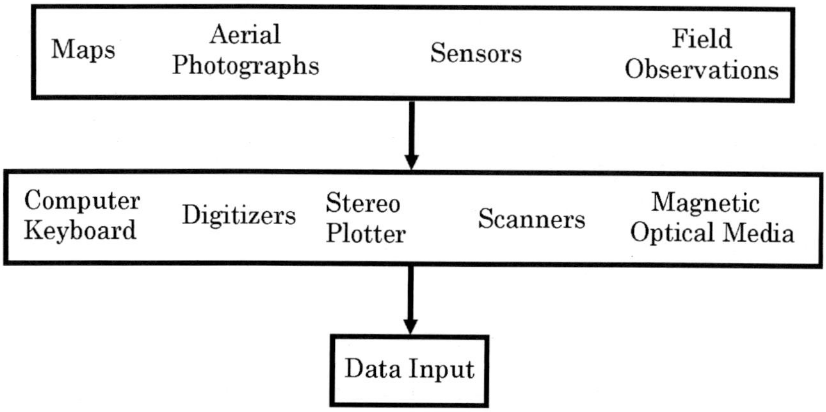

Figure 2.6 : Data Collection and Input

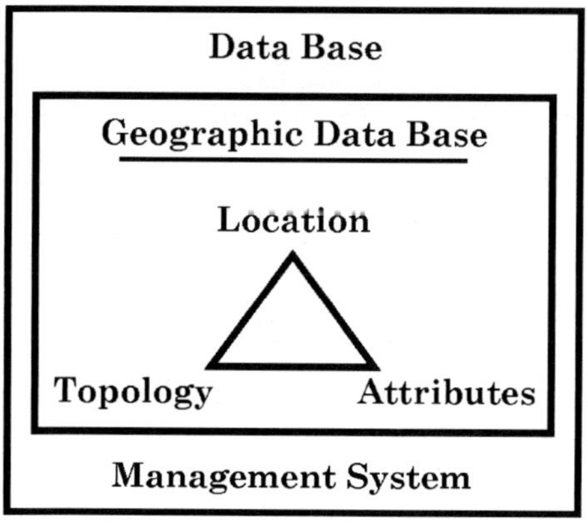

Figure 2.7 : The Components of the Geographic Data Base

2.2.2.2 *Data Storage and Database Management*

It concerns the way in which data about the location, linkages (topology), and attributes of geographical elements (points, lines, areas, and more complex entities representing objects on the earth's surface) are structured and organised, both with respect to the way they must be handled in the computer and how they are perceived by the users of the system (Figure 2.7). The computer program used to organize the database is known as a Data Base Management System (DBMS).

2.2.2.3 *Data Output and Presentation*

Concern the ways the data is to be displayed and how the results of analyses are reported to the users. Data may be presented as maps, tables, and figures (graphs and charts) in a variety of ways ranging from the ephemeral image on the computer screen, through hardcopy output drawn on printer or plotter to information recorded on magnetic media in digital form (Figure 2.8).

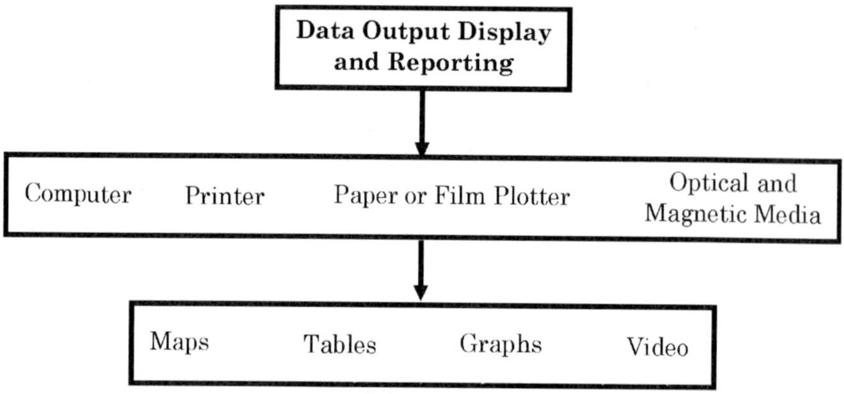

Figure 2.8: Data Output

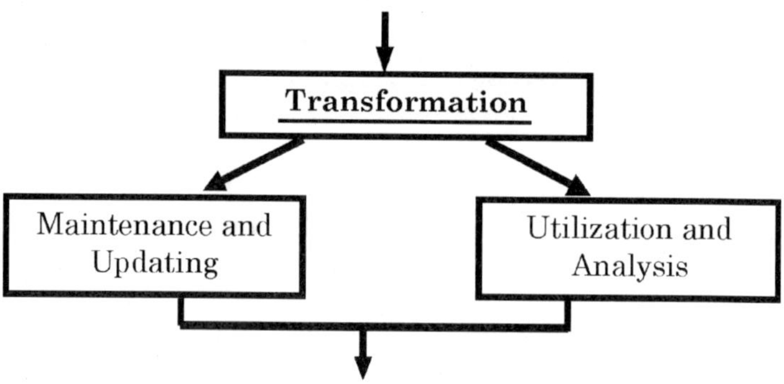

Figure. 2.9: Data Transformation

2.2.2.4 *Data Transformation*

Embraces two classes of operation, namely (a) transformations needed to remove errors from the data or to bring them up to date or to match them to other data sets, and (b) the large array of analysis methods that may be applied to the data in order to achieve answers to the questions asked of the GIS. Transformation can operate on the spatial, topological, and the non-spatial aspects of the data, either separately or in combination. Many of these transformations, such as those associated with scale changing, fitting data to new projections. Logical retrieval of data, and calculation of areas and perimeters are of such a general nature that one should expect to find them in every kind of GIS in one form or another. Other kinds of manipulation may be extremely application-specific, and their incorporation into a particular GIS may be only to satisfy the particular users of that system. The kinds of transformation methods available, their optimum use and misuse, the ways in which sets of simple transformation may be combined in order to achieve types of geographical or spatial modeling (Figure 2.9).

2.3 ADVANTAGES OF GIS

GIS applications provide systematic collection, updating, processing, and distribution of land-related data. Information about zoning, properties, roads, schools and parks, all pertain to geographic location. The capability to handle large quantity of land survey data is a common requirement of a GIS. GIS is able to analyse and manipulate spatial information as well as to integrate different types of data at a high speed. In addition, it is a useful tool for monitoring and management of various information.

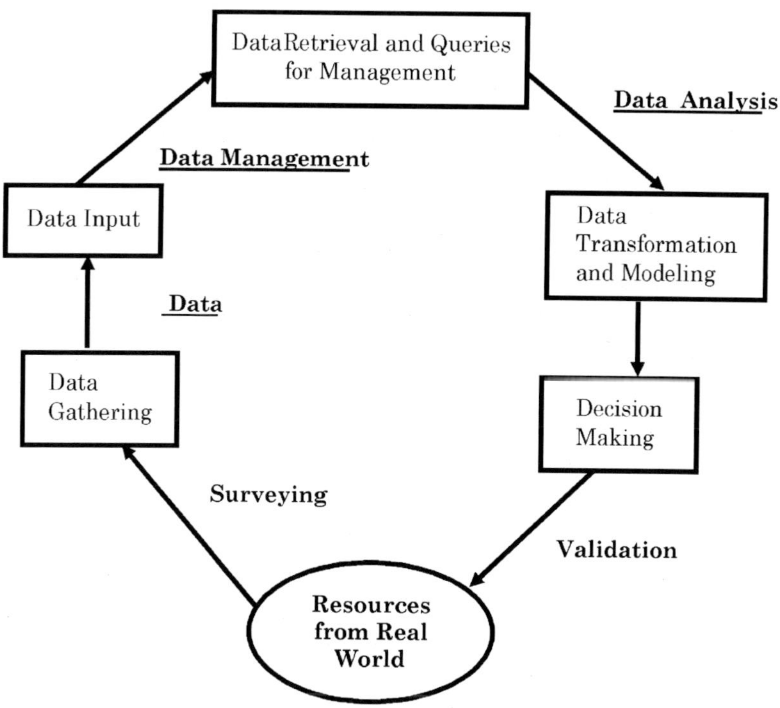

Figure 2.10: GIS as a Management Tool

Overlay of various thematic maps is quickly done using the capabilities of GIS. Thematic maps can be prepared and updated using high-resolution spatial and temporal remote sensing data. The advantages of GIS shift quite rapidly from manual analysis, particularly if the data sets to be analysed are very large. GIS has also been found to be extremely useful for complex modeling or if the data sets of are to be combined using different weightings. Figure 2.10 shows the capabilities of GIS as a management tool.

2.4 ANALYSIS IN GIS

The distinguishing features of GIS systems are their capability of performing an integrated analysis of spatial and attribute data. The data are manipulated and analyzed to obtain information useful for a particular application. There is an enormously wide range of analytical operations available to GIS users, and a number of classifications of those operations have been suggested. Table 2.1 presents some of GIS software and their capabilities in terms of spatial modeling. GIS has a numerous modeling and analysis functions.

Table 2.1 GIS Supporting Spatial Statistics

Software	Developer and URL Address	Software Type	Selected Statistical Analysis Capability
ARC/INFO GRID	Environmental Systems Research Institute, Inc. www.esri.com	Vector and raster GIS	Descriptive statistics, regression and correlation, spatial autocorrelation, multivariate classification, surface interpolation trend surface analysis.
IDRISI	Graduate School of Geography, Clark University www.idrisi.clarku.edu	Raster GIS	Sampling descriptive statistics, regression and correlation, spatial autocorrelation, multivariate classification, surface interpolation, quadrant analysis, trend surface analysis, time-series analysis
Space Star	Anselin (1995) www.rri.wvu.edu/ spacestat.htm	Statistical add-in vector GIS	Exploratory spatial data analysis, Moran scatter plot outliers, scatter plot brushing
S + Spatial Stats	MathSoft, Inc. www.mathsoft.com	Statistical software, limited vector GIS capabilities	Variograms, kriging, spatial autocorrelation, spatial regression modes, trend surface analysis.

2.4.1 Advance Analysis

Reclassification and classification operations transform the attribute data associated with a single map layer. They involve the grouping of objects into classes according to the new values assigned to the objects of the input data according to certain location and non-location attribute values. Classification implies the recognition of pattern and organization in the data being considered for a particular analysis. To this end, the reclassification procedures can be grouped into two categories: reclassification involving thematic attribute values associated with a data layer and reclassification based on topology properties of objects on a particular data layer. In the former type of reclassification, the values on the output layer are assigned as a function of the values on the input layer. In topological reclassification, the new values are assigned as a function of the position, the contiguity, the size or the shape of the spatial pattern represented on the input layer .

The attribute-based reclassification can involve comparison operations. These operations determine whether or not a relationship holds between the value associated with a particular object and a specified or derived constant value (threshold value). The list below specifies the basic comparison operations:

1. Equal to (=): assigns the value of 1 to those objects that are characterized by a specified attribute value; otherwise, the value of 0 is assigned.
2. Greater than (>): assigns the value of 1 to those objects that have an associated attribute value greater than a specific value; otherwise, the value of 0 is assigned.
3. Less than (<): assigns the value of 1 to those objects that have an associated attribute value less than a specific value; otherwise, the value of 0 is assigned.
4. Greater than or equal (≥): assigns the value of 1 to those objects that have an associated attribute value grater than or equal to a specific value; otherwise, the value of 0 is assigned.
5. Less than or equal to (≤): assigns the value of 1 to those objects that have an associated attribute value less than or equal to a specific value; otherwise, the value of 0 is assigned.

2.4.1.1 *Scalar Operations*

This class of operations make use of a single, uniform value, a scalar data layer. It is constructed by assigning the value desired to each location on the data layer. One primary use for scalar operations is to transform all objects of an existing data layer in order to change all values by a given figure. The output data layer contains new attribute values, depending on the type of operation and a constant value (scalar). The fundamental scalar operations include:

1. Addition (+): adds a specified constant to each attribute value on the input layer.
2. Subtraction (-): subtracts a specified constant from each attribute value on the input value layer.
3. Multiplication (x) : multiplies each attribute value on the input layer by a specified constant.
4. Division (/): divides each attribute value on the input layer by a specified constant.
5. Exponentiation (^): raise each value on the input layer to a specified exponent.

2.4.1.2 *Overlay Operations*

An overlay procedure generates a new layer (output layer) as a function of two or more input layers (Figure 2.11). Specifically, the attribute value assigned to every location (or set of locations) on the output layer is a function of the independent values associated with that location on the input layers. If the input data are in the raster format, the output is one raster data layer containing a value in each cell that results from some arithmetic or logical combination a value in each cell that results from some arithmetic or logical combination of the values in the input data sets. If the input layers are in topologically structured vector data, the output is a new set of polygons which are formed by the intersection of all the boundaries in the input data layers. Overlay operations may involve any combination of points, lines, areas, or pixels.

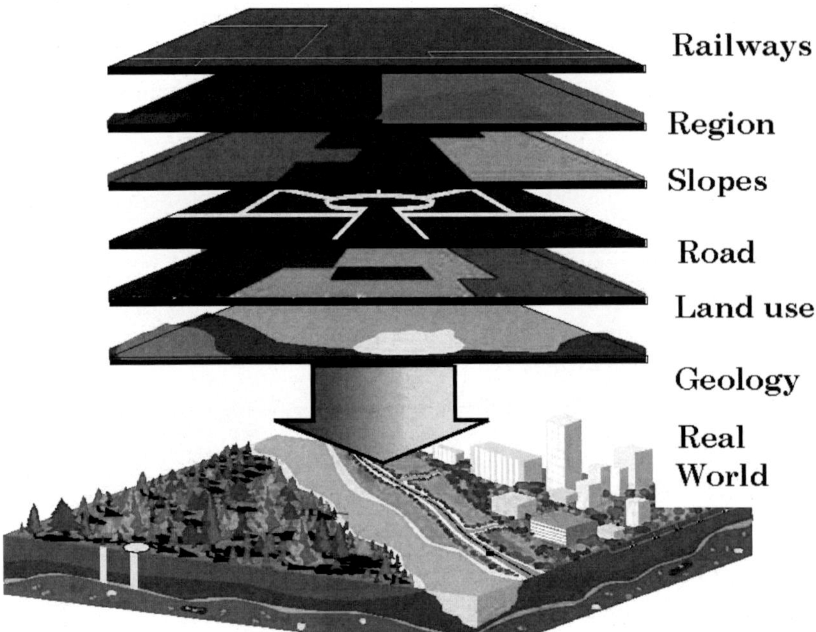

Figure 2.11: Overlay of Thematic Maps in GIS

The preprocessing procedures are also referred to as the geometrical phase of the overlay techniques. The output from this phase is the input to the overlay operations. These operations can be classified into two categories; location-specific overlay operations, which assigns a value (a function of the point-by-point coincidence of the input layers) to each location on the output layer, and category-wide operations, in which the values are assigned to a set of locations (region, zone) as a function of the values that are associated with the corresponding data on a point-by-point bases, the category-wide overlaying summarizes the spatial coincidence. Depending on the nature of the data contained in the input map layers, the overlay operations can be performed in terms of arithmetic overlay includes the following:

1. Addition: The value at each location in input layer A is added to the value at the corresponding location in input layer B.
2. Subtraction: The value at each location in input layer A is subtracted from the value at the corresponding location in input layer B.
3. Multiplication: The value at each location in input layer A is multiplied by the value at the corresponding in input layer B.
4. Division: The value at each location in input layer A is divided by the value at the corresponding location in input layer B.

Beside these four fundamental algebraic operations, a number of algebraic and statistical operations can be performed using overlay procedures. Examples of these types of operations include:

1. Average: The average value at corresponding locations in input layers A and B is calculated.
2. Power: The value at each location in input layer A is raised to the power of the value at the corresponding location in input layer B.
3. Rank: The values at corresponding locations in input layers A and B are ranked.
4. Minimum: The maximum value at corresponding locations in input layers A and B is identified.
5. Maximum: The minimum value at corresponding locations in input layers A and B is identified.

Logical operations are based on Boolean algebra. There are three basic logical operators or Boolean operators or Boolean operators:

1. Intersection: the logical AND
2, Union: the logical OR
3. Complements: the logical NOT.

2.4.1.3 *Connectivity Operations*

The connectivity concept has a different connotation in the context of vector and raster data structures. In the case of vector structure, the term connectivity is used to describe the linkage of points or polygons to each other. It is one of the properties describing topological relationships. For raster data the term connectivity describes the linkage between two pixels, that is, whether or not two pixels are connected, and if connected, the kind of connection involved. The distinguishing feature of connectivity functions is that they accumulate the values over the area being traversed. The connectivity functions include such operations as proximity, buffering, spread and network analysis.

2.4.1.4 *Neighborhood Operations*

This class of operations involves ascribing values to a location according to the characteristics of the surrounding area. Specifically, the value assigned to each location on the output layer is a function of independent values in the neighborhood of that operations, at least three parameters have to be specified. These parameters include one or more target locations, the neighborhood membership (i.e., a set of operations around each target point that are within a specified distance and direction), and the function to be performed on the locations within the neighborhood. Neighborhood membership is also referred to as a window.

2.4.1.5 *AML*

Advanced languages, such as AML (Arc Macro Language of ARC/INFO GIS) can be used to merge the large database and to place the complex query. ARC Macro Language (AML) is a high level algorithmic language for generating the end-user applications. Features include the ability to create on-screen menus, use and assign variables, control statement execution, and get and use map or page unit coordinates. AML includes an extensive set of commands that can be used interactively or in AML programs (macros) as well as commands that report on the status of ARC/INFO environment settings. The AML program, which contains commands of ARC/PLOT in the same sequence and calling this program in ARC/PLOT environment, produces the same results in a more convenient and faster way since this program can be saved. In the present work, a set of AML programs is used in overlaying the various themes.

2.5 FUNCTIONAL ELEMENTS OF GIS

There are five essential elements that a GIS must contain: data acquisition, preprocessing, data management, manipulation and analysis and product generation For any given application of a geographic information system, it is important to view these elements as a continuing process (Figure 2.12). As a guiding principle, the analyst should develop as end-to-end model of the task at hand. Even when the precise details of the steps to be taken may depend on the results of intermediate calculations and analyses, an explicit outlines of the process, like a working hypothesis in a scientific experiment, can be very valuable.

2.5.1 Data Acquisition

Data Acquisition or collection is one of the most time-consuming and expensive, yet important of GIS tasks. There are many diverse sources of geographic data and many methods available to enter them into a GIS. The two main methods of data collection are data capture and data transfer. It is useful to distinguish between primary (direct measurement) and secondary (indirect derivation from other sources) data capture, for both raster and vector data types. Data transfer involves importing digital data from other sources. There are many practical issues associated with planning and executing an effective GIS data collection plan. Following paragraphs reviews the main methods of GIS data capture and transfer, and introduces key practical management issues.

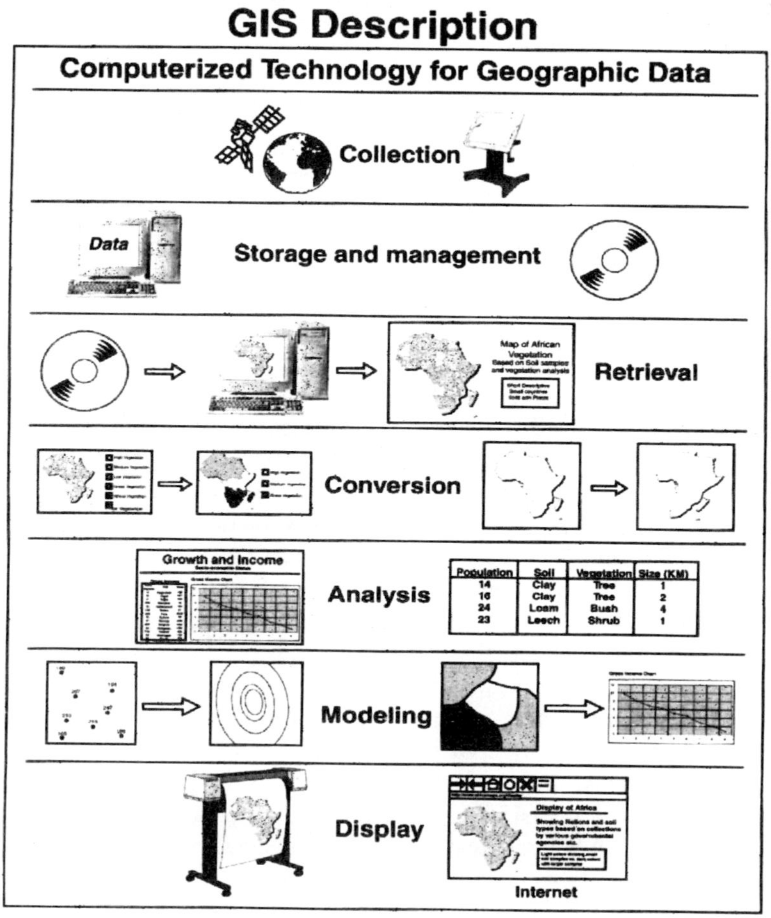

Figure 2.12: Various Functions of GIS

GIS can contain a wide variety of geographic data types originating from many diverse sources. From the perspective of creating geographic databases, it is convenient to classify raster and vector geographic data as primary and secondary (Table 2.2). Primary data sources are those collected specifically for use in GIS. Typical examples of primary GIS sources include raster SPOT and IKONOS Earth satellite images, and vector building survey measurements captured using a total survey station. Secondary sources are those that were originally captured for another purpose and need to be converted into a form suitable for use in a GIS project. Typical secondary sources include raster scanned color aerial photographs of urban areas, and USGS and IGN paper maps that can be scanned and vectorized.

Geographic data may be obtained in either digital or analog format. Analog data must always be digitized before being added to a geographic database. Depending on the format and characteristics of the digital data, considerable reformatting and restructuring may be required prior to import.

Table 2.2. Classification of geographic data for data collection purposes

	Raster	Vector
Primary	• Digital remote-sensing images • Digital aerial photographs	• GPS measurements • Survey measurements
Secondary	• Scanned maps or photographs • Digital elevation models from maps	• Topographic maps • Toponoymy (place name) databases

The processes of data collection are also variously referred to as data capture, data automation, data conversion, data transfer, data translation, and digitizing. Although there are subtle differences between these terms, they essentially describe the same thing that is, adding geographic data to a database. Data capture refers to direct entry; data transfer is the importing of existing digital data across the internet, WANs, or LANs; or using CD ROMs, zip disks, or diskettes. In the early days of GIS, when geographic data were very scarce, data collection was the main project task and it typically consumed the majority of the available resources. Data collection still remains a time consuming, tedious, and expensive process. Usually it accounts for 15-50% of the total cost of a GIS project. Data capture costs can in fact be much more significant because in many organizations (especially those that are public) staff costs are often assumed to be fixed and are not used in budget accounting. Furthermore, as the majority of data capture effort and cost tends to fall at the start of projects, they receive greater scrutiny from senior managers. If staff costs are excluded from a GIS budget then in cash expenditure terms data collection can be as much as 60-85% of costs. After an organization has completed basic data collection, their emphasis moves on to data maintenance. Over the multi-year lifetime of a GIS project, data maintenance often turns out to be a far more complex and expensive activity than initial data collection. This is because of the high volume of update transactions in many systems (for example, changes in land parcel ownership, maintenance work orders on a highway transport network, or logging military operational activities) and the need to manage multi-user access to operational databases.

2.5.1.1 *Primary Geographic Data Capture*

Primary geographic capture involves the direct measurement of objects. Both raster and vector data capture methods are available.

2.5.1.1.1 *Raster Data Capture*

Much the most popular form of primary raster data capture is remote sensing. Broadly speaking, remote sensing is a technique used to derive information about the physical, chemical, and biological properties of objects without direct physical contact. Information is derived from measurements of the amount of electromagnetic radiation reflected, emitted, or scattered from objects. A variety of sensors, operating throughout the electromagnetic spectrum from visible to microwave wavelengths, are commonly employed to obtain measurements. Passive sensors are reliant on reflected solar radiation or emitted terrestrial radiation; active sensors (such as synthetic aperture radar) generate their own source of electromagnetic radiation. The platforms on which these instruments are mounted are similarly diverse. Although Earth-orbiting satellites and fixed-wing aircraft are by far the most common, helicopters, balloons, masts, and booms are also employed. As used here, the term remote sensing subsumes the fields of satellite remote sensing and aerial photography. From the GIS perspective, resolution is the key physical characteristic of remote sensing systems. There are three aspects to resolution: spatial, spectral, and temporal. All sensors need to trade off spatial, spectral, and temporal properties because of storage, processing, and bandwidth considerations. Spatial resolution refers to the size of object that can be resolved and the most usual measure is the pixel size. Satellite remote sensing systems typically provide data with pixel sizes in the range 1m –1 km. The cameras used for capturing aerial photographs usually range from 0.1m – 5m. Image (scene) sizes vary quite widely between sensors – typical ranges include 1000 by 1000 to 3000 by 3000 pixels. The total coverage of remote sensing images is usually in the range 10 by 10 – 200 by 200 km.

Spectral resolution refers to the parts of the electromagnetic spectrum that are measured. Since different objects emit and reflect different types and amounts of radiation, selecting which part of the electromagnetic spectrum to measure is critical for each application area. Remote sensing systems may capture data in one part of the spectrum (referred to as a single band) or simultaneously from several parts (multi-band or multi-spectral). The radiation values are usually normalized and resample to give a range of integers from 0-255 for each band, for each pixel, in each image. Temporal resolution, or repeat cycle, describes the frequency with which images are collected for the same area. There are essentially two types of commercial remote sensing satellite: Earth orbiting and geostationary. Earth orbiting and geostationary. Earth orbiting satellites collect information about different parts of the Earth surface at regular intervals. To maximize utility, orbits are typically polar, at a fixed altitude and speed, and are Sun synchronous. The French SPOT (Systeme Probatoire d'Observation de la Terre) satellite, for example, passes virtually over the poles at an altitude of 832 km sensing the same location on the Earth surface during daylight every 26 days. The SPOT platform carries two sensors: a single band panchromatic sensor measuring radiation in the visible part of the electromagnetic spectrum at a resolution of 10 by 10 m: and a multi-spectral sensor measuring green, red, and reflected infrared radiation at a resolution of 20 by 20m. Each SPOT scene covers an area of 60 by 60 km.

Much of the discussion so far has focused on commercial satellite remote sensing systems. Of equal importance, especially in medium to large-scale GIS projects, is aerial photography. Although the data products resulting from remote sensing satellites and aerial photography systems are technically very similar (i.e. they are both images) there are some significant differences in the way data are captured and can be interpreted. The most notable difference is that aerial photographs are normally collected using analog optical cameras (although digital cameras are becoming more widely used) and then later rasterized, usually by scanning a film negative. The quality of the optics of the camera and the mechanics of the scanning process both effect the spatial and spectral characteristics of the resulting images. Most aerial photographs are collected on an *ad hoc* basis using cameras mounted in airplanes flying at low altitudes (3000–10000 m) and are either panchromatic (black and white) or color, although multi-spectral cameras/ sensors operating in the non-visible parts of the electromagnetic spectrum are also used. Aerial photographs are very suitable for detailed surveying and mapping projects. An important feature of satellite and aerial photography systems is that they can provide stereo imagery from overlapping pairs of images. These images are used to create a 3D analog or digital model from which 3D coordinates, contours, and digital elevation models can be created. Satellite and aerial photograph data offer a number of advantages for GIS projects.

The consistency of the data and the availability of systematic global coverage make satellite data especially useful for large area projects (for example, mapping landforms and geology at the river catchment area level) and for mapping inaccessible areas. The regular repeat cycles of commercial systems and the fact that they record radiation in many parts of the spectrum makes such data especially suitable for assessing the condition of vegetation (for example, the moisture stress of wheat crops). Aerial photographs in particular are very useful for detailed surveying and mapping of, for example, urban areas and archaeological sites, especially those applications requiring 3D data. On the other hand, the spatial resolution of commercial satellites is too coarse for many large area projects and the data collection capability of many sensors is restricted by cloud cover. The data volumes from both satellites and aerial cameras can be very large and create storage and processing problems for all but the most modern systems. The cost of data can also be prohibitive for a single project or organization.

2.5.1.1.2 *Vector Data Capture*

Primary vector data capture is a major source of geographic data. The two main branches of vector data capture are:

— Ground surveying and

— GPS.

2.5.1.1.2.1 *Ground Surveying*

Ground surveying is based on the principle that the 3D location of any point can be determined by measuring angles and distances from other known points. Surveys begin from a benchmark point. If the coordinate system of this point is known, all subsequent points can be collected in this coordinate system. If it is unknown then the survey will use a local or relative coordinate system. Since all survey points are obtained from survey measurements their locations are always relative to other points. Any measurement errors need to be apportioned between multiple points in a survey. For example, when surveying a field boundary, if the last and first points are not identical in survey terms (within the tolerance employed in the survey) then errors need to be apportioned between all points that define the boundary. As new measurements are obtained these may change the locations of points. For this reason it is necessary to store both the measurements and the points inside a GIS database until the survey is complete.

Traditionally, surveyors used equipment like transits and theodolites to measure angles, and tapes and chains to measure distances. Today these have been replaced by electro-optical devices called total stations that can measure both angles and distances to an accuracy of 1 mm. Total stations automatically log data and the most sophisticated can create vector point, line, and polygon objects in the field, thus providing direct validation. The basic principles of surveying have changed very little in the past 100 years. Two people are usually required to perform a survey, one to operate the total station and the other to hold a reflective prism that is placed at the object being measured. On some remote controlled systems a single person can control both the total station and the prism. Ground survey is a very time-consuming and expensive activity, but it is still the best way to obtain highly accurate point locational data. Surveying is typically used for capturing buildings, land and property boundaries, man-holes and other objects that need to be located accurately. It is also used to obtain reference marks for other data capture methods. For example, large-scale aerial photographs and satellite images are frequently georeferenced using points obtained from ground survey.

2.5.1.1.2.2 *Global Positioning System (GPS)*

The Global Position System (GPS) is a collection of 27 NAVSTAR satellites orbiting the Earth at a height of 12,500 miles, five monitoring stations, and individual receivers (Steede-Terry 2000). The GPS was originally funded by the US Department of Defense, and for many years military users had access to only the most accurately data. Fortunately this *selective availability* was removed in May 2000 so that now civilian and military users can fix the *x, y, z* location of objects relatively easily to an accuracy of better than 10m with standard equipment. In many respects GPS has revolutionized primary data capture, especially since the development of Differential GPS, the removal of selective availability, and the creation of low-cost, low-power receivers. Today units costing less than $100 can easily provide locational data at better than 10m accuracy. One of the drawbacks of GPS, however, is that it is necessary to have three or more satellites in unobstructed view in order to collect measurements. This can especially be a problem in forests and urban areas with tall buildings. GPS is very useful for recording ground control points for other data capture projects, for locating objects that move (for example, combine harvesters, tanks, cars, and shipping containers), and for direct capture of the locations of many types of objects such as utility assets, buildings, geological deposits, and stream sample points. Strictly speaking, the term GPS refers, only to the US Department of Defense System. GLONASS is the Russian version of GPS offering similar coverage and accuracy; Galileo is the European Union's proposed equivalent.

2.5.1.1.2.2.1 *Principles of GPS*

GPS works according to a simple principle – the length of time it takes a signal to travel from a satellite to a receiver on the ground. The GPS satellites constantly transmit a coded radio signal that indicates their exact position in space and time. The receiver measures how long it takes the signal to travel from the satellite. By measuring the distance from three or more satellite, the location of the receiver can be obtained by triangulation. If a signal can be obtained from a fourth satellite, then the elevation of the receiver can also be determined. Although standard GPS receivers can provide locations at accuracies of 5–10 m, it is important to understand that there are several possible sources of error inherent in these locations. Some of the errors are random in nature, while others are systematic and can therefore be corrected. Errors arise from signal degradation due to atmospheric effects, minor variations in the location of the satellite, inaccuracies in the timing clocks, errors in receivers, and variations in the reflection of signals from local objects.

A number of techniques are available to improve the accuracy of GPS measurements. Many GPS receivers perform averaging of measurements to improve apparent accuracy. Others snap measurements to map features. So, for example, in-car navigation systems snap the location of the vehicle to a road centerline. The accuracy of measurements can also be improved by using Differential GPS. This technique uses two receivers. One is fixed and the other is used to collect measurements. If the location of the fixed (base) receiver is known accurately, comparing the exact location with the location reported by GPS will provide an estimate of error. This error can be used to correct measurements obtained from the roving receiver provided that it is within about 300 km. In some countries, the differential correction information is broadcast freely over airwaves and can be received using a standard radio receiver. Differential GPS can improve accuracy to allow locations to be determining to better than 1m.

2.5.1.2 *Secondary Geographic Data Capture*

Geographic data capture from secondary sources is the process of creating raster and vector files and databases from maps and other hardcopy documents. Scanning is used to capture raster data. Table digitizing, heads-up digitizing, stereo-photogrammetry, and COGO data entry are used for vector data.

2.5.1.2.1 *Raster data capture using scanners*

A scanner is a device that converts hardcopy analog media into digital images by scanning successive lines across a map or document and recording the amount of light reflected from a local data source. The differences in reflected light are normally scaled into bi-level black and white (1 bit per pixel), or multiple gray levels (8, 16, or 32 bits). Color scanners output data into 8-bit red, green, and blue color bands. The spatial resolution of scanners varies widely from as little as 100 dpi (4 dots per mm) to 1800 dpi (72 dots per mm) and beyond. Most GIS scanning is in the range 400–1000 dpi (16–40 dots per mm). Depending on the type of scanner and the resolution required, it can take from 30 seconds to 30 minutes or more to scan a map. Litton (1998) provides an excellent introduction to scanning for GIS.

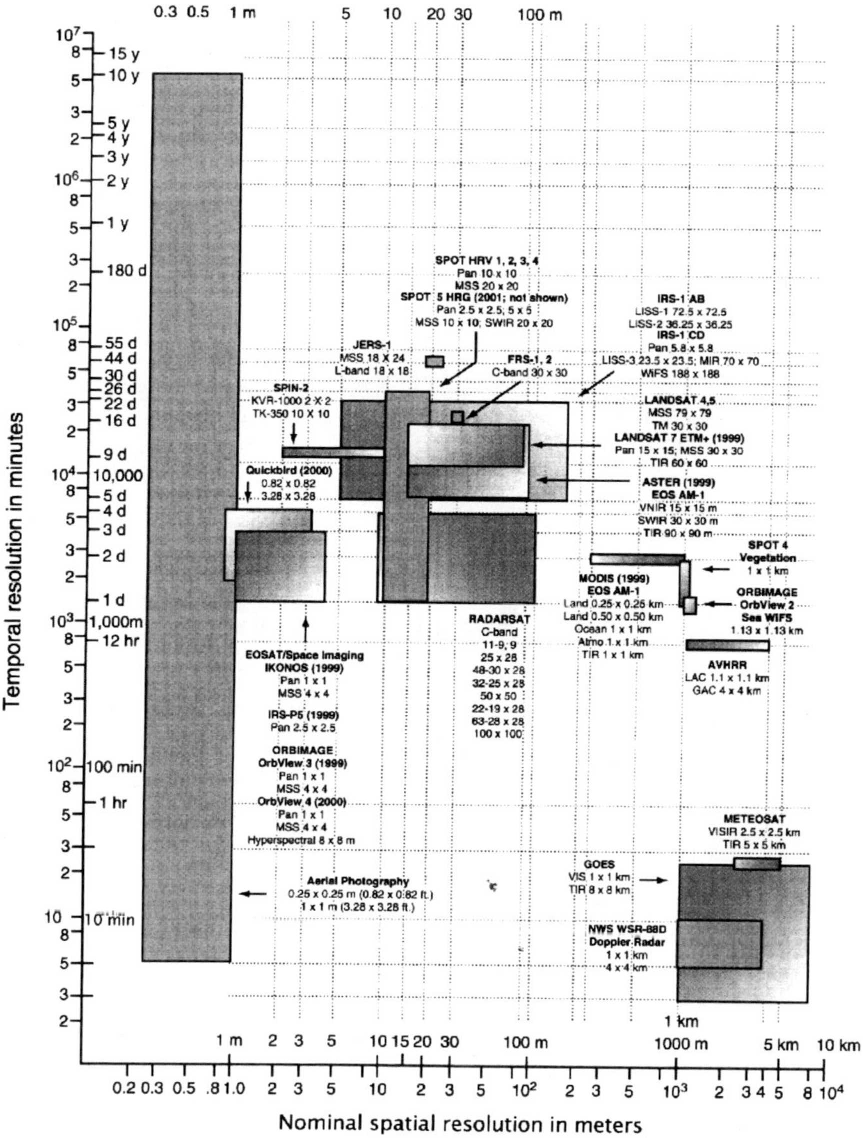

Figure 2.13. Spatial and temporal characteristics of commonly used remote sensing system and their sensors

There are three reasons to scan hardcopy media for use in GIS:

- Documents, such as building plans, CAD drawings, property deeds, and equipment photographs are scanned to reduced wear and tear, improve access, provide integrated database storage, and to index them geographically (e.g. building plans can be attached to building objects in geographic space).

- Film and paper maps, aerial photographs, and images are scanned and georeferenced so that they provide geographic context for other data (typically vector layers). This type of unintelligent image or background *geographic wall-paper* is very popular in systems that manage equipment and land and property assets.
- Maps, aerial photographs, and images are also scanned prior to vectorization (see below).

An 8 bit (256 gray levels) 400 dpi (16 dots per mm) scanner is a good choice for scanning maps for use as a background GIS reference layer. For a color aerial photograph that is to be used for subsequent photo-interpretation and analysis, a color (8 bit for each of three bands) 1000 dpi (40 dots per mm) scanner is more appropriate. The quality of data output from a scanner is determined by the nature of the original source material, the quality of the scanning device, and the type of preparation prior to scanning (e.g. redrafting key features or removing unwanted marks will improve output quality).

2.5.1.2.2 Vector Data Capture

Secondary vector data capture involves digitizing vector objects from maps and other geographic data sources. The most popular methods are manual digitizing, heads-up digitizing and vectorization, photogrammetry, and COGO data entry.

2.5.1.2.2.1 Manual Digitizing

Manually operated digitizers are much the simplest, cheapest, and most commonly used means of capturing vector objects from hardcopy maps. Digitizers come in several designs, sizes, and shapes. They operate on the principle that it is possible to detect the location of a cursor or puck passed over a table inlaid with a fine mesh of wires. Accuracies typically range from 0.003 inch (0.075 mm) to 0.010 inch (0.25 mm). Small digitizing tablets up to 12 by 24 inches (30 by 60 cm) are used for small tasks, but bigger (typically 50 by 32 inches (120 by 80 cm)) freestanding table digitizers are preferred for larger tasks. Both types of digitizer usually have cursors with cross hairs mounted in glass and buttons to control capture.

Vertices defining point, line, and polygon objects are captured using manual or stream digitizing methods. Manual digitizing involves placing the centre point of the cursor cross hairs at the location for each object vertex and then clicking a button on the cursor to record the location of the vertex. Stream mode digitizing partially automates this process by instructing the digitizer control software automatically to collect vertices every time a distance or time threshold is crossed (e.g. every 0.02 inch (0.5 mm) or 0.25 second). Stream-mode digitizing is a much faster method, but it typically produces larger files with many redundant coordinates.

2.5.1.2.2.2 Heads-up digitizing and vectorization

One of the main reasons for scanning maps is as a prelude to vectorization. The simplest way to create vectors from raster layers is to digitize vector objects manually straight off a computer screen using a mouse or digitizing cursor. This method is called heads-up digitizing because the map is vertical and can be viewed without bending the head down. It is widely used for selective capture of, for example, land parcels, buildings, and utility assets. A faster and more consistent approach is to use software to perform automated vectorization in either batch or interactive mode. Batch vectorization takes an entire raster file and converts it to vector objects in a single operation. Vector objects are created using software algorithms that build simple (spaghetti) line strings from the original pixel values. Depending on the size of the raster file, it typically takes 1–100 minutes to complete vectorization.

Unfortunately, batch vectorization software is far from perfect and post-vectorization editing is required to clean up errors. To avoid large amounts of vector editing, prior to vectorization it is useful to undertake a little raster editing of the original raster file to remove unwanted noise that may affect the vectorization process. For example, text that overlaps lines should be deleted and dashed lines are best converted to solid lines. Following vectorization, topological relationships are usually created for the vector objects. This process may also highlight some previously unnoticed errors that require additional editing. Batch vectorization is best suited to simple bilevel maps of, for example, contours, streams, and highways. For more complicated maps and where selective vectorization is required (for example, digitizing electric conductors and devices or water mains and fittings off topographic maps), interactive vectorization (also called semi-automatic vectorization, line following, or tracing).

2.5.2 Preprocessing

It involves manipulating the data in several ways so that it may be entered it into the GIS. Two of the principal tasks of prepossessing including data format conversion and identifying the locations of objects in the original data in a systematic way. Converting the format of the origin data often involves extracting information from maps, photographs, and printed records (such as demographic reports) and then recording this information in a computer database. This process is a time-consuming and costly effort for many organizations. This is particularly (and sometimes painfully) true when one calculates the costs of converting large volumes of data based on paper maps and transparent overlays, to an automated GIS based on computerized data sets. A second key task of the preprocessing phase is to establish a consistent system for recording and specifying the locations of objects in the data sets. When this task is completed, it is possible to determine the characteristics of any specified location in terms of the contents of any data layer in the system. During these processes, it is very important to establish specific quality control criteria for monitoring the operation during the preprocessing phase, so that the databases can be of maximum value to the user.

2.5.3 Data Management

The functions of data management govern the creation of, and access to, the database itself. These functions provide consistent methods for data entry, update, deletion and retrieval. Modern database management systems isolate the users from the details of data storage, such as the particular data organization on a mass storage medium. When the operations of data management are executed well, the users usually do not notice. When they are done poorly, everyone notices; the system is slow, cumbersome to use, and easy to disrupt. Under these circumstances, the smallest human and machine errors create large problems for both the users and the system operators. Data management concerns include issues of security. Processors must be in place to provide different users with different kinds of access to the system and its database. For example, database update may be permitted only after a control authority has verified that the change is both appropriate and correct.

2.5.4 Manipulation and Analysis

They are often the focus of attention for a user of the system. Many users believe, incorrectly, that this module is all that constitutes a geographic information system. In this portion of the system, the analytic operators work with the database contents to derive new information. For example, one may specify a region of interest and request that the average slope of the area be calculated, based on the contours of elevation that have already been stored in the GIS database. Since no single system can encompass the complete range of analytic capabilities a user can imagine, one must have specific facilities to be able to move data and information between systems. For example, there may need to move data from a GIS to an external system where a particular numerical model is available, and then transport the derived results back into the spatial database inside the GIS. This kind of modularity, where other data processing and analysis systems can be linked to a GIS, is very valuable in many circumstances, and permits the system to be easily extended over time by paring it with other analytic tools. When one speaks of geoprocessing, one is often focused on the manipulation and analysis components of a GIS.

2.5.5 Product Generation

It is the phase where final outputs from the GIS are created. These output products may include statistical reports (such as a table or reports, maps and graphics of various kinds (such as a set of bar charts). Some of these products are soft copy images: these are transient images on television-like computer displays. Others, which are durable since they are printed on paper and film, are called hard copy. Increasingly, output products include computer-compatible materials: tapes and disks in standard formats for storage in an archive or for transmission to another system. The capability of taking the output of an analytic process, and placing it back into the geographic database for future analysis, is extremely important.

2.6 MEASUREMENTS IN GIS

The line measurements involve distance operations. These operations include measurement of the distance between two points along straight or curvilinear lines. For example, the length of a bus route or an emergency route can be determined using the line measurement function. There are two basic area measurements: the area of a polygon, measuring the extent of an object, and the perimeter of a polygon area, calculating the distance around the object. A measurement of a flood area is an example of area measurement, while the perimeter operation can be useful in measuring the length of a dike system to contain the food. The fourth category of measurement involves volumetric measurement. This operation is more complex than point, line, or area measurements. It can be performed either by means of a cross-section technique or through overlays of multiple surfaces.

2.7 CAPABILITIES OF GIS

Geographical information system's capabilities could be summarized as follows:

2.7.1 Performing Geographic Queries and Analysis

The ability of GIS to search databases and perform geographic queries has saved many companies literally millions of dollars. GIS have helped reduce costs by

- Streamlining customer service.
- Reducing land acquisition costs through better analysis.
- Reducing fleet maintenance costs through better logistics.
- Analyzing data quickly, as in this example:

A realtor could use a GIS to find all houses within a certain area that have tiled roofs and five bedrooms, then list their characteristics. The query could be further refined by adding criteria - the house must cost less than $100 per square foot. You could also list houses within a certain distance of a school.

2.7.2 Improving Organizational Integration

Many organizations that have implemented a GIS have found that one of its main benefits is improved management of their own organization and resources. Because GIS have the ability to link data sets together by geography, they facilitate interdepartmental information sharing and communication. By creating a shared database, one department can benefit from the work of another - data can be collected once and used many times. As communication increases among individuals and departments, redundancy is reduced, productivity is enhanced, and overall organizational efficiency is improved. Thus, in a utility company the customer and infrastructure databases can be integrated so that when there is planned maintenance, affected customers can be sent a computer-generated letter.

2.7.3 Making Better Decisions

The old adage "better information leads to better decisions" is as true for GIS as it is for other information systems. A GIS, however, is not an automated decision making system but a tool to query, analyze, and map data in support of the decision making process. GIS technology has been used to assist in tasks such as presenting information at planning inquiries, helping resolve territorial disputes, and sitting pylons in such a way as to minimize visual intrusion. GIS can be used to help reach a decision about the location of a new housing development that has minimal environmental impact, is located in a low-risk area, and is close to a population center. The information can be presented succinctly and clearly in the form of a map and accompanying report, allowing decision makers to focus on the real issues rather than trying to understand the data. Because GIS products can be produced quickly, multiple scenarios can be evaluated efficiently and effectively.

2.7.4 Making Maps

Maps have a special place in GIS. The process of making maps with GIS is much more flexible than are traditional manual or automated cartography approaches. It begins with database creation. Existing paper maps can be digitized and computer-compatible information can be translated into the GIS. The GIS-based cartographic database can be both continuous and scale free. Map products can then be created centered on any location, at any scale, and showing selected information symbolized effectively to highlight specific characteristics. The characteristics of atlases and map series can be encoded in computer programs and compared with the database at final production time. Digital products for use in other GIS can also be derived by simply copying data from the database. In a large organization, topographic databases can be used as reference frameworks by other departments.

2.8 APPLICATIONS OF GIS

GIS has been found to be useful to planners, policy makers and decision makers. It enables decision makers, planners, project mangers and development practitioners to improve management of natural resources and overall development of a region. It allows entry of many types of data in a single spatial framework.

A GIS facilitates the following:

1. Overlays of data for purpose of comparison
2. Updating of information to illustrate changes over time
3. Changes of scale for micro-analysis
4. Derivation of non-available data through manipulation of known factors
5. Integration of physical and social science data sets
6. Incorporation of remotely sensed data such as satellite imagery for continuous environmental monitoring
7. Modeling of physical, economic and social processes for the purpose of simulation and prediction
8. Digital format data is compact and large quantities of data can be maintained and retrieved at a greater speed and lower cost
9. Ability to manipulate and integrate different types of data in a single analysis
10. Ability to perform complex spatial analysis rapidly providing both qualitative and quantitative advantage

11. Planning scenarios, decision models and interactive process are normal functions of GIS.

Some of the important applications of GIS are Engineering mapping, Cut/fill analysis, Utility/facility mapping and management, Surface water mapping, Dynamic event mapping, Census and related statistical mapping, Management of watershed and its prioritization, Environmental impact studies, E-business/E-commerce/E-governance, Mobile mapping/WAP application, 3D/4D GIS and Virtual GIS, GPS and GIS integration. Applications of GIS in various disciplines are shown in Figure 2.13. GIS is also being used for a wide range of planning applications for preparing work orders to repair roads and sewers to providing comparative analyses of alternative development plans.

Some of the planning applications include -

1. Cadastral Information System – land ownership database required for tax and revenue planning; demographic analysis; social services planning, etc.
2. Utilities Mapping and Management – electric; water; telecommunication; road; drainage; sewer, etc.
3. Land use Planning – potential for development; suitability assessment for crops, industrial and for a particular land use; sitting of important structures, such as nuclear plants, power stations, etc.
4. Agricultural Development – identify developmental needs in agriculture at village level, etc.
5. Environmental Condition Assessment – environmental impact assessment; environmental sensitive zones, etc.
6. Wasteland Development – suggest an optimum plan considering people's needs, etc.
7. Natural Resource Management – forest; soil; water, etc.
8. Integrated Development Planning - based on resource potential, existing settlement pattern and population; suggest developmental planning in resources and services in a sustainable manner, etc.

2.9 FUTURE OF GIS IN INDIA

The GIS project requires certain objectives to be set, which have to be followed by manuscript preparation of desired layers, decision on attribute information, codification of various polygons, lines and point data, conversion of analogue data into digital format by digitizing/scanning and vectorising, editing, projecting, transforming, edge matching, map joining, dissolving, etc., before data is analyzed and results presented. In developed countries, the main adopters of GIS were planning and development departments, highway and estate departments and other legal services, parks and recreation departments. The basis of acceptance of GIS was its faster information processing capability, better decision support system and savings in cost and time. In India, a major initiative would be required by the different organisations/ departments to initiate applications of GIS.

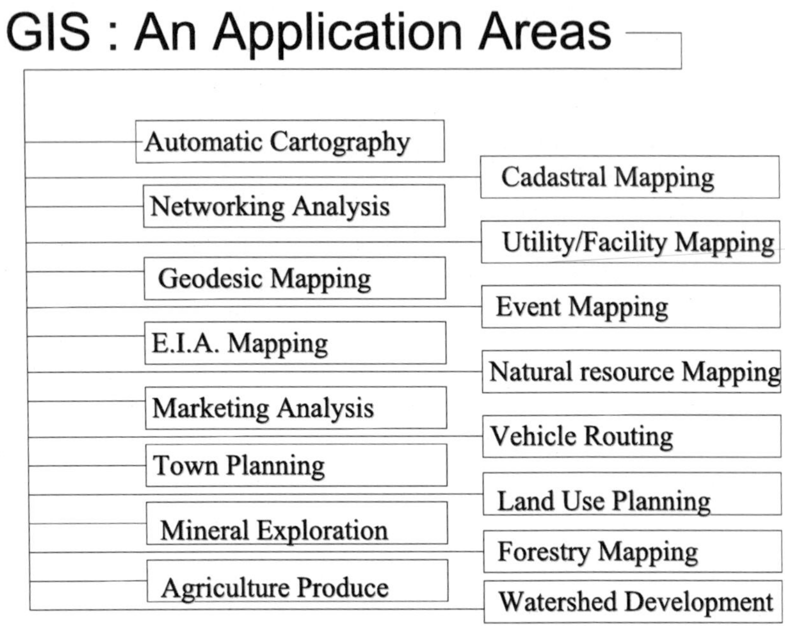

Figure 2.14: Applications of GIS

The information technology plays an important role for the communication of real-time information. In addition, it saves lot of time in duplicating the same information if the information is available in digital form. Like other fields, information technology brings out a revolution in the field of GIS too. Mobile phones, E-mail, Internet, and Global Positioning Systems are some of the examples of information technology which can benefit the direction applications of GIS by providing latest and precise information. The advantage of information technology increases many folds if the data are to be transmitted in real-time to various organisations within India and abroad. Collection of field data for the preparation of GIS data base is a time-consuming task. Since most of the studies involving the use of GIS lack from non-availability of desired information, the information technology can play a key role to update the GIS data base as well exchange of results. It is therefore, extremely necessary to have a nodal point in the country which would be responsible to collect and analyse the information as well disseminate the results to user's organisations. This nodal center may have all the necessary facilities and infrastructure as well as the manpower to take up the responsibility. All the organisations having required data may be electronically networked to the nodal center for flow of real-time information and these organisations may automatically communicate the information with the nodal center as and when there is a change in information. The data format requirement may also be decided before hand, so that the data available from various organisations is in a common format.

Chapter - 3

GLOBAL POSITIONING SYSTEM

The Global Positioning System (GPS) is a burgeoning technology, which provides unequalled accuracy and flexibility of positioning for navigation, surveying and GIS data capture. The GPS NAVSTAR (Navigation Satellite timing and Ranging Global Positioning System) is a satellite-based navigation, timing and positioning system. The GPS provides continuous three-dimensional positioning 24 hrs a day throughout the world. The technology seems to be beneficiary to the GPS user community in terms of obtaining accurate data upto about100 meters for navigation, metre-level for mapping, and down to millimetre level for geodetic positioning. The GPS technology has tremendous amount of applications in GIS data collection, surveying, and mapping.

3.1 GEO-POSITIONING - BASIC CONCEPTS

By positioning we understand the determination of stationary or moving objects. These can be determined as follows:

- In relation to a well-defined coordinate system, usually by three coordinate values and
- In relation to other point, taking one point as the origin of a local coordinate system.

The first mode of positioning is known as point positioning, the second as relative positioning. If the object to be positioned is stationary, we term it as static positioning. When the object is moving, we call it kinematic positioning. Usually, the static positioning is used in surveying and the kinematic position in navigation.

3.2 BASIC FACTS

The GPS uses satellites and computers to compute positions anywhere on earth. The GPS is based on satellite ranging. That means the position on the earth is determined by measuring the distance from a group of satellites in space. The basic principle behind GPS is really simple, even though the system employs some of the most high-tech equipment ever developed. In order to understand GPS basics, the system can be categorised into following five logical Steps:

- Triangulation from the satellite is the basis of the system.
- To triangulate, the GPS measures the distance using the travel time of the radio message.

- To measure travel time, the GPS need a very accurate clock.
- Once the distance to a satellite is known, then we need to know where the satellite is in space.
- As the GPS signal travels through the ionosphere and the earth's atmosphere, the signal is delayed.

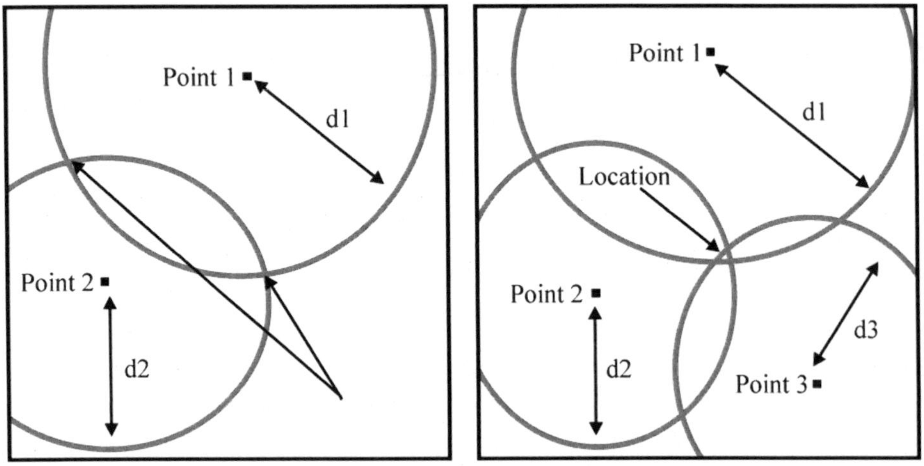

Figure 3.1. Triangulation Method

To compute a position in three dimensions, we need to have four satellite measurements (Figure 3.1). The GPS uses a trigonometric approach to calculate the positions, The GPS satellites are so high up that their orbits are very predictable and each of the satellites is equipped with a very accurate atomic clock.

3.3 COMPONENTS OF GPS SYSTEM

The GPS is divided into three major components (Figure 3.2)

- Control Segment
- Space Segments
- User Segment

3.3.1 Control Segment

The Control Segment consists of five monitoring stations (Colorado Springs, Ascesion Island, Diego Garcia, Hawaii, and Kwajalein Island) as shown in Figure 3.3. Three of the stations (Ascension, Diego Garcia, and Kwajalein) serve as uplink installations, capable of transmitting data to the satellites, including new ephemerides (satellite positions as a function of time), clock corrections, and other broadcast message data, while Colorado Springs serves as the master control station.

The Control Segment is the sole responsibility of the DoD who undertakes construction, launching, maintenance, and virtually constant performance monitoring of all GPS satellites. The DOD monitoring stations track all GPS signals for use in controlling the satellites and predicting their orbits. Meteorological data also are collected at the monitoring stations, permitting the most accurate evaluation of tropospheric delays of GPS signals. Satellite tracking data from the monitoring stations are transmitted to the master control station for processing. This processing involves the computation of satellite ephemerides and satellite clock corrections. The master station controls orbital corrections, when any satellite strays too far from its assigned position, and necessary repositioning to compensate for unhealthy (not fully functioning) satellites.

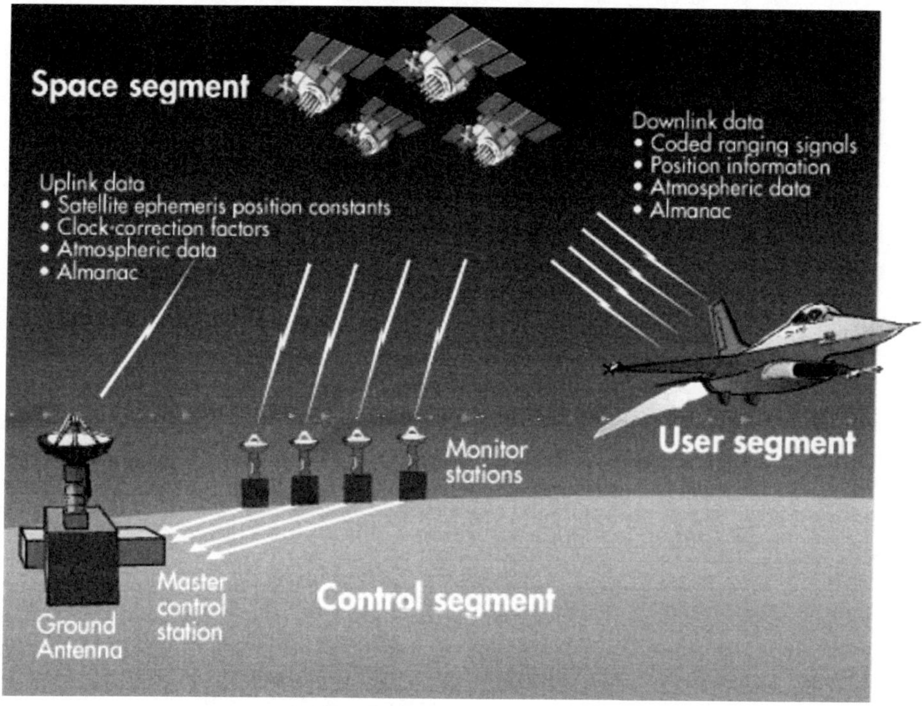

Figure 3.2. Components of GPS

3.3.2 Space Segment

The Space Segment consists of the Constellation of NAVASTAR earth orbiting satellites (Figure 3.4). The current Defense Department plan calls for a full constellation of 24 Block II satellites (21 operational and 3 in-orbit spares). The satellites are arrayed in 6 orbital planes, inclined 55 degrees to the equator. They orbit at altitudes of about 12000, miles each, with orbital periods of 12 sidereal hours (i.e., determined by or from the stars), or approximately one half of the earth's periods, approximately 12 hours of 3-D position fixes. The next block of satellites is called Block IIR, and they will provide improved reliability and have a capacity of ranging between satellites, which will increase the orbital accuracy.

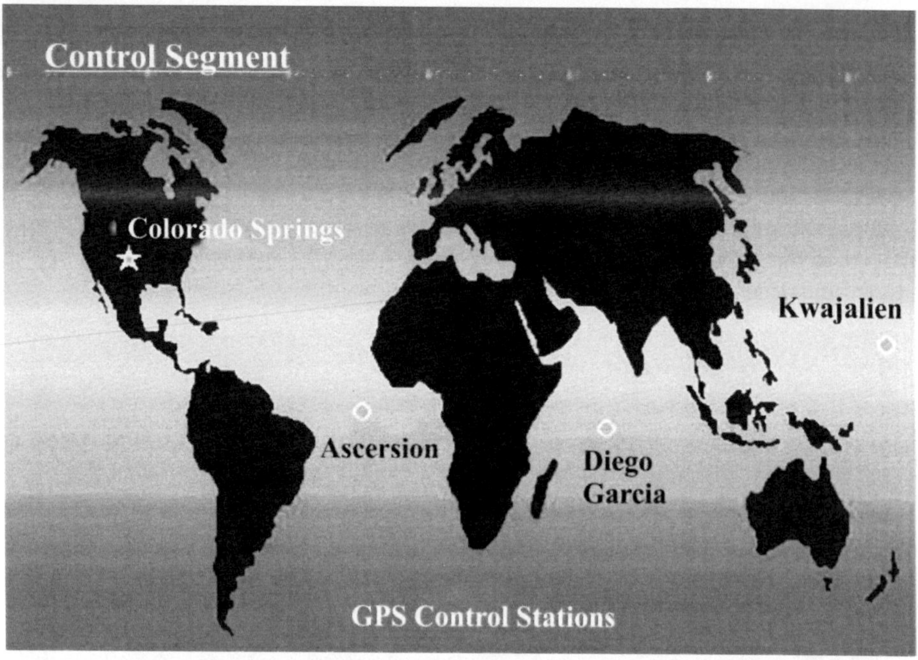

Figure 3.3. Control Segment

Each satellite contains four precise atomic clocks (Rubidium and Cesium standards) and has a microprocessor on board for limited self-monitoring and data processing. The satellites are equipped with thrusters which can be used to maintain or modify their orbits.

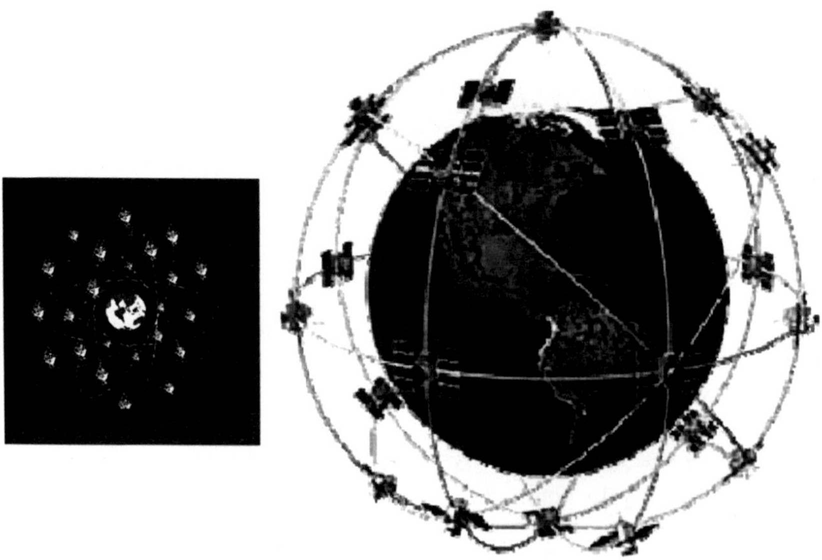

Figure 3.4 Space Segment

3.3.3 User Segment

The user segment is a total user and supplier community, both civilian and military. The User Segment consists of all earth-based GPS receivers. Receivers vary greatly in size and complexity, though the basic design is rather simple. The typical receiver is composed of an antenna and preamplifier, radio signal microprocessor, control and display device, data recording unit, and power supply. The GPS receiver decodes the timing signals from the 'visible' satellites (four or more) and, having calculated their distances, computes its own latitude, longitude, elevation, and time. This is a continuous process and generally the position is updated on a second-by-second basis, output to the receiver display device and, if the receiver display device and, if the receiver provides data capture capabilities, stored by the receiver-logging unit.

3.4 TYPE OF GPS POSITIONING

The various types of GPS positioning (Figure 3.5) could be summarized as follows:

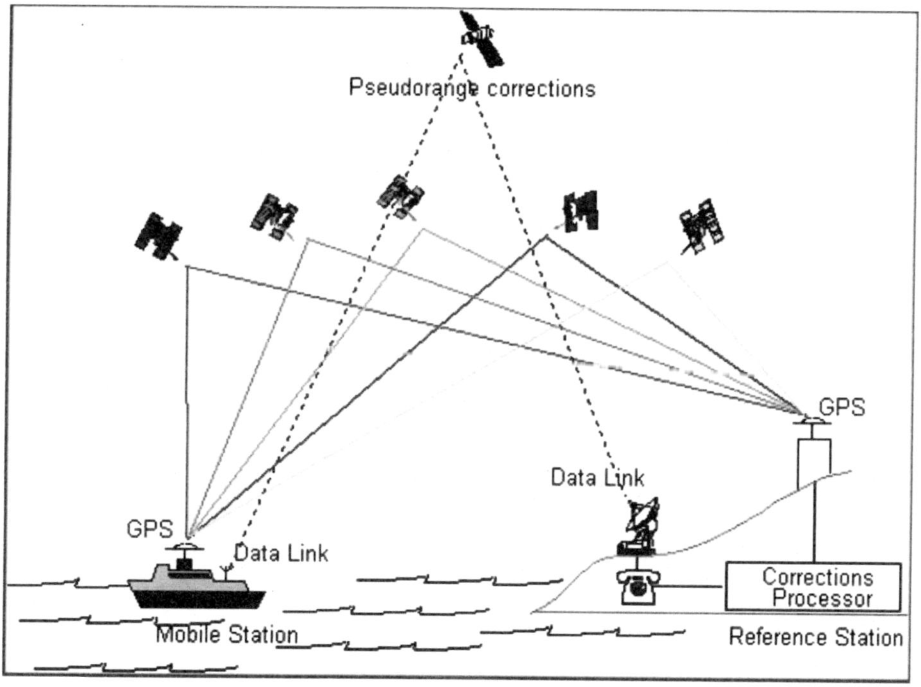

Figure 3.5. Concept of Geopositioning

3.4.1 Absolute Positioning

The mode of positioning relies upon a single receiver station. It is also referred to as 'stand-alone' GPS, because, unlike differential positioning, ranging is carried out strictly between the satellite and the receiver station, not on a ground-based reference station that assists with the computation of error corrections. As a result, the positions derived in absolute mode are subject to the unmitigated errors inherent in satellite positioning. Overall accuracy of absolute positioning is considered to be no greater than 50 meters at best by Ackroyd and Lorimer and to be + 100 meter accuracy by the U.S. Army Corps of Engineers.

3.4.2 Differential Positioning

Relative or Differential GPS carries the triangulation principles one step further, with a second receiver at a known reference point. To further facilitate determination of a point's position, relative to the known earth surface point, this configuration demands collection of an error-correcting message from the reference receiver.

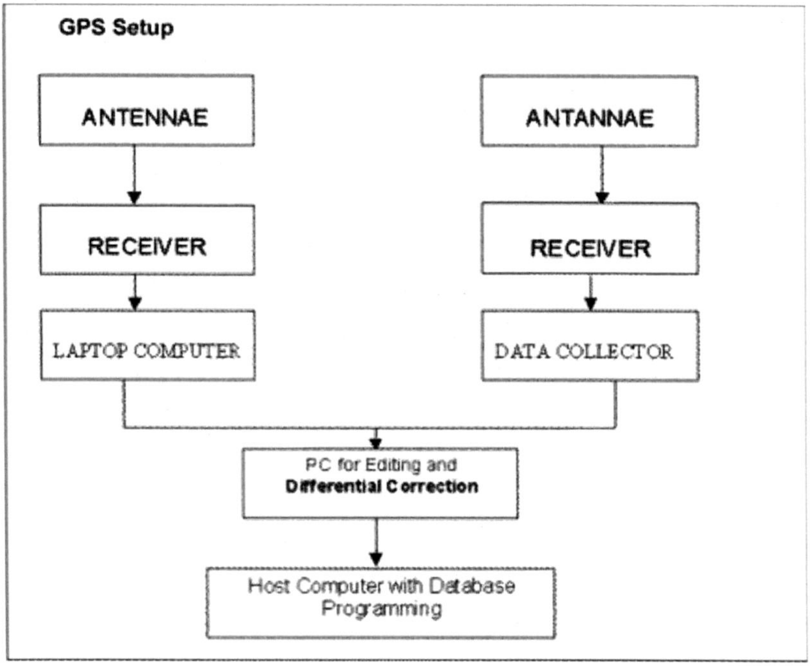

Figure 3.6. Concept of Differential Geopositioning

Differential-mode positioning relies upon an established control point. The reference station is placed on the control point, a triangulated position, the control point coordinate. This allows for a correction factor to be calculated and applied to other roving GPS units used in the same area and in the same time series. Inaccuracies in the control point's coordinate are directly additive to errors inherent in the satellite positioning process. Error corrections derived by the reference station vary rapidly, as the factors propagating position errors are not static over time. This error correction allows for a considerable amount of error of error to be negated, potentially as much as 90 percent.

3.5 ACCURACY OF GPS

There are four basic levels of accuracy or types of solutions with real-time GPS mining system:

Table 3.1. Accuracy in GPS

Autonomous	Accuracy	15 - 100 meters
Differential GPS (DGPS)	Accuracy	0.5 - 5 meters
Real-Time Kinematic Float (RTK Float)	Accuracy	20cm - 1 meter
Real-Time Kinematic Fixed (RTK Fixed)	Accuracy	1cm - 5 cm

GPS satellites broadcast on three different frequencies, and each frequency (or career wave) has some information or codes on it. One can think of it as three different radio stations broadcasting several different programs. The table below lists the signals and the contents:

Table 3.2. GPS Frequencies.

L1 Career	L2 Career	L3 Career
19 cm wavelength	24 cm wavelength	Data not available
1227 M Hz	1775 M Hz	
C/A Code	P Code	
Navigation	Navigation Message	

- *P Code :* Reserved for direct use only by the military
- *C/A Code :* Used for rougher positioning
- For Single frequency use only L1 career is used
- For Double frequency, L1/L2/L3 career is used
- The navigation message (usually referred to as the ephemeris) tells us where the satellites are located, in a special coordinate system called WGS-84. If you know where the satellites are at any given time, then you can compute your location here on earth.

3.6 SOLUTIONS BY A GPS

3.6.1 Autonomous Positions

Uses.......... C/A code only

Requires..... Only one receiver data from at least four satellites

Provides......An accuracy range of about 15 - 100 meters

This solution is designed for people who just need an approximate location on the earth, such as a boat at sea or a hiker in the mountains.

3.6.2 Real-Time Differential GPS (DGPS) Positions

Uses.......... C/A code only

Requires..... Two receivers

A radio link between the two receivers

- Reference receiver at a known location broadcasts RTCM (Radio Technica Commission for Maritime Services) corrections.
- Rover receiver applies corrections for improved GPS positions

Data from at least four satellites - the same four at both the references and rover (common satellites)

Provides...... An accuracy range of about 0.5 - 5 meters depending upon the quality of receiver and antennae used.

This solution gives much better results because here we have a known position at a reference receiver. However it must have a radio link between the reference receiver and the roving (moving) receiver.

3.6.3 Real Time Kinematic (RTK) Float Positions

Uses.......... C/A code and career waves.

Requires..... Two receivers

- Reference receiver at a known location tracks satellites and then broadcasts this satellite data over a radio link in a format called CMR (CMR is a Trimbleâ - defined format).
- Rover receiver receives data from both the satellites and the reference station.

A radio link between the two receivers. Data from at least four common satellites.

Provides An accuracy range of about 20 cm to 1 meters. This solution uses more of the satellite signal than the autonomous or DGPS solution. The CMR data is carrier phase data. The float solution is actually an intermediate step towards the most precise answer, which we'll discuss next.

3.6.4 Real Time Kinematic (RTK) Fixed Solutions

Uses.......... C/A code and career waves.

Requires..... Two receivers

- Reference receiver at a known location tracks satellites and then broadcasts CMR data over a radio link.
- Rover receiver receives data from both the satellites and the reference station.

A radio link between the two receivers.

Initialization, which is achieved most easily with dual-frequency receivers. Data from at least five common satellites to initialize on-the-fly (in motion) Tracking at least four common satellites after initializing.

Provides... An accuracy range of about 1 - 5 cm.

It is noticed that with each increasing level of precision, there are more requirements. The most important unique requirement for the RTK fixed solution is something called an initialization. Here it is not feasible to explain what's happening in an initialization, but it is relevant to mention that initialization is necessary to work at centimeter level accuracy. Dual frequency receivers can perform this process automatically. If the receiver looses the initialization - which can happen if it fails to track enough satellites-then your working accuracy will drop to the float solutions status temporarily. Remember, however, both of these solutions require a radio link to your reference receiver. If, for any reason, you loose your radio link, you will drop back to the autonomous level - the least precise - until the radio link is regained.

3.7 FACTORS AFFECTING GPS

There are a number of potential error sources that affect either the GPS signal directly or your ability to produce optimal results:

- **Number of satellites - minimum number required**

You must track at least four common satellites - the same four satellites - at both the reference receiver and rover for either DGPS or RTK solutions. Also to achieve centimeter -level accuracy, remember you must have a fifth satellite for on-the fly RTK initialization. This extra satellite adds a check on the internal calculation. Any additional satellites beyond five provide even more checks, which is always useful.

- **Multipath - reflection of GPS signals near the antennae**

Multipath is simply reflection of signals similar to the phenomenon of ghosting on our television screen. GPS signals may be reflected by surfaces near the antennae, causing error in the travel time and therefore error in the GPS positions.

- **Ionosphere - change in the travel time of the signal**

Before GPS signals reach your antenna on the earth, they pass through a zone of charged particles called the ionosphere, which changes the speed of the signal. If your reference and rover receivers are relatively close together, the effect of ionosphere tends to be minimal. And if you are working with the lower range of GPS precisions, the ionosphere is not a major consideration. However if your rover is working too far from the reference station, you may experience problems, particularly with initializing your RTK fixed solution.

- **Troposphere - change in the travel time of the signal**

Troposphere is essentially the weather zone of our atmosphere, and droplets of water vapour in it can effect the speed of the signals. The vertical component of your GPS answer (your elevation) is particularly sensitive to the troposphere.

- **Satellite Geometry - general distribution of the satellites**

Satellite Geometry - or the distribution of satellites in the sky - effects the computation of your position. This is often referred to as Position Dilution of Precision (PDOP) (Figure 3.7).

  - PDOP is expressed as a number, where lower numbers are preferable to higher numbers. The best results are obtained when PDOP is less than about 7. PDOP is determined by your geographic location, the time of day you are working, and any site obstruction, which might block satellites. You can use planning software to help you determine when you'll have the most satellites in a particular area.

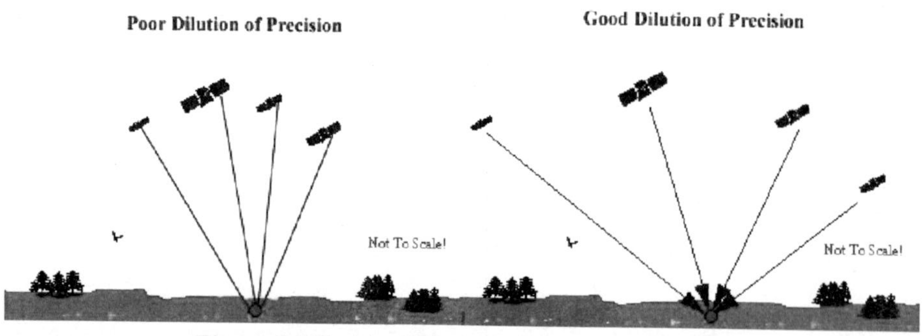

Figure 3.7 Position Dilution of Precision (PDOP).

When satellites are spread out, PDOP is Low (good). When satellites are closer together, PDOP is High (weak).

- **Satellite Health - Availability of Signal**

While the satellite system is robust and dependable, it is possible for the satellites to occassionally be unhealthy. A satellite broadcasts its health status, based on information from the U.S. Department of Defense. Your receivers have safeguards to protect against using data from unhealthy satellites.

- **Signal Strength - Quality of Signal**

The strength of the satellite signal depends on obstructions and the elevation of the satellites above the horizon. To the extent it is possible, obstructions between your GPS antennae and the sky should be avoided. Also watch out for satellites which are close to the horizon, because the signals are weaker.

- **Distance from the Reference Receiver**

The effective range of a rover from a reference station depends primarily on the type of accuracy you are trying to achieve. For the highest real time accuracy (RTK fixed), rovers should be within about 10-15 Km (about 6-9 miles) of the reference station. As the range exceeds this recommended limit, you may fail to initialize and be restricted to RTK float solutions (decimeter accuracy).

- **Radio Frequency (RF) Interference**

RF interference may sometimes be a problem both for your GPS reception and your radio system. Some sources of RF interference include:

  - Radio towers
  - Transmitters
  - Satellite dishes
  - Generators

One should be particularly careful of sources which transmit either near the GPS frequencies (1227 and 1575 MHz) or near harmonics (multiples) of these frequencies. One should also be aware of the RF generated by his own machines.

- **Loss of Radio Transmission from Base**

If, for any reason, there is an interruption in the radio link between a reference receiver and a rover, then your rover is left with an autonomous position. It is very important to set up a network of radios and repeaters, which can provide the uninterrupted radio link needed for the best GPS results.

3.8 REFERENCE STATION

Different levels of accuracy in GPS positions have already been discussed in previous paragraphs. For any level except autonomous (which can have a large amount of error in it), you must have a reference receiver, which is stationary, and a rover, which can be mobile or stationary. The GPS reference station normally operates continuously, 24 hours a day. The coordinates of this station must be known before you can begin using GPS on any of your machines. First a proper site for the reference station is to be selected, then a GPS survey is performed to obtain the known coordinates. This is usually done as part of the installation, either by the installation team or other qualified personnel.

Once it is installed, the GPS reference station can perform two functions simultaneously:

- Receive data from the satellites
- Broadcast GPS data to the rovers in the mine

One reference station can support unlimited rovers. The primary constraint may be distance, because your accuracy may suffer if you're working too far from the reference station. This maximum distance will vary with your accuracy requirements and environment.

3.8.1 Selecting the Reference Station

Some of the features of a good reference site are:

- Clear View to the Sky
- Proximity to your Working Areas: This is both a GPS issue and a radio issue. Remember, RTK is generally limited to about 10-15 Km (6-9 miles)

for reliable initializations, due primarily to potential errors from the ionosphere. Therefore, you should select a reference site that is within about 10-15 Km of where your rovers expect to work.

- Absence of RF Interference: Try to place the reference station away from sources of radio interference, which arise from radio towers, transmitters, television or other satellite dishes, high-voltage power lines, and any other obvious source of interference.
- Minimal Sources of Multipath: Multipath at your reference site can cause inaccurate answers or interfere with your rover's ability to initialize.
- Continuous AC / DC Power Source.
- Stable Monumentation: One should have a stable survey monument or other similarly well-defined physical point at the reference station. Without this, we will have to survey and compute new coordinates to the reference station anytime you move the GPS antennae.
- Stable Antennae Mount: Not only the monument should be stable, but also the GPS antennae itself should be secure and stable to minimize the movement.
- Accessibility of the station.

3.8.2 Reference Station Equipment

- GPS receiver
- GPS antenna
- Radio and antenna, Power supply and Cables

3.8.2.1 Radios

Each GPS rover must receive information from the reference station to achieve accurate positions. To maintain constant communication between your reference station and rover, you need these items at the reference station and at each rover:

- Radio
- Radio Antenna
- Cables

The radios are cabled directly into the GPS receiver. Power may be provided to the radio through the GPS receiver. At the reference site, GPS data is broadcast through the radio. At the rover site, the reference GPS data is received by the radio and routed into the rover receiver, where it is processed together with rover's GPS data/ The rover radio can also draw power from the GPS receiver.

Repeater Radios: If, for any reason, the reference station transmission cannot reach your rovers, then you must use one or more repeaters. A repeater relays the data from your reference or another repeater. The maximum number of repeaters you can use depends on your type of radio. Repeaters differ from your reference and rover radios in two important ways: they must have their own source of power, and they can be moved as the needs change. The radios draw very low power, but they require uninterrupted power. Because repeaters may need to be moved to accommodate your needs, batteries or compact solar power units are normally used.

Frequency and Bandwidth: Most radios used in GPS fall within one of the following frequency ranges:

- 150-174 MHz (VHF)
- 406-512 MHz (UHF)
- 902-928 MHz (spread spectrum)

The lower-frequency radios (150-174 MHZ) tend to have more power, due to design and legal issues (not Physics), However, the bandwidth, which determines the amount of data you can transmit, is narrower in these lower ranges (also due to design, not physics). In the nominal 450 MHz and 900 MHz ranges, the bandwidth is wider. This has positive effects both on the amount of data transmitted and on the number of repeaters possible within the radio network.

3.8.2.2 Radio Range

To guarantee steady, uninterrupted transmission over the radio, one should be aware of some of the factors that affect the radio's effective range.

- *Antenna Height*: raising the radio antenna is the easiest and most effective way to increase range.
- *Antenna design:* radiating patterns vary, depending on the antenna design. For best performance, be sure you understand how yoour antenna transmits signals.
- *Cable length and type:* radio signals suffer loss in cables, so keep the length to a minimum. If you must use long cables, use low-loss cables.
- *Output power:* doubling output power does not double your effective range. Be sure one understands the relationship between power and gain before the best system is decided.
- *Obstructions:* Buildings, walls and even the machines can block or interrupt radio transmission. The repeaters should be carefully used to help minimize the effect of obstructions.

3.8.2.3 Grounding

The radio antenna may be a target for lightning. To avoid damage, you may wish to ground your reference station antenna.

3.9 GPS APPLICATIONS

One of the most significant and unique features of the Global Positioning Systems is the fact that the positioning signal is available to users in any position worldwide at any time. With a fully operational GPS system, it can be generated to a large community of likely to grow as there are multiple applications, ranging from surveying, mapping, and navigation to GIS data capture. The GPS will soon be a part of the overall utility of technology. There are countless GPS applications, a few important ones are covered in the following passage.

3.9.1 Surveying and Mapping

The high precision of GPS carrier phase measurements, together with appropriate adjustment algorithms, provide an adequate tool for a variety of tasks for surveying and mapping. Using DGPs methods, accurate and timely mapping of almost anything can be carried out. The GPS is used to map cut blocks, road alignments, and environmental hazards such as landslides, forest fires, and oil spills. Applications, such as cadastral mapping, needing a high degree of accuracy also can be carried out using high grade GPS receivers. Continuous kinematic techniques can be used for topographic surveys and accurate linear mapping.

3.9.2 Navigation

Navigation using GPS can save countless hours in the field. Any feature, even if it is under water, can be located up to one hundred meters simply by scaling coordinates from a map, entering waypoints, and going directly to the site. Examples include road intersections, corner posts, plot canters, accident sites, geological formations, and so on. GPS navigation in helicopters, in vehicles, or in a ship can provide an easy means of navigation with substantial savings.

3.9.3 Remote Sensing and GIS

It is also possible to integrate GPS positioning into remote-sensing methods such as photogrammetry and aerial scanning, magnetometry, and video technology. Using DGPS or kinematic techniques, depending upon the accuracy required, real time or post-processing will provide positions for the sensor which can be projected to the ground, instead of having ground control projected to an image. GPS are becoming very effective tools for GIS data capture. The GIS user community benefits from the use of GPS for locational data capture in various GIS applications. The GPS can easily be linked to a laptop computer in the field, and, with appropriate software, users can also have all their data on a common base with every little distortion. Thus GPS can help in several aspects of construction of accurate and timely GIS databases.

3.9.4 Geodesy

Geodetic mapping and other control surveys can be carried out effectively using high-grade GPs equipment. Especially when helicopters were used or when the line of sight is not possible, GPS can set new standards of accuracy and productivity.

3.9.5 Military

The GPS was primarily developed for real time military positioning. Military applications include airborne, marine, and land navigation.

3.10 FUTURE OF GPS TECHNOLOGY

Barring significant new complications due to S/A (Selective Availability) from DOD, the GPS industry is likely to continue to develop in the civilian community. There are currently more than 50 manufacturers of GPs receivers, with the trend continuing to be towards smaller, less expensive, and more easily operated devices. While highly accurate, portable (hand-held) receivers are already available, current speculation envisions inexpensive and equally accurate 'wristwatch locators' and navigational guidance systems for automobiles. However, there is one future trend that will be very relevant to the GIS user community, namely, community base stations and regional receive networks, as GPS management and technological innovations that will make GPS surveying easier and more accurate.

■■■■

Chapter - 4

BLAZING TECHNOLOGIES

The importance of E-governance or Information Technology for the modern-day office needs no emphasis, and, offices of government are no exception. While it is agreed that E-governance is not a substitute to Good Governance, it cannot be denied that it is an important tool/aid in that direction. Aware of this aspect, one of the Terms of Reference the Government placed before the Administrative Reforms Committee is:

"To review and make recommendations for the extensive utilization of information technology and all other modern management techniques in Government administration for improving the efficiency and transparency in administration ..." In order to bring about innovation in service delivery to the citizenry, Government must make more and more use of modern technology. This will provide the citizens better access to information and services. It will also help to reduce costs, increase revenues, enhance speed and efficiency and also increase accountability. In short, there is no escape from more and more use of modern technology. In the present study a brief descriptions of two important technologies (E-governance and Geo-informatics) are given in the following paragraphs:

4.1 E-GOVERNANCE

It is well known that finances of the State are under stress and the urgency to improve this financial situation is more relevant now than at any point in the history of the State. E-governance can certainly be one of the steps in that direction because it helps to cut costs by replacing excess manpower, among other things (Figure 4.1). Because of growing awareness, people's expectations have also increased and comparisons with the private sector and developed countries are inevitable. The government must therefore, be seen to be performing and IT is certainly among the better tools for this purpose. Only this can lead to competitiveness, to innovation that is so essential for reforms. In the traditional structure of Government, government departments provide the interface between Information Systems on the one hand and citizens on the other. Ideally, however, it should be the other way round and Information. Systems should provide the means of interface between the Government and the Citizens. These citizens include businessmen and also non-government organizations. Naturally this interface will be non-discretionary and gradually available from any time to anytime and anywhere.

We are still at initial stage of e-Governance. Though lot of work is under progress in G to G application, G to C applications are yet to gain momentum. (G Government, C : Citizens).

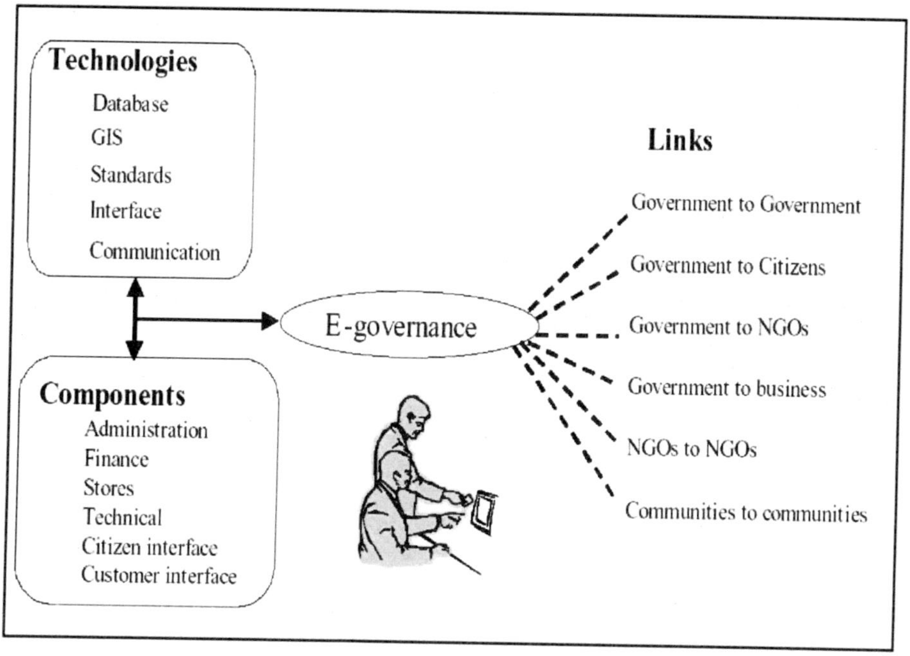

Figure 4.1 Fundamentals of e-governance

E-Governance is a tool for Good Governance not an object in itself. So while declaring agenda for e-Governance, it must be defined what kind of Good Governance will be achieved through e-Governance.

In short role of \`e' is very small or like a catalyst in e-Governance policy and \`Governance' has a prime importance.

4.1.1 Good Governance

To reach at better level of Governance, the system should be effective and result oriented. Because only efficient Governance may be helpful with Government to Government or Government to Employee application but service to citizen may not have precedence in efficiency. Minimum discretion in policy implementation as well as in service delivery is another essentiality of Good Governance. To achieve this it will require high degree of coordination and exchange of information. This coordination would be not only horizontally across the departments but also should include crisis-cross coordination model. This will facilitate exchanging as well as making data available to all levels of department seamlessly and also speedily. This can happen when departmental walls as well as protocol and procedures created for pasting and receiving information should be either torn apart or those should be so transparent that co-ordination, exchange of data becomes effortless.

4.1.2 It Structure of Good Governance

When review is taken of all successful information systems in Government, we observe that usage and benefit of the system is restricted to purpose specific user. Slightest deviation in the requirement is not even entertained by the system. It is therefore necessary to have one common application system spread horizontally across between various Government departments and members of public. Presently Government is known to public by various departments viz. Police, Revenue Administration, Treasuries, P.W.D. etc. This has varied impact on the minds of citizen. In fact Government is a one single entity and hence citizens should also perceive it as a single entity and any transaction with Government. also should be ideally at one point. In today's scenario individuals from life to death activities have a base of paper / certificate / some or other record in printed form. This base boosts requirement of more paper, more authority hierarchy and ultimately entanglement in Red Tape. The conversion of this REDTAPE to ETAPE will be possible. Such vital record can be accessible to the user any where and any time. This will give more power to citizen and eliminate the discretion existing at the level of information providers.

4.1.2.1 *E-Value*

To enable to provide this seamless service efforts are needed to joining up this activity. In this efforts it may not be necessary to create own data or rebuild the information system. But components of data or application can be derived from various departments. This is possible with orientation of computer technology.

4.1.2.2 *Horizontal Integration*

Success of e-Governance system largely depends on how much integrity is possible among various departments. It is observed that 70% of e-Governance initiatives have failed due to lack of integrity. Real benefits of computerization in terms of speed accuracy and volume of service can be achieved only through integration of existing manual or semi manual system. Replacing the existing manual system and introducing new procedures breeds delay or risk of failure.

4.1.2.3 *Alternatives for Integration*

1. *Common Software:* This includes development of common software like ERP package / SAP which can be used by any department with its scaled down or full blown versions. This is feasible when user requirements are fixed and doesn't need any customization.
2. *Co-ordinate Development:* Development of entire IT system is distributed among various user departments. Modules are developed in such a manner that those fit in like jigsaw which makes one Ideal workable system.
3. *City Planning Model:* Each user is allowed to develop applications independently as per his requirements and procedures. However, each user has to follow Master plan of computerization which is like development plan of city.

When developer/user departments adhere to this master plan, connectivity between the software systems and integration of data is ensured. This facilitates horizontal proliferation seamlessly.

4.1.2.4 *Advantages of City Plan Model*

Once the Master plan is made available to user group, they can start the system development immediately. User is not required to wait for clearance and concurrence from other departments and developers. The development is module by module hence costs are spread over the time. With such moderate freedom computerization efforts can be started immediately wherever other administrative machinery is geared up for development and implementation. This also ensures that developmental and maintenance costs are kept at optimum level. It is so possible because these packages are modular and are small in size.

4.1.3 The Road Map Ahead

To have widespread and successful implementation departments and their developers must adopt open standards while designing the system. It is necessary to prepare and maintain common masters with one central agency, which can be used by other system designer and system integrator e.g. Citizens Master, Property Master, Organizations Master, others like Village, Districts, Employee, Accounts Head etc. Simultaneously it is also necessary to create two tier data center at State and District level for sharing, refreshing and updating data generated at resource level. These data centers will also serve for providing master file level data. This common database (expecting to have common data structure) will provide proper interface for web-enabled databases and also service to citizen at their home or kiosks.

4.1.4 Present Status

As a fundamental step towards this aim IT department has issued GR No. COM/1001/CR174/01/39 dt 29/10/2001. In this GR power to sanction IT related projects are delegated to respective Administrative departments. This will speed up the process of computerization projects. In this GR common software developed and made available are also prescribed. This GR also specifies hardware and software platform to be adopted in future by the government departments. Now to build the mechanism to create, update and maintain the master is in process and soon these masters will be available on web site and user can download these masters as well as authentic user can generate code for his employee or relevant data code specific to his software. The impact scenario could be visualized as shown in Figure 4.2. With citizen focused at the center of IT system there are good examples of outstanding individual initiatives implemented as.

- Single window facility in Pune Collectorate
- Citizen Facilitation Centre in Ahmednagar
- Citizen facilities in Nagpur

- 'SETU' in Thane Collectorate with active participation in terms of money and manpower from non government organisation.
- 'Sankalp' in Pune Collectorate.
- Citizen Facilitation Center at Kalyan Dombivali Municipal Corporation.

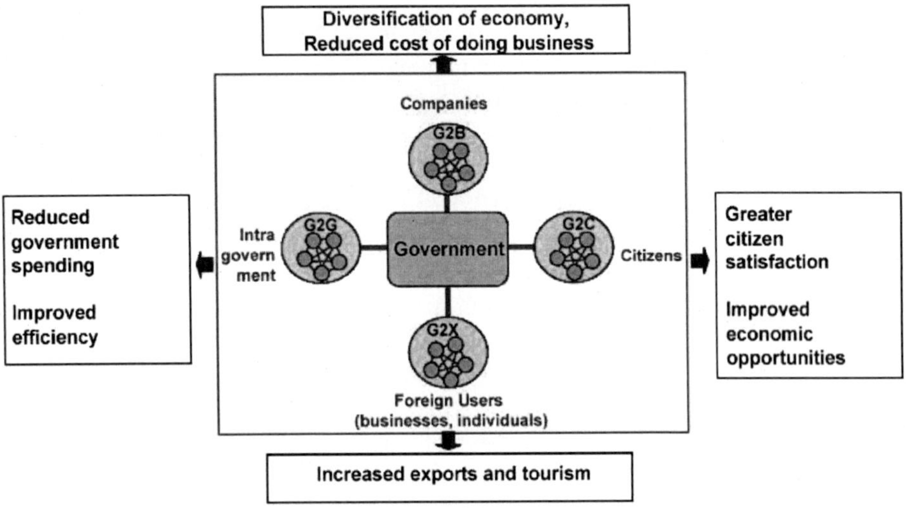

Figure 4.2 Impacts of e-governance

Similar such set ups are coming up in Aurangabad, Nanded, Akola, Latur. Some Zilla Parishads also heading towards such citizen based systems.

4.1.5 Impediment in E-Governance

Though there are good numbers of examples of these successful projects, unfortunately there is a hesitation in replicating these systems by other users. Similarly in absence of clear-cut directives or statutory orders even there is problem in continuing in existing system in the same office. Though objective of the system is identical, each user has either developed different software or used different operating platform. This creates hurdle in exchanging data or information within department itself. Hence, sharing a data with other departments is further more difficult.

There is no awareness of this aspect among staff working on such project. Their mind is not trained to look into such matter. Even operational level formal training modules are also not developed. Even system is running satisfactorily and giving good results, it gets eroded gradually. Single window system established at Pune with touch screen facility came to an halt gradually and nobody can be blamed pointedly. Some systems are implemented without reengineering the existing system suitable to computerization. Wherever reengineering is done, it is not backed by the documents. In some places even there are no formal user manuals or technical manuals.

In absence of proper security features, formal and full proof password system is not existing. Digital signature concept is yet not launched. This has lost credibility of the system. Similarly no orders or GRs are issued to authenticate and support these IT projects. This ultimately boils down as a local venture. When a state level computerization project is implemented and it is monitored centrally at HoD level, there is a possibility of having common strategy but this is not made possible in case of Collectorates, Zilla Parishads, and Municipal bodies. In fact networking and multi user system is ideal environment for such application. But lack of technical expertise to handle network systems, huge capital requirement for initial set up networking couldn't be carried out. Apart from administrative sustainability, financial sustainability is becoming difficult especially funds required for software support, recurring expenditure required for hardware maintenance, stationary and consumables are not available substantially and regularly. Use of all the systems is restricted to office hours and only on working days. Hence, idle time of hardware and software is proportionally much more as compared to its utility and usage. In absence of other infrastructural support like technical man power support, ISDN lines or VSAT facility and computer literacy amongst user the implementation has spread only up to district level though ultimate user or representative citizen of India is in villages.

4.2 GEO-INFORMATICS

Informatics has emerged as the thrust area for the Government as it can enable the administration to re-engineer and improve its processes, connect citizens and build interactions with and within the society by bringing radical changes in its functioning leading to **S**imple, **M**oral, **A**ccountable, **R**esponsive and **T**ransparent **(SMART)** governance. This new way of governance adopted by the public administration for the delivery of services on the Internet and Intranet, constitute the concept of Electronic Governance (E-Governance).

E-Governance leverages on following change potentials for sustainable development:

- Automation replacing existing manual processes, which involve accepting, storing, processing, outputting or transmitting data/information
- Transformation of data into information supporting current processes of decision-making and implementation.
- Transmission of information by extensive use of Electronic Forms and Interfaces through use of web and Internet technology (Paperless Government-On-Line).

With the availability of information technologies including Database Management, Data Warehousing and Data Mining, e-Governance is aimed at converting the transactional data into business relevant information and managing this repository. This asset, in turn, is made accessible to the decision makers by computer based Decision Support System (DSS) using alphanumeric information in various application areas (Figure 4.3).

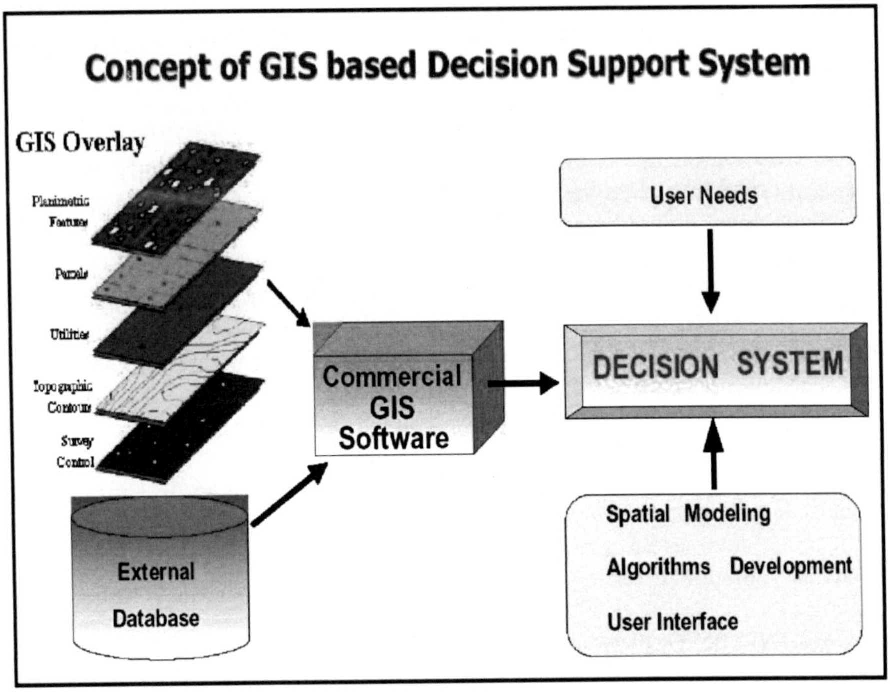

Figure 4.3 Fundamentals of Geo-informatics

The basis for e-Governance is fusion of information with communication technologies (ICT) supporting and transforming the governance by processing and communicating data. Key innovation of e-Governance is computer networks, from intranets to the Internet, creating a digital network between various user nodes

4.2.1 Geo-Informatics in E-Governance

Geo-informatics or Geographic Informatics encompasses all aspects of geo-referenced data viz. acquisition, integration, decision support and dissemination. Its applications are fundamental to population growth, economic development and consumption of natural resources. Geo-informatics addresses the vital elements of any information system viz. geographic measurement, geo-accounting, spatial analysis and integrated decision-making. Geo-informatics comprises of remote sensing, photogrammetry, geographical information system (GIS), cartography and GPS. Currently, it addresses market of ten billion-dollar in the area of services and software products. Functions of e-Governance include planning, preparation and approval of mega-plans, management of existing infrastructure and restructuring of facilities. 80% - 90% of government data is geographic in nature - containing an address, service boundary, pin code, or latitude and longitude co-ordinates. In local government, city planners view maps for development plan; engineers need information on utilities to forecast how serving a new colony will affect overall service; and the estate office updates data with measurements taken from a recent survey.

Table 4.1 Application areas for e-governance using Geo-informatics

Area	Benefits	Beneficiaries
Land records management	• Mapping and updation of land parcels using high resolution remote sensing data and field surveys • Computerized maps of land parcels • Easy updation of changes in land mutations and ownership • Linkages to RDBMS for management of land records	• Land Registration • Revenue Department • Land Settlement
Industrial estate management	• Mapping of industrial plots and infrastructure using high resolution remote sensing data and field surveys • Computerized maps of layout and infrastructure details • On line allotment of lands to customer • DSS for user departments viz. Land, Engineering, Planning departments • Web GIS based application for providing plot level details to potential customer	• State Industrial Development Corporations/ Councils • Industrialists • Potential investors
Disaster Management	• Mapping of disaster prone areas (flood, earthquakes, cyclones, landslides, snow avalanches, fire etc.) at varying scales • Mapping of amenities and infrastructure required for disaster management • Integration with network technology for exchange of information between and among institutions, governments, citizens, NGOs etc. • DSS built around Disaster Management Information System can address needs at all the stages such as preparedness, early warning, relief/response and Mitigation.	• Administration • Rehabilitation Cells • NGOs working in the Disaster Management • Organizations providing scientific databases to study hazard prone areas
Integrated development	• Mapping Landuse/Landcover, Ground Water Potential, and Soil Conditions etc. • Monitoring of landuse/landcover and impacts incl. Salination and land degradation • Implementation of developmental programmes such as Hill Area, Drought Prone Area, Wasteland etc. • DSS can focus on concentration of development activities, feasibility of developmental project to ensure that development goes in right direction.	• Agriculture Department • Ground Water Development Agencies • District Rural Development Agencies

Area	Benefits	Beneficiaries
Telecom network planning and management	• Mapping of components of network w.r.t. land base such as roads, parcels etc. • Asset management for efficient maintenance of components • Routing of the network • Complaint management	• Organizations like BSNL, MTNL and priv-ate telecom companies • Citizens
Electricity distribution and transmission network	• Mapping of components of network w.r.t. land base such as roads, parcels etc. • Asset management for efficient maintenance of components • Routing of the network • Fault management by locating fault location on the map • Designing and optimization of network, load management, load forecast etc.	• SEB • Agencies responsible for Electricity Transmission • Establishments such as industries, townships
Transport network development	• Mapping of roads and other transport network • Identify pocket of poor transport network • Planning new transport network • GIS based Road Information System is effective tool to monitor road development and maintenance.	• PWD • Ministries and Departments dealing with Rural Development • Citizens
Law and order	• Spatial analysis of crime data • GIS based Police Information System • Emergency planning	• Police Departments • Society
Environmental management	• Mapping of environmentally sensitive areas. Issues include water, air and noise pollution, soil salination, oil spills, deforestation, wetland depletion etc. • Modeling for predicting extent of environmental degradation • Inputs for mitigating negative impacts on environment • Environmental Impact Assessment based on spatial data	• NGOs dedicated to envi-ronmental conservation • Ministry of Environment and Forests • PCB • Research Organizations

Geo-informatics link government departments and users to a central network so that everyone has access to the same set of files at the same time for different applications. Easy access to complete information with common geographic-reference using high-speed communication network (including Internet) helps employees to improve their efficiency and government to make better decisions. With the availability of spatial data using Geo-informatics, there is shift from `conventional' DSS to the `spatial' DSS (S-DSS). Till the availability of spatial information from remote sensing data, not much thought was given to the `spatial' dimension to DSS. In fact, any decision involving conservation issues invariably considers geographical parameters such as location, distance, direction, proximity, adjacency, topography etc. and there has been a long pending demand for providing `Spatial Dimension" to alpha-numeric DSS.

4.2.2 Areas for E-Governance using Geo-Informatics

Following areas have been discussed considering the potential of GIS to offer better governance for the benefit of both, government and citizens.

4.2.3 Applications of Geo-Informatics

4.2.3.1 *GIS enabled Land Management System*

Mandates of Industrial Development Agencies/Corporations include planning layouts of industrial sites, attracting entrepreneurs for investment, leasing of industrial plot, construction and maintenance of infrastructure and amenities such as roads, drainage systems, providing streetlights, water supply and other amenities such as built-up accommodation for banks, post offices, telecom facilities, police station, fire station, medical facilities, canteens, shops etc. in the industrial areas in the State.

GIS enabled Land Management System can be developed using Geo-informatics and RDBMS using Oracle for handling transactions of industrial plots (Figure 4.4).

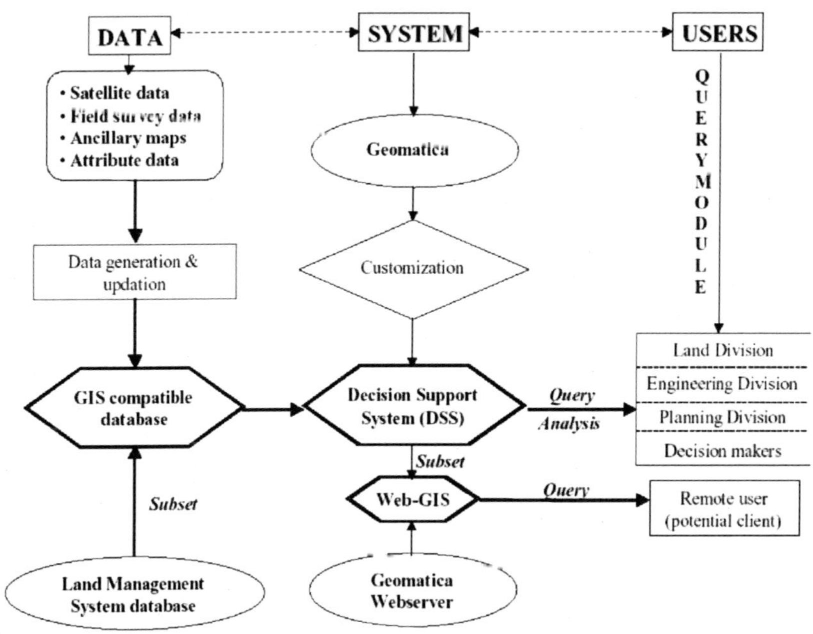

Figure 4.4. Flow Diagram of Land Management System (LMS).

The broad features are:

- Day to day land data creation
- MIS Report Generation (Monthly, Quarterly, Vacant Plot list, etc.)
- Auto Notice generation
- Updation of Regional Office transactions to central database server
- Receipt and Pay order generation.
- On demand auto generation of Offer letter, Lease deed, Scrutiny report, etc.
- Selective dispatch of generated reports (email and / or hard copy)

4.2.3.2 *Spatial Decision Support System (S-DSS)*

S-DSS is developed using commercially available off-the-shelf (COTS) GIS software, Geomatica. S-DSS assists officials to create integrated spatial database consisting of plots, infrastructure and amenities followed by query, analysis, retrieval, and displaying information of industrial sites in the form of maps. The broad features of S-DSS are:

- Geographical database creation, updation (using high-resolution remote sensing data and field surveys) and management
- Decision support using customized query and analysis modules for land transactions, development and maintenance of infrastructure and planning of new sites.
- Linkage with the LMS to provide up-to-date data on the land transactions.
- Display analyzed information in the form of maps

4.2.3.4 *Web GIS based Land Information System*

Web GIS ensures full use of the relevant information generated in GIS and attracts investments from industrial sectors. Potential investors can access this information using conventional Web Browsers (Internet Explorer/Netscape Navigator).

Following are the features of Web GIS:

- The database is kept in the central server and beneficiaries can extract the relevant information using Internet browsers.
- The pre-defined queries are provided so as to allow users to select Industrial sites and industrial plots of their requirement.
- Web site also provide relevant links to appraise on other business aspects

4.2.4 Illustrations

4.2.4.1 *Development of Knowledge Corridor along Mumbai-Pune Express way*

In the wake of global development of Information Technology (IT), Maharashtra government has taken initiative to develop a "Knowledge Corridor" along newly developed Mumbai-Pune expressway. The same is important, as IT industries have been sprawling in the outskirts of Mumbai and Pune, towards the expressway. This has led expressway a potential destination for IT industries. MIDC, which is Special Planning Authority for development of `Knowledge Corridor' has assigned consultancy for developing a GIS of entire corridor along with specific clusters providing information about the most suitable sites for development of various `knowledge based industries'. C-DAC has implemented the project using Geo-informatics. The project envisaged following:

- Developing baseline spatial information for entire corridor along Mumbai-Pune express way at coarser scale
- Developing detailed spatial information for specific clusters
- Providing suitable sites for locating various knowledge based industries such as IT, biotechnology, leisure tourism, corporate offices, meditation centers etc.

To meet the above old topographic maps were updated using the geometrically corrected satellite data as Mumbai-Pune expressway is a recent development and information about which is not available on older topographic maps. Based on selected phenomenon `multi-criteria analysis' was performed in Geomatica to identify the suitable sites for development of `knowledge corridor' along Mumbai-Pune expressway (Figure 4.5). This project has opened new vistas for providing IT enabled solutions for providing crucial inputs in planning the developmental projects of international standard.

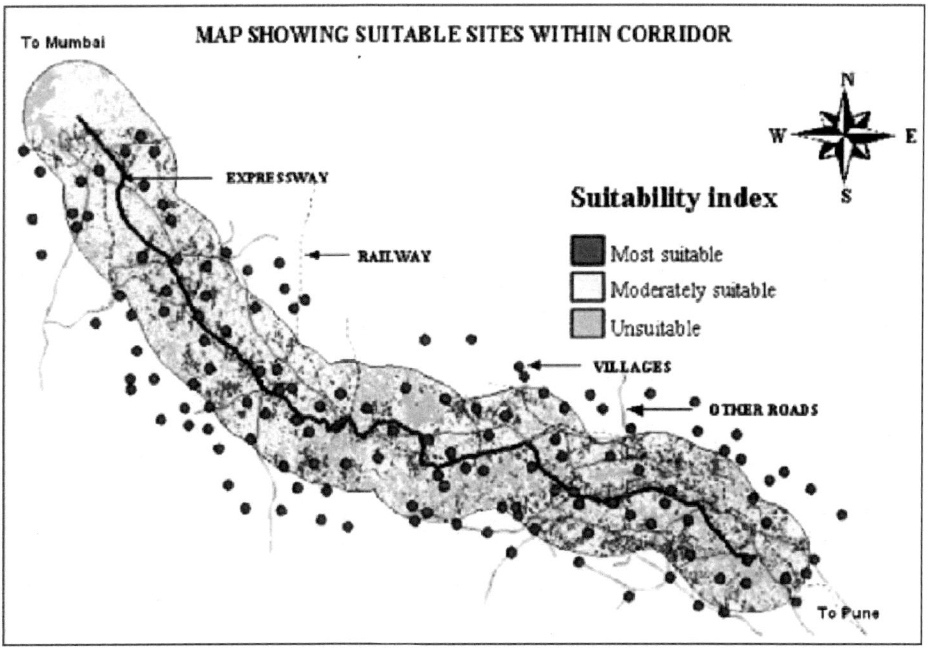

Figure 4.5. Suitable Corridor Sites using Geo-informatics.

4.2.4.2 Site selection for disposal of solid hazardous waste from Industrial Area of Shedre, Aurangabad

The main aim of the project is to identify suitable sites for disposal of solid hazardous waste (SHW) from the industrial area of Shendre in Aurangabad using Geo-informatics. CPCB, New Delhi, has specified several locational criteria for identifying the suitable site. The SHW disposal site shall not be located within a certain distance of geographic features like drainage, lakes and wells. The sites should be preferably located in areas of low population density, low alternative landuse value and low groundwater contamination potential.

During the investigation, spatial and non-spatial parameters required to select suitable sites for disposal of SHW were identified. Relevant spatial and non-spatial data were then collected and various thematic maps were generated in 1:50,000 scale. 1:10,000 scale map of industrial site was registered using GPS readings. Topographic, proximity, overlay and multi-criteria analyses were carried out using Geomatica software to identify the most suitable site for disposal of SHW. Based on the norms specified by CPCB, two suitable sites were identified in the MIDC area. The site, which covers an area of 20.564 hectares, was selected as the most suitable site for disposal of SHW. This site is more or less flat, has a regular geometry, very good accessibility and conducive groundwater conditions.

■■■■

Chapter - 5

FORMULATION OF GIS FOR RESOURCE ASSESSMENT

The Rajasthan GIS Project addresses one fundamental developmental problem of Rajasthan; the problem of easy access to a wide variety of data relating to natural resources, socio-economic factors and physical infrastructure in the state. These data are necessary to understand the extent of development in different parts of the state so that improvements can be planned on a priority basis. Developmental strategies can then be outlined and programmes implemented in specific regions. Any number of queries can be developed when a complete GIS of the resources that exist in the state of Rajasthan is developed, which will be of great help for the developers and planners. To formulate a geographical information system of the Rajasthan Resource Assessment database following process have been followed:

5.1 DATA AND DATA BASE

This study is primarily based on secondary data from various state government and non-government organisations, libraries, national institutions, corporations, and data from primary sources. Some case studies involving field data collections were also carried out to collect the primary information to validate/ replace the collected information. The data were stored, processed and analysed by the computer. Relevant software (depending on the requirement) was used to process the data and prepare Geographical Information System for Rajasthan. The final results were produced in the form maps along with tables, charts, diagrams. Spatial and non-spatial data have been collected and processed separately. They were linked in the GIS environment to prepare the final outcome of the study.

5.1.1 Spatial Data

The prime source of data for the present study was *survey* of India Topo-sheets on 1:250,000 and 1:50,000 scale. Topographical features which exist on them such as state, district, tehsil/block boundary, settlements etc. were digitized from 1:250,000 scaled Topo sheets. Topo-sheets were also used for the demarcation of Drainage-watershed, forest boundaries, surface water bodies transport network including railway, roadways etc., whereas census maps were used to demarcate village locations. Some help was also taken from the secondary sources of data as abstracted from the reports of revenue, irrigation department.

It is evident that the analysis of spatial data is meaningful only when location data are digitized from some large-scale maps. Thus, on the face of it, it would make more sense digitizing the entire study area, the state of Rajasthan and from 1:50,000 maps (maps of scales larger than this are available only for limited parts of the country). The main problem with this, however, besides the time factor, was that it is only for small areas that adjoining map sheets of 1,50,000 scale can be found, because the SOI has not yet printed new editions or imprints for those areas where maps have long since been sold out. Also, all Topo-sheets with grid marking are excluded from free circulation because of military restrictions. It was possible only with great difficulty and help from many different sources to cover just two districts in Rajasthan completely with these maps. Since fewer sheets were needed to cover the whole state with 1:250,000 Topo-sheets, this was the scale chosen for digitizing the whole state. However, even here, ultimately it was not possible to get one out of a total of forty-four. This was the extreme western part of Rajasthan adjoining Pakistan.

5.1.2 Non-spatial Data

Data are the raw material of any database and consequently of a GIS. Much of the early stages of the establishment of a GIS involve the collection of data, their correction and coding, and then their conversion into machine-readable form. The data collected is processed in GIS environment before channeling the output to the user. It is the core of the GIS, and its creation is usually very time consuming. It is imperative that there be no errors in construction (as distinct from errors in the data).

The non-spatial data or attribute data presently available and collected for the Rajasthan Resource Assessment Study are as follows:

1. District level Natural resource data:
   i. Geology
   ii. Geomorphology
   iii. Drainage
   iv. Slopes, levels, contours
   v. Climate
   vi. Soil
   vii. Forest and hills
   viii. Land-use and land cover
   ix. Water bodies
2. Mineral Resources
3. Irrigation and Agricultural Practices
4. Industrial development
5. District level Demographic data:
   i. Population: total, male, female, urban, rural, growth rates
   ii. Population: Schedule caste and tribes, total, male, female, growth rates

   iii. Population: literacy, total, male, female
   iv. Population: language and Religion
   v. Education: enrolment of boys and girls, schools, teachers
   vi. Health: immunization, health centers, hospital
   vii. Agriculture; crop acreage, production and yield.
   viii. Irrigation: area under different sources
   ix. Electricity: villages covered
   x. Telecommunication: exchange, population served
   xi. Banks: branches, villages serviced
   xii. Post offices: numbers
6. Transportation facilities:
   i. Railways
   ii. Roads – All type
   iii. Aerodromes, air routes
7. Communication Network:
   i. Post and telegraph
   ii. Telephone etc.
8. Election Data: for parliamentary constituencies for 1998 and 2003. Data are available for polling stations, winning parties and candidates.

5.2 GEO-CODING AND GIS ENVIRONMENT

A GIS is a whole set of procedure and tools that works within a computer environment, manipulating a database so prepared to produce graphic and non-graphic output. A GIS of resource database is created from data that are collected from various map sources. The important point here is that such data have a definite spatial identity in that each feature on the earth that is recorded has a position on the earth's surface. This position on the earth's surface is commonly identified by some coordinate system, such as latitudes and longitudes. On a paper map, a feature can appear only as a symbol, whereas in a GIS based resource database, 'the computer holds information about position in a "pure" digital form…' This can be stated in simple term as the computer storing a position on the globe (the co-ordinates of a location) in digital form without the requirement of a symbol to represent it. In contrast, on a paper map locations cannot be shown without the use of symbols.

GIS can be an efficient tool for making spatial decisions only if it is built on a well-defined and structured cartographic database. Most people are familiar with databases that hold descriptive texts and numeric data in tables. The GIS is different in that it is primarily concerned with spatial data and all data are geo-coded. Quite simply, this means that every item of data in the GIS is attached to one and only one unique position. This location forms the most crucial part of a Resource database, because without the location data the database would have no spatial component.

The core of the GIS resource database is the location data or the data that are a record of the position of all features in the study area. These features may be hamlets, villages or towns; or perhaps pack tracks, footpaths or roads; or lakes or rives. This is not to imply that a resource database is only concerned with location data. Indeed attributes of locations also form part of the database and it is this combination of location and attribute of a feature that makes GIS such a powerful analytical tool. There is no limit to the amount of attribute data that can be stored. If the locations of features are displayed, this will form the equivalent of a 'base map' on paper. When the attribute of a feature is shown, the 'map' shows a theme and is sometime called a thematic map. But it is not necessary to store all the attribute data as 'map'; it can remain in digital form in the computers' memory. In fact, the data can be analyzed without being presented as a map, but a map often is important, since studying a map carefully can make many inferences.

Rajasthan Resource database creation and storage in computer is done using a number of different techniques, depending greatly on the source of data, the hardware and the software used. As in this study most of the data is in a paper form, a scanner or a digitizer were used to convert map features into digital form. For the Rajasthan Resource database, location features were input only from paper map. All the adjoining map sheets of map scale 1:250,000 that were used to create the Rajasthan Resource database. At present the database does not have any data from remotely sensed sources. It was built using the MAP Info Professional software (GIS). This software packages uses a topological structure and requires all features to be stored as points, lines or polygons. It should be noted that all features on a map can, in fact, be broken down to these three simple geometric figures: the point (with no dimension), the line (with one dimension) and the polygon (with two dimension). The basic units in which all features are stored are, therefore, points (for instance villages or spot heights), or lines (for instance roads and streams) and polygons (blocks, districts and towns).

The first step is to identify every feature on the map as the smallest indivisible unit. This means that a recognizable feature on the map should be recorded as one of its kind. Thus, if a point on the map is to show the location of a town, it can represents only one town. Once the features are identified, then each must be categorized as a line or point or polygon and then coded to how some of its attributes. For example, a line with code 'one' is an unmetalled road, 'two' a metalled road, 'three' a state highway and 'four' a national highway. Reference points on the maps are also marked in advance to enable adjoining map sheets to be topologically joined so that projection could be changed to avoid distortions. The maps are then placed on an electronic table or digitizer and all features that are to be digitized are traced with a cursor. While this sounds fairly simple, the process is actually very demanding and laborious. It requires a high degree of precision. The digital base map is ready for use only after meticulous editing, which may sometimes take more time than digitizing.

The creation of the attribute tables is somewhat simpler, for data need to be input through the computer keyboard. In the case of the Rajasthan Resource database, Microsoft EXCEL and DBASE – IV were used; and the data was then exported to

MAP-INFO. As It is an imperative data, the only way to ensure this is by giving a particular feature the same unique code during digitizing as noted for the corresponding item in the dbase file. For example, if there is a village called Sitapura, which has been given code of 435 when digitizing its location, then in all the files where attribute data for a Sitapura are recorded the code 435 must be use. Great care must be taken when revising data or introducing new items of data.

5.3 HARDWARE AND SOFTWARE

The choice of hardware and software is largely dependent on their capabilities, although to a great extent the nature and source of data also influence the selection. A thorough market survey of what was commercial available to match the specifications required for the purpose of the project, within the limited budget was undertaken. A comparison was made of their capabilities with respect to mode of input, storage, management and exchange of data; the descriptive, analytical and graphic presentation functions, and so on, before the final selection. Following hardware and software were used in the study:

5.3.1 Hardware

System One :	P-IV computers with following specifications:
	Processor – Pentium -IV
	Motherboard – Intel
	RAM – 256 MHz
	HDD – 40GB
	Monitor – 17" coloured
	CD writer – 24x
Digitizer: One	A-3 size Cal comp
Printer Two: -	A-3 size coloured (HP – 1125C)
	A-4 B/W (HP Laserjet 1200)
MODEM One:	33.6 KBPS
Scanner One:	A-4 Umax 1200

5.3.2 Software

Besides the software necessary for running the system and for utilities, other software were used for making graphs, word processing and Relational Data base Management. For The purpose of thematic mapping, a popular package, MAP- Info was used. Different modules of Map-Info were used for their powerful GIS function. Map-Info was used extensively for presenting data using techniques such as located symbols and located graphs and proved efficient for data analysis and data input and exchange. It was also used chiefly for its excellent data integration and analysis, for flexibility and powerful map composition capabilities.

5.4 ORGANIZATION OF THE DATABASE

The Rajasthan Resource database is organized as nine independent databases, which can be grouped into five broad areas of focus: administrative units; electoral units; settlements; communication; and physiography.

The structure of the nine subsystems of the Resource database is:

1. The district database, where the basic unit for collecting data is the revenue district.
2. The urban database consists of all settlements classified as 'urban' in the census. All towns of Rajasthan are part of the database.
3. The communication network database includes railways and roadways. All the railway lines and type of roads are featured.
4. The drainage network database includes both streams and rivers. All major rivers and water bodies and streams have been digitised.
5. The terrain database records height above sea level. These are elevations recorded as spot heights, benchmarks and triangulation points in topographic maps.
6. The land-use and land cover database consist of layers of hills, forestlands and vegetative covers.
7. Geology and soils database is also developed for the entire state of Rajasthan.
8. Irrigation and Agricultural practices have also been recorded to form a part of a Rajasthan resource assessment study database.
9. The Lok Sabha and State Assembly database consists of all parliamentary constituencies and local assembly constituencies of Rajasthan and includes data on polling stations and poll results.

It should be noted that in spite of this logical division of the database, all the data could be used together.

5.4.1 The Data Dictionary

As the volume of data entered into the computer from maps and other sources grows, it becomes increasingly difficult to recollect files, their information and locations in the computer's storage space. A simple way out of this situation is to build and constantly update a data dictionary right from the very beginning. The structure of the data dictionary that is used to organize the Rajasthan Resource Assessment Study database is explained here.

- The data dictionary is the index to the entire database and gives information on every file that forms part of it. It is necessary to maintain a comprehensive data dictionary to ensure that none of the data files are duplicated and at the same time can be located, retrieved and edited by user easily. It also enables user to retrieve data quickly just as the catalogue in a library enables one to identify the exact location of a publication and save time from searching in a random fashion. The data dictionary, therefore, is an essential part of any large database.

- For the Rajasthan Resource database, a two-tier data dictionary has been organized. This includes one master file and number of description files. The master file is, in effect, a simple table with fields, representing sets of information that is considered essential for each data, text, or map file.
- Each file that forms a record in the master file in the data dictionary has a corresponding description file within the dictionary. Each description file is named by the 'code' in the master file. This file has only two fields (i) name and (ii) description. In the case of the data files 'name' represents the exact field names as recorded in the data file and description records an explanation of what this 'name' signifies, including where applicable, the method of computation. In the case of the map data, however, 'name' represents a map file name and 'description' records some significant data regarding this file. No description files are maintained for text files.
- As a general practice, in the Rajasthan Resource Assessment Study it was decided in the earliest days of it's starting that every time when a new file was created, the master file in the data dictionary, would be updated by introducing a new record. Concurrently, the corresponding description file would also be created. At a later date, when a field in the database file was added, deleted or edited these changes were recorded in its associated description file, and also if necessary, in its corresponding record in the Master File. Similarly, when creating, editing, transforming into geographic coordinates or projecting digitized map files, information is recorded in the data dictionary's master file and second level description file. This rule had been strictly followed and ensured at all stages of the study.

5.5 MEASURES OF QUALITY CONTROL

The problem of fragmentary, unreliable and inaccessible data is fairly common in developing nations. But the problem appears to be acute in the case of Rajasthan, and probably many other states of India. Unless the problem of data is solved, there can be no clear understanding of development problems. Consequently, the generation and organization of data becomes the first step in development planning. A very substantial part of this research study was creation of the Rajasthan Resource Database from a large number of sources; published and unpublished, fragmentary and complete, contemporary and dated, some questionable but others of fairly reliable quality. As far as possible, the data has been corrected of outstanding errors. The Rajasthan Resource Database is the most comprehensive repository of digital data on Rajasthan. Following points were kept in mind while carrying out the study:

- Error is unavoidable and often occurred in a significant part of the data recorded in Rajasthan. In all cases obvious or known errors have been mentioned so that any user of the database is warned of its pitfalls and knows to what extent it may be useful or otherwise. This has been done in the spirit that the concerned authorities become aware of the error and correct the relevant section.

- The reliability of data is of utmost importance in a database. In the case of a Resource Assessment Study this applies as much to the accuracy of non-spatial data as it does to spatial data in the form of map. In Rajasthan Resource Assessment Study the good quality of database has been properly maintained through cross checking the resource data by collecting the same information from various sources independent to each other. For an example, with changes in district boundaries there have been consequent changes in the area of the districts that have been subdivided. The areas of all the new and old districts have been computed by totaling the area of the blocks that constitute them. The data on the area of bocks are taken from the urban/village Primary Census Abstract of the District Census Handbooks, village and Town Directories.
- As data processed at different stages in the project, many different problems relating to them were faced. But these problems were usually such that in future they could be quiet easily solved by the concerned departments. Common limitations of data include incomplete reporting, repetition and change in tabulation format. Non-availability of simple data is a common phenomenon.
- A major problem encountered was the incomparability of data over time. Individual departments collect data according to their own needs and update them at different times. Thus it is not possible to collect data for all items for the same point in time. To overcome this problem it was decided that for the Rajasthan Resource Assessment Study any data (and preferably the latest data) available would be collected. The development of infrastructure is fairly slow in Rajasthan, and so there is no great danger of gross misrepresentation of existing conditions, even when the data is five years old. The Census of India, 2001 is the primary source of al population data.
- As the cartographic database is the core of the GIS, the location data should be free of errors. This, however, is easier to say than actually achieved. Errors can creep in, in many different ways as was noticed when digitizing for Rajasthan. Quit often there is more than one settlement of the same name both on the map and in the departmental file. However, there is no way to judge which record in the table applies to which settlement on the map. Sometimes, names change and the change are not reflected in the maps. Occasionally, settlements that appear fairly important on the map do not appear in the census records. At other times, though very rarely, villages on the extreme edges of some SOI map sheets are repeated on the edge of adjoining map sheets, leading to duplication. It is also unlikely that large-scale topographic maps and the District Census Handbook maps correspond entirely.
- The meticulousness, which is necessary for computerizing coded data, is not the same as that needed when dealing with manually computed data. Since

much of the data gathered for the Rajasthan Resource Assessment Study was originally compiled for manual examination, they lack rigorous application of nomenclature and codes. This problem has been overcome by giving fresh codes to all the tabulated data in a scientific nomenclature.

- It is to be mentioned that the SOI maps are also not completely free from errors. However, in the case of these maps, the errors may be attributed to manual techniques of maps production and probably not to surveying. For instance, in a number of map sheets, over mapping lines of administrative boundaries, roads and streams are so untidy than one cannot read them precisely. Railway lines are sometimes marked with incorrect gauge symbols and the same village settlement is shown on two adjoining map sheets. These are, however, not major problems in comparison to the problem of actually gaining access to a maps that are not available at the map sale counters, either because thy are out of print or out of stock, or are restricted, classified or secret. The 'system' can take anywhere between six months and one year to deliver a restricted map if permission is granted by the Ministry of Defense to give it on loan.
- The precision achieved in the analysis primarily depends upon the correctness of the database used in the analysis. Hence to optimize it, the fed information/data was screened and checked thoroughly for its correctness in each stage of the study. The first level of check was introduced by way of involving himself to collect and analyze the data. All the data and information have been collected and gathered personally through visiting and meeting the concern persons of the government, non-governments and institutions. The data collected from the government sources was triangulated from primary sources using FGD and field survey wherever possible. Standard GIS and analysis software (popular brand) were used to prepare GIS and conduct analysis and arrive at conclusion. Ground verification was also conducted after completion of the analysis for few cases to draw and suggest rational interventions.

5.5.1 Spatial Data and Temporal Change

A major part of the Rajasthan Resource Assessment Study was created in the years 2003-2004. The sources of map data were much older and so was one of the most important sets of attribute data: population. Some adjustments were made when recording data as will be pointed out in the following discussion.

- One of the greatest shortcomings of manual cartography is the tremendous time lag between the survey of an area and the production of the paper map. This became very obvious when inspecting the (SOI) map. The most recent 50,000 scale maps of Rajasthan published by the SOI in 1972 were surveyed in 1948. The implication of this is that while the maps were being digitized in 2003, they were, in fact, reflecting ground conditions of a three decade old. Clearly, the map data, for some elements at least, were already questionable. Temporal changes taking place on the ground cannot show on a static map. A number of map frames, depicting the same area at different time phases, could however, be an answer to overcoming this problem. Unfortunately, in most cases, not more than one map of an area could be obtained because only one was printed in the last three decades – and for some places none at all of a particular scale were available. In the present case the incompatibility was that the major source of location data was older than the attribute data.

- In the Rajasthan Resource Assessment Study, the attribute data were all from the post 2000 period whereas the map data were from maps of areas mostly surveyed in the seventies or earlier. With automation in maps production, it is hoped the problem of spatial data being outdated even as the map become available for public use, will be overcome.
- Changes in location data take place over time due to natural phenomena like shifting of riverbeds and flooding. This not only affects the drainage system but also he location of settlements in the flood prone areas. Roads and railways, recently constructed, are also not reflected in an old map. This lacuna in data from Topo-sheets on recently constructed railway lines and roads was filled in/compensated for by consulting current railway timetables and road map of the department of transport, which were updated up to 2002.
- Another point to consider is that maps can hardly keep pace with the changes relating to boundaries of administrative units, especially the revenue district. In 1991 there were 27 districts in Rajasthan and today they are 32. Most of the printed maps of the state show 27 districts.
- The most convenient way of reconciling this problem is of obtaining a list of blocks in each distinct (as at present) from the department of planning and regroups the blocks on the maps to show the new district boundaries. This exercise was done by consulting National Atlas and Thematic Mapping Organization's (NATMO) on 1:1 million administrative plates, since they show all administrative units above that of the revenue village.
- All the location data recorded were digitized from 1:250,000 topographic maps, except the rural units which were digitized from map sheets of the scale 1:50,000.

Chapter - 6

PHYSIO- CLIMO AND BIOTIC RESOURCES

The physiography of Rajasthan is the product of long years of erosional and depositional processes. The present landforms and drainage systems have been greatly influenced and determined by the geological formation and structures. The Aravalli hill ranges, running from north-east to south west, divides the state approximately into the western arid and eastern semi-arid regions. It is also a major water divide. The area, to its east, is well drained by several integrated drainage systems, while the area, to the west, has only one integrated drainage system that is the Luni drainage system in the south eastern part of the desert.

6.1 PHYSIOGRAPHY

Physiographically, Rajasthan is divided into four regions, bearing a close relationship with the geological history of the State, viz., (i) Western desert plains, (ii) Aravalli mountain range, (iii) Eastern plains and (iv) Vindhyan plateau (Figure 6.1)

The western desert plain is a sandy desolate expanse of dune fields and rocky pavements. The plain, lying to the west of Aravalli hills, covers 1,75,000 km2 area of the Rajasthan. It is affected by scanty rainfall. The Luni and Sukri rivers, ephemerally flowing in a southwesterly direction, form the main drainage of the area. A large part of western Rajasthan is characterized by its inland drainage and carries some of the largest salt-lakes, viz., the Sambhar, Didwana, Kuchaman and Talchapar lakes (Figure 6.2). The Aravalli hills of Rajasthan traverse the State in a NNE-SSW direction almost from end to end dividing Rajasthan in two unequal parts, three-fifths of which constitutes the western Rajasthan, two-fifths constituting the eastern Rajasthan. The Aravalli mountain range which exposes principally rocks of the Delhi Supergroup, comprises alternating hill ranges and valleys extending from Delhi on the NE to the plains of north Gujarat on the SW for a distance of about 650 km (Figure 6.3). The entire hill range can be broadly divided into followings:

(i) The north-eastern hill range to the northwest of Jaipur district with an average height of 550-650 m (Alwar hills) and intermittent wide valleys;

(ii) The central Aravalli hill range comprising hills of Shekhawati and Marwar region with an average height of 400 m (Shekhawati hills);

(iii) The Mewar hills covering the area of Udaipur, Dungarpur and Sirohi districts with the relief varying from a plateau between Kumbhalgarh and Gogunda to uniformly rolling country on the south-central part; and

(iv) The Mt. Abu hill range, with an average 1,200 m height, constituted by the Abu batholith of the Erinpura Granite. The highest peak called Guru Shikhar, 1,727 m above MSL, is situated on the Mount Abu.

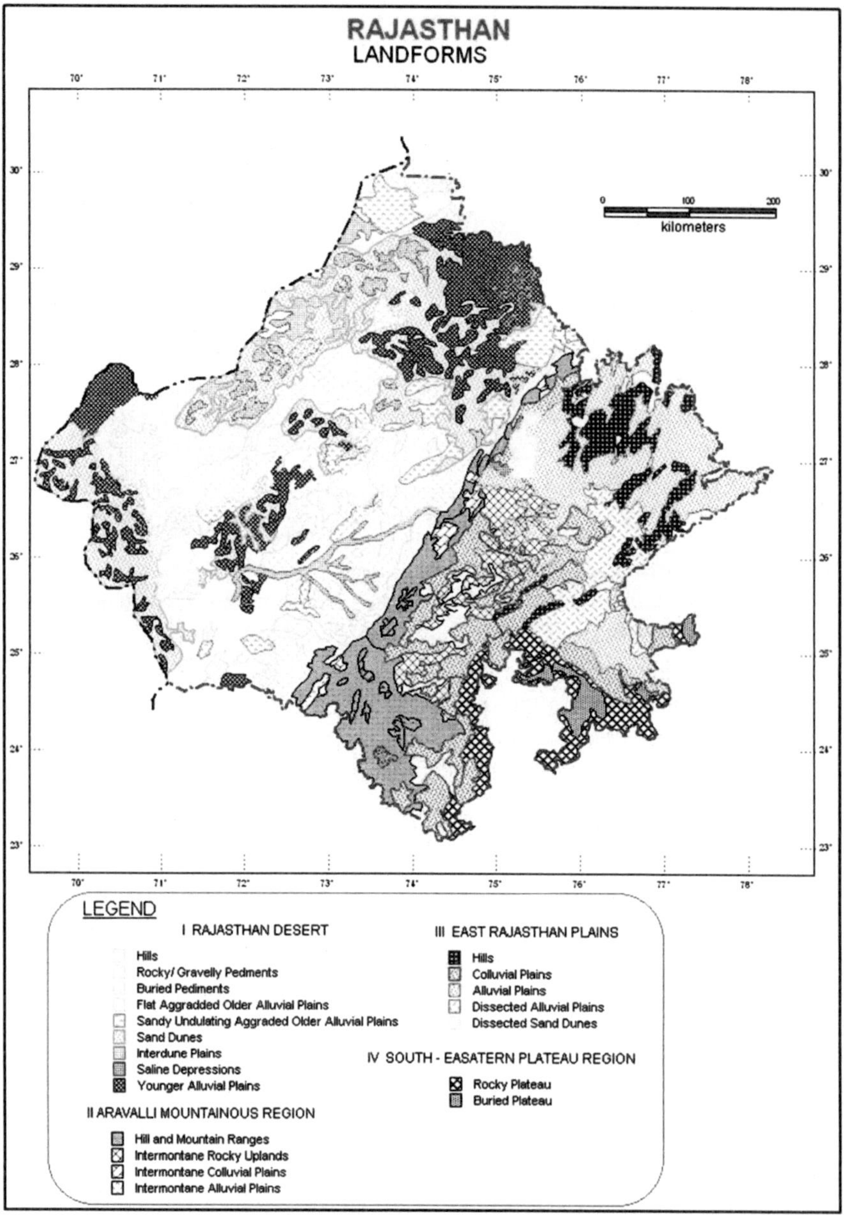

Figure 6.1. Landforms
(Source: Survey of India, Rajasthan Atlas, Resource Atlas, Watershed Atlas, Groundwater Atlas, CAZRI and Agricultural Geography of Rajasthan by Dr. T.S. Chouhan)

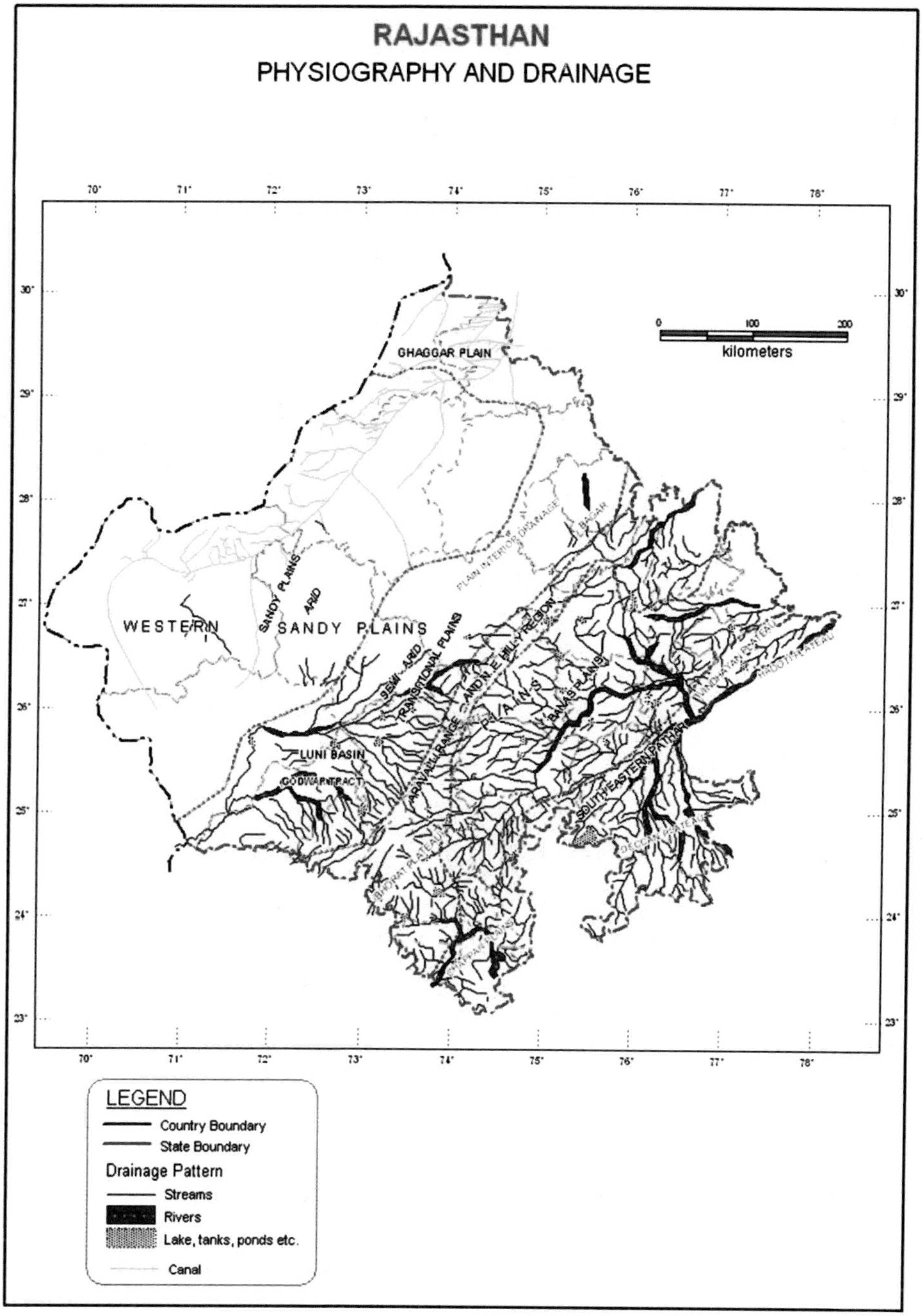

Figure 6.2 Physiography and Drainage
(Source: Survey of India, Rajasthan Atlas, Resource Atlas, Watershed Atlas, Groundwater Atlas, CAZRI and Agricultural Geography of Rajasthan by Dr. T.S. Chouhan)

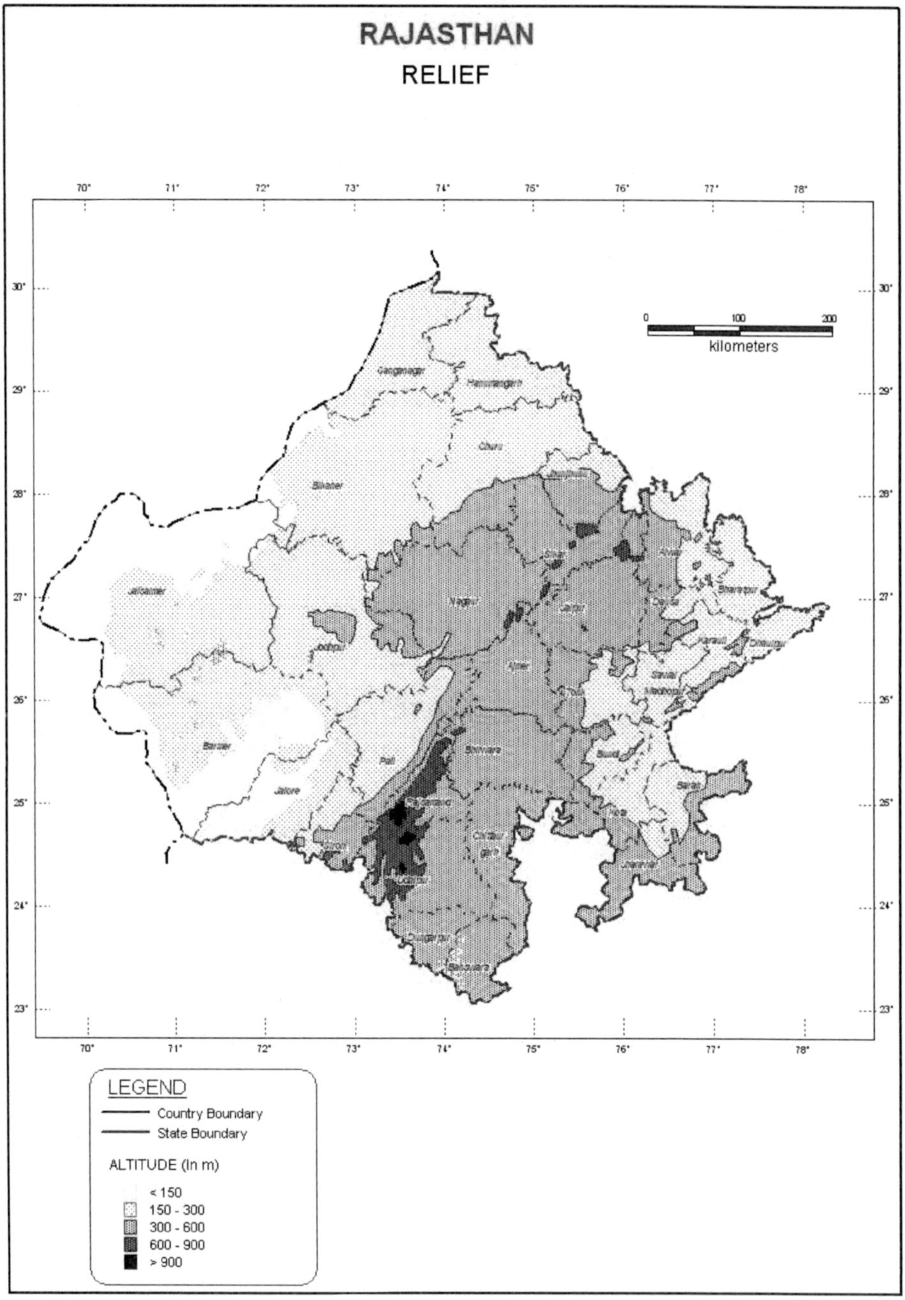

Figure 6.3 Relative Relief
(Source: Survey of India, Rajasthan Atlas, Resource Atlas, Watershed Atlas, Groundwater Atlas, CAZRI and Agricultural Geography of Rajasthan by Dr. T.S. Chouhan)

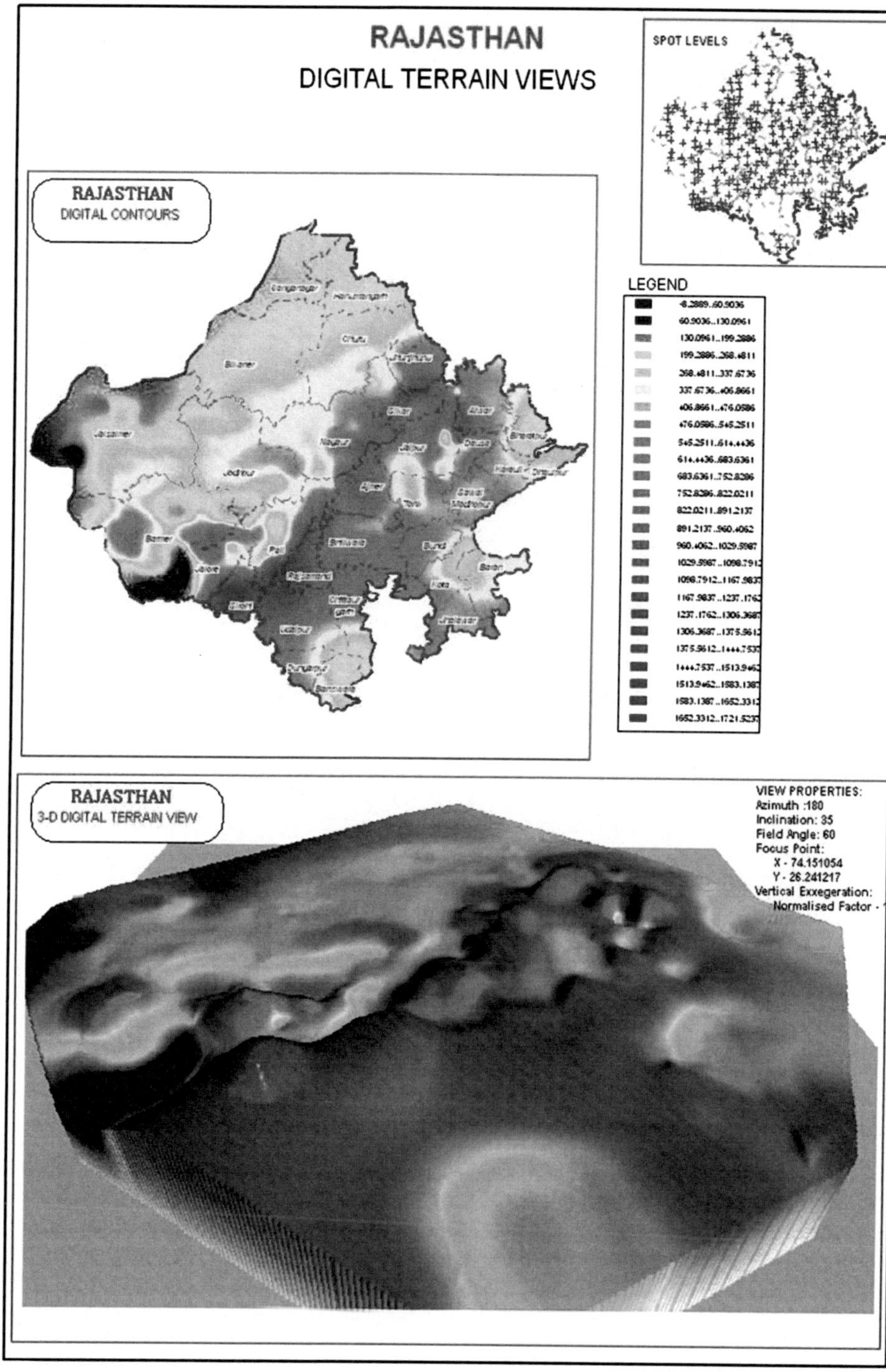

Figure 6.4 Digital Terrain Model of Rajasthan

A Pictorial 3D view of generated from Digital Terrain Model of Rajasthan from different angles is shown in Figure 6.4. The cross-sectional profiles generated from the DTM along major directions are shown in Figure 6.5.

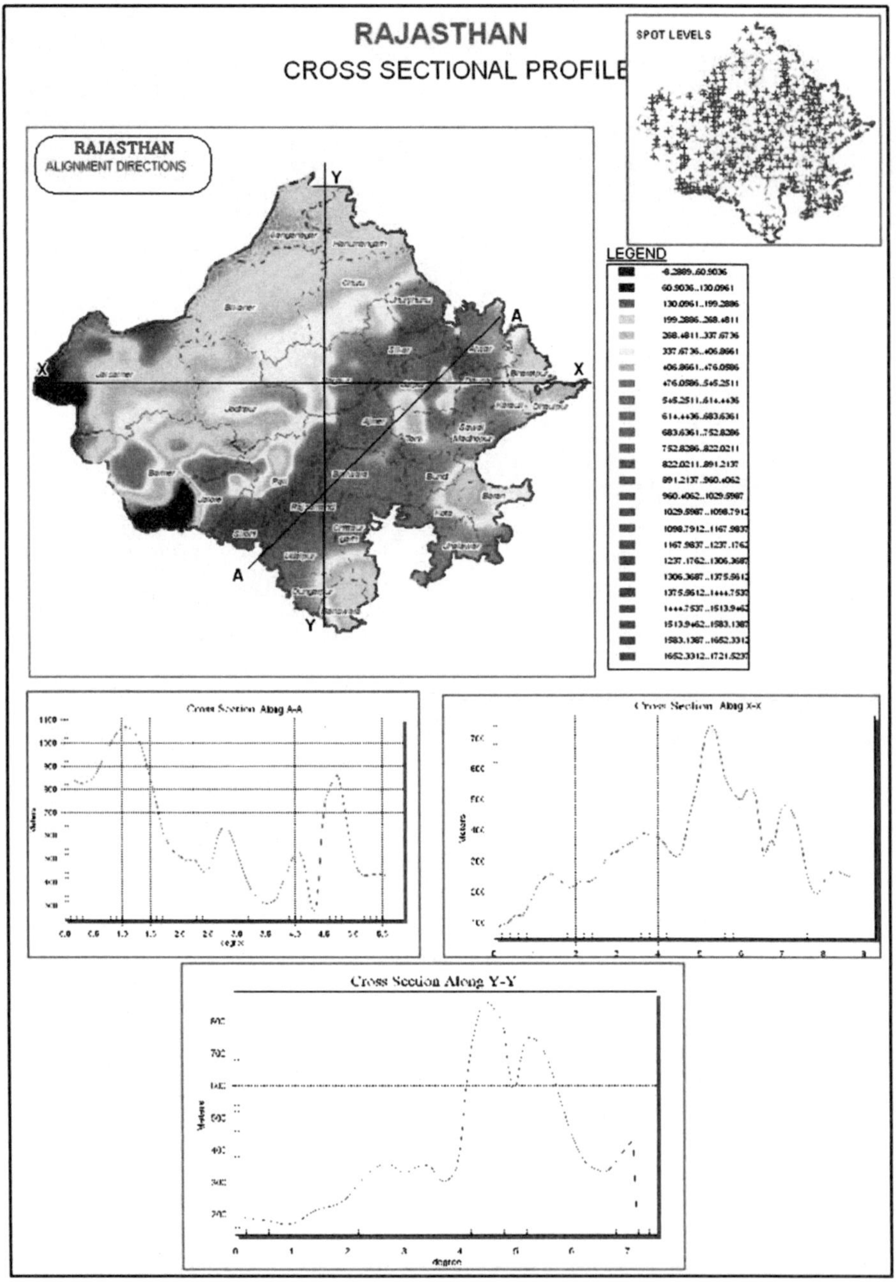

Figure 6.5. X-Sectional Profile of Rajasthan

The Aravalli range partly forms one of the main watersheds of India, which divides drainage to the Bay of Bengal from the Arabian sea. Ajmer city, lying almost at the centre of the range, is in fact on a trijunction of the watershed. The watershed passes along the centre of the range from Ajmer to the southwest up to Kumbhalgarh where it swings to the southeast for some distance to pass ultimately through Udaipur. South of Udaipur, the drainage is to the south into the Gulf of Cambay through the Sabarmati river. The streams emanating from the north and northeast of Udaipur flow into the Banas river through its tributaries. The Sahibi river drains the northern parts of the State and ultimately disappears in the plains of Haryana State. The Banganga river drains the areas in Alwar and Jaipur districts and flows through a large part of Bharatpur district, before joining the Jamuna river. The Eastern plains occur to the northeast, east and southeast of the Aravalli range. The plains cover about 23% area of the Rajasthan. Its northern boundary meets with the Gangetic plains in Bharatpur district; whereas the south-eastern boundary lies in contact with the Vindhyan plateau. Three major basins viz., the Chambal basin, Banas basin and Mahi basin comprise the plains. It constitutes a vast undulating, pediplained country occupied by the gneisses and schists of the Banded Gneissic Complex as well as the phyllites and other rocks of the Aravalli Supergroup. The terrain is covered by a thick mantle of alluvium and wind-blown sand.

The Vindhyan plateau in the southeast, constituted by alternating sequences of sandstone, shale and limestone, forms the western extremity of the main Vindhyan basin. The Vindhyans in this area comprise three plateau - the Bhander plateau to the northeast, followed by the Rewa plateau to the south and west, and the Kaimur plateau encircling the Rewa and Bhander plateaux. The Deccan Traps plateau, further south, continues in Madhya Pradesh. The junction between the Vindhyan plateau and the Eastern plains of Rajasthan is marked by the Great Boundary Fault (GBF). The drainage in this part of Rajasthan is generally to the north and northeast. The Berach river which drains the south-western parts of the Vindhyan plateau, runs northwards, turns to the northeast and flows parallel to and on the northern side of the GBF The Berach river flows into the Banas river which joins the Chambal river in Dholpur area. Elsewhere the rivers and streams flow in different directions and join the Chambal river which flows in a northeasterly direction parallel to the GBF through Kota. The Chambal river draining into the Jamuna river in Uttar Pradesh is the only perennial river in the State.

6.2 GEOLOGY

The Rajasthan State forms one of the geologically important terranes of the Indian Shield, for it contains rock sequences ranging from Archaean to Recent times. The central part of the State constitutes an older Archaean granite-greenstone belt (Mangalwar Complex) which is flanked by a younger Archaean greenstone belt (Hindoli Group) in the east and by the Middle Proterozoic Delhi Fold Belt in the west. The Sand Mata granulite gneisses, which probably represent the oldest rock sequence, occur as tectonic slivers within the Mangalwar terrane. The Mangalwar Complex forms the basement for a number of synformal metasedimentary basins like those of Jahazpur, Gangwara, Sawar, Bharak-Bethumbi-Rajpura-Dariba-Bhinder,

Pur-Banera and Parasoli - Shriramnagar. These sequences are of Early Proterozoic age. In the southern Rajasthan, the Mangalwar Complex forms the basement for the Early Proterozoic Aravalli sequence. In the north-western part of the Indian Shield, the Delhi sequence occurs in the form of two distinct fold belts i.e. North Delhi Fold Belt (NDFB) in Alwar, Jaipur and Jhunjhunu districts and South Delhi Fold Belt (SDFB) in Ajmer, Rajsamand and Sirohi districts. The Delhi Supergroup rocks are followed westwards by the Upper Proterozoic Malani Igneous Suite and the Erinpura Granite batholith. The Malani Igneous Suite comprises bimodal volcancics dominated by felsic volcanics, biotite granite (Jalore Granite) and aegirine-riebeckite granite (Siwana Granite). The Upper Proterozoic sequence in the east and southeast Rajasthan is represented by the Vindhyan Supergroup comprising conglomerate, sandstone, limestone and shale. Marwar Supergroup of rocks comprising Jodhpur sandstone, Bilara dolomite- limestone, Hanseran evaporate and Nagaur sandstone, clay and siltstone. These sequences overlie the basement constituted mainly of the Malani Igneous Suite. The rocks of Marwar Supergroup are unconformably overlain by the Bap Boulder Bed which forms the basement for the Badhaura sandstone, the youngest Palaeozoic sequence. The Mesozoic and the Cenozoic geological sequences are represented by sandstone, shale and limestone in Jaisalmer, Barmer and Palana areas. Several mineral deposits of economic importance occur in association with these geological rock formations, making Rajasthan as one of the leading mineral producing States in India.

6.2.1 Archaean

The Precambrian rocks of Rajasthan were studied in the initial stages by Hacket (1881), Heron (1917, 1936, 1953), Coulson (1927, 1928) and Gupta (1934). Heron (1936) and Gupta (op.cit.) classified the Precambrian sedimentary sequences of Rajasthan in the order of increasing antiquity as the Vindhyan System, Delhi System, Aravalli System and Banded Gneissic Complex (pre-Aravallis). Heron (1953) considered the Banded Gneissic Complex (BGC) as representing the remnants of primordial crust on which the Aravalli sediments, equivalent to the Lower Dharwars of South India, were deposited. However, these pre-Aravalli rocks were not differentiated by the earlier workers. Kurien (1950) advocated that the older schists were granitised to form the gneisses and migmatites (pre-Aravalli) within which occur the quartz-sillimanite schists and calc-granulites as unassimilated lenses of older schists. Raja Rao (1967) and Raja Rao *et al.* (1971) attempted for the first time to classify the BGC and designated these rocks as the Bhilwara Supergroup. Gupta *et al.* (1980) synthesised the voluminous ground data generated over decades of pioneering work in Rajasthan and parts of Gujarat and designated pre-Aravalli rocks as the Bhilwara Supergroup.

6.2.1.1 *Bhilwara Supergroup*

The pre-Aravallis in Rajasthan, hitherto known as the Banded Gneissic Complex, have been regrouped into the Bhilwara Supergroup which form the oldest stratigraphical horizon around/upon which the entire geological succession of Rajasthan has developed. On the basis of tectonoenvironmental setting, magmatic

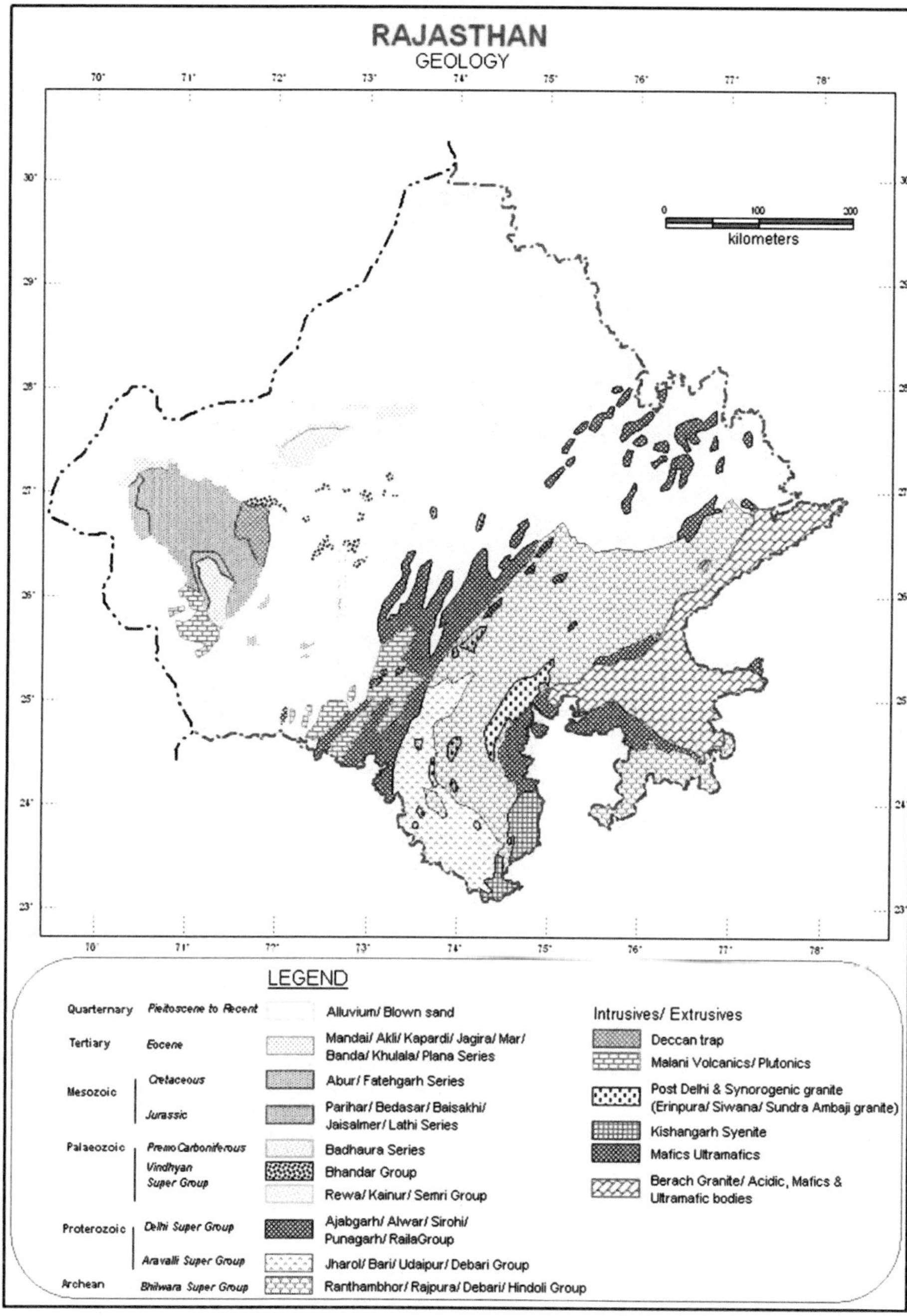

Figure 6.6. Geology
(Source: Survey of India, Rajasthan Atlas, Resource Atlas, Watershed Atlas, Groundwater Atlas, CAZRI and Agricultural Geography of Rajasthan by Dr. T.S. Chouhan)

events, lithological homogeneity and order of superposition, the rocks of Bhilwara Supergroup have been classified into the Sand Mata Complex, Mangalwar Complex and Hindoli Group (Archaean); Rajpura-Dariba, Pur-Banera, Jahajpur, Sawar and Ranthambhor Groups (Lower Proterozoic). Rocky regions of central Mewar, to the east and southeast, comprise essentially igneous rocks of different types and periods of intrusion which have been welded into a streaky and banded composite gneiss, having wide variations of texture and composition. Banded biotite gneiss and foliated gneissic granites are the most common members of the complex which have been intruded by basic dykes and acid sills of widely variable composition and texture. The basement granite gneisses are intruded by many younger granites, basic and ultrabasic rocks, and support many sedimentary cover sequences. Their geology and age relations are complex and the entire assemblage has been further differentiated into the Sand Mata Complex (granulite facies) and the Mangalwar Complex (amphibolite facies) on metamorphic consideration.

6.2.1.1.1 *Sand Mata Complex*

The rocks of the Sand Mata Complex occur in a belt (200 km long and 50 km wide) from Lambori and Amet in the south to Kishangarh and Patan in the north. The western boundary is marked by the Delhi Supergroup of rocks which overlie the Sand Mata Complex with an overlap. The major part of the contact is, however, faulted. The eastern boundary with the Mangalwar Complex lying to the east of Parbati and passing through Agucha and Jatan, described as a faulted contact, has been rendered obscure by high-grade metamorphic effects and migmatisation. The metamorphic contact between the Sand Mata Complex and the Mangalwar Complex practically coincides with the Delwara Lineament. The present spatial disposition of the rocks of Sand Mata Complex may be due to vertical tectonics. On the southeast, the rocks of the Sand Mata Complex are bordered by the rocks of the Aravalli Supergroup. The metasedimentary assemblage of the Sand Mata Complex is heterogeneous in composition, comprising migmatite, composite gneiss, calc-gneiss, biotite schist, garnet-sillimanite schist, sillimanite- staurolitegarnetiferous biotite schist, hornblende schist, biotitechlorite schist, dolomitic marble, conglomerate, quartzite, para-amphibolite, epidiorite and pyroxene granulite and has been regionally metamorphosed to high-pressure amphibolite to granulite facies, and migmatised to different degrees. The group consists of varied igneous suites like ultramafics, mafics and felsics.

6.2.1.1.2 *Mangalwar Complex*

The heterogeneous assemblage containing pelitic rocks deposited during the orogenic phase of the Bhilwara geological cycle, regionally metamorphosed to amphibolite facies and syntectonically migmatised have been included in the Mangalwar Complex. The complex occurs to the east of Delwara Lineament in three main stretches, the first one extending as a 12 to 48 km wide belt from Pipalkhunt to Mangalwar for over 350 km; the second one occurs from west of Amrita to Dadala over 50 km (Sarara-ki-Pal inlier), and the third one extends from Kunda to Pratappur for over 30 km (Mando-ki-Pal inlier) having a NE-SW trend. The eastern boundary of the complex is marked by the Hindoli Group of rocks with an uncertain stratigraphic relationship. The Mangalwar rocks are unconformably overlain by the Aravalli Supergroup rocks to the west and by the Deccan Traps in the south. The Mangalwar Complex is considered representing an Archaean primary granite- greenstone belt comprising of migmatite, gneiss, feldspathic mica schist, garneti-ferous mica schist, sillimanite-mica schist, hornblende schist, quartz-feldspar gneiss, granite gneiss, calc-silicate rock, dolomite, quartzite, impure marble and para- amphibolite. These rocks have been intruded by the Untala-Gingla Granites, Giyangarh-Asind acidic rocks, RaipurJalayan mafic rocks and Ultramafic rocks. The complex has been subdivided into the Lasaria, Kekri, Sarara, Mando-ki-Pal, Suwana, Potla and Rajmahal Formations.

6.2.1.1.3 *Hindoli Group*

The Hindoli Group named after Hindoli village, located 20 km northwest of Bundi, comprises a lowgrade (greenschist), predominantly argillaceous flyschlike sequence deposited during geosynclinal phase of the Bhilwara geological cycle. The rocks of the group occur as a 10 to 50 km wide arcuate belt extending from Gyaspur in the south to Deyi in the north and further north-eastward for 400 km distance, trending NNW-SSE in the southern part, becoming N-S near Chittorgarh, and further northwards veers to ENEWSW in Hindoli-Deyi area. The group represents a secondary granite-greenstone belt and is bounded by the Vindhyans in the east and by the Deccan Traps in the south. In the west its contact with the Mangalwar Complex is not clear. The group comprises shale, slate, phyllite, mica schist with interstratified sequence of metagreywackes, metalvolcanics, quartzite, dolomite and limestone. The grade of metamorphism increases gradually towards west, and greenschist fades rocks grade into intermediate pressure amphibolite fades at the contact with the Mangalwar Complex. The rocks are migmatised in Mangalwar- Deoli area.

6.2.2 **Proterozoic**

The Proterozoic rocks of Rajasthan are included in the Rajpura-Dariba, Pur-Banera, Jahazpur and Sawar Groups (Bhilwara Supergroup); the Aravalli, Delhi, Vindhyan and Marwar Supergroups.

Overlying the Hindoli Group of rocks and the Mangalwar Complex with -an unconformity occur the next younger groups of rocks classified as the Rajpura Dariba, Pur-Banera, Jahazpur and Sawar Groups which are exposed in a series of isolated linear belts. All these synformal metasedimentary basins/ structures of Lower Proterozoic age occur in disjointed belts as outliers. They are mostly composed of dolomite, marble, calc-gneiss, calc-schist, and calcbiotite schist; graphite-kyanite-staurolite schist, garnetiferous mica schist; banded chert, banded ferruginous chert and quartzite. These groups of rocks form important repositorics for deposits of copper, zinc, lead, silver, iron ore, marble, kyanite, steatite, quartz, corundum and garnet.

6.2.2.1 *Bhilwara Super Group*

6.2.2.1.1 *Rajpura-Dariba Group*

The sulphide-bearing predominantly chemogenic rocks, admixed with minor elastics, lying unconformably over the rocks of the Mangalwar Complex and occurring from Bhinder in the south to northwest of Dhor, over a strike length of 100 km as a 2.5 km wide belt have been assigned to the Rajpura Dariba Group. The chief lithounits of the group are sulphide-bearing graphitic mica schist, dolomitic marble, calc-biotite schist, actinolite schist, garnetiferous mica schist and quartzites. The metasediments exhibit polyphase deformation and regional metamorphism corresponding to almandin-eam-phibolite fades. The sulphides of zinc, lead, copper and iron occur as thin bands and layers exhibiting characteristic syngenetic bedded geometry and show relict sedimentary structures. The lithounits have been classified into the Bhinder, Malikhera, Dariba, Sindesar and Satdudhia Formations.

6.2.2.1.2 *Pur-Banera Group*

The dominantly chemogenic sequence with bands of elastics occurs from south of Banera to Samodi in the west, for over 80 km in a 3 km to 12 km wide belt. These rocks overlie unconformably the rocks of the Potla Formation of the Mangalwar Complex and have been included in the Pur-Banera Group. The sequence comprises conglomerate, garnetiferous mica schist, calc-schist, amphibolite schist, calc-gneiss, dolomitic marble, calc-silicate marble, magnetite quartzite (a persistant marker horizon), banded magnetite chert and carbonate rock with associated sulphide mineralization. These rocks exhibiting superposed deformation and regional metamorphism corresponding to amphibolite fades, have been grouped into the Pur, Pansal, Rewara, Tiranga and Samodi Formations.

6.2.2.1.3 *Jahazpur Group*

Jahazpur Group comprises volcanic, elastic and chemogenic package occurring 2 km southeast of Nandrai in Bhilwara district to Naenwa in Tonk district running over 110 km as a 0.5 km to 8 km wide linear belt. The different lithounits occur in three subparallel stretches: (1) extending from Nandrai to Jhikri, (2) from Katralia to Jahazpur and (3) from Utrana to Naenwa, and have been grouped under the Chuleswarji, Jhikri and Umar coeval Formations. The rock types of the group comprise conglomerate, quartzite, gritty quartzite, mica schist, ferruginous chert and breccia, dolomitic marble, carbonaceous phyllite and metavolcanics. The rocks show low-grade metamorphism (greenschist fades), and unconformably overlie the Sujanpura Formation of the Hindoli Group.

6.2.2.1.4 *Sawar Group*

The chemogenic and elastic rocks exposed as isolated outcrops over 13 km long and 3 to 5 km wide area between Bajata in the north and Sawar in the south, have been designated as the Sawar Group. It comprises quartzite, sulphide-bearing dolomitic marble, calc-silicate marble, mica schist, biotite schist, phyllite and carbonaceous phyllite. These rocks unconformably overlie the Kekri Formation of the Mangalwar Complex. The group has been divided into the Morhi and the Ghatiali Formations.

6.2.2.1.5 *Ranthambor Group*

It comprises quartzite and shale with intercalations of slate and phyllite and is considered to be the youngest member of the Bhilwara Supergroup. The molasse-like sediments exhibiting low-grade metamorphism occurring as isolated outcrops over a distance of 200 km from southeast of Bari Sadri to north of Hora and Mandalgarh have been included in the Ranthambhor Group. It comprises quartzite, slate and phyllite intruded by concordant dolerite sills. The rocks occur as synformal hills and overlie the Bhadesar Formation of the Hindoli Group with structural discordance. Three coeval formations viz., The Bari Sadri, Hora and Mandalgarh have been recognised on the basis of their spatial disposition in isolated basins of deposition.

6.2.2.2 *Aravalli Supergroup*

A thick pile, chiefly comprising metamorphosed and complexly folded clastic sediments with minor chemogenic and organogenic assemblages and interlayered basic volcanics, overlying the Mangalwar Complex and the Sand Mata Complex of the Bhilwara Supergroup with an erosional unconformity, has been assigned to the Aravalli Supergroup. This assemblage of stratified metasediments and interlayered extrusives, together with synorogenic and late- to postorogenic acidic, basic and ultrabasic rocks, covers a time span from 2,500 Ma to 2,000 Ma. The rocks of the Aravalli Supergroup are wellexposed over a stretch of 350 km. The width of the belt in the north is about 40 km gradually fanning out to 150 km in the south. The eastern margin, marking the first order erosional unconformity with the underlying Bhilwara Supergroup, represents one of the most ancient strand line. The present disposition of the strand line is arcuate in shape, with a NW-SE trend in the south, gradually veering to NE-SW in the north. On the west, the Aravalli Supergroup is overlain by the rocks of the Delhi Supergroup and towards south, its continuity is lost below the cover of Deccan Traps and Tertiary sediments. The rocks have undergone polyphase deformation and attained progressively higher grade of regional metamor-phism reaching up to amphibolite facies. The rocks also show migmatisation in the vicinity of the synorogenic plutonic activity.

The boundary between the Aravalli and the pre-Aravalli rocks is marked by the development of a thin quartz-arenite bed, overlain by a suite of synsedimentational basic metavolcanics followed up in the sequence by a coarse pebble polymict metaconglo-merate, coarse meta-arkose, feldspathic quartzite, quartzite, dolomite, calcarenite and carbonaceous phyllite. This basal sequence, indicating a near-shore shelf sedimentation along the cratonic border, has been designated to as the *Debari Group.* The basal sequence is followed upward by a flysch (proximal turbidite) sequence designated as the *Udaipur Group,* and this is overlain by a distal shaly flysch-like sequence comprising chloritic phyllite and quartzite, representing the depositional episode in the Aravalli geosynclinal trough, named as the *Jharol Group.* Between the turbidite sequence and shaly-flysch again synsedimentational basic volcanics associated with epiclast are seen along a N-S trending lineament and are designated as the *Bari Lake Group.* A sequence of phyllite, mica schist, metasub-greywacke, quartzchlorite schist, metasemipelite minor dolomitic lenses, manganiferous phyllite, exposed in southern Rajasthan, has been included in the *Lunavada Group.* This assemblage represents a molasse facies developed after the main paroxysm of the Aravalli orogeny when the deeper trough was transformed to shallow-basins.

The dominantly calcareous rocks to the north of the Banas Lineament, representing a shallow-marine environment (epicontinental realm) of deposition and migmatites are included into the *Kankroli Group* and the *Dovda Group.* These two groups, considered to be coeval with the Udaipur and Bari Lake Groups, and the Jharol Group respectively, are exposed in the area south of Banas Lineament. Dolomitic marble and other calcareous units along with conglomerate, quartzite and carbonaceous phulite of ghasar-rama-gurali-nathdwara area, representing shallow-water facies, have been designated to as the nathdwara group. The rocks of the aravalli Supergroups are intruded by the synorogenic (aravalli orogeny) granites in Udipur, Udaidagar, Darwal and Jaisamand areas, in Jaisamand and Banswara areas the rocks of the Udaipur Group are migmatised. Similarly, the rocks of the Lunavada Group also show migmatisation.

6.2.2.2.1 *Debari Group*

Clastic, chemogenic and biogenic coastal and shelfsediments deposited marginal to Mangalwar, Sararaki-Pal and Mando-ki-Pal cratonic masses with first order erosional unconformity and associated synsedimentational shore-line basic volcanics correspo-nding to an early depositional and volcanic episodes of the Aravalli geological cycle, have been included in the Debari Group. It comprises a sequence of conglomerate, meta- arkose, quartzite, phyllite, mica schist, basic metavolcanics with associated pyroclastics, calcareous quartzite, dolomitic limestone, dolomite, calcitic marble, ferruginous chert, algal phosphatic dolomite and chert, and carbonaceous and manganiferous phyllite. The entire sequence has undergone polyphase deformation and regional metamorphism to greenschist facies. The rocks of the Debari Group forming a NW-SE trending chain of high hills in sharp contrast to the peneplained tract of Mewar to the east, can be traced between Pakhand to the north and Waori Khera to the south over a distance of 160 km. Beyond Waori Khera the continuity of these rocks is lost below the Deccan Traps. The Debari Group is further exposed in the Sarara-ki-Pal and Mando-ki-Pal areas. Lithostrati-graphy of the Debari Group has been worked out separately in Debari, Jaisamand, Ghatol and Sarada sectors.

6.2.2.2.2 *Udaipur Group*

A thick sequence of flysch-like coarse- to fine grained rocks associated with minor bands of chemogenic/biogenic rocks representing the main depositional episode during the geosynclinal phase of the Aravalli geological cycle, has been included in the Udaipur Group. Lithologically, the Udaipur Group comprises a thick pile of phyllite, metagreywacke, feldspathic mica schist and migmatite with intercalatory bands of greywacke, conglomerate, quartzite, dolomite, gritty dolomite, dolomitic marble, oligomictic conglomerate, phosphatic and stromatolitic dolomite, carbonaceous, manganiferous and ferruginous phyllite, chert and amphibole schists. The rocks show frequent interfingering and gradational relationships of different litho-units. The argillaceous metasediments constitute the bulk of the group. The rocks have undergone superimposed deformation and show a grade of metamorphism ranging from greenschist fades to amphibolite fades associated with syntectonic migmatisation. The rocks of the Udaipur Group can be traced from northwest of Eklingjee in the north to Santrampur in south, over a distance of 180 km in a NNW-SSE trending broad belt. The outcrop of the group is narrow between Eklingjee and Sapetia from where it widens to 15 km in the Udaipur valley and extends southwards up to Sarara-ki-Pal Inlier where the outcrop bifurcates. The eastern one extends southeastwards through Salumber, Ganora and Talwara to Banswara where its continuity is lost below the Deccan Traps and the Lunavada Group. The western outcrop follows a N-S trend and extends southwards through Paduan, Rakhabdev, Dungarpur, Peit and Vereth to Santrampur where it is overlapped by the rocks of the Lunavada Group. The rocks of the Udaipur Group have been divided into the Sabina Formation, Eklinggarh Formation, Balicha Formation, Neemach Mata Formation and Banswara Formation in the Debari sector while Mandli, Baroi Magra and Zawar Formations are recognised in the Sarada sector.

6.2.2.2.3 *Bari Lake Group*

An assemblage of coarse- to fine-grained elastic sediments, synsedimentational basic volcanics, pyroclastic and volcaniclastic rocks overlying the Udaipur Group has been designated as the Bari Lake Group. This volcano-sedimentary association marks the local upwarping and shallowing of the trough and contemporaneous activation of a N-S trending deepseated fracture (Rakhabdev Lineament) resulting in the outpouring of basic lavas. The Bari Lake Group is principally composed of basic metavolcancis viz., actinolite- chlorite schist and chlorite schist, metamorphosed pyroclastics, quartzpebble- conglomerate, meta-arkose, quartzite, chloritephyllite and metasiltstone with minor bands of dolomite, chert and carbonaceous phyllite. The rocks have undergone polyphase deformation and lowgrade regional metamorphism of greenschist facies.

These rocks occur over a distance of 30 km between Kharpena in the south to Madar in the north through Bari Lake as a 3-5 km wide belt. Further eastwards and northeastwards the group gradually widens to 15 km and can be traced up to Khamnor where it is terminated against the E-W trending Banas Lineament. The coarser clastic rocks have been included in the Sajjangarh Formation while the overlying volcanosedimentary rocks are included in the Varla Formation. A thick sequence of phyllite and mica schist, around Khamnor, Batia, Kadmas and north of Iswal occupying area in continuity of the Bari Lake metavolcanics, and representing the metamorphosed products of fine grained pyroclastic and volcanic rocks, has been designated as the Khamnor Formation and is included in the Bari Lake Group. The

Formation is overlain by a persistent quartzite horizon considered to be the base of the overlying Jharol Group.

6.2.2.2.4 *Kankroli Group*

A complexly folded and metamorphosed assemblages of rocks of argillaceous and calcareous facies developed under shallow-marine conditions around Kankroli has been included in the Kankroli Group. The rocks appear to have deposited on a topographic high developed towards the northern extremity of the Udaipur-Zawar trench due to the reactivation of the Banas Lineament. The bulk of the Kankroli Group consists of garnetiferous muscovite-biotite schist, hornblende schist, hornblende-biotite schist, hornblende gneiss and dolomitic marble with minor intercalations of quartzite. These rocks have undergone polyphase deformation and regional metamorphism to amphibolite facies conditions. The rocks of the Kankroli Group are exposed as a hammer-head shaped arcuate outcrop between Sangat on the west and Pipli on the east. The Kankroli sequence is surrounded on all sides by the younger rocks of the Dovda Group except along its southern margin where it is in contact with the older members of the Debari Group. Upper units of the Kankroli Group show interfingering relationships northeast of Nathdwara with the rocks of the Bari Lake Group. The rocks of the group are divided into the Madra, Morchana, Rajnagar, Puthol and Sangat Formations. The Madra, Morchana and Rajnagar Formations have been considered coeval to the Udaipur Group while the Puthol and Sangat Formations are equivalent to the Bari Lake Group.

6.2.2.2.4 *Jharol Group*

A thick sequence of almost carbonate-free, shaly, flysch-like elastic sequence representing the orogenic phase of the Aravalli geological cycle has been assigned to the Jharol Group. This predominantly argillaceous suite with minor arenites, deposited in the distal axial portion of the Aravalli geosyncline, is designated as the Jharol Group and considered to have developed, perhaps, consequent to the reactivation of the Rakhabdev Lineament. The bulk of the group is made up of phyllite of different types, chlorite schist, garnetiferous mica schist, intercalatory quartzite and minor lenses and bands of calc-schist and marble. Intercalations of quartzite are frequent and thicker in the upper portion of the group. The rocks have undergone polyphase deformation and show a progressive increase in the grade of metamorphism from lower greenschist fades in the east to lower amphibolite fades in the west. Towards the east, the Jharol Group overlies volcanosedimentary association of the Bari Lake Group and is overlain by the rocks of the Delhi Supergroup in the west. The rocks of the Jharol Group are exposed in a N-S trending linear belt, extending from Modasa in the south to Paraoli in the north, over a distance of 190 km. The belt is about 40 km wide in the south gradually decreasing to 10 km near Butala from where it runs as a 2-5 km wide strip up to Paraoli. The Goran Formation, predominantly argillaceous in character, and the overlying Samlaji Formation which is also argillaceous but with frequent thicker intercalatory sandy layers and minor calcareous layers, have been identified in the Jharol Group.

6.2.2.2.5 *Dovda Group*

An association of chemogenic and elastic rocks showing attributes of shelf-type sedimentation in a shallow-epicontinental basin has been included in the Dovda Group. This group is chiefly composed of an interbanded assemblage of hornblende schist, hornblende-biotite schist, biotite schist, garnetiferous biotite schist, amphibolite gneiss, quartz-biotite gneiss and migmatite with numerous intercalatory bands of calc-silicate rock, dolomitic marble and quartzite. The assemblage as a whole has undergone polyphase deformation producing complex fold patterns and regional metamorphism under amphibolite facies conditions accompanying syntectonic migmatisation. The rocks of the Dovda Group are exposed north of the Banas Lineament surrounding the rocks of the Kankroli Group. The outcrop extends from Masind in the southwest to Khandel in the northeast. The Dovda Group overlies the Kankroli Group in the central part, the Bari Lake Group in the southern part and shows interfingering and gradation relationship with the Jharol Group (at Paroti). It is overlain by the rocks of the Delhi Supergroup in Richer and Kuanthal areas. Depti and Devthari Formations have been recognised in the Dovda Group.

6.2.2.2.6 *Nathdwara Group*

Predominantly chemogenic rocks associated with minor clastic sediments deposited in shallow-inland basins have been assigned to the Nathdwara Group. The group is temporally correlated with the Jharol and Dovda Groups and rests on the Debari, Udaipur and Bari Lake Groups. The bulk of Nathdwara Group in the eastern tract comprises phyllite and calcareous phyllite with numerous intercalatory bands of dolomite and calcitic marble. Thick bands of quartzite occur towards the base. The Nathdwara sequence has undergone polyphase deformation and regional metamorphism to greenschist facies.

The rocks of the Nathdwara Group occur in two separate areas. The eastern tract lies between Rama in the south and Berach in the north while the western tract extends from Madar in the south to Kotharia in the northwest through Iswal and Nathdwara. The Kadmal and the Rama Formations have been identified in the eastern tract. The most abundant rock type in the western tract is dolomite/ dolomitic marble with thin intercalatory bands of ferruginous chert, quartzite and mica schist, referred to as the Haldighati Formation.

6.2.2.2.6 *Lunavada Group*

A characteristic assemblage comprising thick accumulation of clastic metasediments with minor bands of chemogenic and biogenic rocks developed in southern Rajasthan during late to post-orogenic phase of the Aravalli geological cycle has been referred to as the Lunavada Group. With excellent preservation of sedimentary structures the rocks represent a shallow-marine environment of deposition perhaps in the intertidal, beach and shelf zones with a regression of coastline from NE to SW. Structural discordance and typical coarser lithology separate this mollas setype assemblage from the flysch-like and volcano sedimentary asemblages o f the Udaipur, Jharol and Bari Lake Groups. The bulk of the Lunavada

Group is composed of a sequence of phyllite, mica schist, quartz-chlorite schist, metasubgreywacke and quartzite with minor bands and lenses of dolomitic limestone, conglomerate, manganiferous phyllite and phosphatic algal dolomite. At least two phases of deformation have affected the Lunavada rocks resulting in the complex folding of the quartzite bands and formation of dome and basin structures. The rocks generally show greenschist facies of metamorphism. The continuity of the group is either terminated by the intrusive Godhra Granite or lost below the Deccan Traps. The rocks of the Lunavada Group have been divided into the Kalinjara, Wagidora, Bhawanpura, Chandanwara, Bhukia and Kadana Formations. The notable igneous events of this period are the intrusion of the Udaipur, Salu-mber, Udaisagar and Darwal granites. The geosynclinal phase of basic and ultrabasic intrusion include metabasalts and ultrabasics. The Rakhabdev ultrabasic suite comprising serpentinite, talc-carbonate rocks and the talc-chlorite schist represents the geosynclinal phase.

6.2.2.3 *Delhi Supergroup*

The metamorphosed terrigenous and marine clastics with subordinate chemogenic sediments and temporally related concordant and discordant intrusives and their extrusive phases unconformably resting over the rocks of the Bhilwara Supergroup in northeast Rajasthan have been included into the Delhi Supergroup. The rocks of the Delhi Supergroup are traceable as a NE-SW trending rectilinear belt from Delhi in the northeast to Himmatnagar in the southwest over a strike distance of about 850 km. These form the Aravalli orographic axis separating the plains of Marwar from the hilly tract of Mewar in southwest Rajasthan. The continuity of the rocks in the northeast is punctuated by the sandy tract of Rajasthan desert where to the north of Sambhar lake these continue as ridges into the main expanse of Delhis in northeast Rajasthan. In the southwest, beyond Himmatnagar, the solid geology is lost under the cover of alluvial blanket.

The Delhi Supergroup comprises dominantly quartzite, biotite schist, calc-schist, calc-gneiss, and marble with subordinate conglomerate and chert. Metavolcanics, granite, ultramafics, basic flows, epidiorite, hornblende schist, amphibolite and gabbro occur as associated intrusives and extrusives. These rocks exhibit polyphase deformation and regional metamorphism varying from greenschist facies to amphibolite facies. In Rajasthan and in contiguous parts of North Gujarat the Delhi sedimentation took place in two depositional regimes. The deposits in the north of the Rajgarh-Pisangan line (Ajmer and northeast Rajasthan) represent beach and shelf-type sedimentation which has been classified into the Raialo, Alwar and Ajabgarh Groups. South of Rajgarh-Pissangan line the deposition took place on shelf-margin to shelf-interior, main geosynclinal trough and in the post-orogenic successor basins. The dominantly arenaceous and argillaceous facies deposited on shelf-margin has been referred to as the Gogunda Group, the mixed calcareous and argillaceous facies of the shelf-interior have been included in the Kumbhalgarh Group. The clastic sediments of trough-fill have been included in the Sirohi Group and the molasse sediments of the successor basins have been accommodated in the Punagarh and Sindreth Groups. Besides the sediments, the magmatic phases of

Phulad Ophiolite Suite, synsedimentational flows and tuffs and synorogenic granite, have also been included in the stratigraphy.

6.2.2.3.1 *Raialo Group*

The name 'Raialo Series' was introduced by Heron (1917) to designate a predominantly calcareous sequence with subordinate quartzite occurring between the Aravalli and the Delhi Supergroups. Heron (1953, p.22) correlated these calcareous formations occurring in widely separated outcrops in parts of Ajmer, Alwar, Udaipur and Bhilwara districts. He assumed that the Raialo rocks, which had no equivalent anywhere else in India, were deposited during the 'Eparchaean Interval'. The correlation of geographically isolated outcrops of the so called 'Raialos' of Heron (op. cit.) is hardly tenable and the type Raialo Group having a restricted geographic spread is developed only in Jaipur, Dausa and Alwar districts. The entire sequence of the Raialos, mainly developed around Baldeogarh and Tehla, gradually thins out and disappears in strike extension. This suggests its sedimentation in a localised basin during the 'Eparchaean Interval'. The Raialo Group, comprising a lower calcareous and an upper arenaceous sequence (viz., the Dogeta and the Tehla Formations respectively) with marble, phyllite and schist is best exposed in the northeast sector around Dogeta, Gola-ka-Bas, Kharar, Jhiri, Tilwarhi, Todi and Kho-Dariba.

6.2.2.3.2 *Alwar Group*

The metamorphosed dominantly arenaceous clastics with subordinate argillaceous and calcareous rocks and interlayered basic volcanics resting unconformably over the Raialo Group have been included in the Alwar Group. Heron (1917) described these rocks as the Alwar Series. The type sections can be examined near Jaipur and at many places in Alwar and Ajmer districts. The generalised sequence of the Alwar Group is a basal conglomerate, arkose and quartzite followed by a sequence of argillaceous and impure calcareous rocks represented by pelitic schists containing different mineral assemblages, impure marble and para-amphibolite which in turn is followed by a sequence of mainly arenaceous rocks with some amount of pelitic sediments. The rocks of the group have been classified into the Rajgarh, Kankwarhi, Pratapgarh, Nithar, Badalgarh and Bayana Formations in northeastern Rajasthan and into the Srinagar and the Naulakha Formations in the Ajmer sector respectively.

6.2.2.3.3 *Ajabgarh Group*

The metamorphosed argillites with intercalated arenites and subordinate carbonates have been included in the Ajabgarh Group. The 'Kushalgarh Limestone' and 'Hornstone Breccia' horizons of Heron (op. cit.) have been grouped under the Ajabgarh Group. These assume the status of formation and have been named as the Kushalgarh and the Sariska Formations respectively. Other rock units of this group include carbonaceous phyllite, interlayered sequence of quartzite, phyllite and schists containing chlorites, garnet, staurolite and andalusite. These are included under the Thanagazi, Bhakrol and Arauli Formations. The rocks of the Ajabgarh Group in the

Ajmer area have been given independent formational status and described as the Ajmer Formation.

6.2.2.3.4 *Gogunda Group*

The metamorphosed terrigenous clastics comprising quartzite and interbanded schist with subordinate impure calcareous metasediments deposited mainly in the shelf zones of the Delhi trough have been included in the Gogunda Group. The quartzite and intimately associated schists of the Gogunda area and its strike extension in Mewar were included in the Alwar Series of the Delhi System by Heron (1953). Even though the continuity of these quartzites is punctuated in the northeast near Badnor, an emperical correlation was made by Heron (op.cit.) between the Delhi rocks of southwest Rajasthan (Mewar) with those of Jaipur and Alwar in northeast Rajasthan. The rocks which were earlier mapped as Lower Delhi Group by Arora and Ramchandran (1971) together with the outliers of calc-schist of Ajabgarh (cf. Heron) have been grouped together under the Gogunda Group. The rocks of the Gogunda Group are traceable from Kishangarh to Richer. These rocks take a sinistral shift near Antalia and then continue southwards (up to Himmatnagar) in Gujarat. The unit attains its maximum width around Gogunda. The rocks of the Gogunda Group have been separated into three depositionally related litho-associations which have been designated as the Richer, Antalia and Kelwara Formations in descending order of antiquity.

6.2.2.3.5 *Kumbhalgarh Group*

The metamorphosed, marine, calcareous sediments with subordinate argillaceous and arenaceous sediments deposited in the shelf interior of the main rift-controlled Delhi geosyncline have been included in this group. The rocks of this group are traceable as a broad, linear belt occupying the western part of almost the entire length of the main Aravalli range. It runs NESW to NNE-SSW from south of Rajgarh-Pisangan area in the north up to Khed Brahma in the south, over a distance of about 290 km. Lithologically, this group is represented by a sequence of calc-gneiss and calc-schist, pure and impure calcitic marble, muscovite-biotite schist, calcbiotite schist, migmatite and gneiss, quartzite, conglomerate and amphibolite. The sequence has a general sharp contact with the rocks of the underlying Gogunda Group and at places with the overlying Sirohi Group. The rocks of the Kumbhalgarh Group have been regionally metamorphosed to amphibolite facies and the effects of contact metamorphism are well-marked towards the western side where they are intruded by acidic and basic rocks. The rocks of the Kumbhalgarh Group have been divided into the Todgarh, Sendra, Basantgarh, Beawar, Kotra, Barr, Ras and Kalakot Formations which do not always indicate the order of superposition. The rocks show a general younging towards west, and the Barr and the Ras Formations may represent the younger part of the group.

6.2.2.3.6 *Sirohi Group*

The flysch-like thick deposits of fine argillaceous sediments viz., phyllite, mica schist and biotite schist along with intercalations of arenaceous and calcareous sediments (quartzite and marble) deposited during or terminal orogenic phase of the Delhi geological cycle, have been designated as the Sirohi Group. The rocks occur in the form of a broad belt occupying the plains of Marwar lying to the west of Aravalli range. They have been traced as outcrops from Reodar Chandrawati area in southwest to Gandoch and Kot in the northeast for a strike length of about 160 km over a width of about 30 km. The continuity of the belt is either lost under a thick soil/sand cover in the area or interrupted by the intrusive Erinpura Granite. The rocks have been deformed and have been regionally metamorphosed to greenschist facies. Lithologically, the group comprises a thick pile of metasediments consisting predominantly of phyllite, mica schist and biotite schist with intercalated bands of calcitic marble and quartzite. The metasediments of the Sirohi Group have been subdivided, in order of antiquity, into the Jiyapura, Reodar, Ambeshwar and Khiwandi Formations.

6.2.2.3.7 *Punagarh Group*

A thick clastic sequence representing a mollasse like association accompanied with basaltic volcanics deposited in a continental basin and developed during the late to post-orogenic phase of the Delhi geological cycle in the Pali district of Rajasthan has been assigned to the Punagarh Group. The rocks of the Punagarh Group occur in roughly oval-shaped outcrop from Sheopura in the northeast to Bhumadra in the southwest and from Jadan in the east to Nimal in the west. Lithologically, the group consists of a sequence of shale, slate, phyllite, mica schist and quartzite, with synsedimentational basic volcanics containing pillow lava, metatuff, bedded chert. The entire sequence is folded to form doubly plunging syncline. The Sojat, Bombolai, Khambal and Sowania Formations have been recognised in the Punagarh Group.

6.2.2.3.8 *Sindreth Group*

Clastic sequence representing molasse-like sediments accompanied with basaltic volcanics, deposited in an isolated successor basin, developed during the late to post-orogenic phase of the Delhi geological cycle in Sirohi district of Rajasthan, has been assigned into the Sindreth Group. The rocks of the Sindreth Group are traceable on a roughly linear outcrop from northeast of Padia in the northeast to north of Mirpur in the south and Gotali in east to Angor in the west over a distance of 20 km. The entire sequence of metasediments is enveloped by intrusive granite and migmatite. Litholog-ically, the group consists of a sequence of conglomerate, gritty quartzite, micaceous quartzite, shale, phyllite and synsedimentational basic flows. The entire sequence appears to have been folded into a synform. The Angor and the Goyali Formations have been recognised within the group.

6.2.2.4 *Vindhyan Supergroup*

The western segment of the Vindhyan basin occupies an area of about 26,000 km' in the south-eastern Rajasthan. The Vindhyan sediments overlie the metamorphites of the Bhilwara Supergroup and the Berach Granite with a sharp, angular unconformity (Great Eparchaean Unconformity) defined by synsedimentary volcanics (The Khairmalia Andesite) and discontinuous bodies of conglomerate of the basal Khardeola Sandstone and Bhagwanpura Limestone Formations. The Vindhyan Supergroup comprises an alternating sequence of sandstone, shale and limestone having a total thickness of about 3,200 m. They are divisible into two broad subdivisions, namely the Lower and the Upper Vindhyans. The Lower Vindhyans are predominantly calc-argillaceous while the Upper Vindhyans are mainly arenaceous. The Vindhyan sedimentation in Rajasthan took place in two different sub-basins, the Chittorgarh-Jhalawar sub-basin in the southwest having a full development of the Lower Vindhyan stratigraphy further classified, in an ascending order, into the Satola, Sand, Lasrawan and Khorip Groups and the Sapotra Karauli sub-basin in the northeast, with only a partial representation of the lower horizons of the Lower Vindhyan sequence equivalent to the Satola and the Sand Groups of Chittorgarh area. The two sub-basins which later came under one cover of the sediments were separated by a median upland near Bundi where the Lower Vindhyans are absent.

6.2.2.4.1 *Lower Vindhyan*

Chittorgarh-Jhalawar sub-basin: The initial phase of the Satola Group sedimentation started with volcanic activity of andesitic affinity (Khairmalia andesite, pyroclastic and tuffs) followed by conglomerate, cross-bedded sandstone and shale of the basal Khardeola Sandstone overlying the metasediments of the Bhilwara Supergroup and the Berach Granite. This is followed by the tidal-flat deposits of the Bhagwanpura Limestone with thin, impersistant conglomerate beds at the base signifying deposition in the coastal regime. The prolific growth of stromatolites belonging to the genera *Collenia, Cryptozoon, Oncolites, Conoplayton* and *Indophyton* in the Bhawanpura Limestone indicate their deposition under stable platform condition. The succeeding Sand Group comprising the Sawa Sandstone and the Palri Shale with procellanite, was deposited in quiet water when there was an incursion of the sea casusing overlaps. The overlying Lasrawan Group of sediments comprising the Kalmia Sandstone and the Binota Shale with glauconitic siltstone interbeds were deposited in deeper water under neritic conditions. The calcargillaceous sediments of the overlying Khorip Group comprising Khori Malan Conglomerate (at the base) grading into the Jiran Sandstone followed by the Bari Shale, the Nimbahera Limestone and the Suket Shale with limestone were deposited in an unstable shelf.

Sapotra-Karauli sub-basin : Sedimentation, contemporary to the Khardeola and Bhagwanpura Formations, also commenced in this sub-basin with the basal conglomerate, glauconitic sandstone and shale followed by the Tirohan Limestone with prolific algal growth of genera *Collenia, Conophyton and Cryptozoon.* The Tirohan Breccia with chert and procellanite, deposited conformably over the Tirohan Limestone, is equated with the Palri Shale of the Sand Group. The bouldery nature of the Basal Conglomerate suggests proximity of the source area and the rate of sedimentation exceeding subsidence of the basin. The succeeding Tirohan Limestone containing algal structures is indicative of its deposition under littoral marine environment. The overlying Tirohan Breccia and porcellanite is autoclastic in origin. The presence of Tirohan Breccia in the vicinity of the GBF zone has been attributed to synsedimentational tectonism. Younger units of the Lower Vindhyans (the Lasrawan and the Khorip Groups) are missing in this sub-basin. Sedimentological studies of the Lower Vindhyan sediments indicate beach to near shore conditions. Sub-rounded to poorly-sorted arkosic sediments of the Khardeola Sandstone, Bhagwanpura Limestone and Sawa Sandstones indicate little transportation compared to rounded and well-sorted sediments of the Jiran Sandstone. The palaeocurrent direction during the Lower Vindhyan time indicates an easterly transport of sediments in the Chittorgarh sub-basin, and NE to ENE in the Sapotra-Karauli sub-basin. The Bhilwara, Aravalli and Delhi Supergroups acted as provenance.

6.2.2.4.2 *Upper Vindhyan*

A well-marked change in the sedimentary environment from predominatly calc-argillaceous (littoral shallow-marine) to clastic (continental fluviatile) took place during the Upper Vindhyan time. During the Kaimur, Rewa and major part of the Bhander time, the two sub-basins in Rajasthan remained unconnected. In the Sapotra-Karauli subbasin the Tirohan Breccia is unconformably overlain by the Badanpur (Kaimur) Conglomerate and the Akoda Mahadev Sandstone suggesting a sharp erosional hiatus between the Lower and the Upper Vindhyans. On the contrary, deposition continued uninterruptedly in the Chittorgarh- Jhalawar subbasin where the Chittorgarh Fort Sandstone (Lower Kaimur Sandstone) conformably overlies the Suket Shale (Khorip Group). Some more part of the basement came under sedimentation and the strand line shifted further to the north and the Akoda Mahadev Sandstone (Upper Kaimur Sandstone) was directly deposited over the older rocks northwest of the GBF, with a distinct basal conglomerate between Akoda and Indargarh. The absence of the Lower Vindhyans in Bundi-Indargarh area is considered to be due to the basement-high known as the Bundi Upland. The Kaimur basin was not so deep initially and it gradually subsided with sedimentation. The orthoquartzitic nature, textural maturity, good sorting and prolific development of crosss-beds and ripple marks at all levels in the Kaimur Sandstone are indicative of its deposition under shallow-marine shelf regime. The presence of glauconite in the upper levels (around Rawatbhata) suggests its disposition under littoral marine condition. Restricted ferruginisation, presence of sun cracks, rain prints and intraformational conglomerate suggest development of local sub-aerial conditions towards the close of the Kaimur sedimentation.

The succeeding Rewa Group of sediments representing alternating argillaceous and arenaceous facies suggest periodic instability of the basin floor. The extremely fine-grained Panna Shale with fine parallel-laminations was locally deposited in a shorelagoon environment. The presence of glauconitic siltstone beds at the base in Rawatbhata-Javda area indicates local reducing condition. The overlying Indargarh Sandstone (Lower Rewa Sandstone) having a restricted development is orthoquartzitic, texturally matured, profusely cross-bedded and rippled, and was deposited in littoral to neritic zones under a shallow marine environment. The Jhiri Shale, covering the older formations, was deposited when there was transgression of the sea. The dominant reddi-shpurplish colour glauconitic siltstone intercalations and proliferation of stromatolitic colonies in the limestone member of the Jhiri Shale indicate that the depositional environments were dominated by warm-water conditions. The overlying orthoquartzitic, texturally matured Taragarh Fort Sandstone (Upper Rewa Sandstone) showing prolific development of ripples and cross- beds, occurring in a blanket-type disposition was deposited under shallow-marine tidal-flat conditions. The presence of glauconite in sandstone and its ferruginous nature locally, suggests deposition under reducing environment. The presence of thin, discontinuous bodies of conglomerate in the upper parts indicates local high-energy conditions.

The orthoquartzite-carbonate-shale association of the Bhander Group of sediments is broadly attributed to stable shelf regime. The reddish-purplish ferruginous Ganurgarh Shale with frequent glauconitic siltstone beds showing ripples, cross-laminations, sun cracks and intraformational conglomerate at the base suggest a shore-lagoon tidal-flat deposition where as the argillaceous limestone and intraformational conglomerate and breccia at the top indicate supratidal conditions. The intraformational conglomerate and breccia, sun cracks and ripples in the lower beds of the Lakheri Limestone suggest their deposition under subaerial conditions. The rest of the formation was deposited under more or less chemogenic environment as suggested by the chert and anhydrite intercalations and almost general absence of detritus. The corresponding limestone unit is missing in the Sapotra-Karauli sub-basin. This condition continued during the Samria Shale. The prolific growth of stromatolites in the limestone interbeds of the Samria Shale favours their deposition in an open, intertidal mudflats, mainly in the protected lagoon. The orthoquartzitic nature, high maturity and good sorting together with profuse crossbeds, ripple marks and current lineations suggest deposition of the Bundi Hill Sandstone under shallow, stable-shelf conditions.

During the deposition of the Sirbu Shale both the sub-basins got united into a single repository wherein the Sirbu Shale and other younger formations viz., the Maihar Sandstone, Balwan Limestone and Dholpura Shale in Bundi-Indargarh area and Upper Bhander Sandstone in Sapotra-Karauli area were deposited in a shallow-marine conditions. The principal mode of palaeocurrent direction in the Maihar Sandstone in Sapotra-Karauli sub-basin is towards north. Uniformity and consistency of the direction of sediment transport signifies stable palaeoslope and general tectonic stability of the basin, especially during the Upper Bhander time. The Balwan Limestone which is stromatolitic limestone, indicates shallow-marine condition of deposition. The youngest Dholpura Shale indicates restricted condition. Thus, it appears that throughout the Vindhyan sedimentation the conditions of deposition were fluctuating due to subsidence and elevation of the basin and consequent transgressions and regressions of the sea. Several evidences of organic life in the form of stromatolite, trace, macro and micro fossils have been reported from the Vindhyans. The stromatolites preserved in the Vindhyans (similar to the Ripheans of the former USSR) indicate an age ranging from over 1,300 Ma to less than 600 Ma. Radiometric dating of basal Khairmalia Andesite (1,250 Ma) and the

Kaimur glauconite (940-910 Ma) is in fair agreement with the results arrived at from the stromatolites.

6.2.2.5 *Malani Igneous Suite*

The Malani Igneous suite encompasses Late Proterozoic plutonic and volcanic rocks. The suite covers an area of about 51,000 km' in parts of Jaisalmer, Jodhpur, Barmer, Churu, Pali, Sirohi and Jalore districts of western Rajasthan and extends further in Sind province of Pakistan. Malani magmatism, the largest anorogenic acid volcanism in India showing little metamorphic effect, is associated with post-Erinpura granitic activity and pre-Marwar sedimentation and covers a span of 850550 Ma. Three phases of igneous activity have been identified viz., eruption of basic flows, followed by voluminous acid rhyolite flows culminating with ashflow deposits (first phase); intrusion of subjacent discordant granites as plutons, ring dykes, bosses and plugs (second phase) and the indiscriminate intrusion of mafic and felsic, coarse- to fine-grained dykes warms (third phase). Malani volcanics have erupted in three stages. The volcanism commenced with eruption of basalt with occasional andesite flows (first stage) followed and covered by the voluminous outpouring of peralkaline and peraluminous rhyolite, rhyodacite and trachyte flows (second stage) and ceased with the outburst of ash-flow deposit (third stage).

The dominant felsic volcanics are rhyolites and rhyodacites having a spread of about 31,000 kin z. The other rock types associated with rhyolite are trachytes, dacites, pitchstone, welded-tuff, vitric, lithic and crystal-ash, ignimbrite, obsidian, pyroclastic slates, agglomerate volcanic breccia and volcanic conglomerates. The cumulative thickness of felsic volcanics exceeds 3-5 km in 95 cumulative flows Bhushan (2001). The final eruptive stage witnessed ash-flow deposits near Pokaran. These ash beds, 50 m in thickness, are composed essentially of light yellow and light brown lithic-, crystal- and vitric-ash having microscopic xenoliths of basalt and rhyolite embedded in it. The mafic-felsic volcanic suite (first phase) is intruded by post-tectonic, anorogenic Siwana and Jalore Granites during the second phase. The Siwana Granite occurs as ring dykes while Jalore Granite attains batholithic dimension near Jalore and also occurs as small bosses near Salawas. The hornblende granite (Malani Granite) occurs as porphyry dykes in Barmer area. The third phase of igneous cycle experienced intrusion of felsic and mafic dykes trending either NNW-SSE or NE-SW direction and the dolerite forming the predominant dyke phase. Five subphases in a dyke-swarm near Sankra have been recorded commencing with felsic intrusions and culminating with the basic intrusions.

The Jalore Granite plots in the granodiorite field. The main difference between the Siwana and the Jalore Granites is the presence of normative acmite and absence of normative anorthite in the former. The Jalore Granite is metaluminous. It has higher SiO2 and A1203. The Siwana Granite which is peralkaline has very high TiO_2 alkalies, REE and iron due to presence of alkali pyroxene and alkali amphiboles. At Siwana, the alkali granites as well as the associated rhyolites (comendites and pantellerites) and trachytes are characterised by the presence of riebeckite, aegirine and arfvedsonite indicating a co-magmatic relationship. Malani volcanics occur towards the western side of tightly compressed Delhi- Aravalli fold belts. A possible sudden release of pressure, causing rifting may have triggered the eruption. NW-SE trending fissures and several calderas and cones were the possible mode of eruptions.

6.2.2.6 *Marwar Supergroup*

The trans-Aravalli sediments, equivalent of Vindhyans deposited on the NW flank of the Aravalli hill range, occurring as numerous scattered flat topped hills and covered under the sands of the Thar desert towards north, have been grouped under the Marwar Supergroup. The general resemblance to the Vindhyan sediments notwithstanding the Marwar beds having deposited in a separate basin, are considered correlatable to the Upper Bhander Group. Laid down in continental seas, upon the Malani volcanics, at a time just before the life forms appeared in abundance, the Marwars are of great interest. Apart from *chordophyceous* tracks and fusiform bodies, bearing some relation of the *Cruziana furcifera,* no determinable organic remains have been found in the Marwar beds. The rocks of the Marwar Supergroup, over 1,000 m thick, are divided into three groups, namely the Jodhpur, Bilara and Nagaur Groups.

The Pokaran Boulder Bed, Sonia and Girbhakar Formations constitute the *Jodlipur Group.* The Pokaran Boulder Bed, about 3 m thick, extending southward from Pokaran in the north to Hariyoli *Nadi* in the south, over a distance of 42 km consists of pebbles and cobbles of granite, syenite, basic rock and quartzite in a red-brown sandy matrix (near Karewa). Cobbles of granite set in a red arenaceous matrix (near Baratka-Gaon). The overlying Sonia Formation, horizontally disposed with rolling dips, commences with a maroon shale with mega ripples which alternates with creamish-buff to reddish, fine- to medium-grained sandstones interbedded with chert, cherty dolomite and jasper. The contact with the overlying Girbhakar Formation is marked by a short break in sedimentation indicated by the presence of pebbly horizon containing pebbles of the Sonia sandstone and chert. The Girbhakar sandstone is reddish and consists of coarse to gritty micaceous sandstone alternating with red and light green shale with pebbly and gritty layers. The Jodhpur Group of rocks shows NW to W palaeocurrent direction. The well-sorted, mature sandstones were deposited in near-shore environment. Heavy mineral analysis suggests that the materials were received from acid plutonic volcanics and low-grade contact metamorphic rocks. The beds are, for most of the part, unfossiliferous.

The calcareous *Bilara Group* which is 300 m thick comprises Dhanapa Formation- chert at the base, followed by well-bedded to laminated cherty dolomite, siliceous dolomitic limestone and stromatolitic limestone; Gotan Formation-dark grey laminated limestone, containing bands of clay; and the Pondlo Formation comprising dolomite, cherty dolomite, claystone, siltstone and reddish sandstone. It is stromatolitic towards top near Shiv Sar and contains siliceous oolites grading into pisolites near Lordiyan. *Weedia, Collenia, Cryptozoon* and algal nodules, illpreserved lamellibranchs and brachiopod shells are reported from the Bilaras.

The *Nagaur Group* unconformably overlying the Pondlo dolomite (the Bilara Group) commences north of Khichan with a 5 m thick conglomerate bed, containing cobbles, pebbles fragments and pieces of dolomite, chert, cherty dolomite, feldspar, quartz and sandstone cemented in a pink-brown calcareous sandy matrix. These are overlain by red, mottled, claystone, siltstone and fine-grained friable sandstone. The Nagaur sandstones have been divided into two formationsthe older unit, designated as the Nagaur Formation, comprises a sequence of brick-red to red claystone, siltstone and sandstone with bands of clay and blotches of green clay and the younger unit, designated as the Tunklian Formation, comprises brick-red claystone, siltstone, calcareous clay and gritty to pebbly sandstone containing pebbles of Malani rhyolite-granite, chert, dolomite, quartz and quartzite near Gudarli.

6.2.3 Palaeozoic

After the deposition of the Nagaur Group (Proterozoic) there was a break in sedimentation as a result of which sediments of most of Cambrian, Ordovician, Silurian, Devonian and Lower Carboniferous periods are not represented in Rajasthan. Upper Carboniferous glaciation is represented in the Nagaur basin by the NE-SW trending remnants of the Bap Boulder Bed-an assortment of pebbles and cobbles with occasional boulders and a few erratics exhibiting well-marked glacial striations, spread over 68 km z from Khirwa to Nokhra. The rock types include gravel/boulder, 5 cm to 25 cm across in size, of rhyolite, tuff, granite, basalt and dolerite of the Malani Igneous Suite; granite, syenite, amphibolite and basic rocks, quartzite, phyllite and slate of pre-Delhi and Delhi age and limestone, dolomitic limestone and sandstone of the Marwar Supergroup. This is overlain by a very small remnant of the marine Badhaura sandstone of 30 m computed thickness. Extending from northwest of Badhaura up to Harbans, these marine sediments, designated as the Badhaura Formation, comprise medium- to coarse grained brownish-yellowish and ferruginous sandstone, clay, shale and siltstone containing Permian fauna and flora. The subsurface continuation of the formation at Pugal between depths 500 and 590 m, and at Shumarwali *Talai* has been established by drilling. The Permian sediments thus have thickened dipwards and represent the post-glacial sediments. The Badhaura sandstone which can be correlated with the sequence of the Indus Basin, implies that the Indus Basin (Salt Range) was connected with the north and northwest Rajasthan.

6.2.4 Mesozoic And Cenozoic

6.2.4.1 *Deccan Traps*

The Deccan Traps (Cretaceous to Eocene) covering an area of about 10,435 km Z in southeast Rajasthan have developed in two separate geographic sectors, namely, the Kota-Jhalawar sector in the southeast and the Chittorgarh-Banswara sector in the southsoutheast. *Kota-Jhalawar sector : In* this sector the Deccan Traps overlying the Vindhyan sediments have attained a cumulative thickness of over 180 m. Twelve flows, varying in thickness from 7 to 56 m are present between 280 and 481 m RL. Out of the 12 flows, five are Pahoehoe-type and the rest are Aa-type. The individual flows are separated either by impersistent red bole bed or by Inter-trappean Beds comprising conglomerate, clay and chert. Both fossiliferous and unfossiliferous clay/chert beds are present among the older flows. *Physa prinsepii* and ostracodes have been recorded near Parbati (24" 23';76°31'), Sohankheri (24°30' : 76°05') and Gadiyo (24°31' : 76°03'). Around Dag (23°56' : 75°50') and Chaumahala (23°5T : 76°36') the flows have undergone widespread laterisation. A gritty, sandy and cherty unfossiliferous friable Infra-trappean clay bed containing angular pieces of sandstone occurs widely in the area around Chippabarod (24°33' : 76"43'), Chabra (24°40' : 76°51') and Chandloi (24°30' : 76°10').

Chittorgarh-Banswara sector : The area is characterised by the N-S trending Deccan Traps plateau with prominent scarp on the west. The Deccan Traps overlie the Lower Vindhyan rocks in the north and northeast, but towards the west and south they rest over the rocks of the Mangalwar Complex and the Aravalli Supergroup. In all, sixteen flows varying in thickness from 11 to 55 m have been identified between 226 and 606 m RL. Out of these, four flows have Pahoehoe characters and the rest Aa-type characters. Fairly widespread unfossiliferous clay bed, as Infratrappean, is present around Kushalgarh (23°12': 74°27') and Barodia whereas fossiliferous Physa *(Bullinus) prinsepii* Infra-trappean limestone bed is reported from Ram-ka-Munna in Banswara district. Unfossiliferous thin Inter-trappean clay beds are noticed around Arnaud (23°53' : 74°49') Dhamotar (24°06' : 74°45'), and Kulmipura (24°07' : 74°45') whereas unfossiliferous chert bed, about 1 m in thickness, is noticed near Acheri (24°26' : 74°50') and Semara (24°26' : 74°42').

6.2.4.2 *Tertiary Alkaline Complex*

The Tertiary magmatic event of the western Rajasthan is characterised by the mildly alkaline and hyperalkaline plutonic volcanic suites of Sarnu Dandali, Karara- Tavidar and Mundwara areas. The hyperalkaline suite of rocks clearly traverses the Malani rhyolites and Cretaceous clastics and, therefore, represents a major Tertiary alkaline magmatic event. The Sarnu-Dandali suite is dominantly nephelinitic in comp-osition, various members being mafic alkaline rocks, ijolites, foidal syenites, phonolites and carbonatites belonging to a carbonatite-nephelinite association. The Mundwara suite is mostly basanitic comprising picrites, foidal gabbros and foidal syenite with carbonatites which intrude the Erinpura Granite.

6.2.4.3 *Sedimentaries*

Triassic rocks do not outcrop anywhere in west and northwest Rajasthan but its subsurface presence, identified on pollen-spore assemblage, at Shumarwali *Talai* borehole, has been established. Land conditions existed during post-Permian to pre- Jurassic period and extensive erosional agencies reigned supreme leaving only traces of these Permian deposits. The Jurassic period commenced with a fluviatile environment which gave rise to the Lathi sandstone. The area north, west and south of Lathi was filled up by the sediments of the Lathi Formation, the southern continuity being checked by the Jodhpur Pokaran-Chotan- Malani ridge. The outcrops of the Lathi sandstone are seen intermittently, separated by the desert sands. The Lathi sandstone, 360 m thick, comprises a sequence of coarse- grained sandstone, gritty sandstone, pebbly sandstone, gritty and pebbly sandstone, conglomerate, arkose, lithic-arenite, siltstone, shale, ferruginous shale, ferruginous sandstone and clay. The Lathis overlie the Malani Igneous Suite of rocks, the Pondlo dolomite, the Nagaur sandstone, the Bap Boulder Bed and the Badhaura sandstone. The contact with the Malanis is marked by a conglomerate bed, grading into pebbly sandstone and grit. A friable, weathered, dirty-white to yellowish sandstone comes in contact with the other older underlying rocks.

The Lathi sandstone has been subdivided into lower and upper members. The lower member comprises a sequence of conglomerate, arkose, lithicarenites and coarse-grained sandstone, indicative of terrestrial deposition under stable delta in fresh water conditions. The upper member comprises bands of sandstone and siltstone, grading into limestone, indicating shallow-marine shelf-sedimentation. The abundance of fossil wood logs is so conspicuous that an area near Akal has been converted into a Fossil Wood *Park.* The fossil wood is of gymnospermous origin and belongs to dicotyledonous stem. The Lathi sandstone has leaf impressions of *Petroph-yllum* sp. *Ptillophyllitm acutifolium, Equisctites* sp. and preserves silicified dicotyledon-ous wood and silicified gastropod shells. The Lathi sandstone contains thin bands of lignite as noted in dug wells near Devikot, Nar Singh-ki-Dhani, Jhinjhinyali and south of Unror. The Mayakor Formation, which overlies the Badhaura Formation with an unconformity has been equated with the Lower Lathis. The deposition of Lathi/Mayakar Formation culminated with the initiation of the marine transgressions in different basins. These include the Jaisalmer basin in the west, the Barmer basin in the south- central and the Palana-Ganganagar Shelf in the north and northwest of Rajasthan.

Jaisalmer Basin

The Jaisalmer basin deepened towards northwest extending into Pakistan. The Jaisalmer-Mari arch, the raised platform, exposes the entire sequence. The Mesozoics are preserved in NE-SW trending strip with their northwesterly boundary being located near Niba, Sam, Kanoi, Mahmud-ki-Dhani, Jaidesar, Sanu and Lila; the Lower Tertiaries are exposed towards northwest of these localities. The Mesozoics are represented by the Lathi sandstone, Jaisalmer limestone, Baisakhi shale, Bedesar sandstone, Parihar (Pariwar) sandstone and Abur limestone, while the Lower Tertiary includes Sanu sandstone, Khuiala limestone and Bandah limestone. Pleistocene Shumar grits overlie the Lower Tertiary sequence as capping of ferruginous grit and laterites. All these units have been assigned formational status.

The Jaisalmer limestone has intertonguing contact with the .underlying Lathis and represents the first marine transgression of the Mesozoic in Rajasthan. Subsurface thickness of this formation is ca 600 m in the Jaisalmer basin. The Baisakhi shale and the Bedesar sandstone occur as mappable rock forma lions of 165 m and 65 m thickness respectively, but in subsurface they occur as intergrading sequences. Each of these formations has been subdivided into units A,B and C. The overlying Pariwar sandstone which represents continental deposition contains angiospermic plant remains and is 300 m thick. The overlying Abur limestone marks the end of the Mesozoic Era in the Jaisalmer basin. The Abur Formation is 66 m thick and in subsurface it is divisible into the Goru Formation and the Parh

Formation of younger ages showing continuity of the marine environments in the deeper parts of the Jaisalmer basin.

The Palaeogene succession is exposed in detached outcrops near Mahmud-ki-Dhani, Sanu, Shera and Lila. It commences with the Sanu sandstone which is 75 m thick and consists of fresh-water arenaceous and glauconitic sandstone, clay, shale and ferruginous shale showing current bedding. The Sanu sandstone appears to grade in subsurface into non-marine and marine bands showing fish teeth. The Khuiala limestone of a total thickness of 100 m overlies the Sanu sandstone and is overlain by the Bandha limestone of 75 m computed thickness marking the close of the sedimentary cycle in the Jaisalmer basin. The Tertiaries occur oblique to the Mesozoic strike and find representation in the Barmer basin and in the Palana-Ganganagar shelf area.

Barmer Basin

Southeast of Jaisalmer a sequence of the Mesozoic and Lower Tertiary rock formations occupies a N-S trending linear tract for about 100 km from Mandai and Fatehgarh in the north to Barmer in the south. This synclinal exposure having a maximum width of 50 km at Akli is known as the Barmer basin. The configuration of the basin is related to Meso-Cenozoic tectonic framework of the Cambay Graben in Gujarat. The rock formation of the Barmer basin include the Fatehgarh sandstone, Akli bentonite, Mandai sandstone and Kapurdi fuller's earth-each unit enjoying a formational status. The Fatehgarh sandstone overlies the Lathi sandstone in the Barmer basin and represents the Cretaceous marine transgression. It occurs along the northern half of the synclinal depression of the Barmer basin and comprises conglomerate, phosphatic sandstone, highly fossiliferous sandstone, ferruginous sandstone and clay in a 50 m thick sequence. These beds are overlain by the Akli bentonite which contains 265 m thick sequence of the bentonite, bentonitic clays, siliceous earth containing phosphatic beds. The continental Mandai sandstone (100 m) overlies the Aklis and is overlain by the Kapurdi fuller's earth (30 m) which marks the closure of Tertiary cycle in the Barmer basin.

6.2.4.4 *Quaternary*

A large track of western Rajasthan constitute the Thar desert. The NE-SW trending Aravalli hill range define its eastern boundary while the flood plains of the Ghaggar-Sutlaj rivers represent its northern periphery. In the southwest the Great Rann of Kachchh is its limit. Average elevation of this undulating desertic terrain varies from about 200 m to 300 m MSL with a few isolated hilly tracts rising up to 500 m above sea level. The Quaternary continental sediments in the Thar desert comprise a multiple episodic succession of fluvial, fluvio-lacustrine and aeolian deposits. A complete succession (reaching up to 300 m) is not developed in any one section. Although the lithofacies show variations in their intensity of development in different sectors, their geological evolution seems to be conformable. The neotectonic movements, forming a series of NE-SW trending horst and graben structures, are considered to have controlled the origin, configuration and developement of depositional basins and their gradual thickening westward. Change in climate from semi-arid to arid resulted in widespread formation of calcrete.

Fluvial and aeolian agencies are the major present day processes which also dominated throughout the Quaternary Era. The fluvial activity prevalent in the eastern margin while the aeolian processes in the west - the relative spatial dominance being a function of the distribution of rainfall. Numerous stable aeolian bedforms inherited from an earlier dry climatic phase have been vigourously acted upon by fluvial processes creating a fine network of rills and gullies along the eastern margins of the desert. The most common landform sequence produced by the fluvial processes within the desert are the desert hill slope, rocky/gravelly pediment, flat buried pediment, older and younger alluvial plains and palaeo-channel signatures. In the Jaisalmer area where near-horizontal, high-level, structural plains (hamada, bounded by steep escarpment) are exposed, the sequence consist of serir, gravel-leg, sandy loam, calcretised conglomerates etc.

A number of saline depressions, ranns or playas, occur within the desert. They are considered to have formed as remnants of a receding sea; as aeolian segmentation of former streams during a drier climate; due to deflation in the wind shadow zone of a hill and stream-trapping by lee-longitudinal dunes etc. The spatial extent of the fluvial landforms has reduced over time due to their progressive burial by the aeolian sands. New units like sand dunes, interdune plains and sandy undulating plains (buried pediment and alluvial plains) have been recognised.

Sand dunes are classified on the basis of their age of formation and stability, and the types recognised are the clustered parabolic, transverse, reticulate, longitudinal and barchans etc. Interdune plains, in between the dune ridges, occur as narrow to wide sandy plain which are flat, sandy arid undulatinghosting gypsite/calcrete beds at places. The calcretes show a large range in the degree of development and morphological appearance depending upon the age of the landform. Soft nodular calcretes have formed within aeolian sandy plains in the 250-400 mm rainfall zone. Nodules, 1-4 cm across, constitute 3-15% of the bulk and occur randomly distributed amidst mildly calcareous sandy mass. Hard nodular calcretes occur extensively in the aggraded older alluvial (Pleistocene) plains or on more ancient surfaces. With depth, the nodules get larger and irregular in shape and are cemented together. These deposits locally called *murnd* find extensive use in construction of roads in sandy dry land terrain.

6.3 STRUCTURE

The Precambrian rocks of Rajasthan exhibit widely varying structural patterns resulting from repeated periods of deformation and accompanying emplacement of igneous rocks. The regional strike of foliation in the Bhilwara and the Delhi Supergroups of rocks is dominantly NNE-SSW to NE-SW, while it is N-S and NW-SE in the Aravallis.

The foliation dips dominantly towards the west at a steep to moderate angle. Local variations in dips are seen towards the east in the rocks of the Hindoli Group. The regional structure is defined by the presence of a series of isoclinal overturned folds and by normal upright folds as in the Delhis of Alwar region. The fold axes in the Mangalwar and the Hindoli Groups of eastern Rajasthan trend dominantly NE-SW swerving to N-S in Rajpura-Dariba and Pur Banera-Bhinder belts of southern Rajasthan. The fold pattern in these rocks indicates en echelon shifts of the fold axes presumably indicating later cross-folding. On the other hand, the fold axes in the Aravallis of Udaipur- Banswara region in southern Rajasthan trend NW-SE, swerving gradually to N-S and NE-SW trends near Udaipur and Nathdwara. There are a series of culminations and depressions, anticlinal culminations in Sarada dome and the Hammer-head syncline of Kankroli as well as other structural complications along the Aravalli belt in comparison with the divergent structural trends in the Bhilwara and the Aravalli Supergroups, the fold axes in the Delhi Supergroup along the main Aravalli mountain range and in Alwar-Jaipur-Sikar region trends consistently in NNE-SSW and NE-SW directions. In the southern part of the Aravalli range, south of Beawar there is a main synclinorium flanked by a series of anticlines and synclines, all plunging towards southwest. Towards the north of Beawar, however, there is a plunge reversal beyond which a series of doubly plunging linear folds (anticlines and synclines) extend towards the northeast as far as the Khetri Copper Belt. The anticlinal culminations within these fold systems are often marked by Alwar quartzites, flanked all around by Ajabgarh schists, phyllites, calcgneiss, calc- schists etc. In additiion to this, the linear anticlinal axes in the Alwar quartzites, in Khetri region of Khetri Copper Belt, also reveal a series of north to south plunging en echelon dextral- and sinistral-type of later sheer folding. After a tight anticlinal arcu-ate swerve, in the Neem-ka-Thana-Babai region, the rocks display a major synclinal structure in the Sikar-JaipurAlwar region. This is marked by a series of sub- parallel upright synclines and anticlines plunging dominantly towards northeast and marked by alternating cluminations and depressions of the axes as one proceeds northward.

Precambrian rocks in Rajasthan have been subjected to many periods of superposed deformation resulting in varying structural trends. The earliest deformation has given rise to tight, isoclinals, reclined folds plunging at about 20°-40° towards WNW. The second deformation which is more intense and prominent has given rise to open to tight upright folds with their plunge varying from 10° to 50° towards SW or SSW in the southern part and NE or NNE in the northern part. The third deformation exhibits NE or SW plunging open, upright folds with sub-vertical axial planes striking WNW-ESE.

The first two generation of folds are noticed in the Bhilwara and the Aravalli Supergroups. The superposition of these two generations of fold has led to different structural trends along NW-SE, E-W and NE-SW axes. The rocks of the Delhi Supergroup, on the other hand, contain the second, third and fourth generation of folds resulting in the formation of tight longitudinal and sometimes isoclinals overturned folds having a dominantly NE-SW trend. The absence of the first generation of folds in the Delhi Supergroup marks a major structural break between the Delhis and the pre-Aravallis/Aravallis of Rajasthan. The other major tectonic element in these rocks is a series of sub-parallel NE-SW trending faults and presumably thrusts in the Delhi and other groups of rock, specially those within the Delhis along the Aravalli range and in Alwar region.

The rocks of the Vindhyan Supergroup are very little disturbed and have gentle dips towards south and south-southeast. The basement rocks beyond Chittorgarh in the west, Bundi-Indargarh in the northwest, and Shivpuri in the east acted as rigid block. The Ramgarh Dome acted as a center from which the stresses were centrifugally radiated resulting in the formation of Bundi-Indargarh and Phalodi anticlines in the northwest; causing a series of anticlines and synclines in Sawai Madhopur, Sapotra and Karauli areas in the north-northwest; generating open rolls in Shabad area in the east; forming Jhalawar anticline in the southeast and developing a series of anticlines and synclines in Chittorgarh in the west. The rejuvenation of adjoining Aravalli-Delhi mobile fold-belt in the west resulted in the Great Boundary Fault (GBF) which trends roughly NE-SW for a distance of 400 km between Chittorgarh and Agra. It started as a gravity-fault along which gradual sinking of the basin floor took place. The present reverse geometry with low-angle thrusting has brought the older basement rocks in juxtaposition with the Vindhyan sequence. The GBF is not a single plane of disruption but it evolved in stages and represented a 10-20 km wide zone (Balmiki Prasad, 1984) comprising a number of parallel and oblique faults showing horst and graben features. Recent studies along the corridor of GBF carried out in Sawai Madhopur and Chittorgarh sectors suggest that it is not a marginal basin fault, but formed during syn- to post-deformation F1 folds. The Marwar Supergroup of western Rajasthan are practically undisturbed except for the presence of a north plunging main basinal synclinal axes trending N-S to NNW-SSE from Sojat to Bilara and Nagaur.

The Malanis were emplaced dominantly as fissure eruptions and accompanying sheet-like intrusions along NW-SE and N-S faults and fissures of post-Delhi age. The rocks of the Mesozoic and Tertiary Era in the western Rajasthan are under formed and constitute very low-dipping strata marked, at places, by gravity faults. The area being mostly covered by desert sand, the subsurface anomalies/ discontinuities, revealed by geophysical and geological exploration by GSI, ONGC and other State agencies, indicate presence of a number of faults which have shaped the present disposition of the various litho units in the area.

6.4 CLIMATIC CONDITIONS

Rajasthan occupies a special position in the climatic history of the Indian sub-continent. The climate of Rajasthan plays a dominant role in structuring its physical as well as biotic environment. The climate is in general varies from hot desert type in the west to subhumid type in the east and south-east. To the west and northwest of the Aravallis, aridity remains a dominant characteristic even during the summer monsoons. The 'temperatures remain high for the greater part of the year and rainfall is scanty, besides being highly erratic and variable. To the east and southeast of the Aravallis conditions improve gradually and there is a considerable variation in the amount of rainfall and temperature distribution. The 500 mm isohyet approximately following Aravallis is the critical dividing line between the arid or hot desert and semi-arid in the west and sub-humid in the east and south-east. Further;

the exist anomalies within these two zones because of differential lateral distribution of climatic components viz. precipitation, temperature, wind velocity and humidity resulting from the state's relief and continental location.

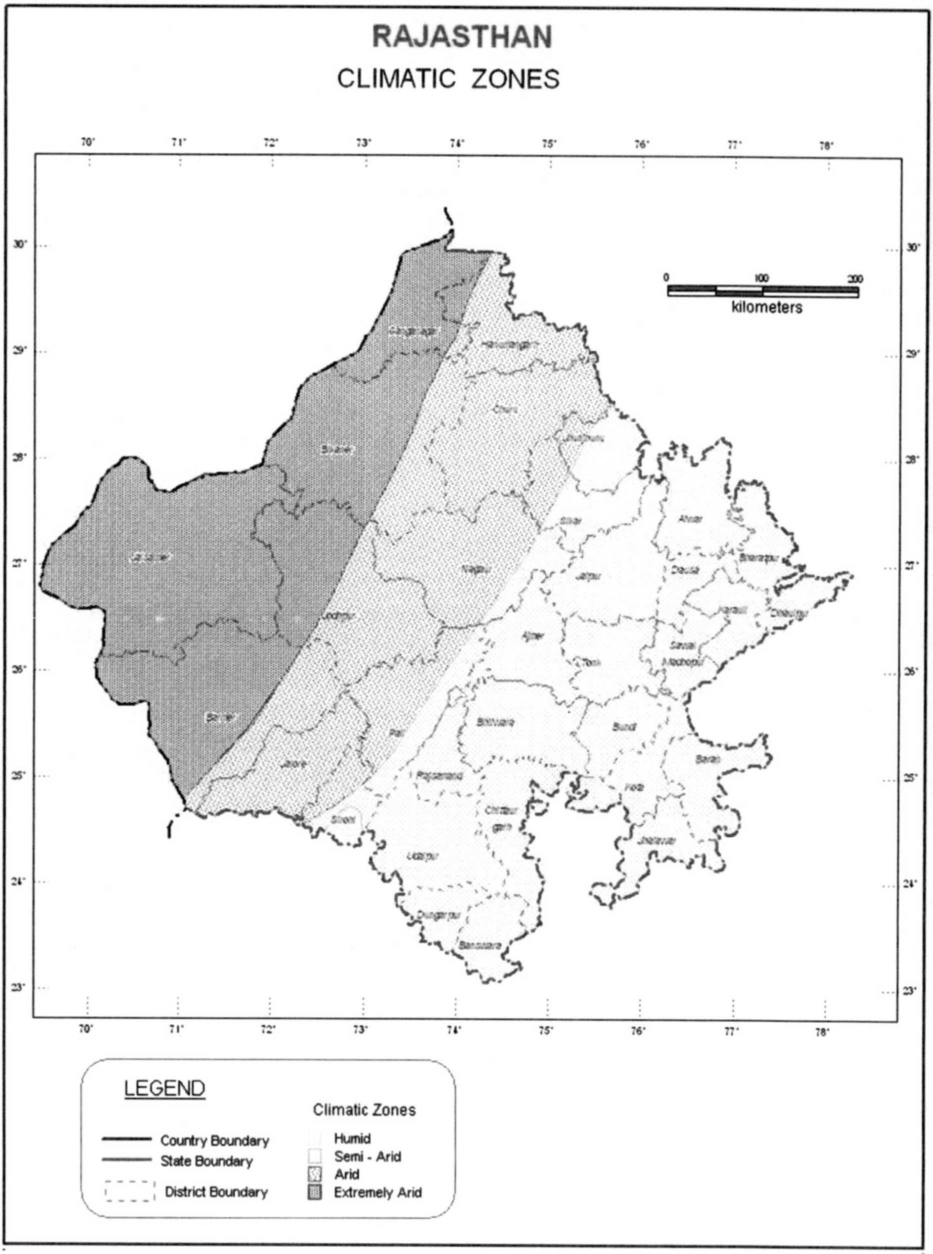

Figure 6.7. Climatic Zones
(Source: Survey of India, Rajasthan Atlas, Resource Atlas, Watershed Atlas, Groundwater Atlas, CAZRI and Agricultural Geography of Rajasthan by Dr. T.S. Chouhan)

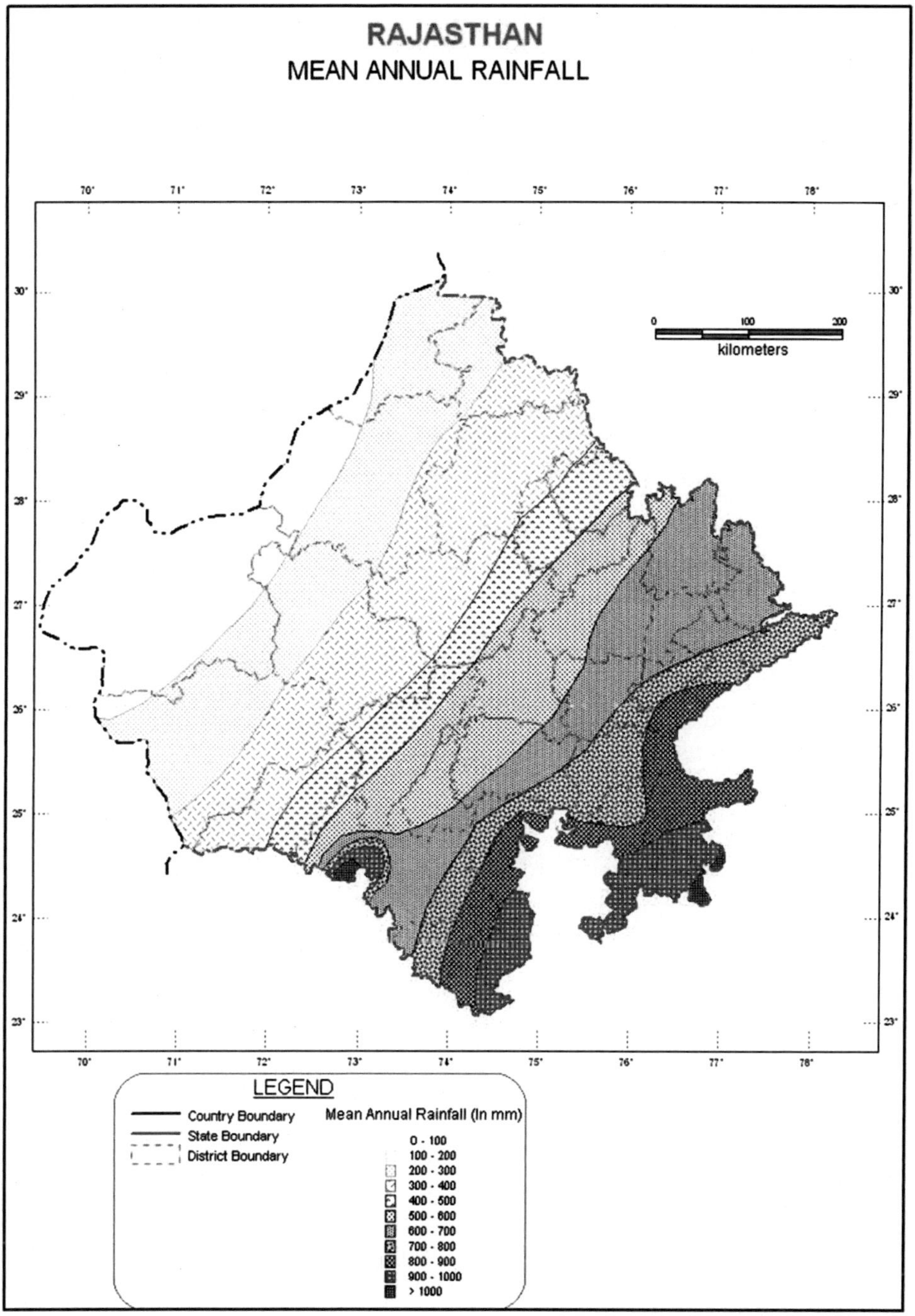

Figure 6.8. Mean Annual Rainfall
(Source: Survey of India, Rajasthan Atlas, Resource Atlas, Watershed Atlas, Groundwater Atlas, CAZRI and Agricultural Geography of Rajasthan by Dr. T.S. Chouhan)

6.4.1. Climate

The climate of Rajasthan, West of Aravalli, is characterized by great extremes of temperatures, long periods of severe drought, accompanied by high winds and low to semi-arid conditions of humidity. The winter is severe and at places the temperature sometimes falls below freezing point. The heat during summer is intense touching 49°C in some parts of the State. In the east and south of the Aravalli there is a considerable variation in the amount of rainfall and range of temperature. Due to the sandy nature of soil, which gets rapidly heated during the day and cools down quickly after dark, difference of as much as 22°C have been recorded between daily maximum and minimum temperature. Based on climatic conditions, Rajasthan is the driest State of the country. Rajasthan is climatically devided in 4 zones based on Moisture Index (After Thornthwaite) as shown in Figure 6.7. Rajasthan receives the rain from both the Arabian Sea and the Bay of Bengal system of the South-West monsoon and about 80% to 90% of the rain in the State occurs from mid-June to September. The 500 mm isohytel line which runs along the western edge of the Aravalli range divides the State into two regions. The rainfall gradually increases towards East and South of this line, and rapidly decreases towards the West. On an average, the State receives 536 mm of rainfall, but there are wide fluctuations in its distribution. Jaisalmer and Bikaner in the west receive hardly 170mm and 390mm of rain respectively where as Banswara and Jhalawar in the South-East receive about 930 mm of rain. On account of temporal and spatial variations in the occurrence of rainfall, a flood-drought syndrome afflicts the State. On the basis of precipitation, temperature, wind velocity and humidity, Rajasthan has been divided into four conventional seasons:

6.4.1.1 *Pre-Monsoon*

This season extends from April to June with temperatures varying between 32°C to 45°C. May is the hottest month, though June is also characterised by high temperature. The mean maximum temperature in May varies from 40°C to 42°C over most of Rajasthan with mean minimum at 23°C to 27°C. The heat during the day is intense and scorching, and Western Rajasthan is the hottest, the mean temperature rising to over 45°C in May and June. Occasionally, the temperatures of the order of 50°C have been recorded here. It will be observed from Figure 6.11 that there is a gradual increase in temperature north and northwest of the Aravallis. Kota and Jhalawar areas in the southeast also indicate a high temperature. The 34°C and 36°C isothermal lines covers greater part of western Rajasthan as well as districts of Kota and Jhalawar. The lowest temperatures prevailing in the hilly Abu region are indicated by an isothermal line of 28°C. The air movement increases in intensity with the advance of the summer season and is strong and vigorous in April and May. The diurnal range of temperature, particularly in the arid west, is of the order of 15°C to 18°C. On account of the dryness of the atmosphere and clear skies and the sandy nature of soil, there is rapid radiation of heat from the earth soon after the sunset and the heat of the day rapidly dies down, with the result that even when the maximum temperature during the day is high, the minimum temperature at night is considerably lower (Pramanik, 1952). Under the spells of heat waves, which sometimes occur over northern India in this season, the day temperatures shoot up considerably above normal. Thus, Ganganagar has recorded a maximum temperature of 50°C, Bikaner and Barmer 49°C, Jaipur and Kota 47.7°C and Jaisalmer 46.6°C. The intense heat of the day is reduced by an occasional dust storm which is sometimes accompanied by a shower. These dust storms are more frequent in the west and northwest decreasing progressively towards the east and southeast with more rainfall. The winds, in this season, are generally from west and southwest. The relative humidity is least during this season, the lowest being in April. It is generally 35 to 60 percent in the morning and 10 to 30 percent in the afternoon hours throughout this season. Values for the month of June are slightly higher due to the occurrence of rains. Thus, the chief features of the weather during this season are dryness of air, high temperatures and hot dry westerly winds.

6.4.1.2 *Monsoon or South-West Monsoon*

This is the main rainy season when over 90 percent of the rainfall of the year is received. The wind flow is mainly southwest or west. Clouding increases with the advance of the season till about August, and spatial distribution is also one of increase towards southeast. The cold, humid winds of the monsoon currents extend over East Rajasthan by about the end of June, and by the third week of July the whole of Rajasthan is under it sway. The advance of the monsoon is generally preceded by thunderstorms and dust storms. The monsoon is most active in July and August. The important synoptic feature of this season is the monsoon depressions, form in the north Bay of Bengal or come from farther east and move westward along the monsoon trough. Some of the depressions go as far as Arabia and The Persian Gulf. The depressions intensity the monsoon activity and sometimes concentrated rainfall in their vicinity, the monsoon withdraws from West Rajasthan early in September and from East Rajasthan by the third week of September.

Normally the rainy season sets in by the end of June over the Aravalli plain and the piedmont zone and by the middle of July in the western and northwestern Rajasthan. The weather cools down with the onset of monsoon showers and brings relief after prolonged heat of the summer season, but in the years of little or no rains, high temperatures persist over vast areas in the State. On an average, the maximum temperatures during this season vary between 29.5°C to 32.2°C in the southern and eastern parts of the state to over 37.7°C in the northern and western parts. The relative humidity is maximum during this season, being highest in August. But it is less marked in the areas to the west and northwest than in the eastern Rajasthan. Winds are generally from west and south west but whenever a depression or low approaches o enters the state; the winds change their direction so as get into the circulation of the disturbance.

6.4.1.3 *Post-Monsoon*

With the withdrawal of the monsoon, clear autumn weather sets in over the area. The skies are generally clear, winds weak and variable over west Rajasthan and north over east Rajasthan. The monsoon is withdrawn from the area by the mid-September after which a second hot season follows which ultimately cools down and merges with the cold season by December. October and November are months of clear and light clouded skies. The temperatures during October, in the whole of Rajasthan, are fairly uniform, the maximum temperature ranging between 33°C to 38°C and minimum between 18°C to 20°C at various stations. The diurnal range of temperature is high; varying from 14°C to 16°C. The relative humidity is quite low, even tower than in the cold season at a number of stations. The winds die down considerably during this season as the anticyclonic conditions succeed to cyclonic, and are generally light and variable.

6.4.1.4 *Winter*

This is a season of cool weather, clear skies, low humidity and low temperature with large diurnal variation. The wind flow is generally weak northerly. The fine weather is occasionally broken by brief spells of cloudy weather, caused by western disturbances which are attended by clouding and rise of temperature in the forward portion, while in its rear western and northwestern winds bring a spell of cold weather. The winter season sets in by the month of December and is marked by a gradual decrease in temperature all over the region. January is the cold set month of the year. The mean maximum temperature during this month ranges from 20°C at Ganganagar to 22°C at Bikaner, 22.8°C at Jaipur and Ajmer 24.4°C at Barmer and Jodhpur to 25°C at Kota. The mean minimum temperature varies from 3.3°C at Ganganagar to 7.7°C at Ajmer, 8.3°C to 10°C at Jaipur, Bikaner, Jodhpur and Jhalawar to 10.5°C at Kota. The isotherms for the north of January show a marked contrast from those of the May isotherms. These isothermal lines approximately run from east to west, with slight variations while crossing the Aravalli hills where these lines dip gradually towards the south, an indication that slight lower average temperature is found over the Aravallis at southern latitude. The incidence of western depressions is a special feature of the weather of the north Indian plains during this season but only a few of them affect and give rain Rajasthan. Only northern and northeastern portions of the state often come under the sway of these disturbances. They bring with them a little rainfall together with significant changes in temperature. The amount of winter rainfall is an insignificant portion of the annual total, hardly 5 to 6 percent. The relative humidity, in this season, is generally between 50 to 60 percent in the morning hours and 25 to 35 percent in the afternoon hours. The winds are generally light and variable, but northeasterly and northerly winds are more frequent than those from the other directions.

6.4.2. *Rainfall*

The annual rainfall in the state varies significantly. The general trend of Isohyets is from north-west to south-east. There is a very rapid and marked decrease in rainfall west of the Aravalli range, making western Rajasthan the most arid part. The average annual rainfall in this part ranges from less than 10 cm in north-west part of Jaisalmer (lowest in the state), to 20 to 30 cm in Ganganagar, Bikaner and Barmer regions, 30 to 40 cm in Nagaur, Jodhpur, Churu and Jalore regions and more than 40 cm in Sikar, Jhunjhunu and Pali regions and along the western fringes of the Aravalli range. On the eastern side of the Aravalli range, the rainfall ranges from 55 cm in Ajmer to 102 cm in Jhalawar as shown in Table 6.1 and figure No.6.8. In plains, Banswara (92.0 cm) and Jhalawar (102 cm) districts receive the maximum annual rain. Mount Abu (Sirohi district) in the south west, however receives the highest rainfall in the state (163.8 cm). The yearly total rainfall is highly variable at different places all over the state and it is most erratic in the western half with frequent spells of drought, punctuated occasionally by heavy downpour in some years, associated with the passing low pressure systems over the region.

Table 6.1: Mean Annual Rainfall

S.No.	District	Mean Rainfall (cm)												
		Jan	Feb	March	April	May	June	July	Aug	Sep	Oct	Nov	Dec	Annual
1	Barmer	0.28	0.36	0.25	0.16	0.66	1.94	8.73	11.31	3.37	0.36	0.06	0.11	27.59
2	Bikaner	0.36	0.62	0.41	0.43	0.99	2.49	7.49	9.12	3.08	0.55	0.09	0.33	25.96
3	Churu	0.86	0.66	0.53	0.42	0.99	3.48	0.10	9.52	4.94	0.52	0.12	0.41	22.55
4	Ganganagar and Hanumangarh	0.97	0.98	0.68	0.53	0.55	2.74	7.74	7.31	3.14	0.32	0.03	0.38	25.37
5	Jalore	0.29	0.45	0.19	0.16	0.55	3.95	15.17	15.52	5.03	0.62	0.08	0.15	42.16
6	Jaisalmer	0.16	0.37	0.13	0.12	0.27	0.84	4.80	6.09	1.14	0.08	0.05	0.04	14.09
7	Jodhpur	0.39	0.53	0.37	0.28	0.93	3.09	9.75	11.93	4.49	0.48	0.11	0.18	32.53
8	Nagaur	0.57	0.55	0.50	0.35	1.18	3.52	12.88	12.66	5.08	0.56	0.11	0.38	38.34
9	Pali	0.35	0.47	0.21	0.20	1.09	4.16	14.48	18.24	7.09	0.65	0.15	0.13	47.22
10	Ajmer	0.51	0.56	0.44	0.30	1.07	4.99	16.20	16.35	7.30	1.01	0.28	0.41	49.42
11	Alwar	1.25	1.09	0.92	0.57	1.26	4.90	17.99	12.77	10.37	1.29	0.24	0.33	52.98
12	Banswara	0.32	0.19	0.14	0.08	0.42	10.97	32.22	29.35	16.13	1.76	0.57	0.62	92.77
13	Bundi	0.54	0.34	0.32	0.25	0.72	6.76	28.10	27.35	10.62	0.78	0.21	0.09	76.08
14	Bharatpur	1.26	1.00	0.74	0.54	1.00	5.11	20.48	20.85	12.10	1.81	0.30	0.42	65.61
15	Bhilwara	0.51	0.22	0.36	0.28	0.68	5.93	25.67	25.30	9.56	0.62	0.19	0.59	69.91
16	Chittorgarh	0.60	0.23	0.23	0.15	0.55	8.55	29.49	30.91	12.50	0.99	0.63	0.58	85.41
17	Dungarpur	0.21	0.19	0.13	0.11	0.72	9.89	28.67	23.32	11.30	1.10	0.43	0.38	76.45
18	Jaipur and Dausa	1.12	0.90	0.59	0.36	0.99	5.13	18.21	18.07	8.50	0.99	0.19	0.10	55.15
19	Jhalawar	1.05	0.54	0.35	0.33	0.92	10.09	33.45	30.01	15.17	1.35	1.29	0.57	95.12
20	Jhunjhunu	1.11	0.94	0.83	0.54	1.35	4.84	14.13	12.57	6.58	0.85	0.19	0.52	44.45
21	Kota and Baran	1.01	0.54	0.34	0.31	0.84	8.34	31.96	28.59	13.48	1.46	0.84	0.57	88.28
22	Sikar	1.22	0.78	0.73	0.50	1.22	4.65	15.08	14.23	6.37	0.59	0.21	0.51	46.09
23	Sirohi	0.42	0.52	0.17	0.25	1.20	5.77	23.81	22.63	7.63	0.92	0.30	0.22	63.84
24	Sawai Madhopur and Karauli	1.04	0.63	0.59	0.38	0.80	5.75	23.39	23.60	10.35	1.19	0.27	0.50	68.49
25	Tonk	0.80	0.47	0.36	0.33	0.77	5.75	21.23	20.75	9.44	0.81	0.21	0.44	61.36
26	Udaipur and Rajsamand	0.80	0.32	0.29	0.17	0.92	6.80	22.89	20.49	10.79	1.00	0.45	0.20	65.12

The south-west monsoon which has its beginning in the last week of June in the eastern parts, may last till mid-September, Pre-monsoon showers begin towards the middle of June and post monsoon rains occasionally occur in October. In the winter season also, there is sometimes, a little rainfall associated with the passing western distribution over the region. At most places, the highest normal monthly rainfall is during July and August. The number of rainy days during this period varies widely in different places, ranging from 10 in Jaisalmer to 40 in Jhalawar and 48 in Mount Abu. Rainfall during the rest of the period ranges from 2.1 cm at Jaisalmer to 7.2 cm at Jaipur, distributed over 2.5 to 6 rainy days (Table 6.2).

Table 6.2: Mean Rainy Days

S.No.	District	Mean Rainy days												
		Jan	**Feb**	**March**	**April**	**May**	**June**	**July**	**Aug**	**Sep**	**Oct**	**Nov**	**Dec**	**Annual**
1	Barmer	0.20	0.30	0.20	0.10	0.50	1.10	4.00	4.40	1.60	0.10	0.10	0.10	12.70
2	Bikaner	0.40	0.70	0.50	0.50	0.90	1.90	4.44	4.60	1.80	0.30	0.00	0.30	16.34
3	Churu	0.80	0.70	0.60	0.40	1.00	2.30	5.30	5.00	2.50	0.30	0.10	0.40	19.40
4	Ganganagar and Hanumangarh	1.10	0.90	0.70	0.50	0.60	1.70	4.20	3.70	1.60	0.30	0.00	0.40	15.70
5	Jalore	0.20	0.40	0.20	0.10	0.40	1.70	6.30	6.40	2.50	0.30	0.10	0.10	18.70
6	Jaisalmer	0.10	0.20	0.10	0.10	0.20	0.40	2.10	2.10	0.50	0.00	0.00	0.00	5.80
7	Jodhpur	0.30	0.50	0.30	0.30	0.90	1.80	5.10	5.50	2.30	0.30	0.10	0.20	17.60
8	Nagaur	0.60	0.60	0.50	0.30	1.10	2.40	6.30	6.30	2.80	0.40	0.10	0.30	21.70
9	Pali	0.30	0.50	0.20	0.20	0.90	2.40	6.70	6.80	3.10	0.50	0.10	0.10	21.80
10	Ajmer	0.50	0.60	0.50	0.30	1.00	2.80	7.80	7.80	3.50	0.60	0.20	0.30	25.90
11	Alwar	1.10	1.00	0.80	0.60	1.10	2.90	8.40	8.30	4.40	0.60	0.20	0.22	29.62
12	Banswara	0.40	0.20	0.10	0.10	0.30	4.40	13.60	12.20	6.30	0.80	0.40	0.50	39.30
13	Bundi	0.60	0.50	0.40	0.30	0.70	4.00	11.80	10.90	5.30	0.50	0.20	0.10	35.30
14	Bharatpur	1.10	0.90	0.70	0.50	1.10	3.30	9.90	10.00	5.40	0.70	0.20	0.20	34.00
15	Bhilwara	0.50	0.30	0.40	0.30	0.60	3.20	11.00	10.30	4.20	0.40	0.10	0.60	31.90
16	Chittorgarh	0.50	0.30	0.20	0.20	0.50	3.90	11.40	10.50	5.10	0.50	0.30	0.30	33.70
17	Dungarpur	0.30	0.10	0.10	0.10	0.60	4.30	12.40	10.70	5.10	0.60	0.30	0.20	34.80
18	Jaipur and Dausa	1.00	1.00	0.70	0.40	1.10	3.40	9.50	9.50	4.50	0.60	0.20	0.10	32.00
19	Jhalawar	1.00	0.50	0.40	0.30	0.90	5.50	12.80	11.50	6.80	0.90	0.80	0.50	41.90
20	Jhunjhunu	1.10	1.00	0.90	0.60	1.50	3.10	7.20	6.90	3.40	0.60	0.20	0.50	27.00
21	Kota and Baran	1.00	0.60	0.40	0.30	0.80	4.50	12.70	11.60	6.30	0.90	0.60	0.50	40.20
22	Sikar	1.20	0.90	0.80	0.60	1.30	2.90	8.40	8.50	3.90	0.50	0.30	0.50	29.80
23	Sirohi	0.50	0.50	0.20	0.20	0.80	2.90	9.90	9.20	3.60	0.70	0.20	0.20	28.90
24	Sawai Madhopur and Karauli	1.00	0.80	0.50	0.40	0.80	3.40	10.70	10.70	5.00	0.60	0.20	0.60	34.70
25	Tonk	0.90	0.60	0.50	0.40	0.80	3.30	10.40	10.40	4.60	0.60	0.20	0.30	33.00
26	Udaipur and Rajsamand	0.80	0.30	0.30	0.10	0.80	3.30	10.30	8.90	4.90	0.70	0.30	0.20	30.90

The rainfall is erratic and is confined mostly to the south- eastern part of the state. There is a rapid and marked decreases in rainfall west of the Aravallis. It varies from above 1500 mm in the south-east to less than 100 mm in the western margin of the state, encircling the Thar Desert or The Great Indian Desert. More than 90 percent of the annual rainfall is concentrated in the rainy months extending from mid-June to mid-September. July and August are the wettest months in Rajasthan. The number of rainy days varies and ranges from less than 15 in western part to more than 55 days in the south-eastern part of Rajasthan.

The distribution of rainfall is very unequal throughout the state. A study of the mean annual rainfall map of the state reveals three distinct rainfall belts :

(i) sub-humid belt on the east and south of the Aravallis, where rainfall varies between 500 to 1500 mm;

(ii) semi-arid belt, from the western hills of the Aravallis in the east to the desert tract in the west, where rainfall ranges between 300 to 500 mm; and

(iii) arid belt in the Thar Desert area with an annual rainfall of less than 100 to 300 mm.

The general trend of the isohyetal lines is from northeast to southwest. On an average, east Rajasthan receives 700 mm of rainfall and west Rajasthan about 300 mm in a year. Of this amount, about 90.0 percent falls during the monsoon season from June to September, and about 5.0 percent during the cold weather period. The seasonal rainfall for various seasons are shown in Figure 6.9. The monsoon pattern in the state from 1983 to 1999 had been most erratic in the state. It can be analysed as lean phase upto 1988 and after 1988 the pattern being alternate wet and lean monsoon conditions upto 1994. After 1994 the monsoon was somewhat generous over the state upto 1997. During last three years i.e.1998, 1999 and 2000 there have been continuous failure of monsoon causing severe draught conditions in the state. It is also known that under normal rainfall conditions, repeated periods of soil moisture utilisation and soil moisture recharge tend to minimize the total drought conditions so well characteristics of the dry years. A comparison of seasonal course of rainfall with course of monthly PE for a station provides much information on many aspects of water relation such as water surplus and water deficit. The water surplus and water deficit areas are shown in Figure 6.10.

6.4.3 Temperature

Marked variations in diurnal and seasonal range of temperatures occur at all places in the state, exhibiting the most characteristic phenomenon of the warm dry continental climate.

The month of March marks the beginning of summer and the temperature starts rising progressively through April, May and June. The temperature rise during this period is almost uniform all over the state. In the western part, mainly at Bikaner, Phalodi, Jaisalmer and Barmer, the maximum daily temperatures varies between 40°C to 45°C (Figure 6.11) Occasionally, it rises to 49°C during the summer months. The diurnal temperature variation is greater in summers. The minimum daily temperature drops down considerably at night and remains between 20ºC to 29ºC. On the eastern side of the Aravalli range, the day temperature are in a similar range but night temperatures hover around 26ºC. At Udaipur and Mount Abu, temperature, however is relatively lower and the mean daily maximum temperature in summer reaches 38ºC and 31.5ºC, respectively.

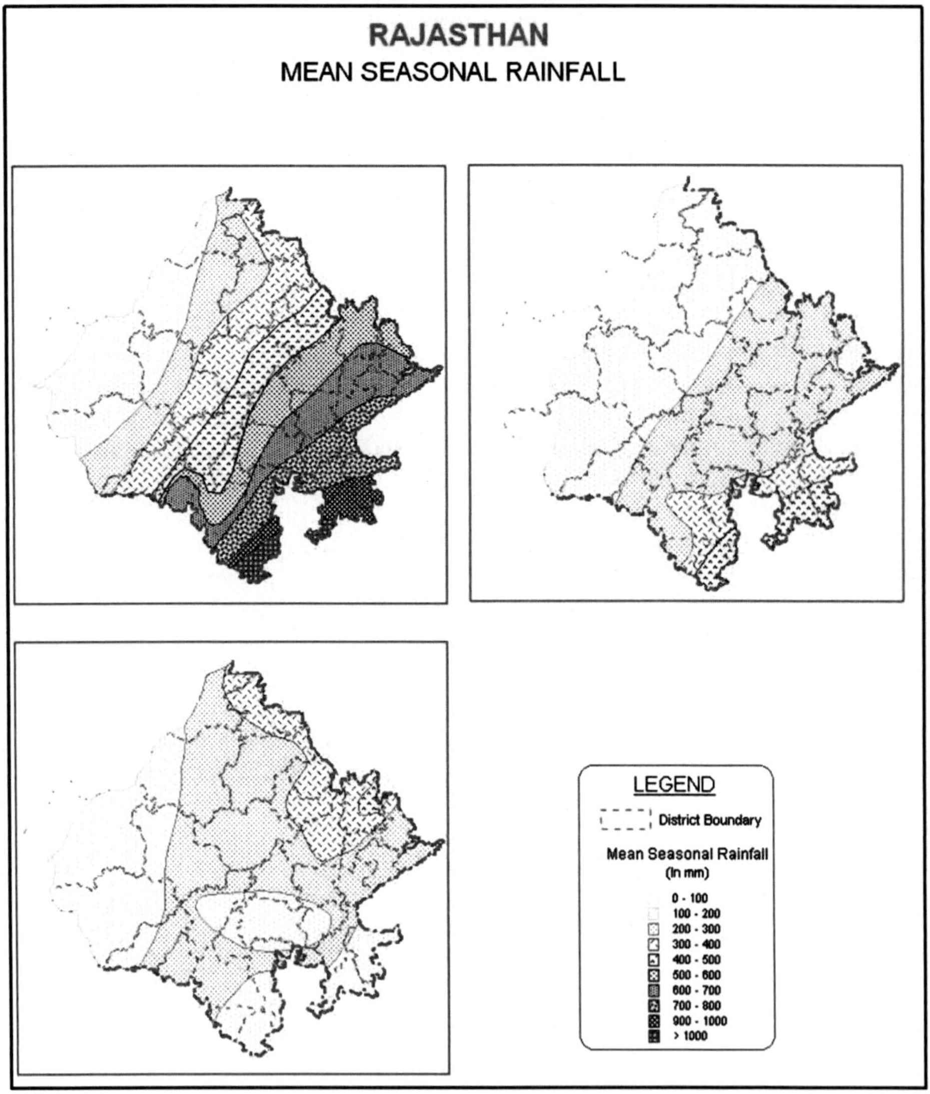

Figure 6.9. Mean seasonal Rainfall (Source: IMD)

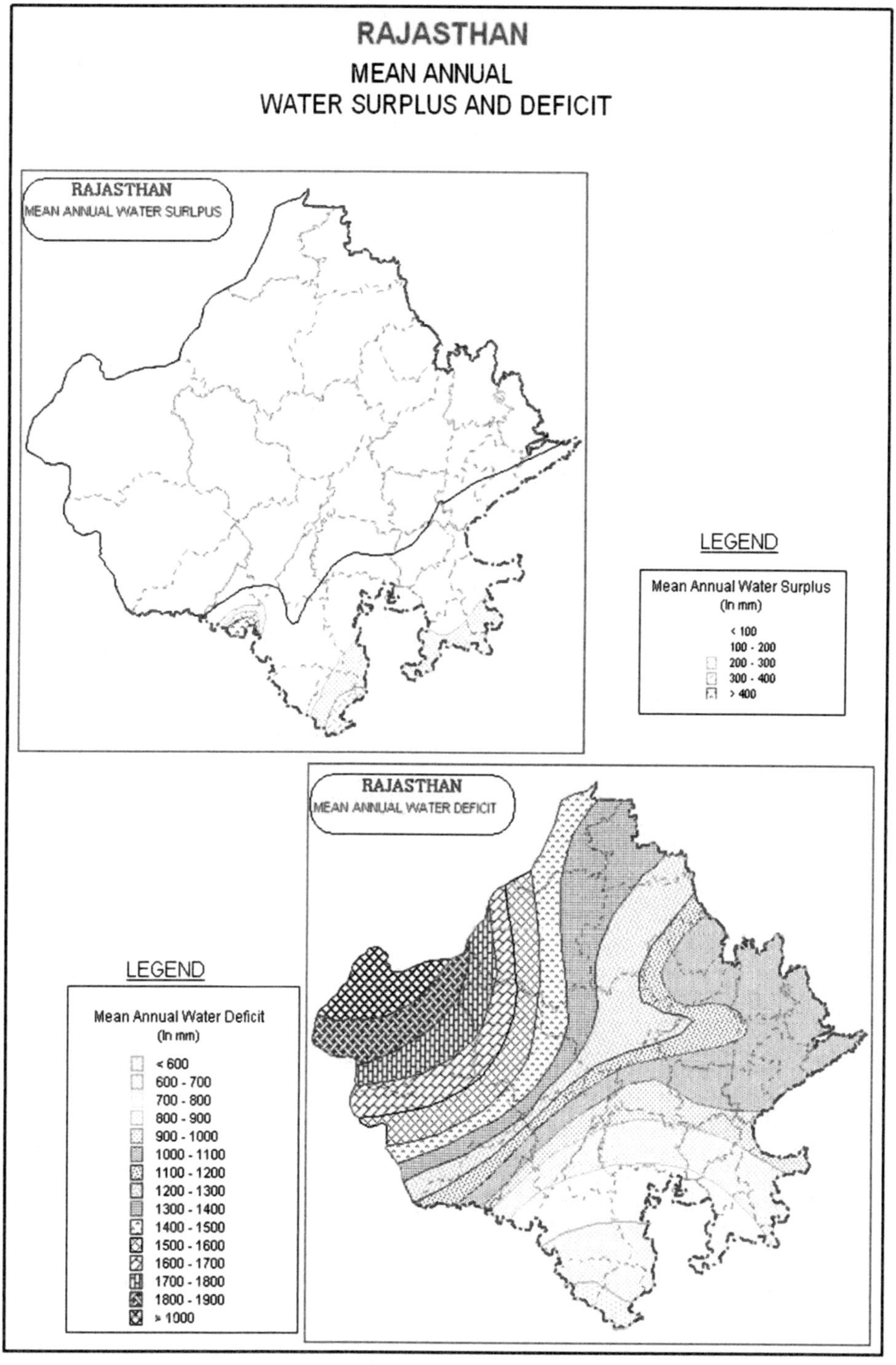

Figure 6.10. Mean Annual Water surplous and Defficit (Source: IMD)

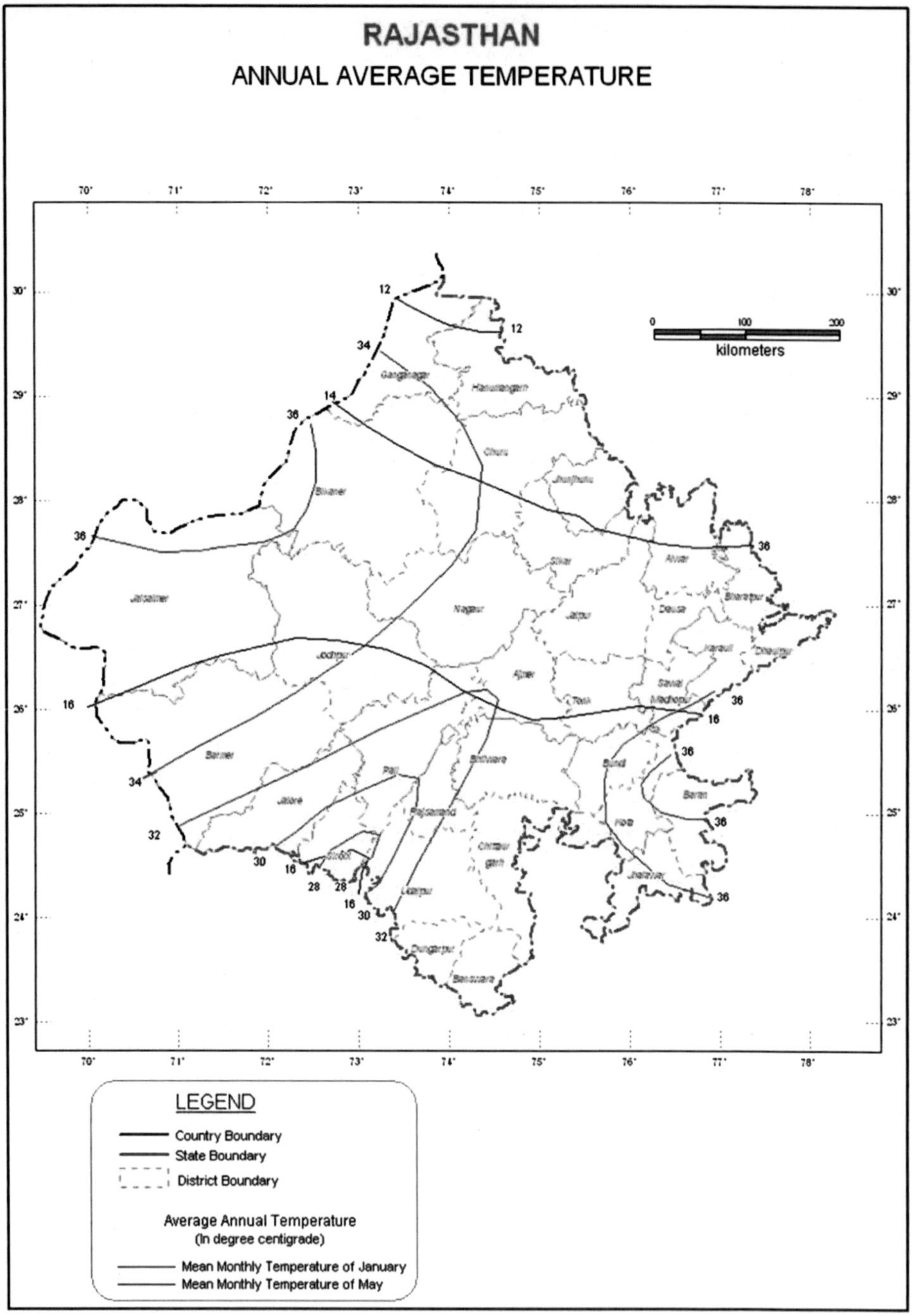

Figure 6.11. Annual Average Temperature (Source: IMD)

The daily minimum temperature for these two stations is around 25°C and 22°C, respectively. The isotherms of summer (June) show variations from 42°C (Sirohi district) to above 47°C (Ganganagar district). A major part of the state, comprising the arid zone of the west and the semi-arid zone of the mid-west, has an average maximum of 45°C in this month. However, in the south-eastern parts, the isotherms show a declining trend from 43°C to 37°C. January is the coldest month of the year. The winter season remains from December through February with marked decline in minimum temperature in December and January. The minimum temperature may fall to minus 2°C in the night at some places like Sikar, Churu, Pilani and Bikaner. A sharp decrease in night temperatures is experienced throughout the arid and semi-arid zone of western Rajasthan on account of quick release of thermal radiations from the sandy soil soon after dusk. Occasional secondary Western disturbances which mostly cross months, causing light rainfall, may also lower the night temperatures and increase wind speed creating wind chill effect.

The isotherms of winter (January), ranging from less than 6°C in the north to above 10°C in the south-east, show few variations in temperatures. The greater part of Rajasthan, except the south-eastern part of Kota, Bundi and Baran and the western district of Barmer exhibits the mean temperatures of more than 10°C. The secondary depressions (western disturbances) accompanied by the ice-cold winds, emanating from the higher latitudes, may bring entire Rajasthan under the spell of the cold wave for the periods of 2 to 5 days during the months. A major feature of distribution in space and time of normal to maximum, minimum and mean temperatures is the small extent of variations over large areas in the monsoon season. One of the peculiar features of Rajasthan is the great extremes of temperature. Variation of as much as 22°C is noted between the mean minimum temperatures due to the sandy nature of the soil. During winter the north-western Rajasthan is colder than the south-eastern region, and at many places the temperature even falls below freezing point and frost occurs. During summer, the westerly or southwesterly winds raise up the temperature making the western region intensely hot and scorching. Due to the dryness of the atmosphere, the nature of the soil, and want of vegetation, the change in the temperature from day to night is sudden, very large and trying. However, in the southern and eastern parts and on the hill tops of Mount Abu and Kumbhalgarh, the speed of the winds is very slow, temperature does not go very high and the nights are cool and pleasant.

The cool season lasts from December to February. January is the coldest month of the year. The Normals of daily maximum, minimum and mean temperatures for a dozen observation stations in the State. Monthly normals of daily minimum temperature vary from 5 to 7.5°C in January to 25 to 30°C in May and June and 25 to 27.5°C in July and August. The annual minimum temperatures vary between 17 to 20'C. Monthly normals of daily maximum, temperature are 20°C to 25°C in January to 40-42.5°C in May and June. The mean annual maximum temperature is 32 to 34°C. Mean daily temperature is 15±2°C in January and 33±2°C in May and June. The highest temperature recorded on any day during the past sixty years or more has exceeded 47.5°C over most of the State and reached as high as 50°C in some areas. The scorching heat and hot 100 winds make the weather during day very trying; while owing to dry atmosphere, clear sky and sandy soil, rapid radiation of heat lowers down the temperature considerably by might.

6.4.4 Evaporation

Evaporation closely follows the march of seasons. When the winds are slight and temperatures are low, evaporation is low. As temperature rises and wind velocity increases, evaporation also increases. Evaporation rates reach their peak in the summer months of April and May generally and display the higher evaporation rates during this period. With the advent of the moisture laden monsoon winds and precipitation, there is a marked fall in evaporation rates.

6.4.5 Evapotranspiration

Evapotranspiration is the combined loss of water vapour to atmosphere in the form of evaporation from soil, water surfaces and transpiration from the plants. This loss is controlled by two factors, namely the availability of moisture at the earth's surface and the ability of the atmosphere to supply energy to vaporise water and transport the vapour. If supply of water is unlimited, loss of water in the form of evaporation and transpiration will depend on the atmospheric conditions and this loss is termed as potential evapotranspiration (PE).

6.4.5.1 *Potential Evapotranspiration*

Potential evapotranspiration is the quantum of water vapour likely to be evaporated from the soil surface and transpired by the vegetation of a certain area under free water availability (moisture) conditions. Increase in evapotranspiration in the summer season is due to higher temperatures, low relative humidity and stronger wind regime. The higher the evapotranspiration, the lower is the spread of vegetal cover on the ground.

6.4.5.2 *Mean Annual Potential Evapotranspiration*

Annual potential evapotranspiration is highest (206.32 cm) at Jaisalmer and the lowest (166.22 cm) at Ganganagar in western part of Rajasthan. In eastern Rajasthan, it varies from 138.12 cm in Udaipur to 174.52 cm in Jaipur. Monthly variations in potential evapotranspiration are marked, making the mean annual values to a limited use as shown in Figure 6.12. Potential evapotranspiration is lowest in the month of December when temperatures are low, but shows a gradual rise from January. From March onwards, this increase remarkably and reaches its highest degree in the month of May and thereafter with the onset of the monsoon, it starts decreasing till the month of August. Then again, there is a slight increase during September and October, owing to rise in day temperature, followed by a gradual decrease from November onwards as winter set in. In Jaisalmer and Bikaner, however, June records the highest rate of potential evapotranspiration.

Potential evapotranspiration is higher than precipitation even during rainy months in the western part of Rajasthan, making these regions arid to extremely arid in climate with perpetual water deficit in all the months. On the eastern side, however, the rainfall in much higher during the rainy season than the potential evapotranspiration demands during that time, thus providing water surplus during the rainy season. The potential evapotranspiration isolines indicate that they are lowest in southern Rajasthan in Dungarpur and Banswara districts (130 cm). The values rise sharply in the western areas, more than 200 cm in Jaisalmer district. In the eastern areas, the values remain around 170 cm in Bharatpur, Dholpur, Jaipur and Tonk districts.

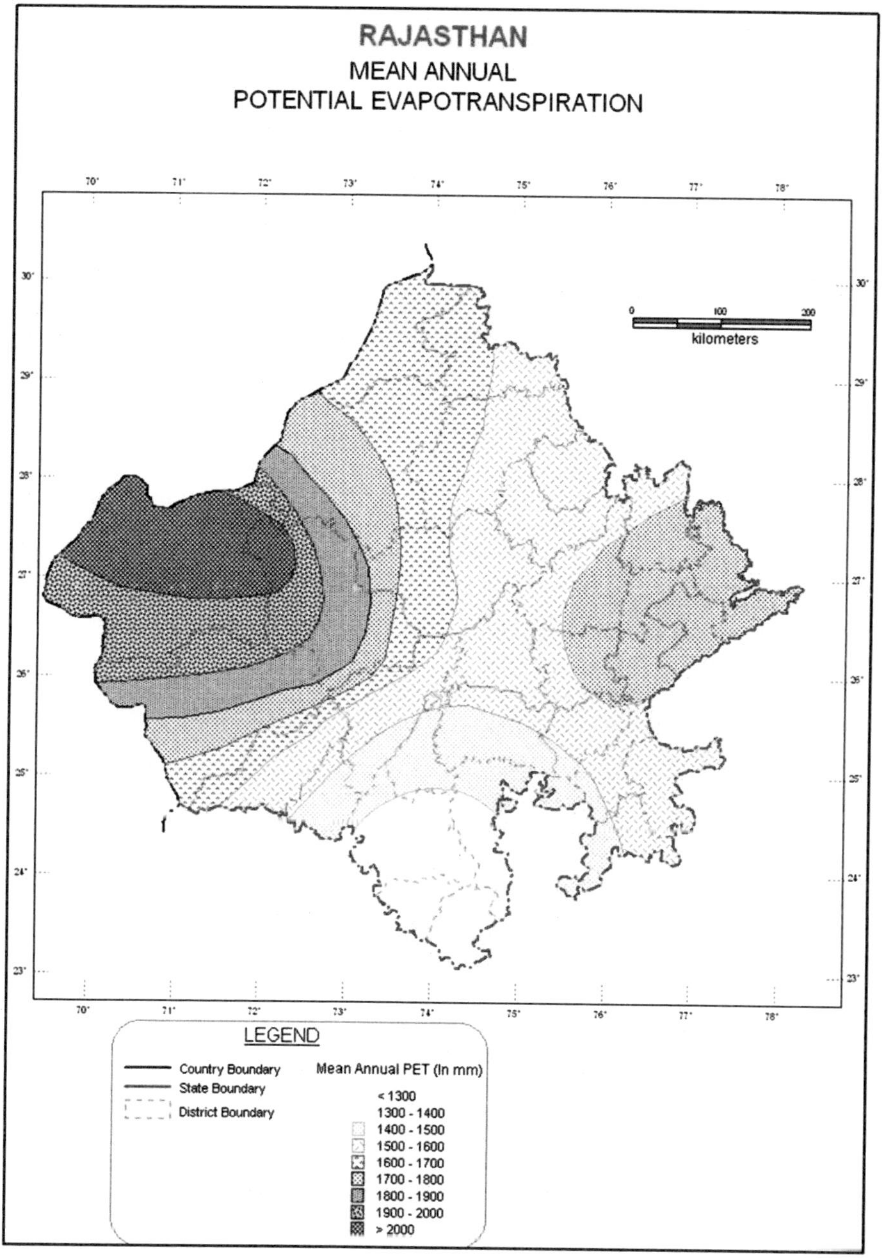

Figure 6.12. Mean Annual Potential Evapotranspiration (Source: IMD and Dr. T.S. Chouhan, Agricultural Geography)

The root system of the plants utilises water. The water available in the root zone of the plants is all that is absorbed by the roots and subsequently transpirated through the leaf system. The amount of water that percolates below the root zone is the excess water over and above that utilised by the plant. The loss of rain water is mainly due to evapotranspiration besides that resulting from surface runoff, percolation and evaporation zone of the soil. The evapotranspiration depends on solar radiation, air temperature, wind velocity and humidity. Based on these factors, a formula to determine potential evapotranspiration is given by Penman. The mean annual potential evapotranspiration varies from a maximum of over 20001 mm and above in northwestern parts of Jaisalmer district to a minimum under 1301 mm in Banswara and Dungarpur districts. This formula was further modified to suit Indian conditions, and on the basis of modified formula, the monthly and annual values of evapotranspiration are worked out.

Potential evapotranspiration (PE) is governed by precipitation, temperature and relative humidity, which in turn determine the water balance of a region. Owing to scanty rainfall, most of the districts in western Rajasthan have very high annual PE rate of above 1600 mm. Highest annual PE is observed in Jaisalmer (2063.2 mm), followed by Barmer (1857.7 mm), Jodhpur (1843.0 mm), Bikaner (1772.4 mm) and Ganganagar (1662.2 mm). In the eastern region, Jaipur district has the record of 1745.2 mm, followed by Ajmer (1566.3 mm), Jhalawar (555.5 mm), Kota (524.7 mm) and Sikar (1503.2 mm). The areas with comparatively higher rainfall such as the south and south-eastern part of the State have annual PE rate below 1400 mm. The annual PE in Udaipur (1381.2 mm) district is the lowest. Near a dam site, in most cases the water remains during the whole year, while in the canal conveyance area and in the command area, the water is made available only during the growing season for the crop. Water is lost by evaporation and percolation in the reservoir and also in are as of the conveyance canal.

The water lost through evaporation may have some effect, on micro climate, but the water that is lost through seepage from the reservoir, canal system and irrigation adds to the water table beneath the ground. The extent of seepage depends on the type of soil. In case of black clay soils, there may be very little seepage; but in case of sandy soils, seepage may account for a substantial part of water loss. Similarly, the lower the extent of cover of soil with vegetation, the lower the moisture content in the soil; and evaporation from such exposed surfaces takes place at a faster rate than from surfaces covered with vegetation. During periods of continuous rainfall, there is substantial loss of water through run off and percolation, while in case of intermittent rainfall there is very little loss due to run off. In case of deep sandy soils, the run off may be low even if the rain at high intensity. In case of clay-soils, where infiltration is poor, the losses due to run off are very high. The construction of proper bunds, may temporarily help storing water, thus reducing losses effected by run off.

The annual PE and its monthly variation for selected stations along with the monthly rainfall values are given in Table 1. It is evident that PE is lowest in December, after which it gradually starts increasing. From March onwards, PE increases remarkably and reaches its highest in the month of May. Thereafter the PE decline during rainy season but once again resumes an upward trend as soon as the monsoon ends. However, in western part of the State the PE rate tends to be high in rainy months, owning to scanty precipitation. In the eastern part, the rainfall exceeds PE during rainy season.

Whenever the moisture available in the soil in any day is less than the PE value, the actual evapotranspiration will decrease depending upon the deficit, and will even cease completely if the moisture level in the top 20 cms is below wilting point. Thus during the years when the rainfall is above normal, there will be a further recharge of water table and increased runoff losses. During the years when the rainfall is below normal and distribution of precipitation is less than evapotranspiration, does the crops may suffer, thus reducing the productivity; but there will be not effect on ground water table level which will decrease only if water is pumped out and used for irrigation.

6.4.6 Wind Velocity, Speed and Direction

The mean monthly and annually wind speed for the year as a whole, winds from south-west or west are most frequent in Rajasthan. Wind velocity is greatest in summer months and various from 10 to 20 km/hour. Wind speed starts increasing from April and reaches its highest in June, after which it gradually declines and comes to its lowest in November. The highest wind velocity is recorded in the month of June at Jaisalmer. (27.2). The minimum wind speed is recorded in the month of November at Udaipur (1.5 km/hour). The mean wind speed is highest at Phalodi station (15.3 km/hour) and minimum at Udaipur station (3.9 km/hour). In the month of April, the wind direction is from west to east. As they come after crossing the heated Thar Desert, they are characteristically dry and warm. During summer, hot and dust raising wind is a common phenomenon. Violent dust storms are most common in the western arid region. Numbers of such dust storms are very few in the eastern part of the state.

Dust storms are more frequent over the *western* than the eastern parts. Their frequency is maximum in the month of June and it is 3 to 4 over the western parts and 1 to 2 over the eastern parts.

6.4.7 Relative Humidity

The relative humidity is minimum in the summer months of March, April and May and maximum during the monsoon months of July, August and September, being lowest in April and highest in August. The moisture content of the air is lowest during the cold season. During the summer season, relative humidity ranges between 35 to 60 percent in the morning hours and 10 to 30 percent in the afternoon hours. In the winter seasons, mainly December to February, relative ranges between 50 to 60 percent in the morning and 25 to 35 percent in the afternoon hours. Ganganagar (68.5 percent) records a slightly higher relative humidity percentage, mainly due to irrigation and appreciably lower dry bulb temperature prevailing during these months of the year. The relative humidity is lowest at Jaisalmer (56.6 percent) and compared with the values of vapour pressure, it will be seen that Jaisalmer is drier than other stations. During the monsoon is less marked increase in relative humidity. This increase is less marked in the west and the northwest of the Aravallis as compared to the areas lying to the east and southeast where rainfall is much higher. After the withdrawal and sometimes the percentage is much lower than that of the winter season.

Table 6.3 District wise Area of Forest According to Legal Status of Rajasthan (as on 31.3.1998) (In Sq. Km.)

S. No.	Districts	Total Forest Area	Reserved Area	Protected Area	Un-Classified Area
1	Ajmer	611.71	194.28	417.32	0.11
2	Alwar	1809.12	1006.06	660.51	142.55
3	Banswara	1236.66		1236.66	
4	Barmer	592.28		558.83	33.45
5	Bharatpur	354.36		353.44	0.92
6	Bhilwara	794.18	434.18	275.57	84.43
7	Bikaner	1077.64		167.13	910.51
8	Bundi	1495.65	801.00	684.42	10.23
9	Chittorgarh	2633.74	1489.58	1120.75	23.41
10	Churu	80.18	7.22	31.78	41.18
11	Dholpur	639.00	7.92	598.33	32.75
12	Dungarpur	646.82	238.73	376.56	31.53
13	Ganganagar	633.44		50.65	582.79
14	Jaipur	944.52	677.73	254.92	11.87
15	Jaisalmer	588.17		101.54	486.63
16	Jalore	545.68	204.68	292.57	48.43
17	Jhalawar	1377.69	453.81	923.45	0.43
18	Jhunjunun	405.31	6.02	392.56	6.73
19	Jodhpur	338.03		180.47	157.56
20	Kota	1391.15	434.00	874.00	83.15
21	Nagaur	235.93	0.80	205.77	29.36
22	Pali	943.82	808.31	124.91	10.60
23	Sawai Madhopur	939.98	790.75	149.23	
24	Sikar	639.30	9.90	629.38	
25	Sirohi	1598.15	611.41	986.74	
26	Tonk	331.56	101.34	228.84	1.38
27	Udaipur	4682.20	3042.00	1604.22	35.98
28	Baran	2202.89		2197.53	5.36
29	Dausa	282.39	133.60	147.19	1.60
30	Rajsamand	392.72	274.53	117.75	0.44
31	Hanumangarh	241.23		113.24	127.99
32	Karauli	1802.57	132.08	1595.42	75.01
	Total	32488.07	11859.93	17651.68	2976.38

Source: Department of Forest, Jaipur

6.5 FORESTS

The forests are the results of the integrated effects of ecological factors viz. climate, topography, soil and past history (including treatment and biotic disturbance). The prevailing locality factors do not permit growth of the high forests of significant commercial value in Rajasthan. The economically productive forest belt in the State is met with in south-east Rajasthan and along Aravalli series of mountains; especially in the districts of Udaipur, Chittorgarh, Banswara, Dungarpur, Jhalawar, Kota, Sawai Madhopur and Alwar. There is hardly any forest growth worth the name to the west of the Aravallis, but for the man-made forests along Rajasthan Canal system.

6.5.1 Forest Cover

The forest area of 31,151 sq.km in the State is only 9.10% of its total area of 342230 sq.km below the norms of 33% prescribed by National Forest Policy of India 1952. Out of total legal forests (31,151 sq.km) hardly one third bears good forest cover which can be termed as potentially productive (Figure 6.13). It can be understood by the following statistics:

(a) Potentially productive forests	10,000 sq.km
(b) Barren hills and grasslands	10,000 sq.km
(c) Rooted Wastes	11,151 sq.km
TOTAL:	31,151 sq.km

6.5.2 Legal Forest Area

The latest area under legal forests in each of the revenue districts is as follows:

6.5.3 Forest Types

Forest type means a unit of vegetation which possesses broad characteristics in physiognomy and structure sufficiently pronounced to permit of its differentiation from other such units (Figure 6.13).

6.5.3.1 *Classification*

C.M. Mathur (1972), described vegetation type for Rajasthan but the most useful classification for management of the forests is still the Forest Types recognized by Champion and Seth (1968) in their book titled "A Revised Survey of the Forest Types of India". According to this, the forest obtained in Rajasthan fall in :

Major Group II-	Dry Tropical Forests
Group 5-	Tropical Dry Deciduous Forests
Group 6-	Tropical Thorn Forests
Major Group III-	Montane Subtrpical Forests
Group 8-	Sub-tropical Broadleaved Hill Forests (only in Mt. Abu)

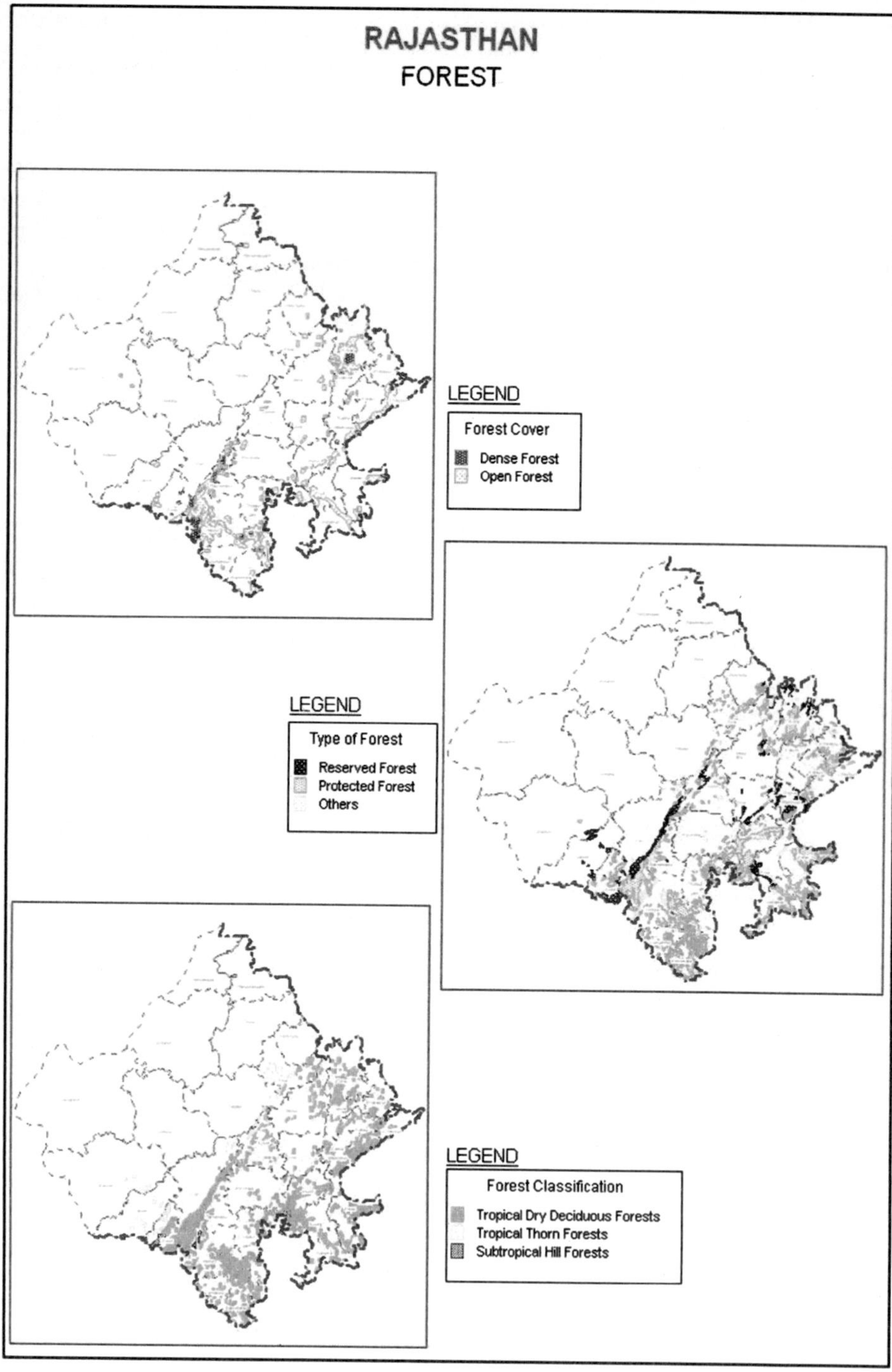

Figure 6.13. Forest Type
(Source: Forest Survey of India and Forest Department of Rajasthan)

Table 6.4. Forest Type in Rajasthan

S.No.	Code	Description Type
1.	5A/C1a	- Very dry teak forests
2.	5A/C1b	- Dry teak forests
3.	5B/C2	- Northern dry mixed deciduous forests
4.	5/E1	- Anogsissus pendula forests
5.	5/E1/D S1	- A. pendula scrub
6.	5/E2	- Boswellia forests
7.	5/E5	- Butea forests
8.	5/E6	- Aegle forests
9.	5/E9	- Dry bamboo brakes
10.	5/DS1	- Dry deciduous scrub
11.	5/DS2	- Dry savannah forests
12.	5/DS3	- Euphorbia scrub
13.	6B/DS2	- Euphorbia scrub
14.	6/E1	- Euphorbia scrub
15.	5/DS4	- Dry grassland
16.	5/IS1	- Dry tropical riverain forests
17.	5/ID3	- Inundation babool
18.	5/2SI	- Secondary dry deciduous forests
19.	6B/C1	- Desert thorn forests
20.	6B/DS1	- Zizyphus scrub
21.	6/E2	- Acacia Senegal forests
22.	6/E4	- Salvadora scrub
23.	6/E4/DS 1	- Cassia auriculata scrub
24.	6/1S1	- Desert dune scrub
25.	8A/C3	- Central Indian subtropical Hill forests

Source: Champion and Seth - 1968

Abbreviations: A. Southern, B. Northern, C. Climatic climax, D. Edaphic climax, DS. DS- Degradation stage

6.5.3.2 *Valuable Forests*

From the point of view of Forestry Management, following Forest types are important in the context of Rajasthan:

6.5.3.2.1 *Teak Forests*

Teak forests extend over an area of about 5120.0Sq.Kms., mainly in Banswara and Dungarpur and sporadically in Chittorgarh, Udaipur, Jhalawar, Baran Divisions, representing the northern most limit (24°42' N. Latitude in Udaipur and 25°12' N. Latitude in Baran) of natural teak forests of India. It is found on geological formations consisting of Aravalli system, pre-Aravalli banded gneissic complex and the Deccan trap on elevations varying between 242 to 485 mts., in yearly rainfall zone of 730mm to 1100 mm, in localities not affected by severe frost. The teak is often pure (50 to 90% in density) and is commonly associated with *Terminalia crenulata* (Sadad), *Diospyros melanoxyplon* (Tendu), *Anogeissus latifolia* (Dhavra), *Lannea coromandelica* (Gurjan of Gol), *Boswellia serrata* (Salar); and Occasionally *Soymida febrifuga* (Rohan), *Miliusa tomentosa* (Umb), *Schrebera swietenioides* (Mokha), *Zizyphus xylopara* (Ghatbor), *Dalbergia paniculata* (Dhoban), *Emblica officinale* (Aonla), *Sterculia urens* (Kadaya), *Dolichandrone atrovirens* (Mendgi), *Mitragyna parvifolia* (Kalam), *Butea monosperma* (Dhak) and *Aegle marmelos* (Bel) are found in tree canopy. The Underwood and undergrowth consists of *Dendrocalsmus strictus* (Bans, in Moister Situations), *Flacourtia indica* (Kakon), *Holarrhena antidysenterica* (Dudhi), *Nyctanthes arbortristis* (Tamat) and in moist locations *Helicteres isora* (Marorphali), Wood-*Fordia fruticosa* (Laldhavri) etc. The Ground-flora consists of *Cassia tora* (punwar), *Ocimum americanum* (bepchi), *Acanthaceae* spp. and grasses like *Themeda* Spp. (Ratedi, Bhedi), Apluda Varia (Bhangta), *Andropogon pumilus* (Masuri), *Heteropogon contortus* (Soorwala), *Eragrostis* (Phuli) *Aristida* (Lapla) etc. along nalas are found *Sorghum halepense* (Barwari), *Ischaemum rugosum* (Toil), *Echinochloa Colonum* (Sawan). Among Climbers Are *Acacia Pennata* (Chillati), *Ampelocissus* Spp., *Discorea* spp. But not profusely. *Dendropthoe falcata* (benda) parasitises salar trees. The teak forests are managed under 'coppice with reserve system' with a rotation of 40 years. Except the reserves of seedling origin, the crop is cut back and tended to yield small timber for agriculture and household purpose.

6.5.3.2.2 *Dhok (Anogeissus pendula) Forests*

Covering about 21,077 Sq. Km in Kota, Tonk, Bundi, Baran, Jhalawar, Jaipur, Ajmer, Pali, Sirohi, Udaipur, Chittorgarh, Alwar and Bharatpur forest divisions in hilly terrain of Aravalli, Delhi system and Vindhyan system, between elevations of 242 to 760 m, Dhok forms the principal forests. Dhok generally forms gregarious stands. Its common associates are *Acacia leucophloea* (Ronj), *A. catechu* (Khair), *Bauhinia racemosa* (Jhinjha), *Wrightia tinctoria* (Khirni) and Tendu. In parts of Jaipur, Ajmer, Alwar and Jodhpur; *Acacia senegal* (Kumat) is common. In upper slopes the main species is replaced by Salar, Kadaya, Gurjan and along foot hills by Dhak. Crop is about 8 m high while in favorable localities like Sawai Madhopur, it may reach upto 12 mts. The understorey consists of *Dichrostachys cinerea* (Kaulasi), *Balanites aegyptiaca* (Hingot), *Rhus mysurensis* (Dansan), *Securinega virosa* (Salepan), *Grewia flavescens* (Firangan or Lodsyali). The undergrowth is sparse and consists of *Achyranthes aspra* (Andhijhara), *Indigofera* Spp. *G. tenax* (Gangan), *Zizyphus nummularia* (Jharberi) etc. and grasss like Bhangta, Soorwal, *Sehima nerovsum* (Sheen), and Aristida. Dhok forests are worked under 'coppice with standard system' with a rotation of 40 years for coppice and 80 years for standards (seed origin). 50-60 standards of seedling origin are retained per Ha. The regeneration period is 10 years.

6.5.3.2.3 *Dry Mixed Deciduous Forests*

Characteristic of hilly region, the dry mixed deciduous forests cover 8960 Sq. Km. Mainly in Udaipur forest division and part of Kota, Baran, Chittorgarh and

Sirohi forest divisions between 330 m. to 450 m above m.s.1. The most dominant species is Anogeissus latifolia (dhavra), often pure. It is usually associated with Tendu, *Albizzia odoratissima* (Kalia Siris), *Dalbergia latifolia* (Kala Shisham), Mokha, *Cassia fistula* (Amaltas), *Pterocarpus marsupium* (Bia), *Buchannia latifolia* (Chironji), Gurjan, Rohan, Aonla, Bel, Dhak Khair, *Cochlospermum religiosium* (Ganiara), *Holoptelia integrifolia*, (Churel, Papdi) etc. Along nalla are found *Mallotus philippinensis*, (Rohini), *Pongamia pinnata* (Karanj), Ficus Glomerata (Gular), Kalam, *Terminalia belerica* (Bahera), *Madhuca indica* (Mahuwa) and some times *Mangifera indica* (Aam), *Terminalia arjuna* (Kohda) etc. Under open canopy, grasses are luxuriant consisting of Bhangta, Ratedi, Soorwala, sheen etc. Bamboos are also found in moist situations. The regeneration is generally deficient. These forests are also managed under 'coppice with standard system', standards of Dhavra and commercially valuable trees are retained

6.5.3.2.4 *Salar Forests*

Boswellia Serrata forests cover 2560.0 sq.kms on upper ridges of Aravallis and Deccan trap between 460 to 950 mts in Alwar and parts of Chittorgarh, Udaipur, Sirohi, Ajmer, Jaipur, Banswara and Dungarpur forests on shallow and skeletal soils, as an Edaphic climax. The common associates are Gurjan, Kadaya and Some-Times Aonla and Dhavra. The Understorey is usually absent but grasses like Bhangta, Soorwala, Aristida and *Chrysopogon montana* (Dhuala) are common. Banda is often found on Salar branches. Tree attains a height of about 14 m with a girth of 90 cm. The seedling regeneration is often adequate in protected localities. The forests are worked under overlapping working circles with 'selection system'; the exploitable girth being fixed at 60 cm at b.h.

6.5.3.2.5 *Bamboo Forests*

Dry bamboo brakes (*Dendrocalamus strictus*) covering about 1000 sq. kms are found practically in all the forest divisions (except the arid north-west portions of Rajasthan) as an under-storey in Dhok, Dharva and teak forests. They occupy sheltered aspects and moist depressions. Good bamboo growth is found in Chittorgarh, Udaipur, Banswara, Kota and Mt. Abu. The clumps may have 10 to 30 culms in height varying from 5 to 9 m. Their quality is under deterioration year after every year. They need rest.

6.5.3.2.6 *Khair Forests*

Occupying an area of about 1120 sq. kms Acacia Catechu (Khair) forests occur as a facie of pure Khair crop in south eastern region e.g. Baran, Jhalawar, Kota, Chittorgarh, Alwar and Sawai Madhopur and are associated with Ghatbor, Kaulasi, Dhok, Bel, Ronj, Jhinjha etc. It is also found mixed with Dhok and Dhavra forests. The natural regeneration is generally deficient. Exploited for Katha (Catechin) manufacture by 'Handi' system (country method of boiling Khair heartwood chips in earthen or metallic pots), the Khair trees are worked under 'selection system' with and exploitable diameter of 10 cms at b.h. and felling cycle of 20 years, with reliance on coppice regeneration.

6.5.3.2.7 *Sub-Tropical Evergreen Forests*

Restricted over about 55.0 sq. kms in Mt. Abu hills between elevations of 1150.0 to 1830.0 mts, these forests support aam, Jamun (*Syzygium cumini*), Kakon, Gadhapals (*Erythrina* Spp.) Bamboo, Rohini, Karonda (*Carissa congesta*), Jasminum Humile, *Rosa moschata*, Strobilanthes Callosus, *Geradiana heterophylla*, (Bichchhu Buti), *Euphorhia neriifolia* (Thor), The Orchids Arides Crispum etc. *Anogeisus sericea* (Androk) is also a marked feature at Mt. Abu and *Phoenix sylvetris* (Khajur) grows even above 1650 mts. To preserve aesthetic beauty these forests are not exploited except for some light dry bamboo removal.

6.5.3.2.8 *Thorn Forests*

Spread over 2000 sq.km the thorn forests are very open. They are mainly confined to arid areas in northwest but also on isolated patches of sandy tract and ravenous land in other parts where soil is generally devoid of humus. The spp. Found are thorny and comprise *Prosopis cineraria* (Kherji), *Maytenus emarginatus* (Bekal), Caparis Decidua (Kair), Hingot, *Acacia jacquemonti* (Baonli), Kumta, *Calotropis procera* (Akda), *Zizyphus* Spp. (Ber, Jharberi), *Mimosa hatata* (Jhinjhini) and *Euphorbia nivulia* (Thor). In more sandy areas *Calligonum polygonoides* (Phog), *Crotolaria burhia* (Senia), *Leptadenia pyrotechnica* (Khimp) are common. The Usual Grasses are *Desmostachya bipinnata* (Dab), *Cenchrus ciliaris* (Dhaman, Anjan), *C. setigerus* (Kala Dhaman), *C. biflorus* (Bhurat), *Lasiurus scindicus* (Sewan), *Panicum turgidum* (Giramana) and *Sporobolus* Spp. These forests are managed only as grazing areas.

6.6 FLORA AND FAUNA

According to Good (1964), the earth has been divided into six floral Kingdoms, Rajasthan falling under Paletropical or Original floral kingdom. He also recognised 37 Floristic Region or Floristic Provinces, out of which Rajasthan is covered under 9-North African-Indian Desert Region and 17-Indian Region. Phyotogeographically the classification of Chatterjee (1939), is still most widely accepted, who divided India into 8 Botanical Regions; Rajasthan falling in Indus Plains. The characteristic spp. or floristic elements of this Botanical division are *Acacia nilotica, A. Senegal, A. Leucophloea, Anogeissus, Albizzia, Dalbergia, Grewia, Prosopis, Salvadora, Capparis, Euphorbia*, and *Calotropis*.

6.6.1 Flowering Plants

Jacquemont (1832) was the first to botanically explore Rajasthan. He was followed by George King (1868), Blatter and Hallberg (1918-21), Parker (1921) Bhandari (1967) and B. Tigai and S. Sharma (1979). The estimated flowering plants of Rajasthan to belong to 1280 species as follows:

Table 6.5. Number of Floral Species

Classification	Dicots	Monocots	Total
Families (After Bentham and Hooker)	92	18	110
Genera	465	128	593
Species	970	210	1280

Source: Jain (1970)

6.6.2 Largest Families

According to number of species the ten largest families in Rajasthan are:-

Table 6.6. Floral Species

S.No.	Families	Species	Genera
1.	Poaceae	180	74
2.	Leguminosae	177	51
3.	Compositae	86	51
4.	Cyperaceae	63	11
5.	Acanthaceae	46	22
6.	Euphorbiaceae	44	14
7.	Malvaceae	40	11
8.	Labiatae	35	13
9.	Convolvulaceae	35	11
10.	Scrophulariaceae	33	19

6.6.3 Largest Genera

The largest genus is Cyperus with 30 spp. While the next is Indigofera with 20 spp. The Acacia occupies 10th position with 11 spp. in Rajasthan.

6.6.4 Endemic Plants

Rajasthan lies in the zone of poor endemism. Of the 134 genera endemic to India, only six namely; *Ouginia, Butea, Caesulia, Glossocardia, Petalidium* and *Haplanthus* are endemic in Rajasthan. Some of the common endemic taxa of Rajasthan are *Cleome gynandra* var. *nana* (Bl and Hall) Bhandari, *Cleome brachycarpa* Vahl ex D.C. var. *glausa* Bl and Hall., convolvulus blatteri Bhandari, *Furestia macratha* Bl. and Hall., *Zizyphus truncata* Bl. and Hall., *Tephrosia uniflora* sub-spp. Petrosa (Bl and Hall) Gillet and Ali, *Lasiurus caudatud, Cenchrus rajasthensis, Tribulus rajasthenisis* Bhandari and Sharma *Anogeissus sericea* brandis var. *nana* King ex Duthie, *Ammannia desertorum* Bl. and Hall., ex Sant., *Euphorbia jodhpurensis* Bl. and Hall., *Puliciara rajputance* Bl. and Hall., *Melhania magnifolia* BI. and Hall.

6.6.5 Fossiliferous Flora

Fossil evidence is extremely scanty for Rajasthan.

6.6.5.1 *Udaipur and Karauli*

Pre-Cambrian plant life of the state is represented by five different types of stromatolites namely *Collenia columnaris, C. symnetrica, Baicalica prima* and *Minjaria calceolate* from matoon formation near Udaipur (Banerjee, 1971), a filament of Eomycetopsis robusta from Aravalli subground of Udaipur (Banerjee, 1974) dated more than 2 bilion years and algal forms Gloeocapsomorpha and Palaeoglaucocystis from the Vindhyan rocks near Karauli.

6.6.5.2 *Jaisalmer*

The lower or Mid Jurassic life recovered from rocks exposed near Jaisalmer (Lathi formation) comprised largely serveral kinds of pteridophytes like *Cyathidites australis rimalis* Bahme *C. minor* Couper, *Dictyotriletes crateris* (Bahme) Srivastva, *Cingulatisporites lathiensis* Srivastava, *Polypodium sporites mathurii* Srivastava, and gymnosperms like *Ginkgocycadophytes deterius* var. *majus* Der, *G. nitidus* (Bahme) Srivastava, *Monosulcites couperi* Der, *Cupresacites ramachandrae* Srivastava, *Araucariacites ghoshii* Srivastava, *Podocarpidites* spp. *Callialasporites dampsieri* (Bahme) Der and other spp. of *Callialasporites* and *Classopollis*. The plant fossils (upper Jurassic) from Habur village (near Jaisalmer) belong to ferns Gleichenites and Phlebopteris, Bennettitales (*Taenopteris, Pterophyllum, Otozamites imbricatus Feistm, Ptilophyllum acutifolium* Morris and conifers *Elatocladus conferta* Oldh and Morris Hall and *Pagiophyllum* spp. In Mudh and Kotri of Palana (Jaisalmer) Formation Pollen grains of Chelianthes Melia and unidentifiable angiosperms of tertiary period are recorded. Now converted as AKAL wood fossil Park near Jaisalmer, huge gymnospermic fossils of large trees of lower Jurassic era estimated to be 180 millions years old are lying.

6.6.5.3 *Barmer*

Barmer area is also rich in fossil finds. The Cretaceous era plant life recorded is comparable with Polypodium, Lycopodium, Lygodium, Coniopteris, Gnetum, Araucaria, Abies, Pinus, Podocarpus and unidentifiable angiosperms. At kapurdi, Fuller's earth deposits (Barmer) contain the Eocene deposits of evergreen trees like *Mesua tertiara*, *Garcinia borooahii, Calophyllum* and fruit of Cocos which are indicative of humid climate with high rainfall

6.6.5.4 *Bikaner*

From the Bap-Bikaner area at Mayakor silicified wood and leaf impressions of *Ptilophylum acutifolium, Pterophyllum* and *Equisetites* of the same age as that of Lathi formation have been recovered.

Table 6.7. List of Rare and Endangered Plants

S. No.	Latin name	Local Name	Family	Habit	Tropical locality	Remarks
1.	*Ammannia desertorum* Blatt. & Hallb.	Moto-Jal-bhangro	Lythraceae	Undershrub	N.W. Rajasthan	Sand stone rocks preferred
2.	*Anogessus Sericea* Brandis var. *numularia* King ex Duthie (=*A.rotundifolia* Blatt. and Hallb)	Indok	Combretaceae	Tree	Ajmer, Jodhpur, Udaipur	
3.	*Anticharis glandulosa* Asch. var. *caerulea* Blatt. and Hallb. Ex. Sant.		Scrophula-riaceae.	Herb	Jaisalmer	Found on rocks
4.	*Bouchea Marrubifolia* Schauer (=*Pleurostigma subrotundum* Hochst. Nomnud)	Bai	Verbenaceae	Herb	N.W. Rajasthan	
5.	*Commiphora wightii* (Arnott) Bhandari (=*C.mukul* Engl. And = *Balsamodendron mukul* Hook. Ex stock)	Gugul	Burseraceae	Shrub atta-ining dend-roid form.	Aravallis	
6.	*Convolvulus scindicus* Stock	Kaland	Convolvulaceae	Herb	N.W. Rajasthan Barmer	
7.	*Crypsis schoenoides* Lamk.		Gramineae	Herb (grass)	Bap, Jaisalmer	
8.	*Dignathia hirtella* Stapf.		Gramineae	Herb (grass)	Jalore	For a short ti-me after rains
9.	*Dipterygium glaucum* Dence		Capparaceae	Undershrub	Jodhpur, N.W. Rajasthan	Sand. Dunes.
10.	*Ephedra ciliata* Fish and Mey ex C.A. Mey (=*E.foliata* Boiss)	Suo-Phogaro	Gnetaceae	Climbing Shrub.	N.W. Rajasthan	Seriously endangered.
11.	*Eriolaena hookeriana* W.&A.	Bhoti dhaman	Sterculiaceae	Tree	Hill	Very rare.
12.	*Euphorbia jodhpurensis* Blatt. and Hallb.	Dudheli	Euphorbiaceae.	Herb.	N.W. Rajasthan	
13.	*Farsetia Macrantha* Blatt. and Hallb.	Motio-hiran-chabbo	Cruciferae	Undershrub	Barmer	Seriously threatened
14.	*Grewia damine* Gaertn (=*G.salvifolia* Weyne ex Roth non Linn.f.)	Phalsa	Tiliaceae	Shrub	Aravalli hills	
15.	*Monsonia heliotropoides* Boiss		Geraniaceae	Herb	Jaisalmer	Endangered
16.	*Melhania denhamii* R.Br.		Sterculiaceae	Herb	Barmer	Sand dune
17.	*M.magnifolia* Blatt. and Hallb.		Sterculiaceae	Shrub	N.W. Rajasthan	On rocks.
18.	*Moringa concanensis* Nimmo	Sarguro	Moringaceae	Tree	Aravalli rocks	Getting very rare.
19.	*Nannca pulla* Lamk. Ex. D.C.		Boraginaceae	Herb	N.W. Rajasthan	Wheat fields
20	*Pulicaria raputanae* Blatt. and Hallb.	Dhola ligru	Compositae	Herb	N.W. Rajasthan	Dry salty bed.

S. No.	Latin name	Local Name	Family	Habit	Tropical locality	Remarks
21.	*Rhynchelytrum villosum* (Parl) Chiov		Graminae	Herb (grass)	Sardar samand (Pali)	Rocky gravel
22.	*Rosa lyellii* Lindl		Rosaceae	Shrub	Mt. Abu Jaisalmer	
23.	*Seetzenia orientalis* Decne	Dhakri	Zygophyllaceae	Herb	Jaisalmer	On rocks.
24.	*Talinum Portulacifolium* Aschers ex Schweinf		Portulacaceae	Shrub	Rocks in Jod-hpur, Barmer.	Very rare
25.	*Taverniera cuneifolia* Arn.	-	Papilionaceae	Under shrub	Pali	-
26.	*Tecomella undulata* (Sm) Seem.	Rohida	Bignoniaceae	Tree	Sandy soil	-
27.	*Tribulus rajasthanensis* Bhandari et Sharma	Gokharu	Zygophyllaceae	Herb	Jodhpur, Jaisalmer and Ajmer	Sand stone ro-cks preferred.
28.	*Tripogon jacquemontii* Stapf.	-	Gramineae	Herb (grass)	N.W. Rajast-han (Osian)	Very rare.
29.	*T. roxburghianus* Bhide	-	Gramineae	Herb (grass)	Jodhpur, N.W. Rajasthan	On moist shady rocks.
30.	*Withania coagulans* Dunal	Paneer bandh	Solanaceae	Under shurb	N.W. Rajasthan	Grassy hills tops.

Source: Forest Survey of India and Forest Department of Rajasthan.

6.6.5.5 *Sambhar-Didwana-Ajmer*

The mixed xeromorphic Thorn Forests with serveral grasses, sedges, composits and forbes (spp. of *Tephrosia, Solanum, Heliotropium, Boerhaavia, Digera* and *Indigofera*) waned about 6000-4000 years ago in the vicinity of Lunkaransar, expanded between 5000-4000 ago in the vicinity of Didwana and appeared at Sambhar between 4500-3000 years ago. Together with this, *Psammophytic scrub* represented by *Calligonum polygon* ides and Aerva grew near Lunkaransar. Phog (*C. polygonoides*) invaded Didwana and Sambhar areas by about 6000 years and 3000 years ago respectively. Ultimately in the northern parts of Rajasthan desert, developed Lithophytic scrub represented by Capparis and Mimosa. Plant life from Pushkar lake near Ajmer is know from about 5000 years. The genus Artemesia which was significantly presented 10,000 years ago is now occasional and even two species *A. scoparia* and *A. nilagirica* are now of rare occurrence in the Rajasthan desert.

6.6.6 Threatened Species

Destruction of natural habitats has resulted in decline of many plant species. Although definite information is lacking, the following plants in Rajasthan are either getting rare or threatened with extinction.

■■■■

Chapter - 7

LAND, WATER AND IRRIGATION RESOURCES

Despite the recent great strides in technology and mechanization agriculture remained the world's most important primary industry, in which the soil plays pivotal role. About 66% of the global population, comprising of farmers derives its living directly from the soil. There are no natural resources more important than soil resource. Whatever its production capacities, whether high in some places of irrigated pockets or low in major areas of dry land and desert, due to inherent limitations, the soil resource of Rajasthan, as a medium of growing crops, has furnished directly or indirectly, a significant share in the income of the state. Rajasthan, being geographically the second largest state in India, has proportionately a greater soil resource. Therefore, the soil resource in the state needs to be used extensively and efficiently so that the state finds an appropriate place in the national food, fiber and fodder production and the state economy is sustained without any depletion through erosion, degradation or overuse. The information and knowledge of soils of the state which could be gained through the study of their physical and chemical properties and their geographically distribution pattern, is an essential prerequisite for their proper utilization, management and conservation. It also helps in proper selection of crops and better land use.

7.1 SOILS

The western, northern and eastern part of Rajasthan is covered with sand sheets of variable thickness, dunes and sand hummocks interspersed. Large variability in the climate, physiography and mineral composition of source rocks have contributed to the variations in the soil characteristics of this region. The influence of past and present climatic gradient and effect of vegetation have also left their mark on the soil attributes. The landscape in the south and south eastern part of the State has remained fairly stable. Soils in this region have been formed due to prolonged effect of climate and vegetation. The soil profile has developed diagnostic horizons. But in the northern and western part large amplitude of variation in the climate has dominated the Aeolian and alluvial processes which have influenced the soil formation processes. In the early and mid Pleistocene, during the humid climate, alluvial plains were formed. Subsequently, during the last 40-50 thousand years the climate has remained arid. High speed desiccating winds in this period resulted in the formation of Aeolian plain of high sand dunes, hummocks and sandy plains. In the northern region the soils have developed from the sediments deposited by the Gagger River and its tributaries. Thus the State is endowed with a variety of soils ranging from desert sand to basaltic silt clay loam.

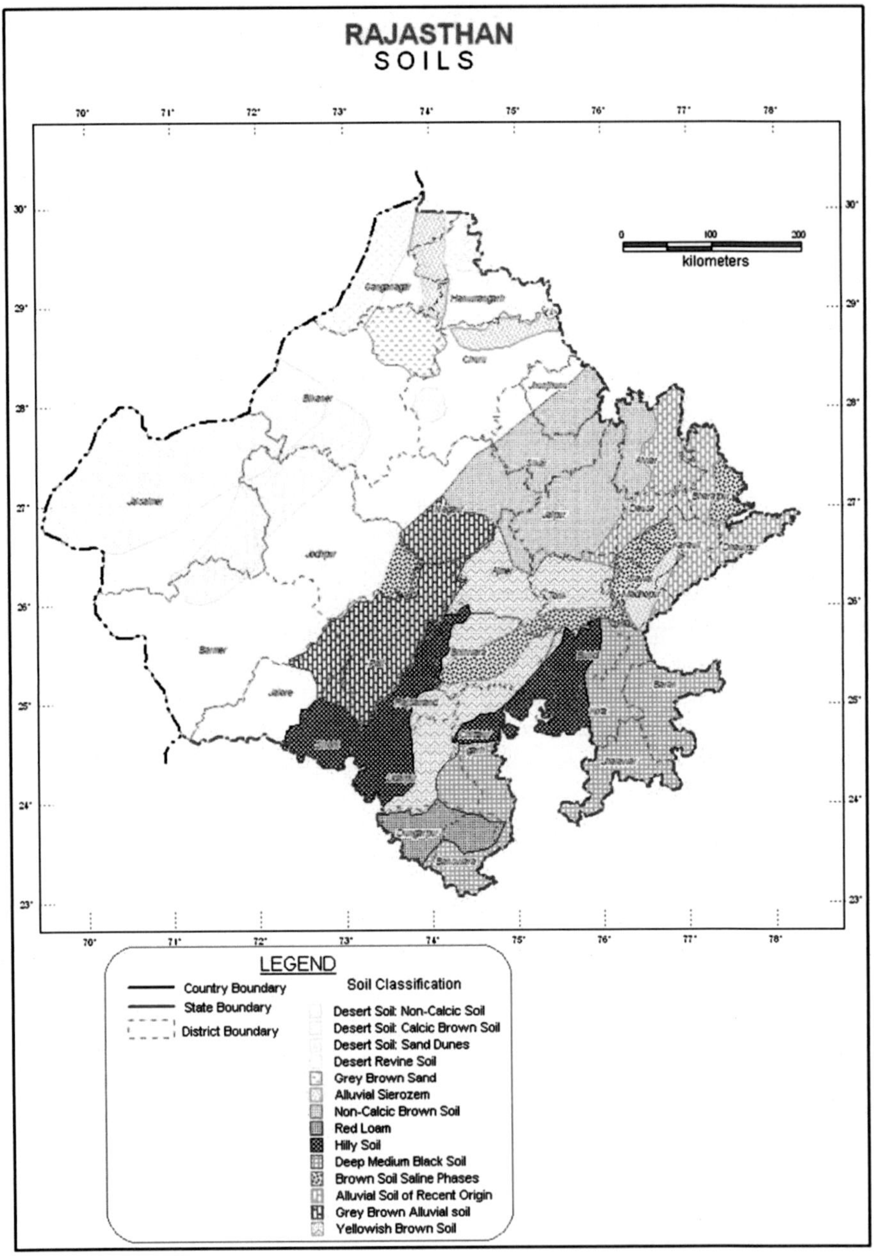

Figure 7.1 Soils
(Source: Survey of India, Rajasthan Atlas, Resource Atlas, Watershed Atlas, Groundwater Atlas, CAZRI and Agricultural Geography of Rajasthan by Dr. T.S. Chouhan)

Rajasthan is one of few States where systematic soil survey investigations have been taken up since beginning. Efforts made by Central and State Government agencies have resulted in valuable information on the characteristics of soils and their geographical distribution. The relationships of soils with the physiography (Lodha *et al.*, 1982), parent material (Mehta *et al.*, 1962; Joshi and Seth, 1972; Dhir and Kolarkar, 1976) and landscape (Khangarol and Mehta, 1976) have been established. Based on these studies, attempts have been made to classify the soils by Mathur *et al.* (1968), Pareek and Seth (1970), Mathur and Shankarnarayan (1975), Dhir (1977), Joshi (1985) and Dhir and Joshi (1991). Based on these reports the soils of the entire State have been classified in 13 broad groups (Figure 7.1).

7.1.1 Major Soil Groups

The physiography -soil association relationship has been presented in Table 7.1 and Figure 7.1. The broad fertility characteristics are presented in Table 7.2.

Table 7.1. Physiography and Soil Association

I. Sandy arid plain	(a) torripsamments/calciorthids/camborthids/ palcorthids
II. Alluvial plain of mid west	(a) Camborithids/calciorthids
	(b) Salorthids
III. Plain of interior drainage	(a) Torripsamments
	(b) Salorthids
IV. Gagger plain	(a) Torrifluvents/torripsamments
	(b) Camborthids
	(c) Torripsamments
V. Aravalli plain	(a) Haplustalfs
	(b) Torripsamments
VI eastern plain.	(a) Haplustalfs
	(b) Ustifluvents
VII. Chappen plain	(a) Ustochrepts
	(b) Chromuster
VIII. North eastern hilly region	(a) Haplustalfs
IX. South-eastern south- eastern hadoti plateau	(a) Haplustalfs
	(b) Chromustert
	(c) Ustifluvent

Salient characteristics of each of the major soil groups have been reported below.

7.1.1.1 *Dune and Inter-Dune Soils*

Dunes are the spectacular features which can be seen scattered throughout arid region. These occur in continuous chain in two -zones. One is in the extreme west in the Barmen Jaisalmer arid Bikaner districts. The another chain of sand dunes runs throughout the eastern part of Bikaner and Churu districts. A discontinuous zone of sand dunes runs from Sanchor, Saila, Phinch, Didwana and Lachmangarh. These three comprises of six types of sand dunes *viz.* obstacle, parabolic, coalesced parabolic, longitudinal transverse, -barchans and shrub coppice (Ghose *et al.,* 1977). The soils associated with the dunes are very deep, line sandy and light yellowish brown to pale brown (Figure 1). The fine sand fraction constitute 88-92%, clay 2.0-5.0% and silt 0.4 to 1.37%. These are generally non calcareous but in Jaisalmer and Barmer districts dunes having 1-2% $CaCO_3$ in powder form or as root cast are also observed. In the associated inter-dunes the soils are similar to the dune except slightly higher amount of silt (2-3%), Clay (5-7%) and $CaCO_3$ (2-5%). The surface of inter-dunes is covered with loose sand and sand hummocks. The dune soil profile does not show pedogenic manifestation. In the profile of inter-dune soil, accumulation of lime or gypsum concretions can be observed. The surface soils are loose and single grained but in the subsoil very weak aggregation is observed. These are coarse loamy, mixed, hyperthermic Typic Tarripsamments.

Table 7.2. Soil Fertility Status

	Districts	Dominant soils	Fertility group		
			N	P	K
1	Ajmer, Jalore and Nagaur	Brown soil, gray brown loam and sandy plain soil	L	M	M
2	Banswara, Chittorgarh, Udaipur, Dungarpur	Medium black, red loam brown soil	M	M	M
3	Bharatpur, Alwar	Soils of younger alluvial plain	L	M	M
4	Barmer, Jaisalmer, Bikaner, Churu	Sandy plain and dune soils	VL	M	M
5	Bundi, Kota, Tonk, Bhilwara	Medium black, brown soil	M	M	M
6	Shri Ganganagar	Indus alluvial soil	VL	M	H
7	Jaipur, Sawai Madhopur	Brown soil	VL	H	M
8	Jhunjhunu, Sikar	Soils of sandy plain	VL	M	M
9	Jodhpur, Pali	Sandy plain, brown light loam, and gray brown loam	L	M	M

The organic matter content of these soils is very low (0.05- 0.10%). The available phosphorus is medium to low. The available potassium is medium to high (Dutta and Joshi, 1989). The available zinc and iron are adequate but the available manganese and copper are below the critical limits (Joshi and Dhir, 1983a,b). The nutrient status is sufficient to meet the requirement of natural vegetation permitted by the low rainfall but upon availability of irrigation water from Indira Gandhi Canal these soils will need besides the macronutrients, also the micro nutrient application for raising the crops. These soils have low water retention capacity and high water percolation rate, within one meter soil depth. 50-80mm of water is retained out of which 80% is available to the plants. Under extreme dry condition also these soils retain water in the subsoil which can meet the requirement of natural vegetation. These soils are prone to wind erosion. Use of heavy machinery for soil operations and uncontrolled grazing result in the loosening of surface soils which lead to large scale sand movement (Figure 2). The dunes should be covered with the perennial grasses/ shrubs and should not be cultivated. In the inter-dunes, pearl millet, moong and moth should be cultivated with strips of grasses.

7.1.1.2 *Brown Sandy Soils*

These soils widely occur on sandy plains in Jodhpur, Barmer, Nagaur, Bikaner, and Jaisalmer districts. The texture of the soil is fine sand to loamy sand. The surface of these soils is covered with loose sand sheeting with occasional hummocks. The surface soil is brown, yellowish brown and pale brown, single grain to granular and noncalcareous. The subsoil is loamy sand having 4 to 6% clay and 3 to 5% silt and weak subangular blocky structure. There is slight increase in $CaCO_3$ content which is present in powdery form. Below 60 to 90cm lime concretionary horizon is commonly encountered. These soils have been classified as coarse loamy, mixed, hyperthermic Typic Calciorthids. Similar soils occurring in Nagaur, Churu, Jhunjhunu and Sikar districts are non calcareous and have been classified as coarse loamy, mixed, hyperthermic Typic Catnborthids. The organic matter content of these soils is low (0.1 to 0.2). The available potassium content is medium to high (Dutta and Joshi, 1988). These soils are fairly provided with the available iron, manganese, zinc and copper (Joshi *et al.,* 1982). The moisture retention capacity of light brown sandy soils varies from 60 to 90 mm per meter depth. These soils are generally cultivated for the rain fed crops like Bajra, guar, moth and moong. Yields of these crops can be stabilized by adopting the dry land practices like inter row water harvesting, field bunding, optimum inter row and plant to plant distance and proper crop combinations. These soils are vulnerable to wind erosion. Use of heavy machinery on these soils should be avoided.

7.1.1.3 *Brown Light Loam Soils*

Soils of this group are typical reddish brown to dark reddish brown. These have developed from the alluvium derived from rhyolite and sandstone. These sails occur in the Jodhpur and Nagaur district. Surface horizon are reddish brown (5 YR 5/4) sandy loam and loam, sub angular blocky and noncalcareous. The subsoil is similar to colour but somewhat heavier with 13 to 15% clay and 8 to 10°h silt and well developed subangular blocky structure. The soil is generally deep to very deep with rock fragments/gravels underneath. The soils are also classified as coarse loamy, mixed hyperthemic Typic Camborthids. The organic carbon content in surface horizons vary from 0.2 to 0.3%. The available phosphorus and available potassium contents arc in medium ranges. The mean values of available iron, manganese, copper and zinc are also fairly within the adequacy limits. The moisture retention capacity varies between 80-120 mm per meter depth. These soils, wherever provided with good quality irrigation water are very productive. High yields of wheat, mustered, cumin and vegetables are obtained on these soils. Under conserved moisture sorghum, pearl millet and pulses are grown in Kharif.

7.1.1.4 *Gray Brown Soils*

These grayish brown soils are loam, clay loam and silt clay loam in texture. The soils have developed from the fine textured alluvium and occur in south eastern part of Jodhpur and Nagaur and in Pali, Sirohi and Jalore districts. The surface horizon is loam and clay loam, grayish brown (10 YR 5/2) and dark grayish brown (10 YR 1/2) and medium subangular blocky. The subsoil is dark grayish brown and very dark grayish brown (10 YR 3/2) clay loam and silty clay loam with 15 to 22% clay and 10 to 15% silt and angular blocky structure. These soils are highly variable in depth (40/150 cm). The $CaCO_3$ content increases with depth. The substrata is formed of lime concretionary horizon. These soils are members of the fine loamy, mixed, hyperthermic family of Typic Camborthids. In surface the soil contain 0.2 to 0.4% organic carbon and medium to high levels of available potassium (Dutta and Joshi, 1989). The micronutrient content of these soils is also adequate to meet the crop requirement (Joshi *et al.*,1981). The soils have high moisture retention capacity (75 to 150mm per metre depth). The soils are free from Aeolian hazards but water erosion is active near shallow gravelly uplands, piedmont zones and hills. At places in depression water logging is also encountered. Under conserved moisture gram, wheat and mustard are grown. With good quality water cotton, condiments, wheat and mustard are grown.

7.1.1.5 *Hard Pan Soils*

These soils are 30 to 40 cm deep underlain by a strongly lime indurated horizon. Hard pan soils occur in Phalodi-Dhawa area of Jodhpur, Jayal-Katoti of Barmer and in scattered areas in Jaisalmer districts. This horizon is formed of lime concretions and gravels/rock fragments cemented together to form an indurated layer which is slight pervious to water but plant roots are unable to penetrate. Surface soils are brown (10 YR 5/3), light yellowish brown (10 YR 6/2) and yellowish brown (10 YR 5/4), fine sand to loamy sand and noncalcareous. The clay and silt content varies from 6 to 10 and 4 to 8%. These soils are the members of coarse loamy, mixed, hyperthermic family of Typic Paleorthids. The content of organic carbon (0.1 to 0.2%), available phosphorus and potassium contents arc quit low. Because of coarse texture and shallow depth the moisture retention capacity is only 40/70mm. These soils are generally unsuitable for rain fed cropping. During normal rainfall year moth and moong can be raise but these soils are better suited for grasses and shallow rooted shrubs. Wind erosion is very active and at places the calcareous substrata is exposed.

7.1.1.6 *Flood Plain Soils*

The landscape is the northern part of the State, covering Ganganagar and part of Bikaner districts is formed of flood sediments deposited by the river Gaggar. The surface soil is brown to gray brown, very fine sand to loamy sand and sandy loam with granular and subangular blocky structure. The content of silt and clay range from 12 to 22 and 11 to 18 per cent respectively. There is wide variation in the morphological characteristics of the soils both within the profile and also between different profiles. These variations are attributed to the varied type of sediments deposited by the river. The sand particles of the Gaggar flood plain soils are much finer compared to the coarse textured soils in rest of the arid region. The moisture retention capacity of these soils is very high ranging from 180 to 231 mm per meter. The content of organic carbon is medium (0.21 to 0.45%). Available phosphorus, potassium and micro nutrient contents is better than other arid zone soils. These soils are fertile and because of availability of good quality water produce bumper crop yields. Sand hummocks and low dunes locally called `Tibbis' are scattered on the surface of these soils. The `tibbis' are 5 to 10 metre high and 2 to 3 meter long. These are members of loamy, mixed, hyperthermic, Typic uslifluvents. The Gaggar flood plain soils are often infested with high concentration of salt. The soil profile has fine textured horizon at some depth due to which drainage is impeded. These soils can be reclaimed with the good quality water available from the Indira Gandhi Canal. Because of good fertility and availability of canal water ibis region is changing in the green belt. High yields of cotton, sugarcane, paddy, wheat, oil seeds and pulses are being realised. With the optimisation of irrigation water and fertilizer application this region has good potentiality.

7.1.1.7 *Light Brown Sandy Soils*

These soils are typical light yellowish brown, fine sand, non calcareous and very deep. These extensively occur in Nagaur, Sikar, Churu, and Jhunjhunu districts. The surface is covered with sand hummocks, loose sand and low dunes. The soil profile is sandy, yellowish brown (10 YR S/4) and dark brown (7.5 YR 4/4) and many meters deep. The soils are free of $CaCO_3$. These soils are very pervious to water. Due to high susceptibility to wind erosion thick sand sheeting and hummocks at fence line and in field proper are spectacular and fields appear in bowl shape. The water retention capacity ranges between 60-70 mm per meter depth. These soils have been classified as coarse, loamy, mixed, hyperthermic Typic Torripsamments. The organic matter content is medium (0.2-0.4%) and soils are also medium with respect to major and micro nutrient convent, but deficiency of Zn and Cu has been reported in some areas. These soils are responsive to fertilizer application and irrigation.

7.1.1.8 *Soils of Younger Alluvial Plain*

These are the soils of vast alluvial plain in the districts of Bharatpur, Alwar and parts of Jaipur district. Soils of these regions are very deep, pale brown and sandy loam to loam. In the subsoil the content of clay and silt ranges respectively from 8 to 15 and 6 to 10 per cent. The $CaCO_3$ content varies from 1 to 2.5% in different horizons which is present in powdery form. In lower horizons soft nodules of $CaCO_3$ may also occur. These soils are fairly resistant to wind erosion because of compact surface. These are the members of loamy, mixed, hyperthermic family of uslifluvents/napluslafs. The organic carbon content of the soils varies from 0.3 to 0.4%. Other major and micro nutrients are fairly well provided. The moisture retention capacity ranges from 200 to 215 mm per meter. Sand hummocks and low dunes may be seen scattered on the fence line. With irrigation available from canal and dug well, winter crops like what, mustard, sorghum, pulses and groundnut are grown. The soils are also responsive to fertilizer application. The problem of salinity/sodicity is wide spread due to water logging in low lying areas.

7.1.1.9 *Brown Soils*

These brown, medium to fine textured alluvial soils occur in Bhilwara, Tonk and Ajmer districts and also scattered in the Udaipur, Chittorgarh and Jaipur districts. These are strong brown (7.5 YR S/8), brown (10 YR S/3) and yellowish brown (10 YR S/4), sandy loam to clay loam at the surface and non-calcareous. The subsoil has clay loam and loam texture and sub angular and angular blocky structure. These soils are slightly calcareous and substrata consist of loosely packed weathered rock fragments. These soils have argillic horizon and exhibit moderate profile development. These are the members of loamy, mixed, Typic Hapluslaf. The organic carbon content varies from 0.5 to 0.8%. The moisture retention capacity varies from 110 to 250 mm per meter. At places these soils are associated with rocky undulating uplands. In such situations water erosion is very active. Natural salinity in soil is the major problem. In canal command areas and saline water irrigation secondary salinisation dominates. Otherwise these soils are very productive both in the irrigated and rain fed areas.

7.1.1.10 *Red Loam Soils*

These soils generally occur in narrow valley fills or in the flat intermountain areas. These have developed from the sediments deposited from the adjoining hills. Such soils are widespread in southern part of Rajasthan covering Dungarpur, Udaipur and Chittorgarh districts. These soils are string brown (7.S YR S/6), yellowish brown (10 YR S/6) and pale brown (10 YR 6/3), sandy loam and loam devoid of CaC03. Such soil is slightly darker with somewhat more of clay and silt content. These are also member of the loamy mixed, family of Typic Ustalft (Joshi and Seth, 1972).

The organic carbon content in the surface soil is medium (0.4 to 0.6%). Deficiency particularly of Zn has been reported. The moisture retention capacity varies from 75 to 80 mm per meter. During rains water is pounded in low-lying areas where paddy is cultivated. on upper slopes soils are shallow and suffer from moderate to severe water erosion. Contour bunding, field bunding and terracing is required to reduce soil loss due to erosion. Perennial shrubs and trees also help in reducing soil losses. In the depression areas, the soils are very productive and cash crops like cotton sugarcane, wheat and grain are raised with very low irrigation requirement.

7.1.1.11 *Black Soils*

These dark grayish brown and very dark grayish brown, very deep soils occur in south and south eastern part of Rajasthan State. In the Kota, Bundi and Sawai Madhopur districts the black soils have developed from the Vindhyan rock and Deccan trap sediments deposited by the river Chambal and its tributaries, the Kali Sindh and the Parvati. In Banswara, Jhalawar and Pratapgarh - Dhariyavad area of Udaipur- Chittorgarh districts the black soils have developed from the Deccan trap. In this region the black and red soils occur at juxtaposition. The black soils are clay loam, silty clay loam having angular blocky structure. The different horizons of the profile do not show variation in the silt and clay content 20-25% and 35-40% respectively. The substrata have lime concretionary. Sharma and Kant (1976) have reported slight translocation of finer particles and their subsequent concentration in alluvial horizons. The soils have characteristic property of swelling when wetting and shrinking in dry condition. These soils develop cracks of 50-100 cm depth after drying. These soils have been classified as Typic Chromusterts and Typic Pellusterts.

The organic carbon content of the soil ranges between 0.4 to 0.6%. Deficiency of copper is common in these soils. In the black soils of Jhalawar deficiency of Zn has been reported. These soils have very high moisture retention capacity (145-200 mm/meter). Because of very low water percolation in these soils the rainwater remains ponded and water logging is common. In absence of good drainage system these soils develop problem of salinity. There are two big dams for irrigation on the Chambal and the Mahi rivers which supply water for irrigation. Higher yields of crops like cotton, sugarcane, sorghum, wheat, mustard, pulses and oil seeds can be obtained by adopting efficient water management systems. Along the banks of the Chambal and its tributaries, very deep ravines have been formed. These are further extending towards the cultivated fields. Proper treatment in the catchment area, by adopting suitable mechanical means, soil conservation measures and growing perennials shrubs and grasses, is required.

7.1.1.12 *Soils of Highly Region*

The southern part of the State in Udaipur, Dungarpur, Banswara and Sirohi districts is occupied by the Aravalli hill range, similarly the south eastern part of Rajasthan covering Sawai Madhopur and Bundi districts is also hilly. The landscape of this region is very uneven, sloppy, covered with stones and rocks. The soils, which occur in pockets, are very shallow gravelly, coarse sandy, reddish brown to grayish brown and non calcareous. These suffer due to severe water erosion. Cultivation sloppy and undulating lands are not possible and good grasslands can be developed. In the depression valley areas where sediments are deposited soils are deep to very deep and fine textured. These are highly variable in texture (sandy loam to silty clay loam) and colour (reddish brown to dark grayish brown). Mehta *et al.* (1962) have observed that these soils have moderate illuviation and have monogenic character of the profile. These are very productive and with well irrigation good yields of paddy, cotton, sugarcane, wheat, pulses and oil seeds are harvested. Hilly soils are the member of Typic ustochreps. The organic carbon content of soils in depressions varies from 0.8 to 1.5%. These soils also contain good amount of available phosphorus, potassium and micronutrients (Joshi et al., 1973; Lodha and Baser, 1971). The shallow soils have very low moisture retention capacity but in deep soils 150-200 mm water can be retained per meter depth.

7.1.1.13 *Salt Affected Soils*

Salt affected soils are of common Occurrence throughout the State. In Rajasthan 1 million hectare area is affected with salinity. Causes of high salinity in soil may be as follows:

7.1.1.13.1 *Natural Salt Affected Soils*

Because of arid climate in western Rajasthan, the salinity in sail and groundwater is quite high. Saline water lakes occur near Sambhar, Pachpadra and

Didwana. In Pali, Jalore and south eastern part of Jodhpur and Nagaur districts vast stretches of natural salt affected soils occur. Due to weathering of rocks salts are released the catchments of ephemeral streams which is carried through water and salinity developed along the banks and in depressions where these streams are lost. Thus large area turns saline. The profile of the salt affected soils is charged with very high concentration of salts. The pH and Ec values (1:2) respectively ranges from 8.2 to 8.7 and 20 to 82 dsm-1. The salinity decrease with depth. These soils are sandy loam to loam and clay loam, dark grayish brown (10 YR 4/2) to very dark grayish brown (10 YR 3/2) with angular blocky structure. The solum is very hard and compact and do not allow rainwater to percolate. These soils are devoid of vegetation except the salt tolerant bushes like *Prosopis juliflora*. The soils are moderately deep to deep (30-50 cm) underlain by a compact lime concretionary horizon. Inspite of high salinity these soils are well provided with the available iron, manganese, zinc and copper (Joshi, 1988a, 88b). In the flood plain of Gaggar in Bikaner and Ganganagar district large natural saline areas occur. These are highly saline throughout the soil profile. Wherever, the canal water is available, these areas are being reclaimed and brought under cultivation. The problem of natural salinity is also common in parts of Bhilwara, Ajmer, Chittorgarh and Bharatpur districts. In low-lying areas salinity is commonly encountered. In the Shahpura and Banera tehsils of Bhilwara districts the problem of slick spots is common.

7.1.1.13.2 *Anthropogenic Salinity*

Irrigation with saline/high residual sodium carbonate water, construction of dams and in the command area of canal is resulting in the salinization of soils. The problem of brackish ground water is common in western part of the State but it is also encountered in Bhilwara, Ajmer, Bharatpur, Alwar, Jaipur and Chittorgarh districts. Due to increasing build up of salinity in soil the crop yields decrease. Farmers are left with the choice of salt tolerant crops and their varieties. The use of water containing high residual sodium carbonate has resulted in severe degradation of soil productivity (Plate 5). Due to development of high sodicity (Table 5) the soils become very hard and compact and nutrient availability has also been reduced (Joshi, 1988, 90). In such areas gypsum application is essential to rehabilitate the land productivity (Joshi and Dhir, 1988, 1992). The construction of irrigation dams results in the increased salinsation in command area of the Chambal and the Indira Gandhi Nahar Projects. Similar problem has also developed with other irrigation dams in the semi arid region in Ajmer, Bharatpur, Alwar, Bhilwara and Pali districts. These salt affected soils can be managed by growing salt tolerant shrubs and grasses.

7.2 LAND USE

The various parts of the state present a variety of land use patterns which exhibit, to a large extent, the availability of soil and water resources in the area and the human endeavors to harness them. In the areas having greater concentration of population, the significance for agricultural cropping of the land increases and therefore, the net shown area in these districts is relatively high. On the other hand, in the districts where soil is less fertile, extensive farming is needed to feed even a smaller population, so that the net area shown again increases. The common classification of land use are forest, land put to non-agricultural uses, permanent pastures, net sown area, fallow land (including current and other fallow lands) and cultivable wasteland (including barren and uncultivable land, and land under miscellaneous tree crops). Land use/land cover mapping of all districts of Rajasthan has been completed under a National Land use/Land cover project, coordinated by National Remote Sensing Agency, Hyderabad. This work has done on 1:250,000 scale district map by two work centers, namely, State Remote Sensing Application Center (SRSAC), Jodhpur and Central Arid Zone Research Institute (CAZRI), Jodhpur. Data of different categories of land use/land cover has been compiled by National Remote Sensing Agency, Hyderabad.

Table 7.3. Land Use Classification and Utilisation

S.N.	Classification	Statistic based on First- Plan period (1951-1956)		Statistic based on VIII-Plan period (1993-1994)	
		Area in Lac ha	perce-ntage	Area in Lac ha	percen-tage
1	Geographical area	342.438	100.00	342.438	100.00
2	Forest land	13.030	3.80	24.257	7.08
3	Not available for cultivation				
(a)	Under non-agriculture use			16.612	4.85
(b)	Barren and unculturable land			26.847	7.84
	Sub-Total (a) + (b)	74.460	21.74	43.459	12.69
4	Other uncultivated land (excluding fallow land)				
(a)	Permanent pastures and other grazing land			17.632	5.15
(b)	Land under misc, tree, crops and groves			0.160	0.05
(c)	Culturable waste land			52.822	15.42
	Sub-Total (a) + (b) + (c)	89.250	26.06	70.614	20.62
5	Culturable area [1-2-3-4(b)]			274.562	80.18
6	Fallow land				
(a)	Old fallow land			19.845	5.79
(b)	Current fallow land			21.943	6.41
	Sub-Total (a) + (b)	57.420	16.77	41.788	12.20
7	Net area sown [5-4(a)-4(c)-6]	106.190	31.00	162.320	47.40
8	Area sown more than once	7.030	2.05	30.226	8.83
9	Gross cropped area (7+8)	113.220	33.06	192.546	56.23
10	Gross un-irrigated area			136.593	39.89
11	Gross irrigated area(9-10)			55.953	16.34
12	Net un-irrigated area			116.346	33.98
13	Net irrigated area (7-12)			45.974	13.42
14	Cropping intensity (%)		106.62		118.62

Source: " Agriculture Department Statistics"

Table 7.4. Land Use Pattern

S. No.	District	Forest	Not Available for Cultivation		Other Uncultivated Land		
			Land put to non-agriculture use	Barren and uncultivated land	Permanent pasture/ grassing land	Land under miscellaneous use	Salt affected
1.	Ajmer	39,375	53,041	36,475	41,590	—	7,685
2.	Alwar	121,78	38,259	23,490	6,570	—	7,437
3.	Banswara	122,190	43,910	—	8,440	—	—
4.	Barmer	10,996	118,262	49,375	94,437	—	31,000
5.	Bharatpur	10,680	30,408	12,450	8,220	—	4,560
6.	Bhilwara	44,500	60,017	62,500	37,500	—	6,250
7.	Bikaner	7,745	85,085	—	166,875	—	1,367
8.	Bundi	103,750	44,546	19,690	1,560	—	315
9.	Chittorgarh	281,560	65,422	13,310	7,235	—	315
10.	Churu	3,535	85,144	351	84,237	—	2,006
12.	Dholpur	30,080	16,536	1,003	—	—	—
13.	Dungarpur	37,815	23,388	1,360	625	—	—
14.	Ganganagar Hanumangarh	2,561	150,813	—	—	—	—
16.	Jaipur Dausa	59,352	123,147	32,375	46,625	—	5,250
17.	Jaisalmer	6,200	43,008	7,620	333,750	—	20,880
18.	Jalore	20,468	86,918	15,375	61,484	—	27,265
19.	Jhalawar	102,977	28,908	—	—	146	—
20.	Jhunjhunu	35,086	34,829	3,284	27,136	—	—
21.	Jodhpur	4,336	104,803	9,219	99,219	—	17,383
23.	Kota, Baran	314375	78,040	8,750	18,125	—	315
24.	Nagaur	15,083	99,872	970	33,935	—	18,155
25.	Pali	84,500	53,653	15,140	—	—	2,190
26.	Sawai Madhopur Karauli	175,200	61,290	101,325	12,516	—	—
27.	Sikar	56,881	45,866	4,963	14,310	—	2,405
28.	Sirohi	141,000	22,330	25,050	4,065	—	—
29.	Tonk	28,250	48,486	2,188	57,624	—	3,938
30.	Udaipur Rajsamand	401,925	80,494	1,875	28,125	—	—

Land Use Pattern ---- Continue

S. No.	District	Cultivable waste					Fellow land	Net area sown
		Water logged area	Marshy/ swampy area	Gullied ravineous area	Sandy area	Land with or without scrubs		
1.	Ajmer	—	—	7690	3315	73910	100168	484851
2.	Alwar	—	—	47778	—	67185	—	525803
3.	Banswara	—	—	—	—	105910	—	223250
4.	Barmer	—	—	—	655313	94562	1634275	150480
5.	Bharatpur	—	—	5310	—	32910	—	404662
6.	Bhilwara	—	—	12500	—	20315	140719	661199
7.	Bikaner	2148	—	726	1122890	143213	296498	891553
8.	Bundi	—	—	44060	315	40625	73920	226219
9.	Chittorgarh	—	—	625	—	19125	55870	642138
10.	Churu	—	—	—	28945	4147	174333	1300302
12.	Dholpur	—	—	52721	—	68495	12766	119149
13.	Dungarpur		—	—	—	99340	—	214472
14.	Ganganagar Hanumangarh	3098	—	—	290993	46334	33835	1536216
16.	Jaipur Dausa	—	—	39250	23963	16462	53556	1006820
17.	Jaisalmer	620	—	—	2143500	728810	366586	189126
18.	Jalore	—	—	8437	35429	9218	50720	748704
19.	Jhalawar	—	—	10951	5588	148825	71281	259224
20.	Jhunjhunu	—	—	5344	5494	6461	35339	440827
21.	Jodhpur	—	—	7383	236600	168867	609484	1027706
23.	Kota, Baran	—	—	29375	—	48125	36219	710276
24.	Nagaur	—	—	5602	210078	67157	218835	1291113
25.	Pali	66560	3125	16560	7315	9375	38000	942282
26.	Sawai Madhopur Karauli	—	—	76584	2938	56850	123014	442983
27.	Sikar	—	—	7563	3420	5121	60124	572547
28.	Sirohi	—	—	—	7185	26300	41857	245813
29.	Tonk	—	—	18825	5250	7875	20662	506302
30.	Udaipur Rajsamand	—	—	—	—	493780	—	721701

Source: Statistical Abstract

7.2.1 Land-Use Classification

The common nine-fold classification of land use has been generalized here into six broad groups:

- Forest,
- Land put to non-agricultural uses,
- Permanent pastures,
- Net sown area,
- Fallow land (including current and other fallow lands) and
- Cultivable wasteland (including barren and uncultivable land, and land under miscellaneous tree crops).

The percentage of cultivable waste and the land put to non-agricultural use is highest in parts of Bhilwara, Chittorgarh, Udaipur, Kota and Baran districts, owing to their hilly terrain. The percentage of forest areas is noticeably high in Kherwara (Udaipur), Dungarpur, Banswara, Sirohi, Kota, Jhalawar, Baran and Sawai Madhopur districts. Forest areas, in general, are conspicuously absent in the entire western Rajasthan. Higher percentages of pasture land, which are often seasonal, exist in Barmer, Jodhpur, Jalore and Nagaur districts of state.

7.2.2 Land Use Pattern

Land Use Pattern in the State has been developed depending on availability of both soil and water resources and human endeavors to harness them. In the western districts of Jaisalmer, Barmer and Bikaner the percentage of uncultivated and fellow land is maximum due to sandy nature of soil and non-availability of water in larger area. In the eastern Rajasthan, the net sown area is highest on account of extensive agriculture due to better rainfall and growing demands of high-density population. The percentage of cultivable waste is the maximum in hilly districts of Bhilwara, Chittorgarh, Udaipur, Kota and Baran. Forest areas are high in the Kherwara (Udaipur district), Dungarpur, Banswara, Sirohi, Kota, Jhalawar and Baran districts.

7.3 WATER RESOURCES

Water is the most critical input for habitation and development of the region. In a development plan, improvement in water resources has an important place in increasing the production of food and fiber, which are basic requirements of human life. As a result of increasing urbanization and other connected developments, there has been shrinkage with the passage of time, in the culturable area available for agricultural development. Rajasthan being a desert State receives about 1% of the country's water in the form of annual rainfall. A good part of it gets absorbed due to dry soil conditions further restricting surface runoff. The water getting absorbed should normally recharge ground water; however, part of it does not reach the deep ground water table while some part goes to ground water which is already brackish leaving comparatively a small part potable enough to be used as drinking water or fit for agricultural use. The surface water resources in the State besides being scarce, are mainly confined to South and South-eastern part of the State. There is no perennial river in the State and it has to depend upon its share of water from interstate-river basins and rain water availability in seasonal rivers of State basins during monsoon period.

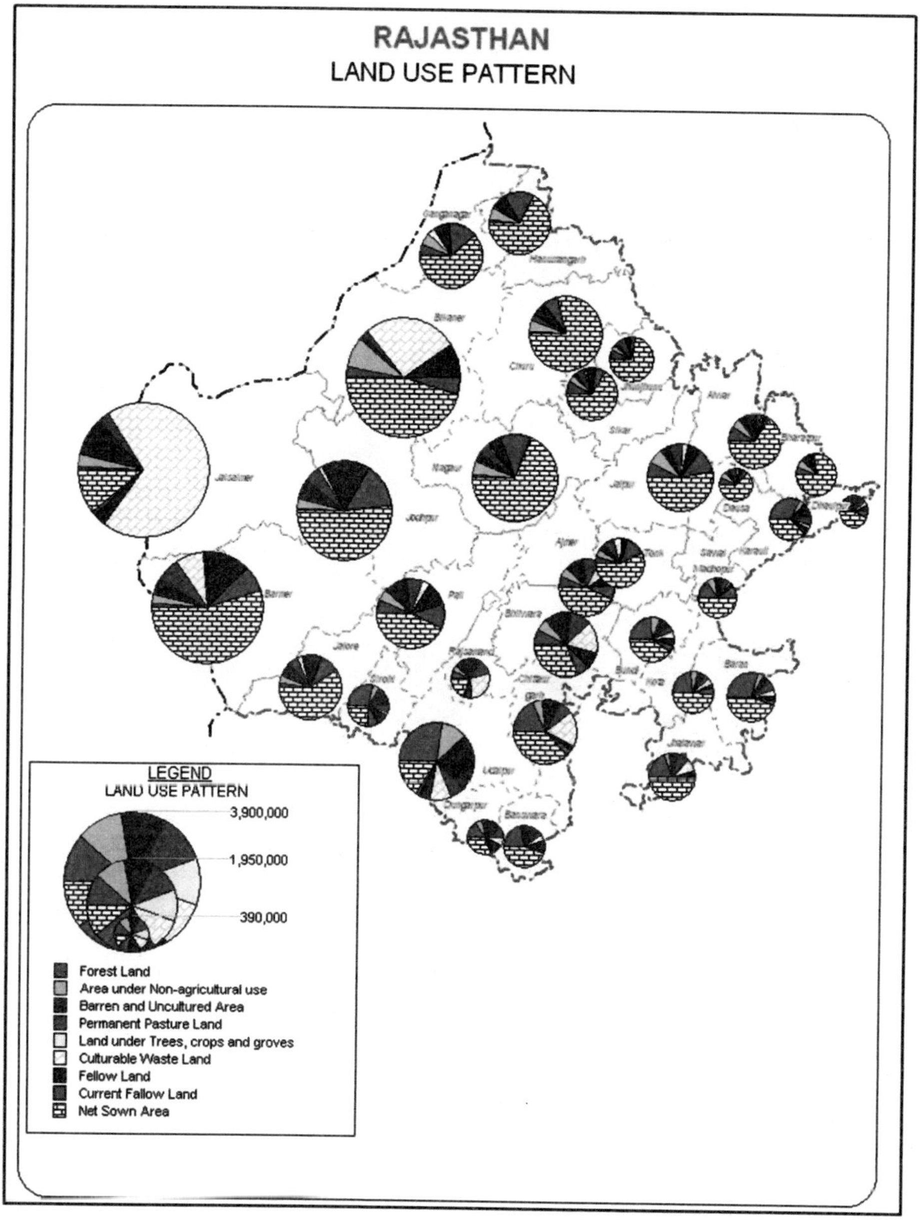

Figure 7.2. Landuse Pattern

The arid to semi-arid climatic conditions of Rajasthan are characterized by many rainless months and a high probability of dry years with very little surface water flow. The average annual rainfall in Rajasthan ranges from 200 mm in the North-west to nearly 950 mm in the South-west. More than 90% of rainfall occurs during monsoon, but the distribution is unreliable in terms of space and time. Ground water under these conditions, is a reliable source in many cases also having the advantage of better quality source of water for drinking, industrial as well as for agriculture use. Ground water, therefore, plays an important role in the development of water resources in Rajasthan. More area in Rajasthan is irrigated by wells compared to all flow Irrigation Schemes put together. Over exploitation of Ground water is resulting in white, grey zones being converted to dark zones and therefore its impact on ground water quality calls for immediate action to check this problem. Use of ground water and surface water in an integrated manner can help in checking the problem of water logging and salinity in irrigated areas.

7.3.1 Ground Water Behavior

Water demand in major part of the state is mainly met from ground water resources because of scarce surface water resources, which endorses ground water to assume the role of supplying water for almost all the development activities. Crystalline, semi consolidated and unconsolidated aquifers are the potential source of water in Rajasthan. In desert region generally the potential aquifer systems often lie adjacent to saline aquifers and receive insignificant replenishment due to water deficit climatic conditions whereas in other parts of the state the rate of withdrawal is so high that the effect of good rainfall is also neglected and the water levels continuously shows a declining trend. Thus in both desertic and non-desertic region trend of water level is towards declining side. Ground water exploitation without adequate recharge has caused depletion in water levels, deterioration in chemical quality of ground water by ingression of saline water and reduction in yield of wells has become common. Water levels vary with rainfall, physiography, drainage and climatic environment. These parameters show wide variation in Rajasthan. Eastern Rajasthan has more rainfall and it decreases westward. From the above analysis it is inferred that in spite of number of wet spells in last 16 years there is a continuous decline in ground water levels in the state which is ultimate result of heavy ground water withdrawals for domestic, irrigation and industrial purposes. If the same trend of exploitation continuous without adequate recharge (natural or artificial) the situation will become so worse that all the productive ground water potential zones will become dry resulting in the permanent damage to the aquifer.

7.3.1.1 *Depth to Water Level*

Depth to water table on the eastern side of Aravalli hill range is shallower than that in the western side. The depth to water table in the State varies widely i.e. less than 5m to over 130m. The depth to water level map is presented in Figure 7.3. In the areas west of Aravalli hill ranges, deep water table i.e. over 60m is recorded in the zone of inland drainage, in north and southwestern part of Jaisalmer district, southern part of Bikaner district, west central parts of Churu district and north eastern part of

Jodhpur district. Locally such areas also exist in the southwestern part of Barmer district, northwestern part of Nagaur district and northwestern part of Jodhpur district. In major portion of western Rajasthan depth to water table is 20 to 60m. Shallow water tables (less than 5m) are noticed locally in the Ghaggar River Basin in Ganganagar district and along Luni River in Jodhpur, Pali and Barmer districts. In the areas on the eastern side of Aravalli hill ranges the depth to water level is less than 5m to over 20m BGL. Shallow water level less than 5m was recorded around Sambhar lake, northern part of Jhalawar district, major part of Kota and Bundi districts, locally in the eastern part of Bhilwara district, in the Mahi-Bajaj canal command area of Banswara district and locally in Dungarpur, Udaipur, Chittorgarh, Bharatpur and Dholpur districts.

7.3.1.2 *Water Level Fluctuation*

The fresh water potential aquifers are under tremendous pressure of exploitation. The State is also prone to frequent droughts combating of which promotes construction of more and more ground water abstraction structures. As per ground water resource estimation as on 01.02.2001, out of 237 blocks in the State, 86 blocks have been categorized as over exploited and 80 blocks as critical. Ever-increasing withdrawal of groundwater for irrigation, industrial and domestic purposes have resulted in declining trends in water levels. The deeper water levels, inferior chemical quality and availability of IGNP water restrict ground water development in westernmost parts of the State and as such the area do not exhibit declining water level trend. The decline in water levels is observed in remaining larger parts (except Mahi and Chambal canal command areas) of the State. Severity of rate of decline in water level also depends on availability of saturated thickness and static ground water resources in the area. In hard rock areas saturated thickness is generally restricted to weathered and fractured zones and hence decline in water level is rapid and impact of which is of serious nature. In alluvial areas having limited saturated thickness (say 10 to 50m) decline in water level of few centimeters to 1m per year may be taken as serious. However, decline in water level of few centimeters or even over 1m may not be considered significant for aquifers having saturated thickness of more than 100 to 500m as found in Lathi basin area of western Rajasthan.

7.3.1.2.1 *Long Term Water Level Fluctuation*

Decline in water level is a countrywide phenomenon. Rajasthan witness that decline in water levels by more than 8m in parts of Jodhpur, Nagaur, Jalore, Jhunjhunu, Jaipur and Dausa districts. Decline in water levels between 4m and 8m has been found in parts of Sikar, Jhunjhunu, Jalore, Nagaur, Ajmer, Jodhpur, Jaisalmer, Barmer, Alwar, Jaipur, Dausa, Rajsamand, Karauli, Jhalawar and Bhilwara districts. In major parts of the Bikaner, Churu, Barmer, Baran, Kota, Sawai Madhopur, Dholpur, Bhilwara, Udaipur, Bundi and Tonk districts decline in water level is mostly less than 4m.

Table 7.5. Change in Water Level

S. No.	District	Average Rise/ Drop (m)	Max. Drop (m) (Spot Value)
1	Ajmer	-5.59	-19.03
2	Alwar	-5.12	-19.45
3	Banswara	0.00	-5.80
4	Baran	-2.01	-17.90
5	Barmer	-1.33	-19.38
6	Bharatpur	-1.11	-7.45
7	Bhilwara	-4.71	-12.60
8	Bikaner	2.47	-12.85
9	Bundi	-3.30	-18.75
10	Chittorgarh	-4.09	-19.70
11	Churu	0.04	-13.56
12	Dausa	-2.18	-16.50
13	Dholpur	-1.35	-7.75
14	Dungarpur	-2.02	-8.20
15	Ganganagar	5.62	-1.25
16	Hanumangarh	4.42	-3.67
17	Jaipur	-5.74	-21.25
18	Jaisalmer	0.23	-7.98
19	Jalore	-8.00	-21.45
20	Jhalawar	-1.22	-6.60
21	Jhunjhunu	-6.22	-14.81
22	Jodhpur	-6.60	-54.26
23	Karauli	-3.43	-29.75
24	Kota	-1.51	-13.17
25	Nagaur	-6.50	27.50
26	Pali	-6.69	-28.86
27	Rajsamand	-4.69	-16.80
28	Sawai Madhopur	-1.74	-9.00
29	Sikar	-4.77	-16.82
30	Sirohi	-6.49	-24.99
31	Tonk	-2.37	-11.60
32	Udaipur	-2.67	-13.33

Source: GWD and CGWD

7.3.1.2.2 *Annual Water Level Fluctuation*

The overall rainfall in the State was more by 22.9% in 01-02 (June – May) as compared to 2000-01. Thirteen districts received normal rainfall (+20% to –20%).

Seventeen districts received above normal rainfall (>20%) Out of these, fifteen districts received rainfall in the range 20% to 60% above normal rainfall and remaining two districts rainfall more than 60% above normal rainfall. This has resulted in rise of water levels. The deficit rainfall (-20% to –60%) was observed in two districts.

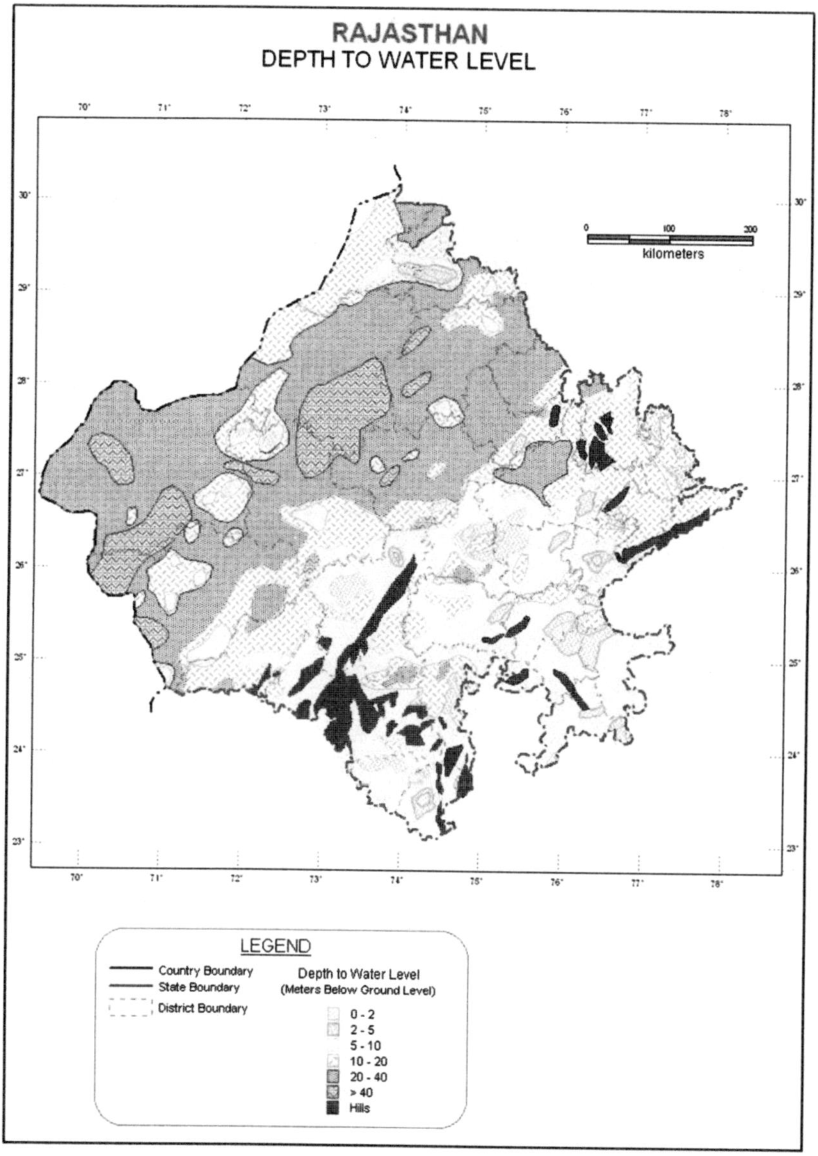

Figure 7.3 Depth to Water Level
(Source: Survey of India, Rajasthan Atlas, Resource Atlas, Watershed Atlas, Groundwater Atlas, CAZRI and Agricultural Geography of Rajasthan by Dr. T.S. Chouhan)

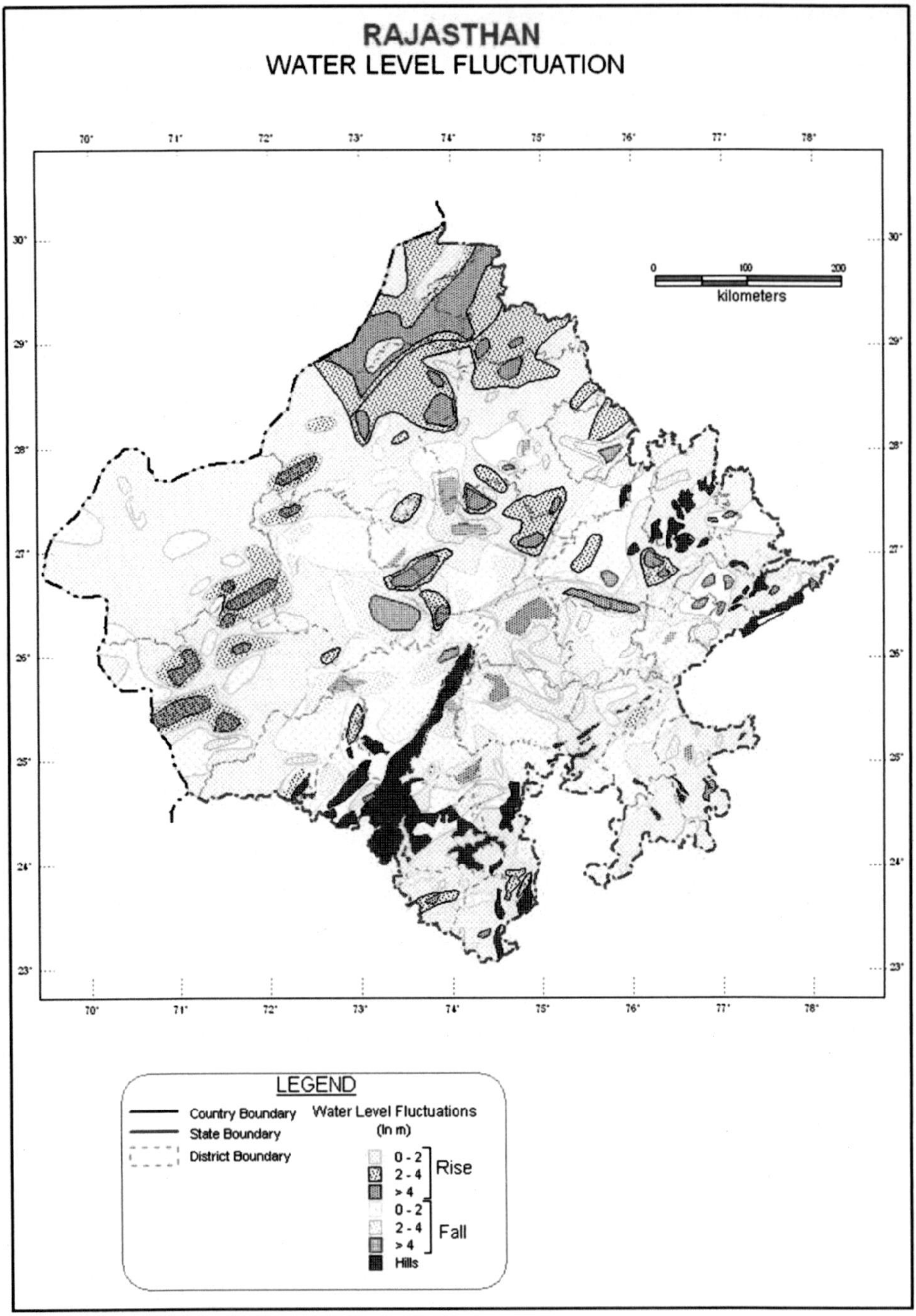

Figure 7.4 Water Level Fluctuation
(Source: Survey of India, Rajasthan Atlas, Resource Atlas, Watershed Atlas, Groundwater Atlas, CAZRI and Agricultural Geography of Rajasthan by Dr. T.S. Chouhan)

7.3.1.2.3 *Decline in Water Level*

Decline in water level less than 2m was observed in 33% of total NHS monitored in the State mainly falling in the districts of Hanumangarh, Dausa, Ganganagar, Alwar, Bharatpur, Karauli, Jaipur, Dholpur, Churu, Bundi, Sikar, Pali, Sirohi and Dungarpur. In other districts decline of this range was observed at isolated stations only. In 4% of NHS, decline in water level in the range of 2m to 4m was observed at isolated locations in Alwar, Karauli, Pali, Nagaur, Bharatpur, Sirohi, Kota, Jodhpur, Sawai Madhopur, Dholpur, Dausa and Jaipur districts. Fall in water level of more than 4m was observed in 5% of wells monitored and was noticed at isolated patches in Karauli, Sawai Madhopur, Bhilwara, Barmer, Sirohi, Alwar, Udaipur, Ganga-nagar, Hanumangarh, Dholpur, Jalore, Dausa, Bikaner and Jaipur districts.

7.3.1.2.4 *Rise in Water Level*

Rise in water levels, less than 2m, was observed in 40% of wells monitored in the State falling in all the districts. Rise in water levels between 2m to 4m was recorded in 13% of NHS falling mainly in Banswara, Rajsamand, Baran, Udaipur, Chittorgarh and Dungarpur districts. Isolated locations in Ajmer, Jodhpur, Barmer, Nagaur, Bhilwara, Jhalawar and Jalore districts exhibited situation. In 5% stations, rise in water level more than 4m was observed in isolated stations in the districts of Rajsamand, Ajmer, Bhilwara, Kota, Jodhpur, Udaipur, Jalore and Bundi.

7.3.2 Ground Water Utilisation

The water balance computations use the computed figures of natural ground water replenishment as an input to the ground water storage and additional inputs come as return flows, thus increasing the supply capacity of aquifers. This increase is growing with time, as more return flows come from the growing domestic, industrial and agricultural water utilization. Drafts from the ground water storage are restrained to the limits of aquifer storage and monthly withdrawal capacity. Thus the water balance computations invariably results in a balanced exploitation of aquifers on average, with over-utilization and under utilization in the high and low dependability climatic states respectively, when higher or lower than average ground water withdrawals were necessary to meet demands, because surface water availability was reduced or increased in these climatic states. This reflects the concept of conjunctive use of surface and ground water which may be encouraged through the development of artificial recharge works in the identified areas. The assessed mean natural replenishment of all the aquifers in Rajasthan is about 13.14 BCM (10.64 MAF) per year. Acute over exploitation of aquifers is evidenced by constantly falling water levels and increase in ground water salinity in the northern basins (Shekhawati, Ruparail, Sabi), in parts of Luni and Banas basins and in part of the outside basin. Exclusively high ground water levels, water logging of soils and salinity are a growing concern in the parts of IGNP and Chambal Projects, as a result of high percolation from leaking canals and in places the presence of a shallow hard pan beneath the surface.

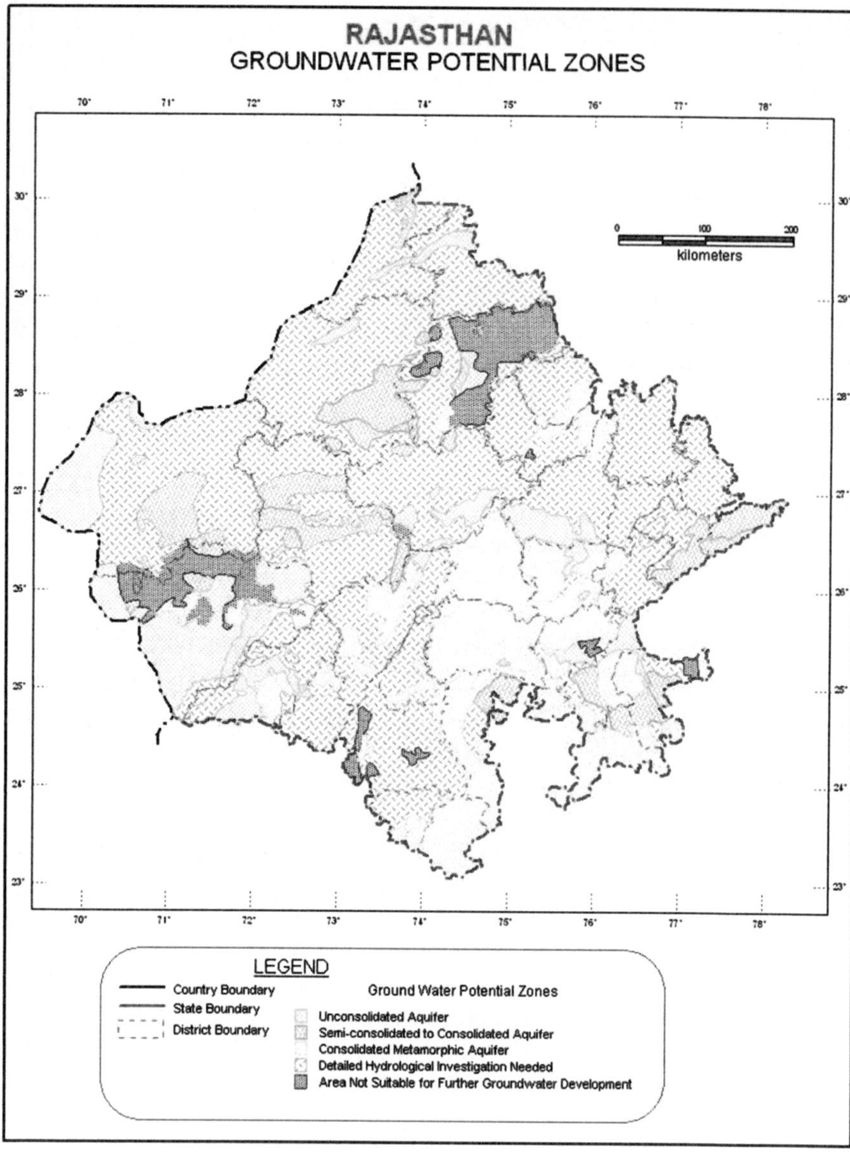

Figure 7.5 Groundwater Potential Zones
(Source: Survey of India, Rajasthan Atlas, Resource Atlas, Watershed Atlas, Groundwater Atlas, CAZRI and Agricultural Geography of Rajasthan by Dr. T.S. Chouhan)

7.4 IRRIGATION

Rajasthan has a coverage of 53.50 lakh hectare irrigated area which is proportionately much less compared to areas of the adjoining States like Uttar Pradesh, Punjab and Haryana. Out of the total culturable area of the 256,56 lakh hectares; the area under net irrigated agriculture is only 43.43 lakh hectares. This irrigated area is 16.91% of the total cultivable area of the State. Surface water irrigation has made good progress all these years. The total canal irrigated area has increased from 3.20 lakh hectares in 1951 to 24.06 lakh hectares in 1993. At present it covers about 35.24% of the total irrigated area of the State. Since 1951, the area under irrigated by wells and tube wells have remarkably been increasing, and at present 60.22% of the total area exist under irrigated farming under this system in the State.

Table 7.6 Irrigation Facilities in Rajasthan

S. No.	District	Total irrigated area		Well irrigation
		Sq. Km.	%	(%)
1.	Ajmer	1247.80	14.11	90.33
2.	Alwar	3744.94	44.89	98.70
3.	Banswara	843.64	16.75	13.00
4.	Baran	2219.57	27.76	59.14
5.	Barmer	1025.80	3.61	99.50
6.	Bharatpur	2104.33	41.53	95.30
7.	Bhilwara	2122.59	20.30	80.72
8.	Bikaner	1805.87	6.62	26.25
9.	Bundi	2273.68	40.75	43.00
10.	Chittorgarh	1916.90	17.65	94.25
11.	Churu	255.42	1.51	100.00
12.	Dausa	1545.87	45.54	97.40
13.	Dholpur	902.46	23.93	94.75
14.	Dungarpur	297.21	7.88	59.00
15.	Ganganagar	8034.85	13.19	0.44
16.	Hanumangarh	4989.12	51.66	3.14
17.	Jaipur	3837.69	35.37	94.53
18.	Jaisalmer	238.95	0.62	7.50
19.	Jalore	3330.04	31.29	93.80
20.	Jhalawar	1568.09	25.12	91.67
21.	Jhunjhunu	1399.50	23.60	99.60
22.	Jodhpur	1388.17	6.07	99.16
23.	Baran	908.45	17.91	80.60
24.	Kota	2289.57	44.04	24.60
25.	Nagaur	2537.88	14.32	99.75
26.	Pali	2824.02	22.79	83.57
27.	Rajsamand	714.81	15.24	85.00
28.	Sawai Madhopur	1443.41	28.62	73.60
29.	Sikar	1777.97	23.00	98.33
30.	Sirohi	1152.38	22.43	91.40
31.	Tonk	1794.51	24.94	65.00
32.	Udaipur	874.27	7.71	78.90

Source: Irrigation Department Progress Report

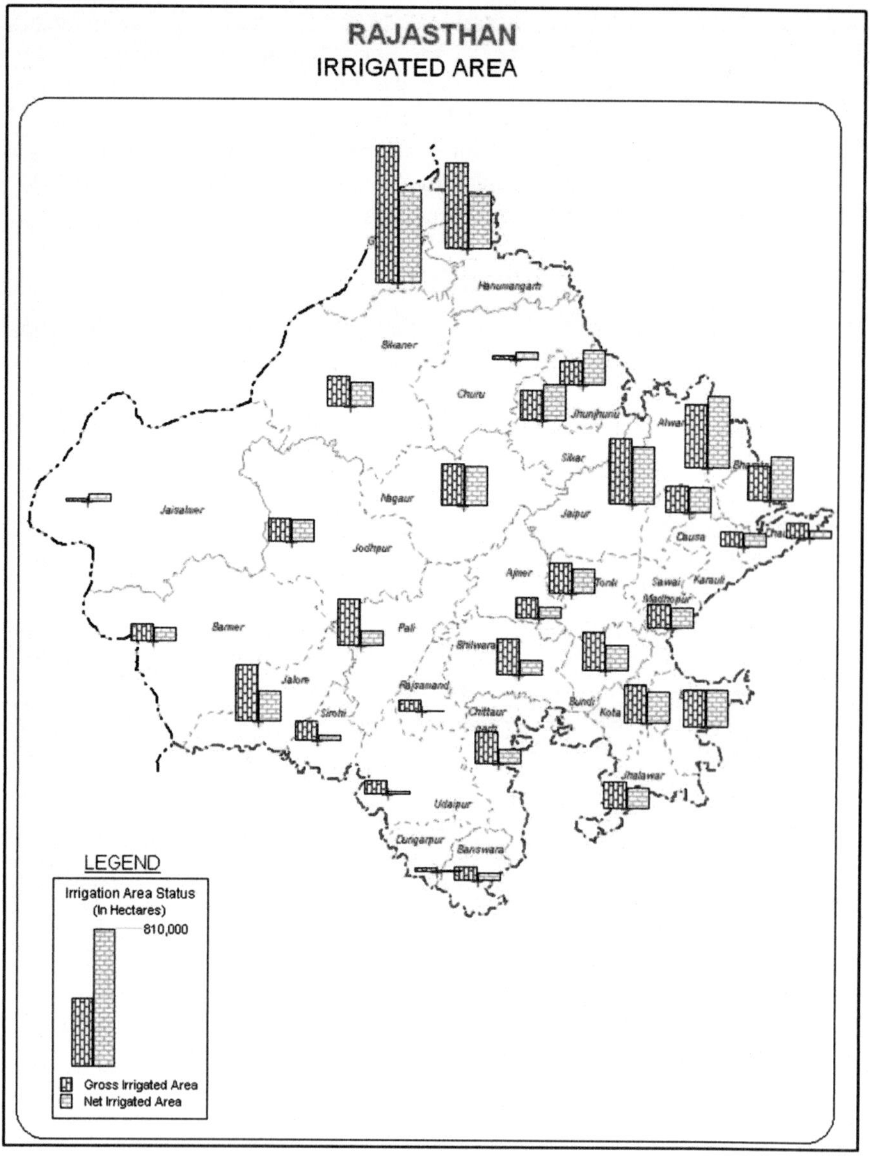

Figure 7.6. Gross and Net Irrigation Areas

7.4.1 Gross Irrigated Area

Gross Irrigated area includes the double – cropped or triple – cropped area, which is irrigated in Kharif, Rabi and summer-seasons in the same field. Such areas are found more in canal irrigated and well irrigated regions. Irrigation in the Kharif season is usually required for cotton, sugarcane, paddy and oil seeds. In Rabi-season, irrigation is mostly used for wheat, mustard, barley, gram, vegetables and fruit plants. The gross irrigated area in the State varies from year to year, depending upon the quantum and pattern of precipitation. The gross irrigated area varies from district to district. It is highest in Ganganagar district. In Kota, Bundi, Jaipur, Dausa and Alwar also, the percentage of irrigated area is observed more. In Jaisalmer and Churu districts irrigation by wells and canals is on a much lower scale. In Dungarpur district also, it is on the lower side, owing to poor aquifers in the hard metamorphic rocks but the situation is much better in Banswara district, where the Mahl-Bajaj canal has now started covering more area under irrigated agriculture.

7.4.2 Net Irrigated Area

The net irrigated area is the part of land, which receives irrigation by any source during any one season i.e. Kharif, Rabi or summer. The net irrigated area is a good indicator of agriculture production potential of the area. With the increase in additional source of irrigation, the net irrigated area is increasing every year. The percentage of net irrigated area varies from under 15 to more than 60%. Almost the entire district of Ganganagar and parts of Alwar, Bharatpur, Dholpur, Jaipur, Kota, Bundi, Bhilwara, Chittorgarh and Udaipur districts are better endowed with irrigation systems, mainly through canals and wells. Elsewhere in Rajasthan, the situation is dismal, with the percentage of net irrigated area being under 30%. As regards tehsils of the southern region in general, have far better net irrigated area than tehsil of northern and western Rajasthan. Ganganagar tehsil has the maximum percentage of net cultivated area (83.5%) followed by Anoopgarh (75.3%), Jaipur (75.2) and Rajsamand (71.4%). Many tehsils of the desert districts including Jaisalmer, Barmer, Jodhpur, Bikaner, Nagaur, and Churu, etc. have very little area under irrigation.

7.4.3 Irrigation Resources

Irrigation Commission (1972) had assessed the ultimate Irrigation potential of the country as 113.5 million ha comprising 58.5 million ha by Major and Medium Irrigation schemes, about 15 million ha by Minor surface water schemes and about 40 million ha by ground water. Indications are that minor surface irrigation potential could be about 17 million ha and ground water potential could be around 64.5 million ha taking the ultimate Irrigation Potential to about 140 million ha. This potential does not include additions due to inter basin transfers.

The mean annual surface and ground water resources of the State are as under:

		Mcum	MAF
(a)	States internal resources:-	19580	15.86
	(i). Already harnessed	8210	6.65
	(ii) To be harnessed by on going schemes	2100	1.70
	Sub Total (I +II)	10310	8.35
	Balance to be harnessed	9270	7.51

(b)	External surface resources:-		
	(i) Surface water allocated under various inter states agreements	17901	14.50
	(ii) Contemplated allocations from other inter state sources viz Ganga water 45800 Mcum (37.10 MAF), Mahi (ex-kadana) water 1926 Mcum (1.56 MAF)	47726	38.66
	Sub Total (I+II)	65627	53.16
	Total long term available surface water resources from internal and external sources	85207	69.02
	Ground Water Resources	13142	10.64
	Grand Total	**98349**	**79.66**

The total culturable area in the State is about 157 lac ha (out of total geographical area of 342 lac ha) which can be used for agriculture production if water required for irrigation is available. The State of Rajasthan is a water deficit State and its present total surface (both external as 14.5 MAF and internal as 15.86 MAF) water resources is estimated as 37481 Mcum (30.36 MAF). An integrated and comprehensive water resources planning study for the State for a 50 year plan horizon i.e. upto 2045 has been under taken through consultants. As per the preliminary report of the study, the likely utilizable resources will be only 16432 Mcum (13.31 MAF) out of the total internal surface resources of 19580 Mcum (15.86 MAF) of the State. Thus, the total utilizable water including all losses in the State with external allocated surface water of 17901 Mcum(14.5 MAF) shall be 34333 Mcum (27.81 MAF). The demand for various sectors has also been tentatively estimated by the consultants, though they are to be further refined during the course of the study. According to the present study the demand of all other sectors combined, except the Agriculture sector, for the year 2045 works out to 8864 Mcum (7.18 MAF) out of which about 2628 Mcum i.e.2.13 MAF (20% of 13142 Mcum) is assumed to be from ground water sources, while the balance 6236 Mcum (5.05 MAF) shall be from surface water resources. This means that the water available for agricultural shall be only 28097 Mcum (22.76 MAF) as per details given in next page.

(a) Surface Water	**Mcum**	**MAF**
Internal sources	16432	13.31
(Utilizable as per Consultants)		
Inter State Agreement	17901	14.50
Sub Total	**34333**	**27.81**

(b) Ground Water	13141	10.64
(c) Demand for other user Departments except Agriculture	8864	7.18
(d) Demand met from ground water (20% of b)	2628	2.13
(e) Balance to be met from surface water (c-d)	6236	5.05
(f) Net surface water available for Agriculture (a-c)	28097	22.67

The ultimate irrigation potential with the utilizable surface water (28097 Mcum) for the State with the constraint of water availability is estimated as 47 lac ha (considering required water depth as 0.60 m including storage and all other losses). The ultimate irrigation potential through contemplated allocations from other inter basin sources viz. Ganga water (45800 Mcum) and Mahi water at Kadana (1926 Mcum) is estimated to be 41.5 lac ha and 2.20 lac ha respectively, which add up to 43.70 lac ha. The ultimate irrigation potential from ground water resources has been estimated as 29.13 lac ha. The combined ultimate irrigation potential through surface water and ground water is likely to be as under:-

(i) With existing internal and external resources 76.13 lac ha

(ii) With existing internal, external and future external resources 119.83 lac ha

7.4.4 Basin-Wise Water Resources and Irrigation Development

The State has fourteen major river basins which are further sub-divided in 59 sub-basins. Irrigation Department, Rajasthan prepared a Draft Master Plan for all Basins and published Surface Water Resources of Rajasthan during 1987-88. according to this Master Plan the State has been divided into 14 River Basins viz Chambal, Banas, Luni, Mahi Shekhawati, Sabarmati, Parwati, Sabi, Banganga, Ruparail, Gambhiri, West Banas, Sukli and other nallahs of Jalore Districts. Surface Water- According to the total Surface Water Resources of Rajasthan prepared during 1987-88, the total surface water available in Rajasthan is 15.86 MAF. Ground Water- Based on hydro-geological investigations, the ground water recharge i.e. replenis-hable ground water resources in all river basins is estimated as 13142 Mcum/year (10.64 MAF). The average ground water resources for irrigation is 10957.68 Mcum/ year (8.875MAF), and the net irrigation draft is 6336.38 Mcum/year (5.375 MAF).

The ultimate ground water irrigation potential is estimated as 2.913 m.ha. and the irrigation potential created up to December, 1994 is 1.807 m.ha. Leaving a balance of 1.106 m.ha. As per estimates as on 1.1.95, 2128.90 Mcum (1.726 MAF) ground water has been earmarked for domestic use. The present draft for domestic use is 689.27 Mcum(0.558 MAF), leaving a balance of 1439.63 Mcum (1.168 MAF) for future domestic and industrial uses. Of the resources available for irrigation, present draft is 6322.68 Mcum (5.130 MAF), leaving a balance of 4950.02 Mcum (4.01 MAF). A brief summary of the State of surface and ground water resources, irrigation and culturable land use from existing, ongoing and proposed surface water schemes, basin wise so far identified, is described as under-

7.4.4.1 *Shekhawati Basin*

The mean annual surface water potential has been estimated as 183.49 Mcum (0.148 MAF), and mean gross recharge from ground water is 275.90 Mcum (0.224 MAF). While the net irrigation draft from ground water is about 385.38 Mcum (0.313 MAF). The total area of basin is 1.269 mha and culturable area of 0.356 mha. The average irrigated area is 0.284 mha (1990-94).

7.4.4.2 *Ruparail Basin*

The mean annual surface water potential has been estimated as 183.49 Mcum (0.148 MAF), and mean gross recharge from ground water is 270.05 Mcum (0.219 MAF). While the net irrigation draft from ground water is about 285.88 Mcum (0.232 MAF). The total area of basin is 0.302 mha and culturable area of 0.212 mha of the basin. The average irrigated area is 0.109 mha (1990-94). The details of surface water schemes in the basin are as under:

7.4.4.3 *Banganga Basin*

The mean annual surface water potential has been estimated as 448.25 Mcum (0.148 MAF), and mean gross recharge from ground water is 843.95 Mcum (0.219 MAF). While the net irrigation draft from ground water is about 1034 Mcum (0.840 MAF). The total area of basin is 0.972 mha and culturable area of 0.652 mha of the basin. The average irrigated area is 0.219 mha (1990-94). The details of surface water schemes in the basin are as under:

7.4.4.4 *Gambhir Basin*

The mean annual surface water potential has been estimated as 531.36 Mcum (0.430 MAF), and mean gross recharge from ground water is 468.20 Mcum (0.379 MAF). While the net irrigation draft from ground water is about 394 Mcum (0.320 MAF). The total area of basin is 0.439 mha and culturable area of 0.263 mha of the basin. The average irrigated area is 0.107 mha (1990-94). The details of surface water schemes in the basin are as under:

7.4.4.5 *Parbati Basin*

The mean annual surface water potential has been estimated as 248.53 Mcum (0.201 MAF), and mean gross recharge from ground water is 223 Mcum (0.181 MAF). While the net irrigation draft from ground water is about 131 Mcum (0.106 MAF). The total area of basin is 0.248 mha and culturable area of 0.132 mha of the basin. The average irrigated area is 0.061 mha (1990-94). The details of surface water schemes in the basin are as under:

7.4.4.6 *Sabi Basin*

The mean annual surface water potential has been estimated as 213.85 Mcum (0.173 MAF), and mean gross recharge from ground water is 313 Mcum (0.253 MAF). While the net irrigation draft from ground water is about 421 Mcum (0.342

MAF). The total area of basin is 0.505 mha and culturable area of 0.243 mha of the basin. The average irrigated area is 0.132 mha (1990-94).

7.4.4.7 *Banas Basin*

The mean annual surface water potential has been estimated as 3451.50 Mcum (2.795 MAF), and mean gross recharge from ground water is 2792 Mcum (2.261 MAF). While the net irrigation draft from ground water is about 2003 Mcum (1.626 MAF). The total area of basin is 4.705 mha and culturable area of 2.183 mha of the basin. The average irrigated area is 0.834 mha (1990-94).

7.4.4.8 *Chambal Basin*

The mean annual surface water potential has been estimated as 8059 Mcum (6.528 MAF), and mean gross recharge from ground water is 3381 Mcum (2.739 MAF). While the net irrigation draft from ground water is about 1141 Mcum (0.926 MAF). The total area of basin is 3.136 mha and culturable area of 1.478 mha of the basin. The average irrigated area is 0.75 mha (1990-94).

7.4.4.9 *Mahi Basin*

The mean annual surface water potential has been estimated as 4529.52 Mcum (3.669 MAF), and mean gross recharge from ground water is 988 Mcum (0.800 MAF). While the net irrigation draft from ground water is about 326 Mcum (0.265 MAF). The total area of basin is 1.661 mha and culturable area of 0.588 mha of the basin. The average irrigated area is 0.217 mha (1990-94).

7.4.4.10 *Sabarmati Basin*

The mean annual surface water potential has been estimated as 513 Mcum (0.415 MAF), and mean gross recharge from ground water is 145 Mcum (0.117 MAF). While the net irrigation draft from ground water is about 47 Mcum (0.038 MAF). The total area of basin is 0.419 mha and culturable area of 0.093 mha of the basin. The average irrigated area is 0.025 mha (1990-94).

7.4.4.11 *Luni Basin*

The mean annual surface water potential has been estimated as 790 Mcum (0.640 MAF), and mean gross recharge from ground water is 1601 Mcum (1.297 MAF). While the net irrigation draft from ground water is about 1727Mcum (1.402 MAF). The total area of basin is 3.827 mha and culturable area of 1.419 mha of the basin. The average irrigated area is 0.476 mha (1990-94).

7.4.4.12 *West Banas Basin*

The mean annual surface water potential has been estimated as 280 Mcum (0.227 MAF), and mean gross recharge from ground water is 91 Mcum (0.073 MAF). While the net irrigation draft from ground water is about 59 Mcum (0.048

MAF). The total area of basin is 0.187 mha and culturable area of 0.069 mha of the basin. The average irrigated area is 0.037 mha (1990-94).

7.4.4.13 *Sukli Basin*

The mean annual surface water potential has been estimated as 87 Mcum (0.070 MAF), and mean gross recharge from ground water is 53 Mcum (0.043 MAF). While the net irrigation draft from ground water is about 42 Mcum (0.034 MAF). The total area of basin is 0.096 mha and culturable area of 0.025 mha of the basin. The average irrigated area is 0.013 mha (1990-94).

7.4.4.14 *Other Nallah of Jalore District-Basin*

The mean annual surface water potential has been estimated as 55 Mcum (0.045 MAF), and mean gross recharge from ground water is 57 Mcum (0.046 MAF). While the net irrigation draft from ground water is about 57 Mcum (0.047 MAF). The total area of basin is 0.186 mha and culturable area of 0.022 mha of the basin. The average irrigated area is 0.099 mha (1990-94).

7.4.4.15 *Area falling out-side Basin*

This area is situated west of Aravali and occupies about half the area of the State. There are no streams and rivers in this area, and except for external sources of surface water of 13728 Mcum (11.22 MAF) being proposed to be utilized through Gang Canal System, Bhakara canal System and Indira Gandhi Nahar Project, and meager ground water sources estimated as 2300 Mcum (1.865 MAF), it is wholly dependent on rainfall. Out of the total area of 16.273 mha of this region, the total culturable area is 8.009 mha. The average irrigated area is 1.657 mha (1990-94). The prestigious Indira Gandhi Nahar Project has been taken up for utilization of the allocated share of the Inter – State Ravi Beas river to Rajasthan which shall be creating to a command of 1.269 mha.

7.4.5 Imported Water

According to existing Inter-state agreements, Rajasthan has the following share in the out of State rivers-

(A)	**Ravi-Beas-Sutlej**		**Mcum**	**MAF**
1.	Pre-partition share through Gang Canal		1370	1.11
2. (a)	Through Indira Gandhi Nahar Project (Stage-I)			
	i. Irrigation	-3.37 MAF |	4432	3.59
	ii. Drinking Water	-0.22 MAF |		
(b)	Through Indira Gandhi Nahar Project			

	(Stage-II)			
	i. Irrigation	-3.35 MAF |	4938	4.00
	ii. Drinking Water	-0.65 MAF |		
(c)	Through Other Sources			
	Gang Canal	-0.30 MAF |	1247	1.01
	Bhakra System	-0.24 MAF |		
	Sidhmukh-Nohar	-0.47 |		
	Sub total (a)+(b)+(c)		10617	8.60
3.	From Bhakra (Sutlej)		1741	1.41
	Total (A)		**13728**	**11.12**
(B)	Narmada Water through Narmada Canal		617	0.50
(C)	Yamuna Water		1119	0.91
(D)	Mahi Water		457	0.37
(E)	Chambal Water		1980	1.60
	Total (B+C+D+E)		4173	3.38
	Total (1+2+3)		**17901**	**14.50**
	Contemplated Imported Water			
(a)	Mahi Waters (ex-Kadana)		1926	1.56
(b)	Ganga Waters		45800	37.10
	Total (a+b)		**47726**	**38.66**
	Grand Total		**65627**	**53.16**

■■■■

Chapter - 8

MINERAL, POWER AND INDUSTRIAL DEVELOPMENT

Mineral resources are the backbone for the industrial and socio-economic development of a country. Rajasthan ranks seconds only to Jharkhand in mineral deposits. Some minerals produced like Wolastonite Jaspar are found only in Rajasthan. Important minerals produced in the State are basemetals viz. copper, lead and zinc; industrial and other minerals like soapstone, calcite, baryte, wolastonite, clays, silica sand, fluorite, cement grade limestone, garnet, gypsum, rock phosphate etc. The State also produces a variety of dimensional and decorative stones viz. granite, marble, slate sandstone, quartzite etc. In fuel minerals, the State has vast resources of lignite suitable for power generation. Oil and Natural Gas has also been found at some places in Western Rajasthan. The State has virtual monopoly in country in the production of wolastonite. Jasper and zinc while it is a producer of gypsum, asbestos, soapstone, rock phosphate, ball clay, calcite, feldspar, lead etc. accounting for more than 70 per cent of country's production of these minerals. The state occupies an important place in the field of dimensional and decorative stones. 95 per cent of country's production of marble is from the State of Rajasthan. There has been a steady increase in production of sandstone, slate and granite in recent years.

8.1 MINERAL RESOURCES

Rajasthan, endowed with a rich mineral wealth, occupies a significant position on the mineral map of the country. In the area of metallic and non-metallic minerals, Rajasthan, by possessing 19% of working mines in the country, leads the country. It possesses a wide range of mineral resources and offers considerable potential for the future. The important metallic minerals with which the name of the State is intimately associated are: lead, zinc, copper and tungsten. It is also richly endowed with a variety of non-metallic minerals. Rajasthan has also enormous deposits of building stones like marble, granite, sandstone, limestone, slate etc. It is a leading producer of asbestos, soapstone /steatite, gypsum, rock phosphate, calcite, feldspar, clay etc. Presently, the State enjoys a monopoly in the production of wollastonite, emerald, jasper and semi-precious garnet. In addition, the State has good deposits of lignite in Bikaner, Nagaur and Barmer districts.

8.2 TYPE OF MINERALS

Occurrences of minerals in Rajasthan is shown in Figure no. 8.1. Brief descriptions of minerals with their occurrence areas in Rajasthan are described below:

8.2.1 Apatite

Apatite occurs as a primary constituent of igneous rocks, but only as accessory mineral. It also occurs as a vein mineral. It is a natural phosphate-bearing mineral having a definite chemical composition. Apatite is mainly of two types. One is chlorapatite $3Ca_3(PO_4)_2CaCl_2$ and the other is fluorapatite $3Ca_3(PO_4)_2CaF_2$. Its main use is as fertilizer. In Rajasthan, occurrences of Apatite have been reported from Sikar and Udaipur districts.

8.2.2 Asbestos

Rajasthan accounts for about 96% of the country's total production of asbestos, whereas Andhra Pradesh and Karnataka, the other producing States contribute the remaining 4%. The asbestos deposits are mainly located in the southern parts of the State, confined to the altered ultramafic intrusives within the rocks of the Bhilwara, Aravalli and Delhi Supergroups. Rajasthan possesses crysotile and amphibole varieties of asbestos.

8.2.3 Barytes

Barytes, Barite or the 'Heavy Spar' are different names of the same mineral having the composition $BaS0_4$. The mineral is very important and vital to the petroleum industry which consumes more than 80° of the world's Barytes production. Barytes finds it principal utility in drilling mud for exploration of because of its high specific gravity, inertness to acids insolubility in water and above all its lower cost compared to many other available heavy materials. In Rajasthan, Barytes mineralization is associated with igneous, metamorphic and sedimentary rocks. It occurs as veins, cavity-fillings, bedded and residual deposits. Occurrences of Barytes are known in the districts of Alwar, Bharatpur, Bhilwara, Bundi, Chittorgarh, Jalore, Pali, Sikar, Sirohi and Udaipur, whereas important deposits are limited to Alwar, Bharatpur, Pali and Udaipur districts.

8.2.4 Bauxite

Bauxite is used mainly for the extraction of aluminum metal, in the refractory and chemical industries, manufacture of emery powder, cement paint and in refining petroleum products. India is bestowed with large deposits of bauxite. The known bauxite reserves of the country are of the order of about 200 million tones, of which 140 million tones are of high-grade. Rajasthan, however, contributes very insignificant quantity of bauxite. In Rajasthan bauxite occurrences in Kota district have been reported by the GSI. Bauxite in these areas occurs in the form of blanket capping as a product of in situ alteration of the Deccan Traps.

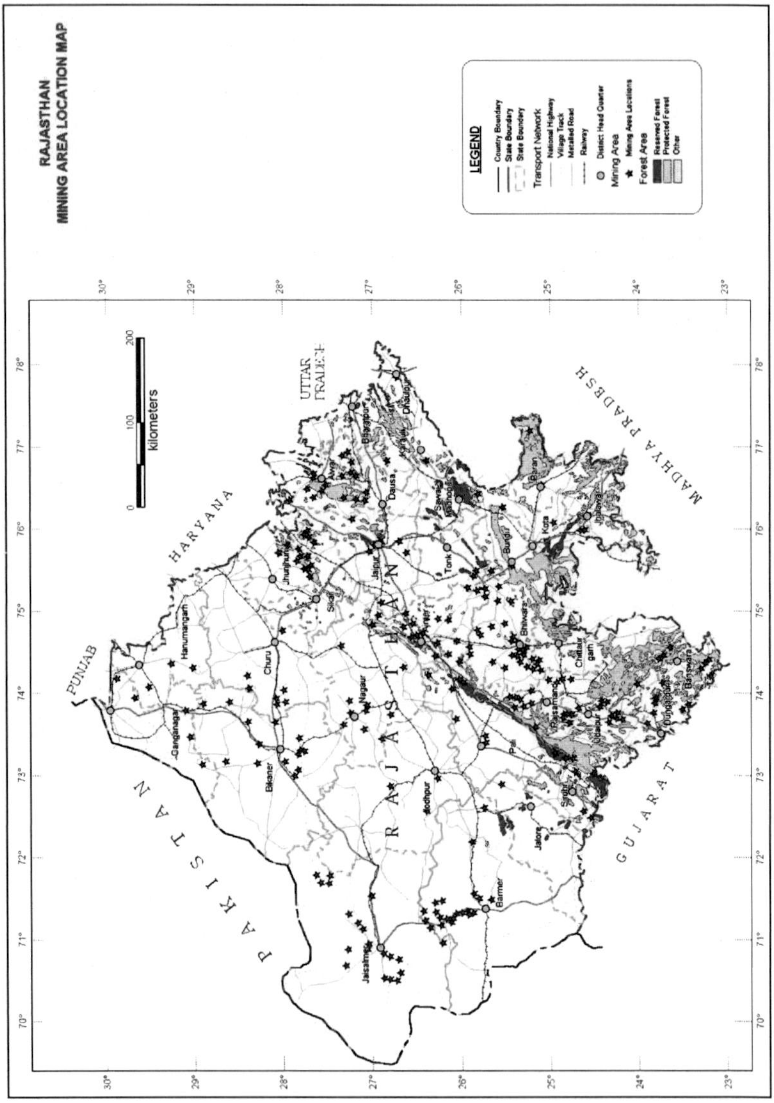

Figure 8.1 Mining Area Locations Map
(Source: Survey of India, Rajasthan Atlas, Resource Atlas, Watershed Atlas, Groundwater Atlas, CAZRI and Agricultural Geography of Rajasthan by Dr. T.S. Chouhan)

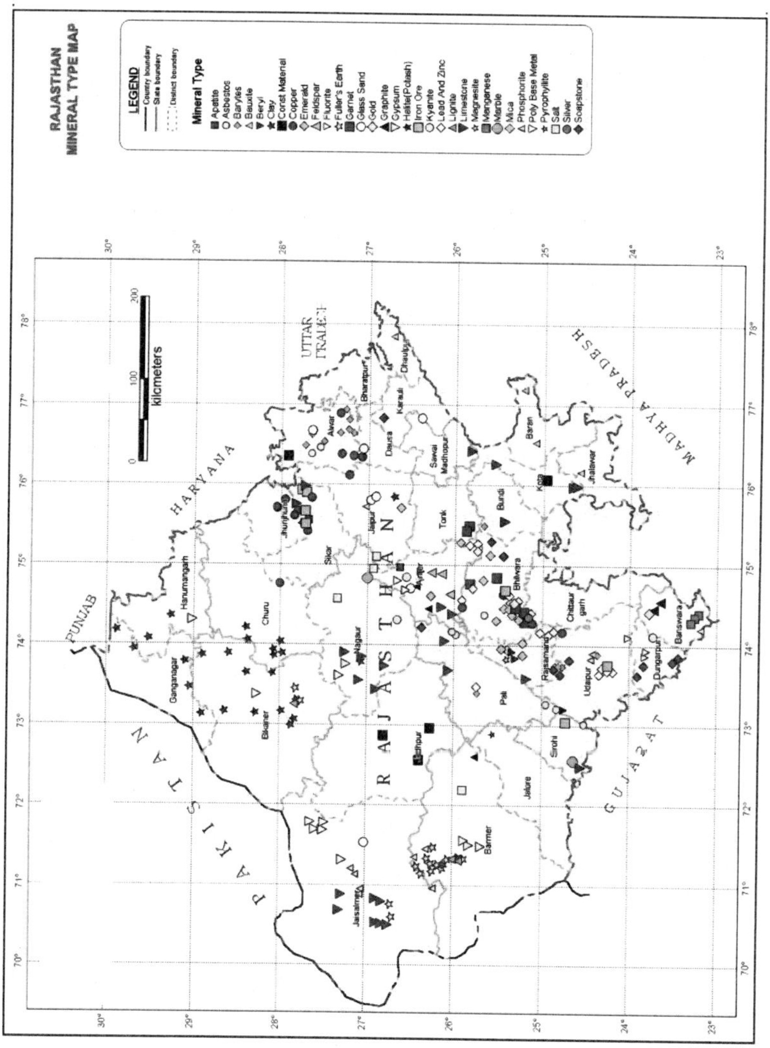

Figure 8.2 Mineral Types Map
(Source: Survey of India, Rajasthan Atlas, Resource Atlas, Watershed Atlas, Groundwater Atlas, CAZRI and Agricultural Geography of Rajasthan by Dr. T.S. Chouhan)

8.2.5 Bentonite

Bentonite is one of the two naturally occurring clay minerals of montmorillonite group with decolorizing and absorbing properties. Almost all Bentonite assume a slightly darker shade due to absorption of moisture. Two types of Bentonite viz., swelling and non-swelling type, are commonly known. Sodium-Bentonite is of swelling type which swells up to 15 times while calcium-Bentonite is of non-swelling type. Commercial importance of Bentonite depends on its swelling index, base exchange properties and setting time. The Bentonite is used as grouting material in engineering constructions, in drilling mud's and as carrier in insecticides, paint and pharmaceutical industries. It is also used for bleaching vegetable oils and petroleum. Rajasthan possesses inexhaustible reserves of both swelling and non-swelling types of Bentonite. Deposits of Bentonite are mostly located in Barmer and Jaisalmer districts.

8.2.6 Beryl

Beryl, a silicate of beryllium and aluminum, is the principal source of beryllium metal which is lighter than aluminum. When alloyed with copper and after suitable heat treatment the alloy develops high-tensile strength and ability to withstand repeated stress. Beryl has become very important in recent years because the metal is used as moderator in nuclear reactors. A small quantity of beryllium is used in the chemical industry in the manufacture of beryllium salts and in ceramic glaze. The green transparent variety of beryl is emerald, which is a precious stone. Rajasthan is the principal beryl producing state in India. Beryl occurs in association with mica pegmatite in the mica mines of Ajmer, Udaipur and Bhilwara districts.

8.2.7 Bismuth

Bismuth is principally used in medical preparations. However, it has some other important uses for making fusible alloys in combination with tin, lead, mercury and cadmium, aluminum alloys (Cerro base alloys), production of atomic bomb, in radar equipment and in nuclear energy projects. Bismuth occurs in nature either in native form or as sulphide, bismuthinite Bi2S3. This mineral is generally covered by thin film of the yellow oxide, bismuth ochre and is frequently associated with the ores of tin, tungsten and other base metal sulphides. Most of the bismuth metal is won as a by-product in the smelting of other ores. In Rajasthan, a very localized and small lensoid occurrence of bismuth ore from an unzoned pegmatite at Narda, Neem-ka-Thana tehsil, Sikar district has been reported with anomalous bismuth (84.20%), arsenic (1.08%), gold (70 ppm with bismuthinite, 1.7 ppm in host rock), lead (0.17%) and copper (0.11°%).

8.2.8 Basalt / Rhyolite

South-eastern parts of Rajasthan are partly occupied by the Deccan Traps, which constitute the main building stone in this area. The rhyolite is mostly used for railway ballast and other construction purposes in western Rajasthan.

8.2.9 Flaggy Stone (Kota Stone)

Upper Vindhyan limestone, quarried in Kota and Chittorgarh districts, occurs in shades of yellow, green and blue. The quarries are located at Ramganj Mandi, Chechat, Suket, Manpura, Morak and Mur-ka-Kheri in Kota district. Flaggy limestone also occurs near Aroutiya, Paroliya and Kishanpura in the Jhalawar district.

8.2.10 Granite

Granite and associated rocks form excellent dimensional and decorative stones. Durability, hardness, capability of taking mirror polish and fascinating colours makes the granite most distinctive stone among marble, limestone and sandstone. It is widely used as polished tiles, slabs and monumental stone all over the world. The granite occurrences in Rajasthan are distributed over wide areas in Ajmer, Alwar, Banswara, Barmer, Bhilwara, Chittorgarh, Dungarpur, Jaipur, Jaisalmer, Jalor, Jhunjhunu, Jodhpur, Nagaur, Pali, Rajsamand, Sawai Madhopur, Sikar, Sirohi, Tonk and Udaipur districts. A district-wise brief account of important localities along with nature, colour and commercial names of the granite is given below

8.2.11 Marble

In Rajasthan, marble and marble based industries are very well developed. The marble and marble products are used in house construction, statues, marble chips, marble/dolomite powder and plaster making; as tyre detergent; and in gur, sugar, and paper industries. Marble is mainly mined from Makrana, Raj Nagar, Rikhabdev, Ajmer, Andhi and Bhaislana areas. Abur limestone and Jaisalmer limestone take good polish and are used as decorative stones. These are marked as 'Supari stone' and 'Yellow-marble' respectively. Morwar in Rajsamand district has a single mine (out of numerous mines) which produces 300,000 metric tones of blocks-slabs of white marble per year creating a world record which find a place in the *Guinness Book of World Records*. Rajasthan possesses the largest reserves exceeding 1,100 million tones accounting for about 95% of the total production of marble in India.

8.2.12 Sandstone

Sandstone is mainly found in the Vindhyan and the Marwar Supergroup sequences of Bharatpur, Tonk, Sawai Madhopur, Bundi, Chittorgarh districts and jodhpur, Bikaner, Pali, Jaisalmer districts respectively. Rajasthan has been the main center of export of sandstone to other parts of the country. The important occurrences of sandstone are found in Bharatpur, Bikaner, Bundi, Jaipur, Jodhpur, Kota, Tonk, Sawai Madhopur districts.

8.2.13 Slate / Schist

Slate finds its use as material for roofing, paving and shelves. Slate deposits are found mainly in the Delhi and Aravalli rocks in parts of Ajmer, Alwar, Bharatpur, Bundi, Pali, Sawai Madhopur, Tonk, Udaipur and Chittorgarh districts. Quarries in Jaipur, Udaipur, Alwar, Dungarpur and Banswara are providing schist and phyllities as roofing material. Alwar slate has acquired a reputation in the world market for its quality and has left China and Spain far behind in the international market. Alwar district has enormous reserves of slates and has monopolized the Indian export Alwar slates are used in the preparation of school slates, as roof tiles, and tabletop of the Billiard table. The slates are available around Bahrod, Maton and Neemrana in huge quantity. Mines are also situated around Jakharana, Bassai, Kalor, Adind Bhaiyasar, Mahatavas and Khundrot villages. RSMDC is engaged in slate mining at Maton.

8.2.14 Calcite

Calcium carbonate (CaC03) is an important constituent of sedimentary rocks, next to quartz and clay minerals. In the form of limestone it is extensively used in the manufacture of lime, cement, ceramic products, calcium carbide and bleaching powder. In the pulverized form, it is used in paint, rubber, textile, paper, leather and sugar industries. Limestone of high-purity is used for imparting glaze in pottery and as a flux in metallurgical operations. Pure, crystallized, transparent variety of calcium carbonate, known as "Iceland Spar" is used in optical instruments. The total reserves of calcite in India are estimated at 10,573 thousand tones. Rajasthan is the leading producer of calcite in the country. In Rajasthan, occurrences of calcite have been recorded from Sikar, Pali, Udaipur, Sirohi and Jaipur districts.

8.2.15 China Clay

Extensive deposits of china clay are found in India. With the increasing demand from industries like ceramic, refractory, textile, paper, rubber and pesticides, the production of the mineral has increased to a very large extent. Almost the entire production of china clay is consumed indigenously. A small quantity of superfine variety is also imported for specialized uses. Rajasthan contributes about 50% production of china clay, 40% of ball clay and 10% of fire clay in the country. China clay is found to occur in Bhilwara, Bundi, Chittorgarh, Jaipur, Jhalawar, Sawai Madhopur, Sikar, Pali districts etc. Ball clay occurs in Alwar, Bikaner and Bundi districts.

8.2.16 Copper

Copper is the most essential metal used by the man since the 'Bronze age'. In spite of its part substitution by aluminium, the demand of copper is ever increasing. At present the production of copper metal in the country is only 15,000 tones as against the annual demand of about 0.5 million tones. A major part of the demand for this metal is, therefore, met by imports causing heavy drain on our foreign exchange. Substantial reserves of copper ore have been established in Rajasthan, which now hold a significant place in the country in this important base metal resource. The main occurrences of copper ore are located along the prominent mineralized belts viz, Khetri Copper Belt and Alwar Copper Belt in the northeast part of the State. Besides, a number of small copper deposits associated with other base metal occurrences, which are under various stages of exploration, have been brought to light in other parts of the State.

8.2.17 Corundum

Corundum, due to its hardness, is used mainly in manufacturing abrasive material. In sintered form, it is used for special refractory crucibles. Synthetic corundum is in great demand as the production of natural corundum is insufficient to meet the demand. Ruby and sapphire are the transparent gem verities of corundum. Ruby is rose-red, carmine or purple. True sapphire is blue. Other colours in sapphire include white, yellow, green or pink.

8.2.18 Dolomite

Dolomite occurs in massive or bedded form. Its main industrial application is as a flux material in the iron and steel industry, which accounts for 91% of the total consumption in the country. Besides, it is used in glass, ferroalloys, alloy steel, fertilizer, ceramics, chemicals, electrode, rubber, building construction etc. Though there are many occurrences of dolomite in the various geological formations of Rajasthan, production of dolomite is very small as there are no metallurgical and other industries in the State. The dolomite deposits occur in Ajmer, Alwar, Bhilwara, Jaipur, Jaisalmer, Jhunjhunu, Jodhpur, Nagaur, Pali, Sawai Madhopur and Udaipur districts. The total reserves in the State are placed at 1,62,644 thousand tones.

8.2.19 Feldspar

Feldspar is an alumino-silicate mineral of potassium, sodium and/or calcium. Though feldspar is among the most widely distributed rock-forming minerals and occurs as a constituent of most of the rocks, its commercial deposits are mainly confined to pegmatite. Feldspar is either quarried from the pegmatite veins or is recovered as by-product to the mining of mica and beryl from the pegmatite. Feldspar is chiefly used in the ceramic and glass industries and in insulator making. Rajasthan is the principal feldspar producing State in the country and its reserves are about 10.13 million tones which constitute 72% of the total Indian reserves. In Rajasthan, feldspar occurrences have been reported from Ajmer, Alwar, Jaipur, Pali and Sikar districts.

8.2.20 Fluorite

Fluorite (CaF_2) is generally colourless to white and purple in colour. The finest-grade of fluorite is used for enameling iron, manufacture of opaque and opalescent glasses and for the production of hydrofluoric acid. The inferior-grade fluorite is used as a flux in foundry work for the manufacture of steel. Transparent fluorite is used in the manufacture of lenses. Use of fluorite in the steel metallurgy and in the manufacture of artificial chryolite makes it an important mineral. Rajasthan has a large number of occurrences and small deposits of fluorite and is only next to Gujarat. Occurrences of fluorite are reported from Ajmer, Alwar, Bhilwara, Dungarpur, Jaipur, Jalore, Jhunjhunu, Nagaur, Sikar and Sirohi districts. Workable deposits occur in Dungarpur, Jalore, Jhunjhunu, Sikar and Sirohi districts. The total reserves of fluorite from

8.2.21 Fuller's Earth

Fuller's earth, also known as *Multani Mitti is* a clay of sedimentary origin composed essentially of attapulgite, montmorillonite and kaolinite. It is used as cosmetic and for cleaning woolen clothes. It has absorbent properties which enable it to remove greasy and oily matter. It is also used in refining oils and animal fats. Major production of fuller's earth in India comes from Rajasthan. Annual production is about 5,0006,000 tones. Barmer is the main producing district contributing about 77% of total production of the State. The rest of the production comes from Bikaner (18%) and Jaisalmer (5%) districts. With the increase in the output of mineral oil and vegetable oil products in India, a progressive increase in demand for fuller's earth is apprehended.

8.2.22 Aquamarine

Aquamarine, a beryllium aluminium silicate, is a transparent variety of beryl. Aquamarine is mainly recorded from Toda Raising and Hanotia in the Tonk district. It occurs in association with mica and pegmatite. Aquamarine is usually cut in pendant form.

8.2.23 Amethyst

Amethyst, a purple to violet-coloured transparent variety of quartz, occurs in well-crystallized form. When heated strongly it looses its colour. Under moderate heating it turns yellow and is sold as citrine. Though faceted, the steep brilliant cuts are common. It is also carved into attractive forms and shapes. Amethyst is reported to occur near the Kishangarh railway station in Ajmer district and from Bidera, Girota, Didwana and Choup in Jaipur district.

8.2.24 Chrysoberyl

Chrysoberyl (oxide of beryllium and aluminium) is green in colour and differs from beryl in its composition. Chrysoberyl occurs in association with mica and aquamarine - bearing pegmatites of Ajmer and Bhilwara districts.

Dumortierite-quartz *(blue aventurine)*

A massive opaque blue variety of crystalline quartz of semi-precious quality is reported to occur at Dewal near Shahpura, Jaipur distrct. The intergrown crystals of dumortierite impart blue colour to quartz.

8.2.25 Emerald

Transparent variety of beryl possesses attractive colours and lustre. The colour varies from colourless to green, blue and yellow. Green coloured stones are known as emerald *(Panna),* while aquamarine has yellow colour. Emerald is a very highly priced precious stone. In India emeralds are available mostly in Ajmer and Udaipur districts of Rajasthan.

8.2.26 Fuchsite quartzite *(green aventurine)*

Fuchsite quartzite is reported from Bhinder area Bhilwara district. It occurs in high-grade psammites of the Pur-Bunera Group. It is used in statue building.

8.2.27 Garnet

Garnet is used in the manufacture of coated abrasive like paper, cloth and discs for grinding and finishing glass, wood, hard rubber etc. In powder form, it is used for making lapping and buffing compounds. Garnet is generally obtained from small pits and shallow underground mines. It is also recovered as co-product in processing of beach sands. Rajasthan is the only State producing gem variety. The garnet occurrences in Rajasthan are recorded from Ajmer, Bhilwara and Tonk districts.

8.2.28 Rock Crystal

Transparent crystal of quartz is used for necklaces and other ornaments. Associated with pegmatite it is available in Ajmer and Tonk districts.

8.2.29 Ruby / Corundum

Pegmatites intruding ultrabasic rocks give rise to development of corundum by desilication of feldspars. A small deposit of corundum/ruby at Jual in Tonk district and an occurrence associated with muscovite at Galwa in Bhilwara district are reported. The corundum is mainly used for manufacture of emery cloth.

8.2.30 Gold

Gold, the 'king of metals', has always attracted mankind for its untarnishable yellow glitter for making the brilliant ornament par excellence and its relative scarcity and physical properties render it to this day the acknowledged international measure of material values. The usages of gold are as the standard for monetary system in the form of coinage or as bullion, for ornamentation especially in jewellery, in making gold leaf, gold lace and various fabrics, gold plating and in some chemical and photographic preparations. It also has some medicinal values. Incidences of gold in multiple environments in several part of Rajasthan are well known. The occurrences are found in the Bhilwara, Aravalli and Delhi Supergroups of rocks of Archaean to Proterozoic age. Gold mineralization is hosted by a variety of rocks types. It is invariably associated with massive to semimassive, disseminated sulphides comprising pyrrhotite, pyrite, chalcopyrite, arsenopyrite, sphalerite, bismuthinite etc.

8.2.31 Graphite

Graphite, also known as 'black lead', is a soft crystalline form of naturally occurring carbon. Graphite crucible industry and foundry units are the major consumers of natural graphite. Refractory industries consumes graphite for manufacturing magnesium-carbon bricks for use in steel making in basic oxygen and electric arc furnaces. There are numerous reported occurrences of graphite deposits in the country. The graphite mined is not of high-grade. About 50% of our requirements are met by import. Domestic graphite is mostly used in the metallurgical industry for steel foundry and graphite crucibles. The mineral occurs in lump, flaky or powdery forms which together with the total fixed carbon (F.C.) content, determines the uses. In Rajasthan, graphite occurrences have been located in Ajmer, Alwar, Banswara and Jodhpur districts.

8.2.32 Gypsum

Gypsum, a hydrous calcium sulphate containing 79% $CaSO_4$ and 21% water, is a widely distributed mineral and is generally transparent or white in colour. It occurs both as massive-type or as fibrous lumps and transparent plates. The varieties of gypsum are selenite (finely crystalline), stainspar (fibrous with silky lustre), alabastar (fine-grained, massive and compact) and gypsite-porous gypsum of inferior quality mined with sand and grit. Gypsum is mainly used in Portland cement to retard and control the time of setting and is also used in the manufacture of ammonium sulphate for artificial fertilizer, plaster of Paris, distemper, 'burtonizing' water for beer making, potteries and as a flux in smelting nickel ores. Its uses in paper industry and as filler in insecticides are also important. Selenite is used for making gypsum plates for petrological microscopes. Alabaster is used for ornamental purpose.

Rajasthan accounts for about 90% of total production of gypsum in the country having a total reserves of 1,013.70 million tonnes as estimated by GSI. This constitutes a lions share of the total proven reserves of 1,150 million tonnes of gypsum in the country. Production of gypsum in Reserves of gypsum have been located mainly in Bikaner, Nagaur, Barmer, Jaisalmer, Pali and Ganganagar districts. The most important deposits occur in Nagaur district.

8.2.33 Ilmenite

Ilmenite is an ore from which titanium metal is derived. It is used in the manufacture of pigments. India possesses rich deposits of beach sands of ilmenite mainly along the coast of Kerala and at a few places along the east coast of India. In Rajasthan, ilmenite is found to occur as scattered disseminations in the country rocks in various parts of Ajmer, Nagaur, Pali, Jodhpur and Sikar districts but these are of only academic importance.

8.2.34 Iron Ore

Iron is extracted from its oxide ores such as hematite and magnetite which when smelted with coke and limestone yield metallic iron. Iron ores are needed to meet the demands of modern iron and steel industry. Iron deposits of Rajasthan are of minor importance to meet the demands of any large-scale iron and steel industry within the State. The iron ore deposits are located mainly at Morija, Rampura, Neemla-Raialo, Kotputli and Dabla in Jaipur district; at Tonda, Sior, Zamalpur and Kali Pahari in Jhunjhunu district and at Narda, Manowas, Bagoli and Sarai in Sikar district. These occurrences are found associated with the Delhi Supergroup of rocks.

8.2.35 Kyanite

Kyanite is an important high-alumina refractory mineral. Kyanite, sillimanite and andalusite are a closely related trio of aluminium silicate (A1203. Si02) minerals having the same chemical composition but differing slightly in physical properties. In Rajasthan, occurrence of coarse-bladed crystals of kyanite and sillimanite in small veins and lenses are reported from Ajmer, Bhilwara, Dungarpur, Jaipur, Pali and Udaipur districts.

8.2.36 Lead and Zinc

In India, Rajasthan and Gujarat have been centers for Pb and Zn mining since centuries Zawar, Udaipur many mines bear ancient world Zawar is the place recognised as first ancient smelting site in the world. In south-eastern Rajasthan Pb-Zn deposits o within the Bhilwara Supergroup of Archaean to Lo Proterozoic age (e.g. Agucha, Dariba-Rajpura, Banera), Lower Proterozoic Aravalli Supergroup (zawar) and Lower to Upper Proterozoic Delhi S pergroup (e.g. Kayar-Ghugra). In terms of tonnage a d grade the deposits occurring within the Archaean b sement rocks far exceed those occurring in the main Aravalli and Delhi basins.

8.2.37 Lignite

Lignite, a low-grade fossil fuel, represents an intermediate stage in the alteration of wood into coal. It is dark-brown to blackish-brown in colour carries a good amount of moisture and often crumbles into powder on drying. In Rajasthan, large deposits of lignite occur in Tertiary formations of Middle Eocene age in the Barmer, Jaisalmer and Nagaur basins falling in Barmer, Jaisalmer and Bikaner-Nagaur districts, respectively.

8.2.38 Limestone

Limestone is one of the most important industrial mineral required in the manufacture of lime, cement, chemicals (soda-ash, caustic-soda, bleaching powder, calcium carbide etc.), fertilizer (ammonium nitrate etc.) and as a flux in iron and steel, ferro-alloy and other metallurgical industries. Limestone occurs in almost all the districts of the State, but important deposits are located in Ajmer, Bundi, Chittorgarh, Jodhpur, Nagaur, Pali, Sirohi and Jaisalmer districts.

8.2.39 Megnesite

Megnesite, a hard white mineral, is the chief source of magnesium. Its importance lies in its property to withstand high temperature. It is extensively used in the manufacture of refractory bricks used in steel-making furnaces, in the manufacture of some special types of cement (sorel cement) and in glass and rayon industry. Megnesite deposit near Sarupa-Chhaja in Ajmer district is known to have reserves of about 1.5 million tones of magnesite with an average grade of 50% $MgCO_3$ (24% MgO). Occurrences have also been reported from Bhimana in Pali district and Lewaka-Gurha area in Udaipur district.

8.2.40 Manganese

Manganese finds its main use in the manufacture of pig-iron and steel, Ferro-manganese (a chief ingredient for production of manganese-steel), chemicals and in dry battery cells. India possesses vast resources of manganese ores and is ranked third in the world in the production of manganese. It is second to iron ore in terms of export. In Rajasthan manganese ore deposits of significance occur in Banswara district.

8.2.41 Mica

Mica is one of the important and indispensable minerals used in electrical and electronic industry due to its excellent dielectric strength and insulating properties. India, once enjoyed the status of being a leading mica exporter but with mica-substitution coming in a big way, the export-oriented mica mining industry is facing a great slump forcing many mica mines to close down.

In Rajasthan, five distinct mica-pegmatite belts have been delineated: Beawer- Ajmer belt, Bhilwara belt, Dungarpur-Banswara belt, Tonk belt and Kaunthal belt. Mica is generally of ruby variety but mostly stained and spotted due to mineral inclusions.

8.2.42 Molybdenum

Molybdenum is mainly used as an alloying constituent in steel industry for imparting strength, elasticity and resistance to shock and fatigue. Metal molybdenum finds application in electrical industry where it is used for contact points, thermocouples and electrodes. Molybdenum-tungsten alloy is used as filament support, in incandescent electric lamps, wireless-receivers and transmitter valves. In India the entire requirement of this metal is met by imports. Molybdenum does not occur in nature in its free or native state. In India molybdenum is generally found associated with Pb, Zn and Cu ores and is recovered as by product concentrates. All India recoverable reserves (as on 1.4.93) are placed at 8.4 million tonnes of ore. In Rajasthan molybdenite (the sulphide of molybdenum) is found in the sodalite-cancrinite pegmatite traversing the nepheline syenites at Manderia near Kishangarh in Ajmer district. Minor occurrences of academic interests are also reported from Khandela area of Sikar district. Molybdenum associated with the copper sulphide ores is also reported from the Khetri Copper Belt of Jhunjhunu district.

8.2.43 Nickel

Nickel is an important ingredient in the production of stainless steel. It is also used as a catalyst in making alloys and coins and in the storage of batteries. Nickel-compounds are widely used in electroplating, chemical and ceramic industry. The requirement of the country in respect of nickel is met solely by imports.

8.2.44 Ochre

Ochre are mineral-pigments with iron as the colouring material either in the form of hematite or limonite or both. These earth pigments are insoluble and are used directly as paint material or to give colour, body and opacity to paint, cement, linoleum, rubber, glasses, enamels, plastics etc. The maximum production is reported from Chittorgarh district (69.5%) followed by Bikaner district (17.9%).

8.2.45 Oil and Natural Gas

The Eocambrian Hanseran Evaporite Group (HE of the Marwar Supergroup contains foetid dolomite a dolomitic limestone at four levels below halite beds N 2,3,4 and 5. It is gray to dark and black carbonaceous fine-grained, massive to vuggy dolomite. It stains pap and smells strongly of petroleum.

Foetid dolomite from Lakhasar contains 30.40 m kg of ethane and hydrocarbon, and is considered t source rock for the petroleum. The hydrocarbon migrated and got trapped in structural traps around Devikot-Nachana subsurface high separating at Jaisalmer basin from the Nagaur-Bikaner basin.

8.2.46 Potash

Potash is one of the key elements used in fertilizers. India produces a meager quantity of potash in Gujarat as by-product in manufacture of salt from sea water. In Rajasthan, GSI has established by detailed geophysical investigation aided by extensive drilling, occurrences of subsurface halite-bearing evaporates over an area of 30,000 km z covering parts of Ganganagar, Hanumangarh, Bikaner, Churu and Nagaur districts.

8.2.47 Pyrite – Pyrrhotite *(Sulphur)*

India does not possess any significant deposit of native sulphur. The entire sulphur requirement of the country is met from imports. The uncertainty of obtaining native sulphur in the international market and its exceedingly high prices has necessitated the search for alternate indigenous sources such as pyrite (iron sulphide). The economic value of pyritepyrrhotite lies in its being a source of sulphur. The sulphur as well as the pyrite itself constitutes the principal raw materials for the manufacture of sulphuric acid, which is use mainly in the manufacture of superphosphate fertilizer and also in several other chemical industries. Pyrite-pyrrhotite deposit near Saladipura in Sikar district is of paramount importance.

8.2.48 Rock Phosphate

Rock phosphate, popularly known as Phosphorite, is an essential ingredient in the manufacture of super phosphate, a fertilizer, whose paramount importance in boosting the green revolution in our country needs no emphasis. Rajasthan contributes a major share in the production of rock phosphate in India. In Rajasthan, phosphorites occur in diverse lithological formations ranging in age from Lower Proterozoic to Tertiary. Of these, the discovery of the Precambrian phsophorites in the Aravalli rocks of Udaipur district, in particular, has opened up new vistas in the sphere of exploration of rock phosphate in similar metasediments in India. The phosphorite deposits in Udaipur and Banswara districts in Rajasthan are located within the rocks of the Aravalli Supergroup of Lower Proterozoic age while those of Alwar and Jaipur districts occur in the rocks of Delhi Supergroup of Lower to Middle Proterozoic age and that of the Chittorgarh district occur in the rocks of the Lower Vindhyan of Middle Proterozoic age. Phosphorite occurrences are also reported from the Mesozoic Formations of Jaisalmer.

8.2.49 Salt

Salt is one of the most common substances used in our daily life and is an essential constituent of human food. Salt is mainly prepared by evaporating seawater. Salt also occurs in nature as beds of rock salt, which are in fact, evaporates derived from ancient basins of highly saline seas. In dry regions like Rajasthan, salt is precipitated from saline ground water at the ground water surface in the form as efflorescence. Extraction of salt from efflorescence forms local cottage industries. Considerable quantities of common salt are prepared from the brines of saline lakes in Rajasthan. Salt lakes of Rajasthan, vary in characters are important sources of salt production in the country. The chief salt producing areas are in Nagaur, Barmer and Bikaner districts.

8.2.50 Silica Sand (Quartz and Glass Sand)

Quartz is an important constituent of almost all rock-forming minerals. Free silica occurs in forms such as pure quartz (crystalline or massive), opal, agate and chalcedony. The main source for silica sand and quartz are sandstones, quartzite, pegmatite, quartz veins etc. In Rajasthan, glass sand occurrences suitable for utilisation in glass industries are known from Jaipur, Bundi, Dausa, Sawai Madhopur, Tonk, Jaisalmer, Bikaner and Barmer districts. The other use of silica sand is in acidic Remi-mass in foundry units.

8.2.51 Silver

Silver is valued as a medium of exchange and also for its use in jewellery. In recent years, its application in industry has increased rapidly. In Rajasthan silver is associated with the Pb-Zn ores of Zawar, Rajpura-Dariba and Bharak areas of Udaipur, Rajsamand and Bhilwara districts respectively and is extracted during the smelting of Pb from the Pb concentrates.

8.2.52 Siliceous Earth

Siliceous earth composed mainly of colloidal silica and other siliceous matters contains chiefly 69-81% SiO_2 and 4-12% Al_2O_3 with very little CaO, MgO and iron oxides. It is homogeneous, porous, easily breakable and light in weight. It has high absorption power. Its main use is as filters and refractory silica bricks. Siliceous earth occurrences are known from Barmer and Jaisalmer districts.

8.2.53 Steatite

Steatite, talc and soapstone are the three terms used in trade for a hydrous silicate of magnesium. A purer variety of compact and massive talc is steatite. Rajasthan is the prime producer of soapstone in India. Rajasthan with 6.3 million tones of reserves accounts for about 41% of the all India total reserves of all categories. It produces about 86% of the national output. Occurrences of soapstone have been reported from Jaipur, Bhilwara, Dausa, Sawai Madhopur, Udaipur, Dungarpur and Banswara districts.

8.2.54 Tin

Tin is recovered from its oxide cassiterite. Metal tin is widely used in alloys and in the chemical industry. Tin oxide is used as polishing material for marble and granite. India does not possess workable deposits of tin and the entire requirement is met by imports. In Rajasthan, Bhilwara district holds some promise of holding a deposit.

8.2.55 Tungsten

Tungsten is a metal of strategic importance. It is used in the manufacture of high-speed alloy and special steels, and is essential for machine tools, Armour plates and other military equipment. It is also used in various electrical appliances. Tungsten carbide, much harder than the hardest steel, is used in making drill bits, cutting tools etc.

The best known deposit of the country is that of Degana in Rajasthan.

8.2.56 Uranium

Uranium is a valuable metal for atomic energy and holds important place both in war and peace.

Occurrences of uranium ores are known to occur at several places in Sikar, Bhilwara and Udaipur districts.

8.2.57 Vermiculite

Vermiculite, yellow and brown in colour, is an alternation product of mica and retains the micaceous cleavage. When vermiculite is heated, its volume increases considerably (12 times or more). The mineral is used where some extreme lightness and low-heat conductivity are required.

Deposits of vermiculite have been recorded near Gudas in Ajmer district'

8.2.58 Wollastonite

Wollastonite, a calcium metasilicate ($CaSiO_3$) is one of the latest to enter in the field of industrial minerals, due to its fast growing industrial applications. It is used in the ceramic, enamel, glass, matches, paints, paper, plastics and plywood industries; in the preparation of art ware, ceiling tiles, floor tiles, insulators, as an extender of short-fibred asbestos or as a replacement for non-fibrous materials. Rajasthan is the only wollastonite producing State in India. Large deposits of wollastonite have been located at Pali, Sirohi, Udaipur and Ajmer districts in Rajasthan.

8.3 STATUS OF RAJASTHAN IN OCCURRENCE OF MINERAL AND ITS PRODUCTION

The status of Rajasthan with regard to category-wise recoverable reserves of minerals and production of minerals is shown in following Table 8.1 and Table 8.2.

Table 8.1 Category-Wise Recoverable Reserves of Minerals in Rajasthan (million tones)

Minerals	Rajasthan			
	Proved	**Probable**	**Possible**	**Total**
Apatite	0.006	0.030	-	0.036
Asbestos	0.265	0.556	1.040	1.861
Barytes	0.004	0.726	1.628	2.358
Bauxite	-	-	0.318	0.318
Bentonite	33.452	107.538	136.116	277.107
Build.stone:				
(a) granite		0.007	-	0.007
(b) marble	0.048	3.172	733.707	736.927
Calcite	0.0403	10458	0.368	2.229
Copper				
(a) ore	24.327	26.037	38.972	89.336
(b) metal	0.204	0.268	0.436	0.908
Dolomite	6.391	84.253	44.777	135.421
Feldspar	231.844	305.476	9239.754	9867.074
Fire clay	10.817	4.812	198.438	214.068
Fluorite	0.412	0.218	0.329	0.959
Fuller's earth	-	0.300	189.700	190.00
Garnet	0.010	0.039	5.110	5.159
Graphite	-	0.048	0.421	0.469
Gypsum	13.650	37.788	430113	93.952
Ilmenite	-	-	-	-
Iron	-	6.000	3.000	9.000
Kaoline and other clay	11.066	13.974	189.829	214.875
Kyanite and sillimanite	-	-	.001	.001
Pb+ zn(ore)	64.819	30.893	75.913	171.625
Lignite	-	-	-	1020.000

Minerals	Rajasthan			
	Proved	Probable	Possible	Total
Limestone	1,310.579	3272.122	5210.568	9793.269
Megnesite	1.336	1.353	2.798	5.487
Manganese	-	-	0.409	0.409
Nickel (ore)	-	-	-	-
Ochre	1.175	3.166	6.099	10.980
Potash	20.398	-	-	20.398
Pyrites	10.900	21.250	2.200	34.350
Rock phos.	59.966	6.848	6.925	73.739
Silver(ore)	52.203	49.101	31.418	132.723
Quartz+ silica sand	20.632	30.620	85.748	136.999
Soapstone	5.317	8765	30.250	44.332
Tin :				
(a) ore	-	-	-	-
(b) metal	-	-	-	-
Tungsten	-	0.722	12.669	13.393
Wo3	-	0.000624	0.058502	0.59426
Vermiculite	-	0.02600	0.004278	0.006878
Wollastonite	0.545	1.551	0.964	2.791

Source: IBM Indian Minerals Year Book,Vol.2, 1998, p.222-614

Table 8.2. Mineral Production in Rajasthan (In Million tones)

Mineral	Unit	All India rank of Rajasthan	Rajasthan			
			April 97-March 98		April 98-March 99	
			Qty	Value	Qty	Value
ALL MINERALS	-	11	-	7747773	-	8309323
FUEL MINERALS	-	13	-	272000	-	290000
Natural Gas (utilized)	mcm	6	136	272000	145	290000
METALLIC MENIRALS		2	-	3758287	-	4053045
Copper ore	T	2	1471049	844292	1376787	791385
Iron ore (total)	x1000T	8	7	599	7	620
Iron ore (lumps)	x1000T	8	7	599	7	820
Lead concentrate	T	1	48820	471174	51697	464353
Silver	Kg	1	27705	204601	34542	255093
Zinc concentrate	T	1	292235	2237621	349890	2541594
NON METALLIC MENIRALS		2	-	3717486	-	3966278

Mineral	Unit	All India rank of Rajasthan	Rajasthan			
			April 97-March 98		April 98-March 99	
			Qty	Value	Qty	Value
Phosphorite	T	1	952732	1282726	1407273	1422137
Asbestos	T	1	24820	3504	19109	3072
Ball clay	T	1	370669	43869	314144	13539
Barites	T	2	3431	1692	4252	1658
Calcite	T	1	50373	17818	57629	18819
Chalk	T	1	15	2	0	0
Dolomite	T	9	7190	600	35253	2843
Feldspar	T	1	60069	6349	57346	5789
Fire clay	T	1	106090	16517	77259	10241
Fluorite (graded)	T	2	1205	4525	1100	4128
Garnet (abrasive)	T	2	92	51	349	172
Garnet (gem)	Kg	1	742	70	889	83
Graphite (R.O.M)	T	4	1061	298	0	0
Gypsum	T	1	2179978	309322	2302530	304432
Jasper	T	1	6118	1483	5592	1063
Kaolin (total)	T	4	156704	15320	153880	17727
Kaolin (natural)	T	1	156704	15320	153880	17727
Kyanite	T	3	538	242	211	119
Sillimanite	T	3	49	49	106	103
Limestone	x1000T	3	15920	1568077	17969	1755083
Megnesite	T	4	6	6	578	504
Mica crude	T	3	23	723	18	453
Mica waste and scrap	T	2	247	0	341	0
Ochre	T	1	282256	141469	313233	12575
Pyrophyllite	T	5	3883	2473	1133	273
Quartz	T	2	60827	6005	46355	4653
Quartzite	T	3	0	0	2522	253
Silica sand	T	3	163406	25644	153645	18790
Steatite	T	1	401643	300704	388176	260447
Vrmiculite	T	4	16	1	0	0
Wollastonite	T	1	97743	92330	94642	85451
Laterite	T	4	65112	2917	87656	3889

Source: IBM, 1999. Monthly Statistics of Mineral Production Vol. 31, no. 3, March 1999

8.4 MINERAL DEVELOPMENT

The National Mineral Policy announced in 1993 has opened the doors for both domestic and foreign direct investment in the Indian Mineral Industry. This is expected to give a fillip to the domestic mineral industry in furthering its growth and making the entire gamut of minerals and mineral based products domestically and internationally competitive. Sweeping changes in the national Mineral Policy has been recently made by Government of India making it adequately liberalised investor-friendly through amendments to MMRD act (now MMDR act). The economic liberalisation, which has been followed by promulgation of the new economic policy and subsequent General Agreement on Tariffs and trade (GATT) pose a new challenge before the nation's mining sector. The mining operations in the State are carried out either by opencast method or buy underground method. The underground mining in the State is mostly in the organised sector where public sector organisations like M/s Hindustan Zinc Limited. M/s Hindustan Copper Limited, M/s RSMM Limited and M/s RSMDC are operating where due consideration is given to the protection of environment. However in the open-cast mining that too, mostly in stone mining, by and large are in unorganised sector, are discriminately dumping the waste and are not maintaining the proper environmental problem of stone industry in the state is waste generation (both during quarrying and processing).

8.4.1 Fuel Minerals

The State Directorate of petroleum has identified large of lignite at depths varying from 200-600 m. in Barmer-Sanchore basin. Recently a team from Australia namely M/s Line Energy has shown interest to develop Underground Coal Gassification (USG) project in koslu area of Barmer district. The State Government has also undertaken Randd project for Coal bed Methane (CBD) in Barmer and Bikaner districts. New Delhi has sanctioned Rs.4027 Crores for drilling. The exploration works has been entrusted to M/s Mineral exploration Corporation Ltd. (MECL). Recently, Government of India has released 7 blocks of Coal Bed Methane for competitive bidding in which one blocks of Rajasthan (north of Barmer-Sanchore basin) has been included. Although lignite deposits have been proved at 33 locations, favourable geological environment spread over a large area in parts of Bikaner, Nagaur and Barmer districts provides further scope for lignite exploration both shallow as well as deeper levels.

8.4.2 Gem Minerals

A long belt has been identified for exploration of emerald in parts of Udaipur, Rajsamand and Ajmer districts. Emeralds of Rajasthan has a good market value. Besides, State also has potential of semi-precious stones like Garnet, Aquamarine and precious stone like Diamond.

8.4.3 Lime Stone

This mineral has wide ranging applications and it has been located over 26 districts of the State. About 9000 MT reserves have been established. At present 15 major and 4 medium cement plants with an installed capacity of 16 MT of cement per annum are already in operation and about 10 cement plants each of over 1 MT capacity are likely to come up in near future on the areas already explored by the department in various districts. Apart from cement industry, the State is also producing large quantity of SMS grade limestone in Jaisalmer districts which otherwise would have been imported.

8.4.4 Arial Prospecting for Basemetals and Noble Metals

With the liberalisation of economy, the department look lead in inviting foreign collaboration in the mineral sector. Rajasthan is the first State in the country to grant reconnaissance permits over large areas for aerial survey for basemetals. In all 54 permits have been sanctioned to multinational companies covering almost 74000 sq. km area. The data generated by these surveys will be very useful for planning detailed ground prospecting for noble metals and basemetals.

8.4.5 New Prospecting Policy

For promoting scientific mining, bringing transparency, the State Government has farmed separate policies for granite, marble and gypsum.

8.4.6 Mineral's Waste

The dimensional stone industry has high percentage of waste generation. Solid stone waste, powder and slurry are the major source of environmental degradation in the mining belts of marble, sandstone, granite, flaggy limestone, state stone etc. The marble waste generation varies widely from 30 per cent by weight in mechanised quarrying to 65 per cent by weight in manual mining. In granite mines, waste generation is even on higher side.

8.4.7 Reclamation of Mined Out Land

Reclamation of the broken land concurrent with the excavation is an integral part of a properly planned and implemented mining strategy. The Indian Bureau of mines has laid down regulations in this regard and every lessee has to submit short term and long term plans for the land reclamation before commencing the mining activities. The IBM monitors the implementation of the proposed plan periodically. The large scale mines owners and the government undertaking like M/s HZL, HCL, RSMML, RSMDC, Wolkem (India) Ltd., ASD, ASI, ACC and the major cement looks plants have a functional environmental management department. which looks after land reclamation in their respective mining leases. The mines and geology department in collaboration with the District Collector and District Industries Department has initiated the work of reclamation. The department has carried out a study to suggest measures for land reclamation in the Kotastone mining belts of the Kota and Jhalawar district using remote sensing data.

8.4.8 Dumping of Marble Slurry

A large quantity of marble slurry is being generated by the thousands of the marble cutting and polishing units installed in various parts of the State. Unplanned disposal of this slurry results in chocking of the natural drainage, corruption of the fertile top soil and silting in the water bodies, besides several other deleterious effects to the natural environment like air pollution. In districts like Udaipur and Rajsamand, the District Collectors have identified lands(s) for dumping marble slurry in the industrial areas. The mineral based industry owners also have a growing awareness to pour this fluid waste to dry up in their factory premises and the resulting solid waste is being dumped by them in the nearby low lying areas. There is a need to identify such lands in all the industrial areas where stone cutting and polishing units are under operation.

Structural use of Marble Powder Slurry: At present, 7-8 hundred marble cutting industries in Rajasthan are producing 2000 MT of marble powder slurry per day as waste product. Set up of more such industries is increasing at a fast rate. The slurry is being dumped in open and big heaps and dumps lying adjacent to the road side are spread in several Km. Disposal of marble powder slurry is a big problem, causing serious environmental hazard and pollution problem. Malaviya National Institute of Technology (MNIT), Jaipur has studied the physical and chemical properties of marble powder slurry and concluded that marble powder slurry can be used for preparing compressed bricks, tiles and decorative tiles. To disseminate the outcome of the study, a workshop was organised in August, 1998 which was attended by the number of entrepreneurs, scientists and concerned officers of various departments.

8.4.9 Studies for the Solid Waste Management

The biggest challenge in the environment management of the mining belts in Rajasthan involves waste rock disposal. An unprecedented growth in the mining activities in last two decades has resulted in generation of huge quantities of waste rock in the mining belts, most of which is being dumped at the quarry-head or in the close proximity, without any scientific considerations. This cam be potentially hazardous from both environmental and mineral conservation points if view. The unplanned and unscientific dumping of the waste rock results in disturbing the regional environment and natural water regimes. An additional problem is posed by the land degradation by the mine waste, which at places, is dumped directly over the cultivable lands. Moreover, in absence of adequate dumb-stabilisation measures, the waste rock spills over the cultivable land and the slurry generated by the rainwater deposit fine-grained mud over the soil in the vicinity of the waste dumps. This results in disturbing the chemical and physical balance of the natural soil. Expedient dumping of the waste rock over the quarry head leads to repeated handling of the waste rock, and occasionally the process becomes too expensive to mine out the mineral economically, leading to the wastage of economically viable minerals.

The problems are especially acute in the dimensional and decorative stones mining belts, where the waste rock problem is exaggerated by the small size of the leasehold areas, which gives the mines owners very little space for systematic dumping of the mining waste. In view of the above stated problems, the departmen carried out several studies using remote sensing data in ecologically the wors affected mining belts, viz. the marble mining belts of the Rajsamand district limestone (Kota stone) mining belts of the Kota and Jhalawar districts and the sandstone mining belt of the Jhalawar districts. In these belts, the remote sensing data were interpreted to prepare land-use maps, which were subsequently used for identifying non-mineralised fallow land use in vicinity of the mining areas, which could be used for dumping the mining waste. Several such areas have been tentatively identified in the mining belts of the Rajsamand district, Work is now continuing using GIS techniques for delineating such areas. Similar studies have been carried out in the Bijoliya sandstone mining belts also. However, in these parts dumping in the mined out parts of the quarries is not possible because of almost horizontal dips of the sandstone horizon. In this belt, the non-mineralised wastelands need to be identified in the vicinity. However, the waste rock in this belt usually comprises fractured sandstone, which can be used as building stone and road material.

8.4.10 Afforestation in Mining Clusters

The department has established an environment and development cell in the Directorate, Udaipur in August, 1995. Since then, all the mining officers in different districts of the state have been instructed to encourage and monitor afforestation and plantation activities in all the mining leases falling their jurisdiction. This has yielded encouraging results and the leases have started plantation and afforestation activities in their leases.

8.5 NON-CONVENTIONAL SOURCES OF ENERGY

State Government created a autonomous agency in 1985 with the name of Renew-able Energy Development Authority (REDA), since then most of the REDA programmes were only subsidy programmes though REDA installed variety of systems but their long term impact was not seen. From 1999 onwards agency modified its approach and an effort was made to off-load the subsidy and make these programmes people's programme. The State of Rajasthan has a strong resource of Non-conventional Energy comparing the lesser availability of electricity, coal, petroleum products and other energy sources. State has adopted a policy with a view to promote generation of grid-grade power through Non-conventional energy sources in the context of non polluting renewable and decentralised availability of inputs like solar, wind, mini-small hydel, biomass and other wastes. However, in view of immense potential of power generation through exploitation of wind resources in Rajasthan and special incentives required because of special nature of wind as resource, it has been felt necessary to adopt a scientific promotion policy for generation of power by use of wind resources. The objectives of this policy is to support a programme of creating 100 MW power generation capacity based on exploitation of wind resource in the state, largely in private sector.

8.5.1 Wind Power Generation

Rajasthan Energy Development Agency had carried out monitoring of wind in past 13 years under a MNES sanctioned project in association with Indian Institute of Tropical Methodology, Bangalore. It has been assessed that State has a gross potential of 1200 MW power generation from wind with the available technology, out of which 400 MW can be harnessed immediately. Agency has already installed 2 MW wind power demonstration farm at Jaisalmer, 2.25 MW at Devgarh and 2.10 MW Phalodi with the help of MNES. In private sector four wind farms has been set up with a total capacity of 9.76 MW. It is expected that by 31.3.2004, State will be able to achieve its target of 100 MW power from wind with capital investment from private sector. Agency has received proposals for projects for a combined of 237 MW from available industry in the wind energy sector. Investments from profit making companies avails 100 per cent depreciation from its profits IT Act.

8.5.2 Solar Photovoltaic Power Plants

Solar Photovoltaic technology is being utilised in the State for providing stand alone domestic_and street lighting systems in the rural areas and have proved to be very successful as this has obviated drawings long transmission/distribution lines which would have called for massive investment. MNES provides 50 per cent of financial assistance for the projects. The main advantage of grid interactive power plants being their maintenance free operating capacity. Three grid interactive power plants of 55 Kwp capacity have been installed so far.

8.5.3 Biomass Power Projects

There are now technologies available where surplus biomass such as mustard stalk, rice husk, wheat stalk and other agro wastes can be run power plants. REDA has got studies done for such availability of biomass in ten blocks in the State. It is estimated that power plants based on biomass totalling to a capacity of about 300MW can materialise. REDA have already received applications from private parties proposing power plants based on biomass totalling to a capacity 30 MW. Ten more blocks have been taken up for assessment of surplus biomass.

8.5.4 Micro Hydel Projects

MNES has now designed small hydel projects also as non-conventional energy sources based power projects. State Government has authorised RSPCL for executing projects based on non-conventional Energy Sources, to facilitate setting up of following small hydel projects:-

1. Jakham Dam (Chittorgarh)
2. Alina RMC at head (Kota)
3. Bisalpur dam RMC (Tonk)
4. Som kamala Amba, RMC head (Dungarpur)

There are other places also on the IGNP where suitable water falls are available for putting up small hydro power stations. It is estimated that if all such falls on the canals in the State are utilised a further 10MW electricity generation capacity can be added and local area can be served. At such falls small capacity plants can be taken up with short gestation period.

8.5.6 Solar Power Project

140 MW Integrated Solar Combined Cycle Power at Mathania is in the advanced stage of implementation with an investment of Rs. 871.74 Crores. $ 45 million GEF grant from World Bank will be available as 100 per cent grant for the project. It is expected that Ministry of non-conventional Energy Sources. Government of India and Government of Rajasthan will provide Rs. 50 Crores each as equity. M/s KFW, Germany have completed the appraisal of the project. Loan agreement in respect of KFW composite loan of dm 250 Million has been signed on 29.10.2001 between Government of India and KFW Germany, Besides this, Global Environment Fund (GEF) is also providing $ 7.50 lac for conducting various studies and technical guidance. It is expected to finalise EPC cum O and M contract upto August/ September, 2002. The ground work at site may start in the mouth of October, 2002.

8.5.7 Promotional Programme for Consumer Products

With the modified strategy of creating mass awareness about promotion of non-subsidised non-conventional energy consumer devices. REDA has found a huge market for promotion of varied products some can be listed as Solar Photovoltaic Home Lighting System, Solar Water Heating System, Solar Cookers, Solar Lantern, SPV Pumps, Compact Fluorescent Lamp, Pressure Cookers, Nutan Stoves etc. Agency has authorised about 180 Aditya Solar Shops in private sector who are doing necessary supporting work for demonstration off the shelf availability, sale, repair and providing guarantees for customer satisfaction without any subsidy.

Chapter - 9

HUMAN AND LIVESTOCK RESOURCES

Advancement in human civilization are marked by improvement of man's role as producer and consumer of resources, progress of civilization marks a gradual shift of his role from the field of physical tail to specialize himself in mental work. By this specialization he builds up culture, invent more effective ways and means they in turn bring him richer harvest of material good and longer leisure. Resource development starts from man, it is implemented by man and it ultimately used for satisfying mans wants. Thus Human resource is most precious of all resources.

9.1 HUMAN RESOURCES

9.1.1 Jurisdictional Changes

After the 1991 Census, the Government of Rajasthan has effected inter-district and inter-tehsil jurisdictional changes by creating 5 new districts and 28 new tehsils. Thus the total number of districts and tehsils has gone up to 32 and 241 respectively. The constitution of new districts is given in Table 9.1 and Figure 9.1.

The 1991 Census data for the new and the affected tehsils has been re-casted to make 1991 and 2001 Census data comparable.

The basic unit for rural areas is the revenue village which has in most of the cases definite surveyed boundaries. The revenue village may comprise several hamlets (Dhanis) but the entire village will be one unit for presentation of data.

There are 38 Census Towns i.e. places considered otherwise rural being villages but treated as urban because of satisfying the above established criterion by them in the state in 2001 Census. District wise details of such Census Towns are as under:

Total No. of Census Towns in 1991	29
Total No. of Census Towns declassified in 2001	07
Total No. of Census Towns merged in 2001	01
Total No. of Census Towns in 2001	38
Number of Census Towns newly created	17

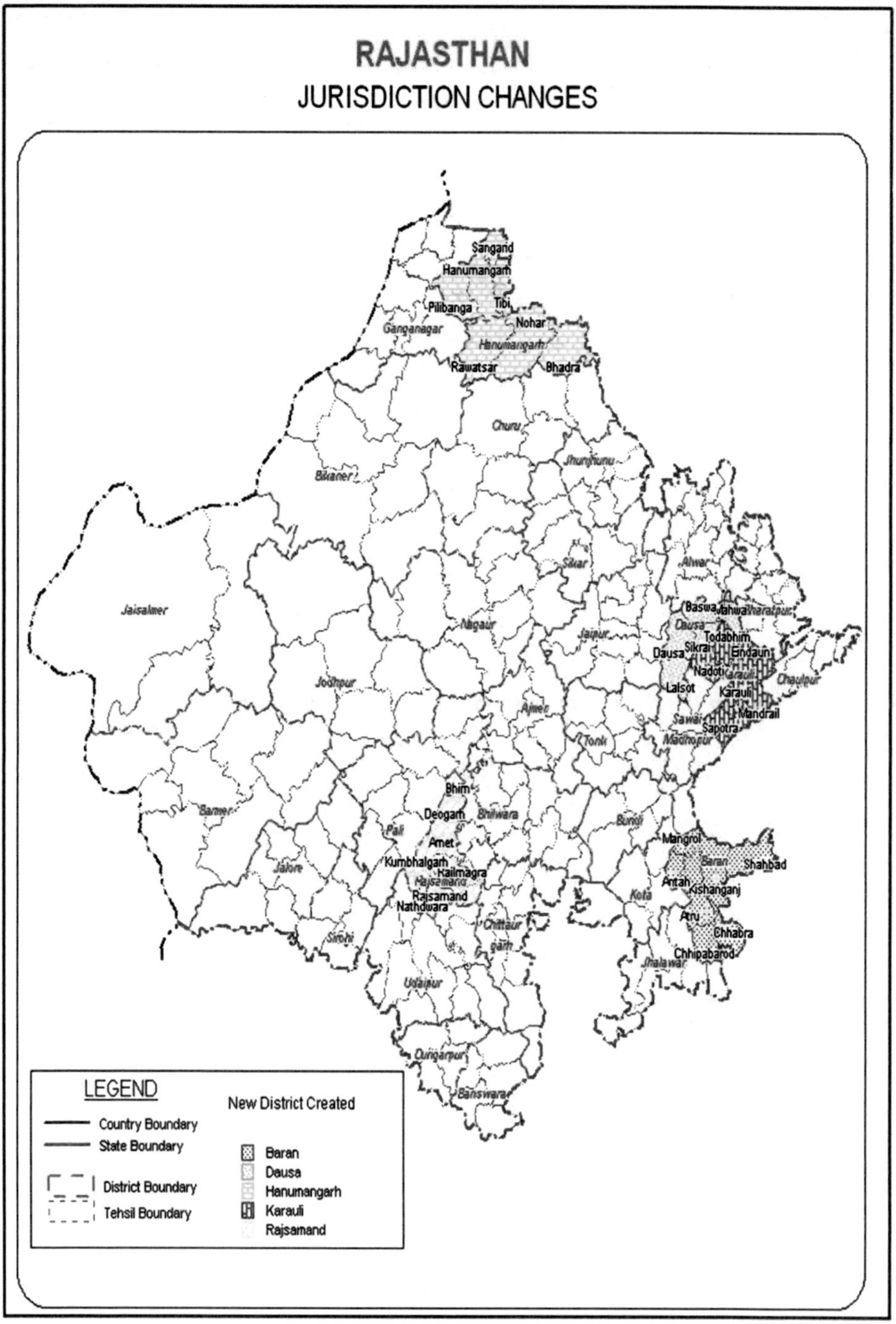

Figure 9.1. Jurisdiction changes (Source: Census of India)

Table 9.1 Details of Districts created after 1991 Census

S.N.	Names of districts created after 1991	Names of the transferred constituent tehsils	Names of the existing districts in 1991 from which the tehsils transferred to new districts	Date of Gazette notification
1.	Dausa	Dausa	Jaipur	31.3.91
		Baswa	Jaipur	
		Lalsot	Jaipur	4.8.92
		Sikrai	Jaipur	
		Mahwa	Sawai Madhopur	
2.	Rajsamand	Nathdwara	Udaipur	31.3.91
		Railmagra	Udaipur	
		Rajsamand	Udaipur	
		Kumbhalgarh	Udaipur	
		Amet	Udaipur	
		Bhim	Udaipur	
		Deogarh	Udaipur	
3.	Baran	Baran	Kota	31.3.91
		Mangrol	Kota	
		Chhabra	Kota	
		Atru	Kota	
		Chhipabarod	Kota	
		Shahbad	Kota	
		Kishanganj	Kota	
4.	Hanumangarh	Hanumangarh	Ganganagar	5.7.94
		Pilibanga	Ganganagar	
		Tibi	Ganganagar	
		Sangaria	Ganganagar	
		Nohar	Ganganagar	
		Bhadra	Ganganagar	
		Rawatsar	Ganganagar	
5.	Karauli	Karauli	Sawai Madhopur	15.7.97
		Sapotra	Sawai Madhopur	
		Hindaun	Sawai Madhopur	
		Todabhim	Sawai Madhopur	
		Nadoti	Sawai Madhopur	

9.1.2 Population Distribution

According to 2001 Census the total population of Rajasthan is 56,473,122. Out of Total Population 43,267,678 persons were living in rural areas and the rest in urban areas of the state (Figure 9.2). The proportion of the population living in rural and urban areas comes to 76.62 per cent and 23.38 per cent respectively.

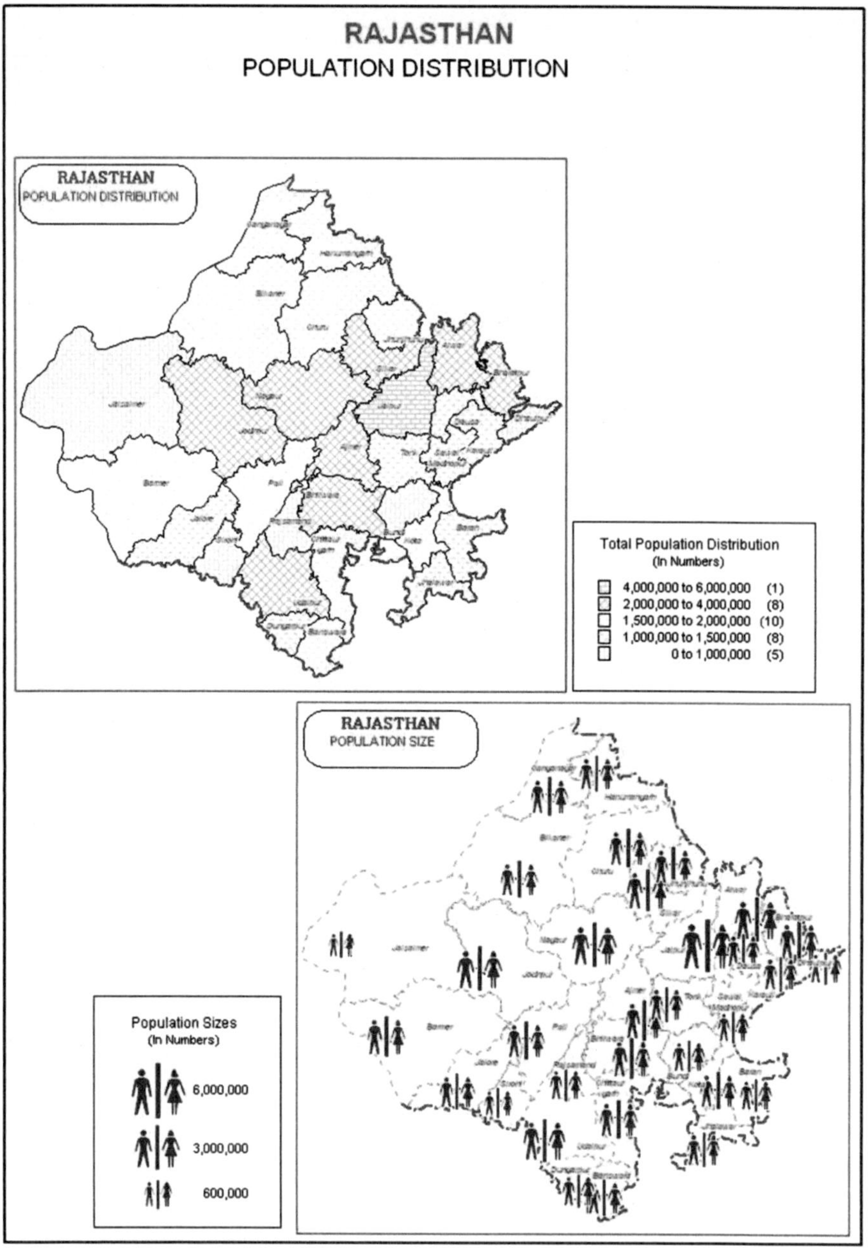

Figure 9.2. Population Distribution (Source Census of India)

9.1.2.1 *Rural-Urban Composition since 1901*

The urban component of population of Rajasthan i.e. the proportion of urban population to its population had been higher than that of India from 1901 to 1951. After that the proportion of urban population of Rajasthan had been lower than that of India. As regards to the trend, the proportion of urban population after declining from 15.06 percent in 1901 to 13.44 percent in 1911, indicated an increasing trend upto 1951 reaching at 18.50 percent. Again, after declining to 16.28 percent in 1961 it registered a continuous increase. As regards to India, after declining from 10.84 percent in 1901 to 10.29 percent in 1911 it has shown a continuous increasing trend which can be seen from the Table given below :

Table 9.2 Rural-Urban Composition of Population in India and Rajasthan

Year	Proportion of Population (%)			
	India		Rajasthan	
	Rural	Urban	Rural	Urban
1901	89.16	10.84	84.94	15.06
1911	89.71	10.29	86.56	13.44
1921	88.82	11.18	85.67	14.33
1931	88.01	11.99	85.28	14.72
1941	86.14	13.86	84.73	15.27
1951	82.71	17.29	81.50	18.50
1961	82.03	17.97	83.72	16.28
1971	80.09	19.91	82.37	17.63
1981	76.69	23.31	78.95	21.05
1991	74.27	25.73	77.12	22.88
2001	72.22	27.78	76.62	23.38

It is correct to visualise in the context of Rajasthan that 'India lives in villages'. Out of 32 districts, the percentage of rural population to total population in 25 districts is higher than that of the state average of 76.62. Kota, Jaipur, Ajmer, Bikaner, Jodhpur, Churu and Ganganagar are the districts where the percentage of rural population to total population is lower than that of state average. Kota is the only district in Rajasthan which can claim to be an urban as only 46.58 percent of population lives in rural areas of the district. Jalore, Barmer, Dungarpur and Banswara are the districts where more than 90 percent of the population lives in villages. On the other hand, Kota, Jaipur and Ajmer are the districts where the proportion of the rural population to total population is less than 60 percent. The proportion of urban population to total population has decreased in 11 districts viz Ajmer, Bikaner, Jodhpur, Churu, Pali, Udaipur, Sirohi, Jaisalmer, Jhalawar, Barmer and Banswara. In other words, the proportion of rural population to total population has increased in these districts. It is very interesting to note that inspite of being Divisional Headquarters and the top cities as their constituents, the proportion of rural population has increased in Ajmer, Jodhpur, Bikaner and Udaipur districts during the decade 1991-2001. Conversely at the urban front, Jaipur tehsil has the highest proportion of urban population of 96.31 percent to its total population in the state. It is followed by Ladpura tehsil (83.49 percent), Jodhpur tehsil (80.40 percent), Bikaner tehsil (75.07 percent) and Ajmer tehsil (73.22 percent). All these tehsils have major cities as their urban constituents.

9.1.3 Decadal Growth of Population

The population of Rajasthan has increased from 10,294,090 in 1901 to 56,473,122 in 2001, registering a population growth of 448.60 percent in 100 years (Figure 9.3). While analysing the growth of population in last 100 years in rural and urban segments of the state, it is found that the growth of population in urban areas is much higher than that of rural areas which stood at 751.60 percent and 394.86 percent respectively. The growth of population in rural areas of the state since 1901 has shown a fluctuating trend. However, it has registered a higher growth rate of 27.49 percent during the decade 1991-2001 as against 25.46 percent registered during 1981-91. Similarly it has shown a fluctuating trend in urban areas also since 1901. It registered the highest growth rate of 58.69 percent during the decade 1971-81. Contrary to the growth rates in rural areas it has shown a declining trend from 39.62 percent registered in 1981-91 to 31.17 percent during 1991-2001. Decline in the decadal growth rate of urban population is mainly attributed to declassification of 7 Municipalities into Gram Panchayats.

Table 9.3. Percentage decadal variation of population from 1901 to 2001 by Rural-Urban Composition in Rajasthan

Year	Total	Rural	Urban
1901	-	-	-
1911	+6.70	+8.74	-4.83
1921	-6.29	-7.26	-0.03
1931	+14.14	+13.63	+17.21
1941	+18.01	+17.25	+22.43
1951	+15.20	+10.80	+39.59
1961	+26.20	+29.65	+11.04
1971	+27.83	+25.77	+38.47
1981	+32.97	+27.47	+58.69
1991	+28.44	+25.46	+39.62
2001	+28.33	+27.49	+31.17
1901 to 2001	+448.60	+394.86	+751.60

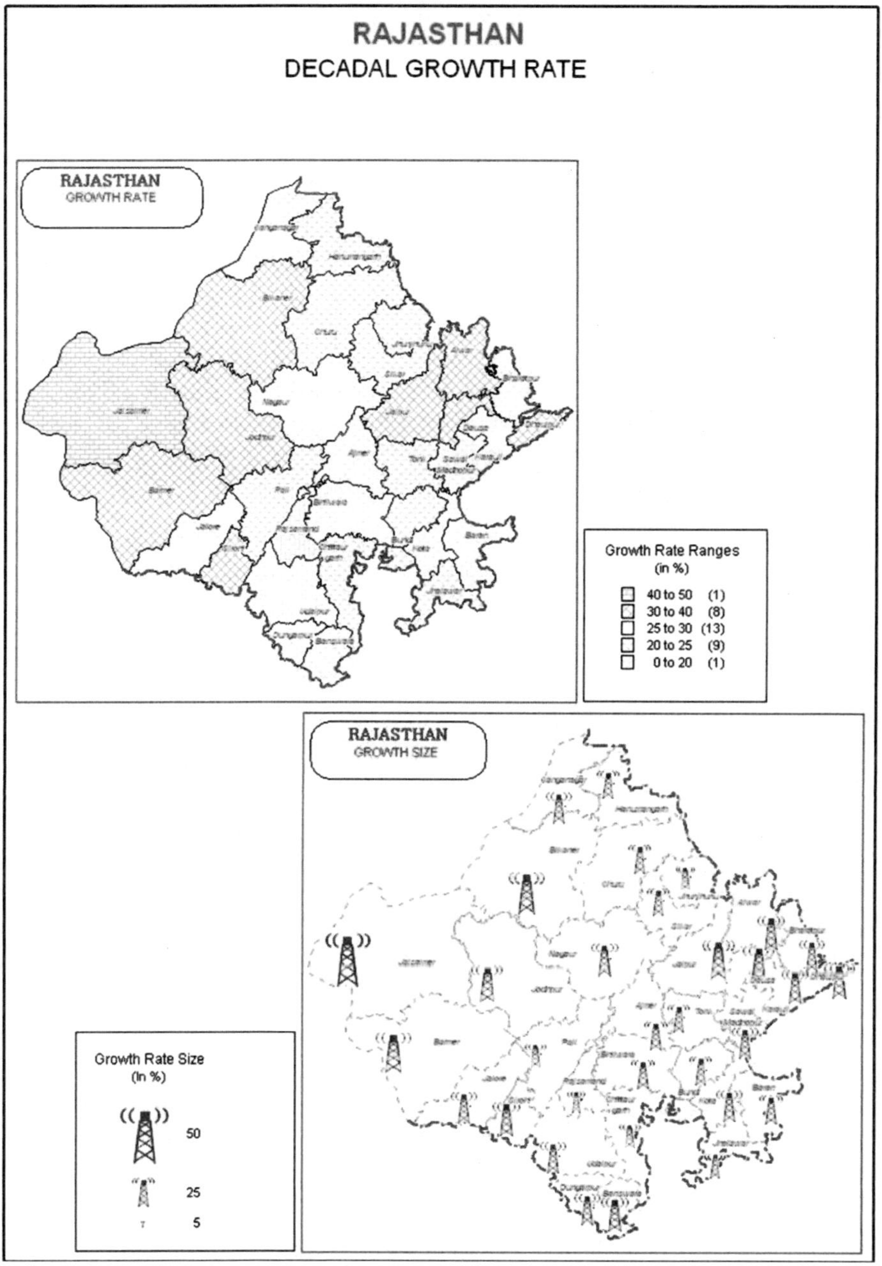

Figure 9.3. Decadal Growth Rate (Source : Census of India)

9.1.3.1 *Percentage Decadal Growth at District level*

Jaisalmer, Bikaner, Barmer, Jaipur and Jodhpur are the districts which have attained the highest percentage growth rates of 47.45, 38.18, 36.83, 35.10 and 33.77

respectively during the decade 1991-2001. Whereas, contrary to it, Rajsamand, Jhunjhunu, Chittorgarh, Pali and Jhalawar are the districts which have the lowest percentage growth rates of 19.88, 20.90, 21.46, 22.39 and 23.34 respectively during the same period.

9.1.3.2 *Percentage Decadal Growth in Rural areas*

The state has registered percentage decadal growth rate of 27.49 during the decade 1991-2001. There are 11 such districts which have attained higher growth rates and the remaining 21 have attained lower than that of state's growth rate. The following table shows the districts which have registered higher/lower growth rates than that of 27.49 i.e. average of the state. Highest growth rate in rural areas have been attained by the districts adjoining to international border i.e. Jaisalmer, Bikaner and Barmer. Apart from these, Jodhpur and Sirohi have also attained higher growth rates. On the other hand, the five districts which have recorded the lowest percentage growth rates in rural areas are Rajsamand, Jhunjhunu, Chittorgarh, Kota and Hanumangarh.

Table 9.4a First five and last five districts of Rajasthan as per their rural decadal growth rate.

First Five Districts		Last Five Districts	
Jaisalmer	(47.99%)	Rajsamand	(18.38%)
Bikaner	(47.82%)	Jhunjhunu	(20.74%)
Barmer	(40.85%)	Chittorgarh	(20.84%)
Jodhpur	(37.41%)	Kota	(21.02%)
Sirohi	(32.97%)	Hanumangarh	(21.08%)

Table 9.4b First five and last five districts of Rajasthan as per their Urban decadal growth rate.

Highest			Lowest		
1.	Jaipur	(46.17)	1.	Barmer	(0.75)
2.	Jaisalmer	(44.54)	2.	Jhalawar	(11.43)
3.	Baran	(40.73)	3.	Sirohi	(18.15)
4.	Sawai Madhopur	(40.17)	4.	Churu	(20.10)
5.	Hanumangarh	(39.33)	5.	Banswara	(20.32)

9.1.3.3 *Percentage Decadal Growth in Urban areas*

As compared to the decadal growth rate of 27.49 in rural areas of the state, a higher growth rate of 31.17 in urban areas has been recorded during 1991-2001. At the district level, 14 have recorded higher decadal growth rate than that of state average whereas 18 have recorded lower rate. The following are the five districts each having highest and lowest decadal growth rates in urban areas. Jaipur district has the highest decadal growth rate of 46.17 in urban areas of the state whereas Barmer district has recorded the lowest growth rate of 0.75. Jaisalmer is the district which has higher growth rates in both rural and urban areas of the state as compared to other districts.

9.1.4 Sex Ratio

The sex ratio is defined as the number of females per thousand males. In Rajasthan, as in other parts of the country the sex ratio has been adverse to females i.e. the number of females per thousand males is significantly low. As per the provisional results of Census 2001, 922 females have been recorded per thousand males in Rajasthan (Figure 9.4). A look at the sex ratio recorded for India and Rajasthan over a period of last 100 years shows that it has always been on a lower side in Rajasthan as compared to that of India.

Table 9.5. Sex ratio in India and Rajasthan 1901-2001

Year	1901	1911	1921	1931	1941	1951	1961	1971	1981	1991	2001
India	972	964	955	950	945	946	941	930	934	927	933
Rajasthan	905	908	896	907	906	921	908	911	919	910	922

Table 9.6. Sex ratio for Rural-Urban areas at state level for 1901-2001 (Rajasthan)

Census year	Total	Rural	Urban
1901	905	898	947
1911	908	904	936
1921	896	896	897
1931	907	908	903
1941	906	907	897
1951	921	919	958
1961	908	913	882
1971	911	919	875
1981	919	930	877
1991	910	919	879
2001	922	932	890

The above table shows that the sex ratio in urban areas was higher than that of rural areas upto 1921. After that, the sex ratio in rural areas was higher than that of urban areas till date except in 1951.

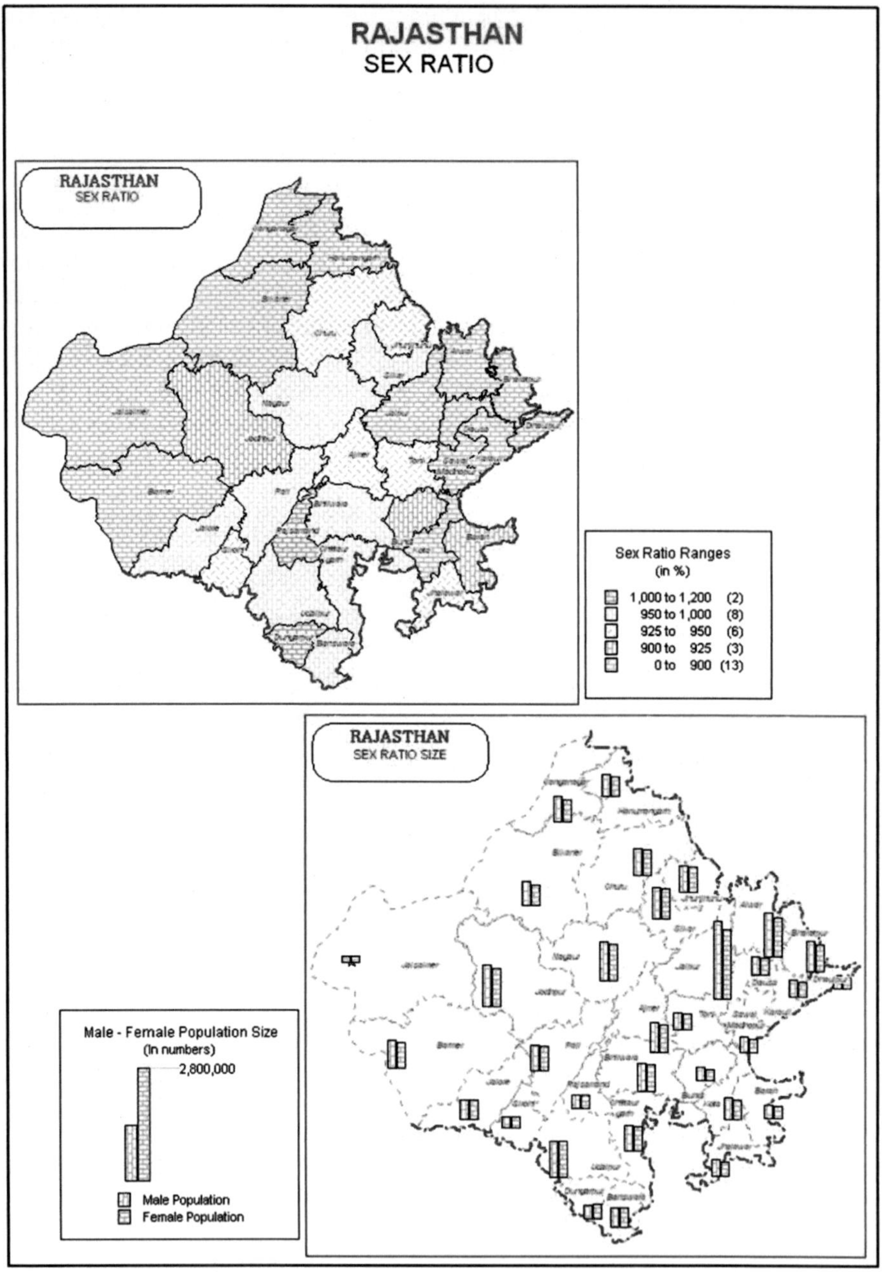

Figure 9.4. Sex Ratio (Source: Census of India)

9.1.4.1 *Sex ratio in Rural and Urban areas at District level*

The sex ratio for rural and urban areas of Rajasthan has been recorded as 932 and 890 respectively in 2001. The sex ratio at district level in Census 2001 is universally higher in rural areas of all the districts than in urban areas except in Bharatpur, Dholpur, Karauli, Tonk and Baran districts. The sex ratio in the rural areas of the districts varies from 821 in Dholpur district to 1,035 in Dungarpur district. Rajsamand (1,013) and Pali (1,003) are also the districts besides Dungarpur of which the sex ratio favours females. In 17 districts sex ratios in rural areas are lower than that of state's average of 932. None of the districts in the state is such where sex ratio favours females in the urban areas. The sex-ratio in urban areas of the districts of Rajasthan varies from the lowest of 764 in Jaisalmer to the highest of 945 in Tonk. In urban areas of 15 districts of the state, sex ratio is lower than the state average of 890.

The five districts each for highest and lowest sex ratio in rural and urban areas are shown to the following Inset Table 9.7.

Table 9.7 Highest and Lowest Sex Ratios in rural and urban areas of Rajasthan.

Rural		**Urban**	
Highest	**Lowest**	**Highest**	**Lowest**
1. Dungarpur (1035)	Dholpur (821)	1. Tonk (945)	Jaisalmer (764)
2. Rajsamand (1013)	Jaisalmer (832)	2. Churu (936)	Alwar (838)
3. Pali (1003)	Bharatpur (854)	3. Dungarpur (935)	Ganganagar (848)
4. Udaipur (988)	Karauli (855)	4. Banswara (930)	Dholpur (857)
5.. Banswara (982)	Ganganagar (881)	5. Rajsamand (927)	Barmer (858)

9.1.4.2 *Sex ratio at Tehsil level*

All the 241 tehsils of the state have been classified into 5 ranges of sex ratio. These ranges are (i) Below 850 (ii) 850 to 899 (iii) 900 to 949 (iv) 950 to 999 and (v) 1,000 and above. While classifying these tehsils into above five ranges of sex ratio we find the maximum number of tehsils i.e. 94 are those which have sex ratio in between 900 - 949.1. It is followed by 63 and 39 in the ranges 950-999 and 850-899 respectively. It is a matter of great satisfaction that 29 tehsils in the state are such of which sex ratio favours females i.e. above 1000. Sex ratio varies from 788 in Jaisalmer tehsil of Jaisalmer district to 1,077 in Aspur tehsil of Dungarpur district.

9.1.5 Child Population (0-6 Age Group)

Child population in the age-group 0-6 constitutes an important segment of the population structure of the society. It helps to compute the effective literacy rate at all levels, be it at National or State or district or city/tehsil level etc. It may also be a trend setter for future growth of population.

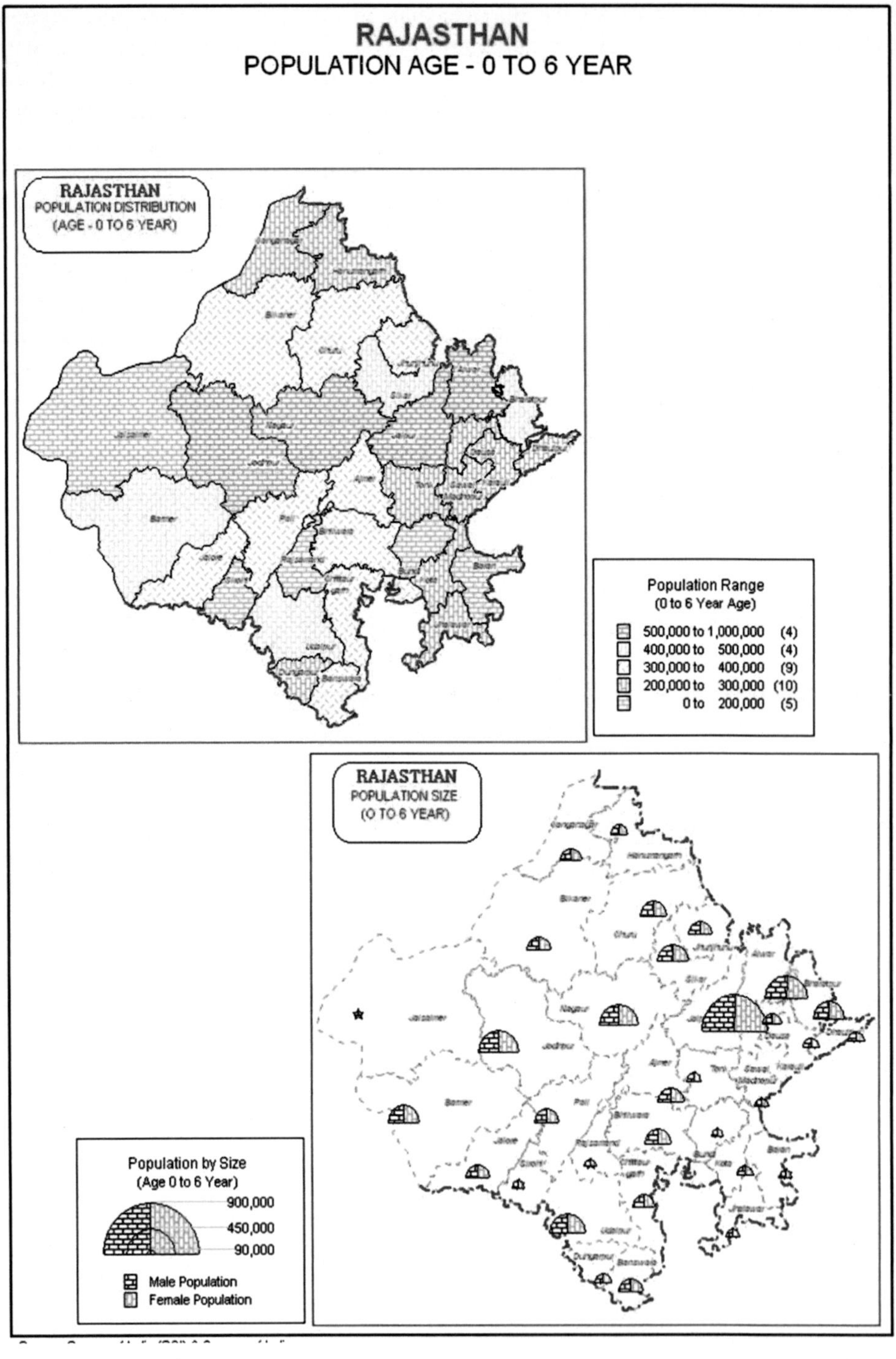

Figure 9.5. Population Age 0 to 6 years (Source: Census of India)

Table 9.8 Population in the age-group 0-6 in Rural and Urban areas by sex in 1991 and 2001

T/R/U	1991			2001		
	Persons	**Males**	**Females**	**Persons**	**Males**	**Females**
Total	8,859,492	4,622,750	4,236,742	10,451,103	5,474,965	4,976,138
Rural	7,013,109	3,655,361	3,357,748	8,450,272	4,414,223	4,036,049
Urban	1,846,383	967,389	878,994	2,000,831	1,060,742	940,089

According to 2001 Census, a total number of 10,451,103 children are found in the age-group 0-6 which constitutes 18.51 percent of the total population of the state (Figure 9.5). In rural areas of the state, the proportion of 0-6 population to its total population comes out to be 19.1.53 percent which is significantly higher than that of urban areas which stands at 15.15 percent. If we analyse the proportion of 0-6 population sex-wise, we find that the proportion of males is higher than that of females in both the rural and urban areas of the state as can be viewed from Table 3. Table 3 also presents the percentage of child population in the age group for total, rural and urban areas of each district.

9.1.5.1 *Sex-ratio in 0-6 age group of population*

The sex ratio for the population in the age-group 0-6 has been recorded as 909, as against 922 for total population in Rajasthan. Similar trend is observed for the sex-ratio in the age-group 0-6 for rural and urban areas of the. None of the districts has recorded sex-ratio in this age-group 0-6 population favoring females. The highest sex ratio of 972 in Banswara and the lowest of 852 have been recorded in Ganganagar district. Out of 32 districts, 17 are such where sex ratio in the said age-group is lower than that of sex ratio of total population. Karauli and Sikar are the only districts where sex ratio in age-group 0-6 in urban areas is higher as compared to that of rural areas. This is evident from the fact that the sex ratio in urban areas of Karauli and Sikar in the said age-group stood at 894 and 891 as against 873 and 880 respectively in rural areas.

9.1.6 Literacy

Rajasthan has marched ahead in literacy by achieving literacy rate of 61.03 in 2001 as against 38.55 in 1991 (Figure 9.6). Rajasthan has got a distinction in achieving the highest decadal difference of literacy rate of 22.48 among the other states of India. Chhattisgarh is near to it with a difference of 22.27. If we consider the literacy rates by rural-urban composition of the state, we find that literacy rate for urban areas is much higher i.e. 76.89 percent as against 55.92 percent for rural areas of the state.

9.1.6.1 *Literacy at District level (Rural-Urban)*

At the district level for rural areas, following are five districts, each the best and worst performers in literacy:

	Best performers (Rural)			Worst performers (Rural)	
	Name of district	Literacy rate		Name of district	Literacy rate
1.	Jhunjhunu	73.24	1.	Banswara	40.78
2.	Sikar	70.39	2.	Bhilwara	44.59
3.	Kota	67.34	3.	Jalore	44.81
4.	Churu	65.29	4.	Dungarpur	45.69
5.	Hanumangarh	63.65	5.	Bikaner	46.33

Likewise following are the five districts, each for the best and worst performers on literacy in urban areas:

	Best performers (Urban)			Worst performers (Urban)	
	Name of district	Literacy rate		Name of district	Literacy rate
1.	Udaipur	86.19	1.	Jalor	66.33
2.	Banswara	84.80	2.	Dholpur	67.48
3.	Alwar	82.27	3.	Nagaur	69.1.37
4.	Ajmer	81.69	4.	Tonk	69.1.57
5.	Chittorgarh	81.01	5.	Karauli	70.22

9.1.6.2 *Literacy by sex*

The male literacy rates have always been higher than the female literacy rates at all the levels i.e. district/tehsil for rural and urban areas. as can be seen from Table 9.9 given at the end of this volume.

Table 9.9 Literacy rates by sex at state level for rural and urban areas in 1991 and 2001

Year	**Total**			**Rural**			**Urban**		
	P	**M**	**F**	**P**	**M**	**F**	**P**	**M**	**F**
1991	38.55	54.99	20.44	30.37	47.64	11.59	65.33	78.50	50.24
2001	61.03	76.46	44.34	55.92	72.96	37.74	76.89	87.10	65.42

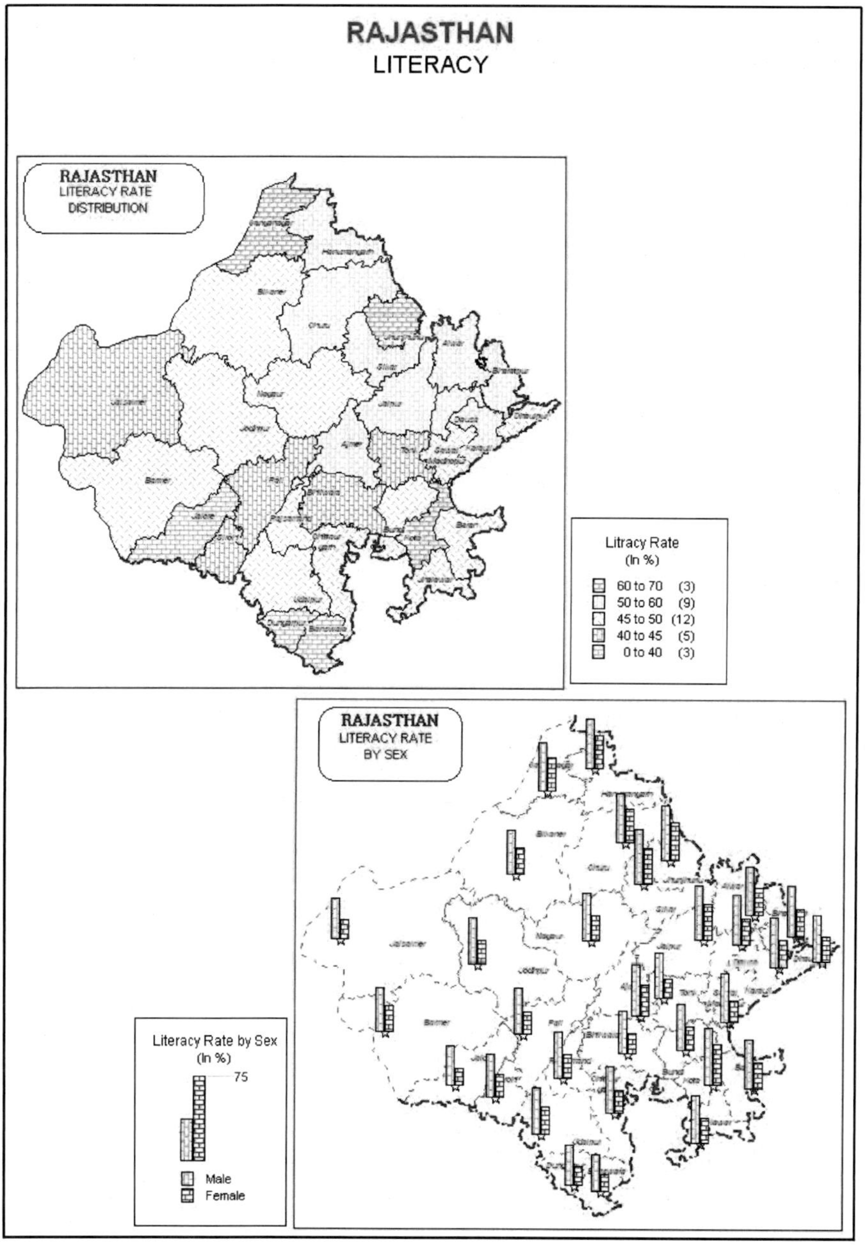

Figure 9.6. Literacy (Source: Census of India)

Among males, the literacy rate has increased from 54.99 percent in 1991 to 76.46 percent in 2001, which is higher than that of all India average i.e. 75.85. In case of females, it has increased more than double from 20.44 to 44.34 percent from 1991 to 2001 respectively, which is still lower than all India average of 54.16.

It is a matter of great significance for Rajasthan that the female literacy rate of 11.59 percent in rural areas which was the lowest at the national level in 1991, has increased more than threefold i.e. 37.74 percent in 2001. The male literacy rate has also progressed well, rising from 47.64 percent to 72.96 percent during last decade.

9.2 ANIMAL HUSBANDRY

The importance of livestock in Rajasthan agriculture is of great significance. The entire field operation from ploughing to the harvesting of crops is carried out by the draught animals mainly bullock and camel. They are the keystone in farming. In addition to crops, draught and milch stock are raised and maintained by the individual farmer. They are his constant companions in the field by day, and live beside his house or even under his roof. The maintenance of livestock and their proper care and treatment closely affect the prosperity of the Rajasthan cultivator. India supports the largest cattle population, which including buffaloes, is 20% of the world's estimated 1115 million bovine populations. In spite of this large number of cattle, the average per capita daily consumption of milk in India is only 130 gram as against 741 grams in Switzerland etc. It is considerably small on account of the very low milk yield capacity of the Rajasthan cattle. The Rajasthan agriculture differs fundamentally from that in the western countries in that it is firmly dependent on the use of draught force. Paradoxically enough, no branch of Rajasthan agriculture is so much neglected as animal husbandry.

In Rajasthan animal husbandry occupies an important role in the state economy. Since more than half of the area is arid or semi-arid and is not suitable for traditional crop farming. Livestock raising has superiority over crop cultivation with respect to growth, stability and resource conservation. The natural forces like soil climate-compels have less impact on livestock farming. Animal husbandry is not a mere adjunct to crop agriculture, but a major economic activity contributing over 12 percent of the total net cosmetic income. The state produced 2.8 million tones of milk and over 10 thousand tones of wool annually which account for over 10 percent and 40 percent of the total milk and wool produced respectively in the country. The State has only 7 percent and 21 percent respectively of the total cattle and sheep population of India. There are significant regional differences with respect to animal husbandry practices and the role of livestock farming in the State economy. The livestock in the State are of better quality in comparison to those found in other parts of the country.

There are seven well defined breeds of cattle namely Haryana, Gir, Nagauri, Kankrej, Tharparker, Rathi and Malvi in Rajasthan. Out of these Tharparker and Rathi are generally classified as dairy breeds, Narauri and Malvi for producing draught animals and Haryana, Gir and Kankrej for producing bullock and milch animals. The important breeds of sheep in the State are Chokle. Magra, Marwari, Nali, Pugal, Jaisalmeri, Malpura and Sonadi. Chokla sheep produce medium apparel and excellent carpet quality wool while Magra, Marwari, Nali Pugal and Jaisalmeri produce medium to good carpet quality wool. The Sonadi and Malpura produce poor quality wool but are suitable for mutton production. The important breeds of goats are Marwari, Thakrana, Sirohi, Jamnapari, and Barbari. The Barbari is highly prolific and produce twins and triplets. Thakrana is reputed for high mild production.

Jamnapari is a hardy and heavy animal and can adapt to almost all types of climate and food.

9.2.1 Livestock and Cattle Regions

There are significant regional differences with respect to animal husbandry practices and the role of livestock farming in the economy of the state.

9.2.1.1 *Arid Region*

The arid and semi-arid areas covering over half the State are agro-climatically not suitable for crop farming. Here animal husbandry is evidently the only source of regional economy which not only support the natives of the area but also constitute an important item of sale to other regions. The natural forces which reduce the productivity and lead to instability in agriculture have much less impact on livestock farming. This is primarily due to the ability of the desert animals to withstand the extreme conditions and the trans-migratory system of husbandry followed which imparts greater adjustability. Further, the region is endowed with some of the best breeds of cattle, sheep and draught animals. Hardy perennial grasses like Sewan (*Lasiurus sindicus*), Ajjan (*Cenchrus setigerus*), Dhaman (*Cenchrus ciliaris*) and also abundantly available. A comparatively low investment and marginal improvement in management and conservation of fodder and water resources can bring remarkable change in returns from animal husbandry whereas in returns from animal husbandry whereas in agriculture even higher input may not have desired impact. Stability in the case of crop farming would need adequate irrigation facilities, soil conservation and other capital investment methods which are precarious in arid areas.

Arid regions are characterized by a stagnant economy and absence of industrial infrastructure. The existing cropping pattern in these areas indicates that it is mainly designed to serve the subsistence needs of the human and cattle population. Low yield coupled with recurrent crop failures necessitates hoarding of the production achieved during a favorable year. This reduces the chances of marketable surplus from crop products. On the other hand, in the case of livestock farming the situation is different. Because local needs of the livestock products are limited, the major share of the produce is surplus. The yields are relatively profitable and are not unstable. About half the milk production and almost the entire quantity of wool and mutton produce in the area are surplus to local requirement. Livestock farming, therefore, in the region by its very nature offers scope for commercialization. This also provides scope for livestock based industries. Unfortunately, the importance of livestock farming vis-a-vis crop farming has not been fully realized and attempts were made to bring increasing areas under crop farming. Since the area was unsuitable for such activity, stress on crop farming brought about faster deterioration of soil as vegetation cover and wasteful use of scarce water resources. It is felt that the resource depletion trend can be checked by reverting all the sub-marginal land to natural plant cover of grasses and shrubs, which will support livestock and cattle development.

9.2.1.2 *Eastern and Southern Regions*

In the eastern districts of the State small and marginal farmers predominate. Most of them derive part of their income from products of animal origin. But in the absence of any organized market for fluid mild, they do not have any incentive for increasing its production. Here the number of animals is quite high as compared to the western Rajasthan but the production of milk, wool and mutton is frustratingly low. No surplus has ever been achieved due to high local consumption. The livestock in the southern districts of the State are of poor quality. The fodder resources though available in plenty are nutritionally poor. However, the number of stock is so large that small improvement in their production is likely to raise the overall production of the State considerably. The areas coming under the command of the Chambal and Mahi projects offer great potential for livestock farming along with cultivation.

9.2.2 Growth of Livestock and Cattle Population

In Rajasthan the livestock and cattle population is already high and has shown a further increase during the present century. At present there are more than 54 million and 12 million of total livestock and cattle population in the State respectively. The growth rate of different categories of domestic animals is not uniform. The differential growth rate of domestic animals is and indication of changing emphasis of cattle and livestock in the rural economy of Rajasthan.

9.2.3 Livestock Distribution Pattern

The district-wise population distribution is represented in Figure 9.7. It is clear from the figure that Barmer, Jodhpur and Nagaur districts leads in the total live stock population while Sirohi and entire eastern Rajasthan is having least population of the livestock. Brief Description of the individual major component of Livestock population is given in following paragraphs.

9.2.3.1 *Cattle*

The percent share of cattle to total Livestock population has decreased from 35.02% in 1966 to 22.1% in 1997 (Figure 9.8). The very low and low cattle distributions were observed in Sawai Madhopur, Dausa Karauli, and Jhunjhunu and Bharatpur districts during 1997. Medium ranges were registered in Kota, Jalore, Sirohi, Alwar and Sikar districts. High ranges were found in Ajmer, Nagaur, Bikaner, Barmer and Banswara district, while very high ranges were located in Udaipur, Jodhpur, Bhilwara and Chittorgarh district of Rajasthan. However, during 1972, ten districts namely Barmer, Bikaner, Churu, Jaisalmer, Jalore, Jhunjhunu, Jodhpur, Nagaur, Sikar and Sirohi indicated change in lower categories. This might be due to severe drought during that interval of livestock census. None of the districts indicated change in higher level of category.

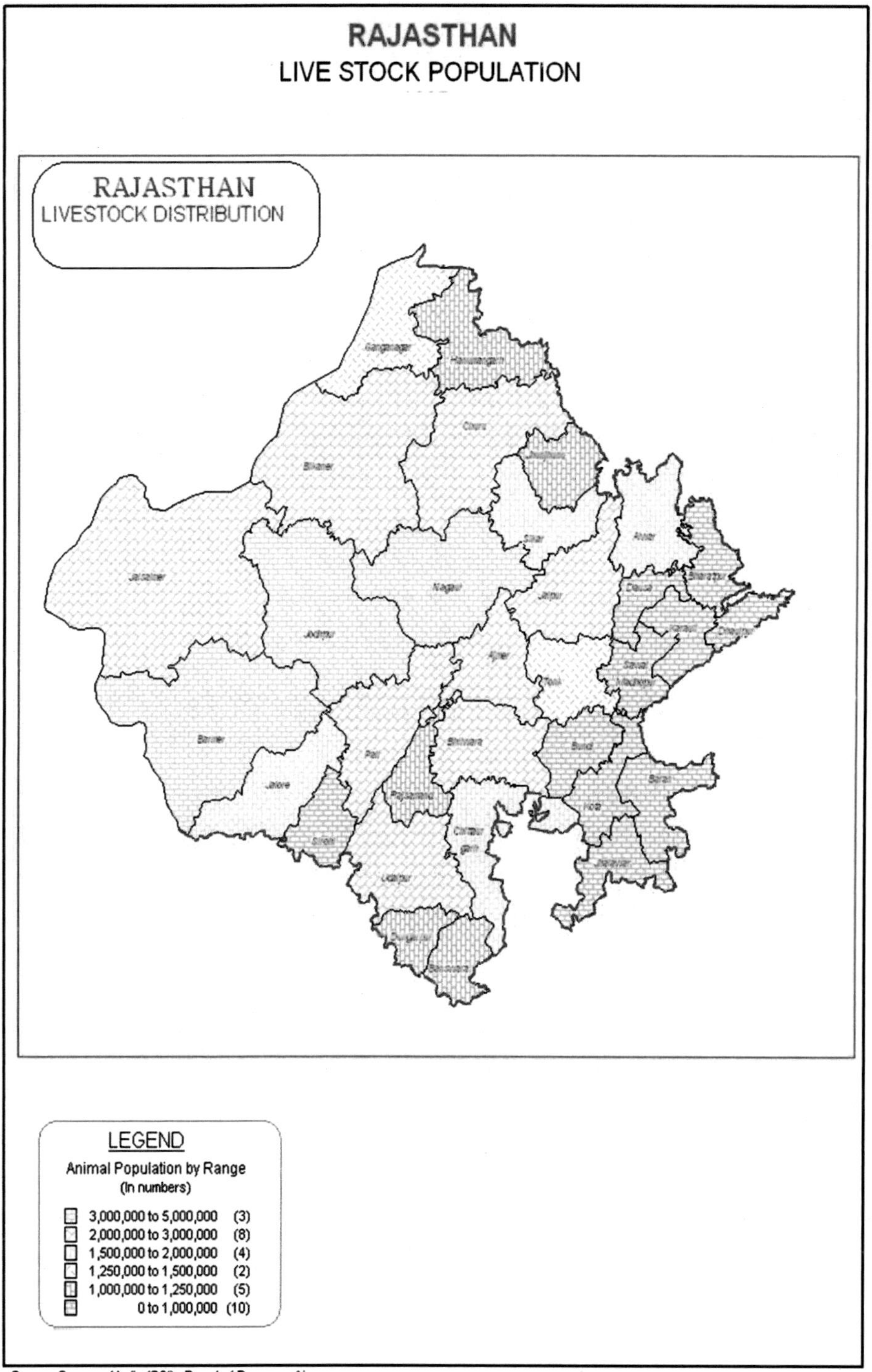

Figure 9.7. Livestock Population (Source: Board of Revenue, Ajmer)

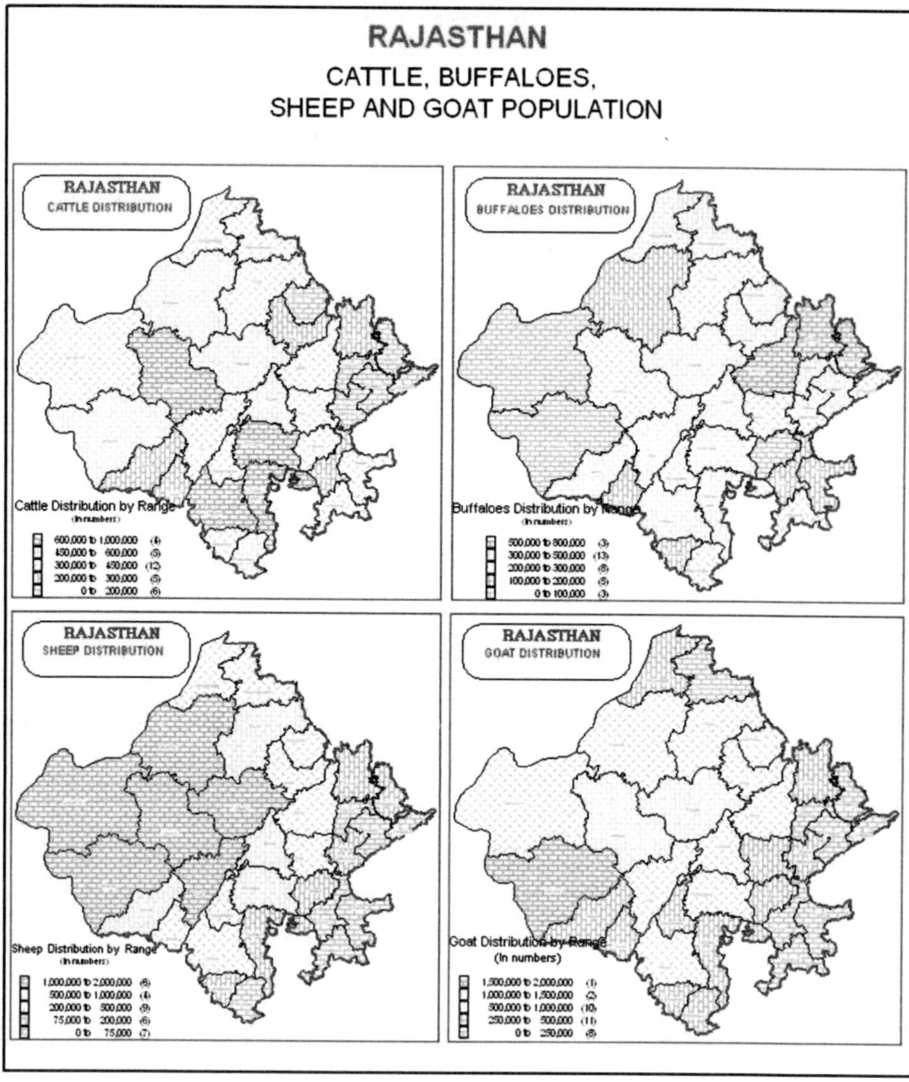

Figure 9.8. Cattle, Buffaloes, Sheep and Goat Population
(Source: Board of Revenue, Ajmer)

9.2.3.2 *Buffaloes*

The total number of buffaloes was arranged in five ranges and mapped as shown in Figure 9.8 namely, very low, low, medium, high and very high. Very low categories was found Jaisalmer, Barmer and Bundi district. Low ranges were identified in Baran, Kota, Dungarpur, Bikaner districts while Medium ranges were traced in Udaipur, Jalore, Pali, Bhilwara, Chittorgarh, Bhilwara, Ajmer, Nagaur, Sikar, Dausa, Karauli, Jhunjhunu, Hanumangarh and Ganganagar district of Rajasthan. Very high ranges were found in Jaipur, Alwar and Bharatpur district of Rajasthan.

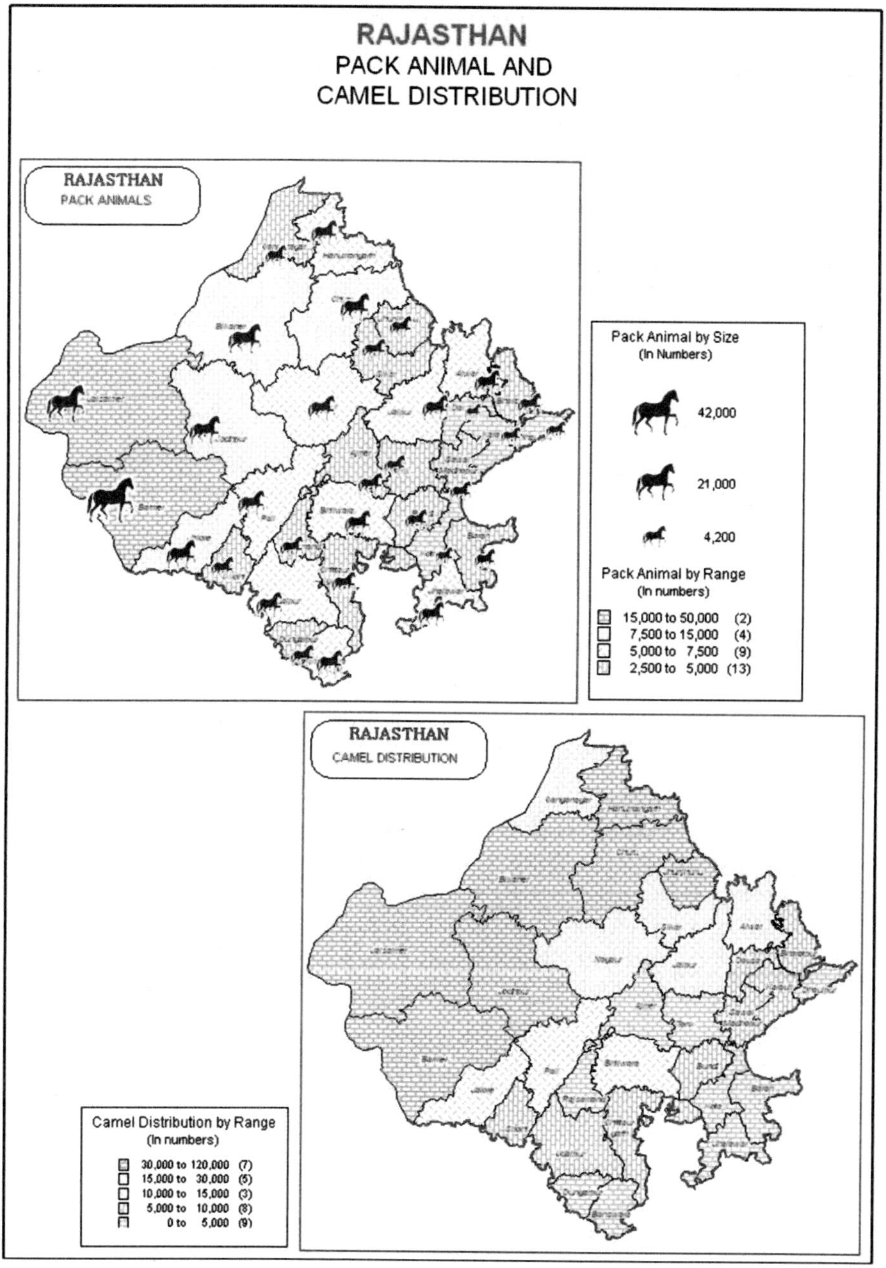

Figure 9.9. Pack Animal and Camel distribution (Source: Board of Revenue, Ajmer)

9.2.3.3 *Sheep*

The sheep population was arranged and mapped in group of five different categories namely very low, low, medium, high and very high (Figure 9.8). Very low category was found in Easter region namely Dausa, Karauli, Dholpur, Baran, Kota, Jhalawar and Banswara. Medium category is found in Udaipur, Rajsamand, Tonk, Jaipur, Jhunjhunu, Sikar, Hanumangarh and Ganganagar districts. Highest concentration of population were found in Desertic regions of Rajasthan like, Jaisalmer, Bikaner, Barmer, Jodhpur, Nagaur, Pali district.

9.2.3.4 *Goat*

The percent share of goat in was worked out and grouped into five different categories (Figure 9.8). Very low category was observed in Eastern districts like Bharatpur, Karauli, Dausa, Dholpur and Hanumangarh districts.

Highest Goat concentration was found in Barmer district of Rajasthan followed by Jodhpur and Nagaur districts.

9.2.3.5 *Camel*

The contribution of camel to Rajasthan's livestock population was worked out and arranged into five different categories namely very low, low, medium, high and very high (Figure 9.9). It indicated that camel population is maximum in desert region of Rajasthan. It is lowest in eastern region of the state. Nagaur, Pali, Jaipur and Alwar districts along with Sikar and Ganganagar is having second highest population of camel.

9.2.3.6 *Pack Animal*

The pack animal consists of horses, ponies, mules and donkeys. The population contribution of pack animals was worked out and grouped into five categories namely very low, low, medium, high and very high (Figure 9.9). It is observed that population of pack animals was very low in Dausa, Dholpur, Karauli and Kota districts. The highest concentrations were observed in Barmer and Jaisalmer districts followed by Jalore, Jodhpur, Bikaner and Churu districts. The entire Aravalli belt was found to have medium to low concentration of Pack Animals in the state of Rajasthan.

Chapter - 10

TRANSPORTATION AND COMMUNICATION

A State's sinews are roads, railways and communications. The effective transport and communication systems accelerate the growth of the economy by integrating markets, reducing the cost of production, enhancing the overall productivity and also helping in attracting large scale industrial investment.

10.1 ROADS

The State has a net work of 1.51 lakh km (151914 km) of roads consisting 2964 km as national highway, 9966 km as state highway, 5947 km as district main roads, 66395 km as other district and village roads and 2239 km of border roads constructed by PHED and 64403 km by other departments in the state respectively. The road density as measured by length of roads per 100 sq. km of geographical area of the State was 44.42 km. The State had accounted for 6.12% of road net work at the all India level of 24.66 lakh km. Total network of roads is encompasses 11 national highways and 519 state highways covering the entire state. Rajasthan State Road Transport Corporation is the main state government department for providing transportation facilities with about 4258 vehicles operating through 47 depots on 2624 routes (including 519 state highways) in the state of Rajasthan. From the Table 10.1 and Figure 10.1 it is clear that Jaipur has the highest National Highway Road length followed by Bikaner and Ajmer district. Sawai Madhopur, Jhunjhunu, Banswara and Hanumangarh are the districts having negligible National highway length in their districts. Jodhpur has the highest all type road length while Dholpur district hast the lowest share in the Rajasthan state. Ganganagar district has the highest road connectivity to its villages followed by Jaipur and Alwar districts. Lowest village connected to roads were found in Jaisalmer and Dholpur district of Rajasthan.

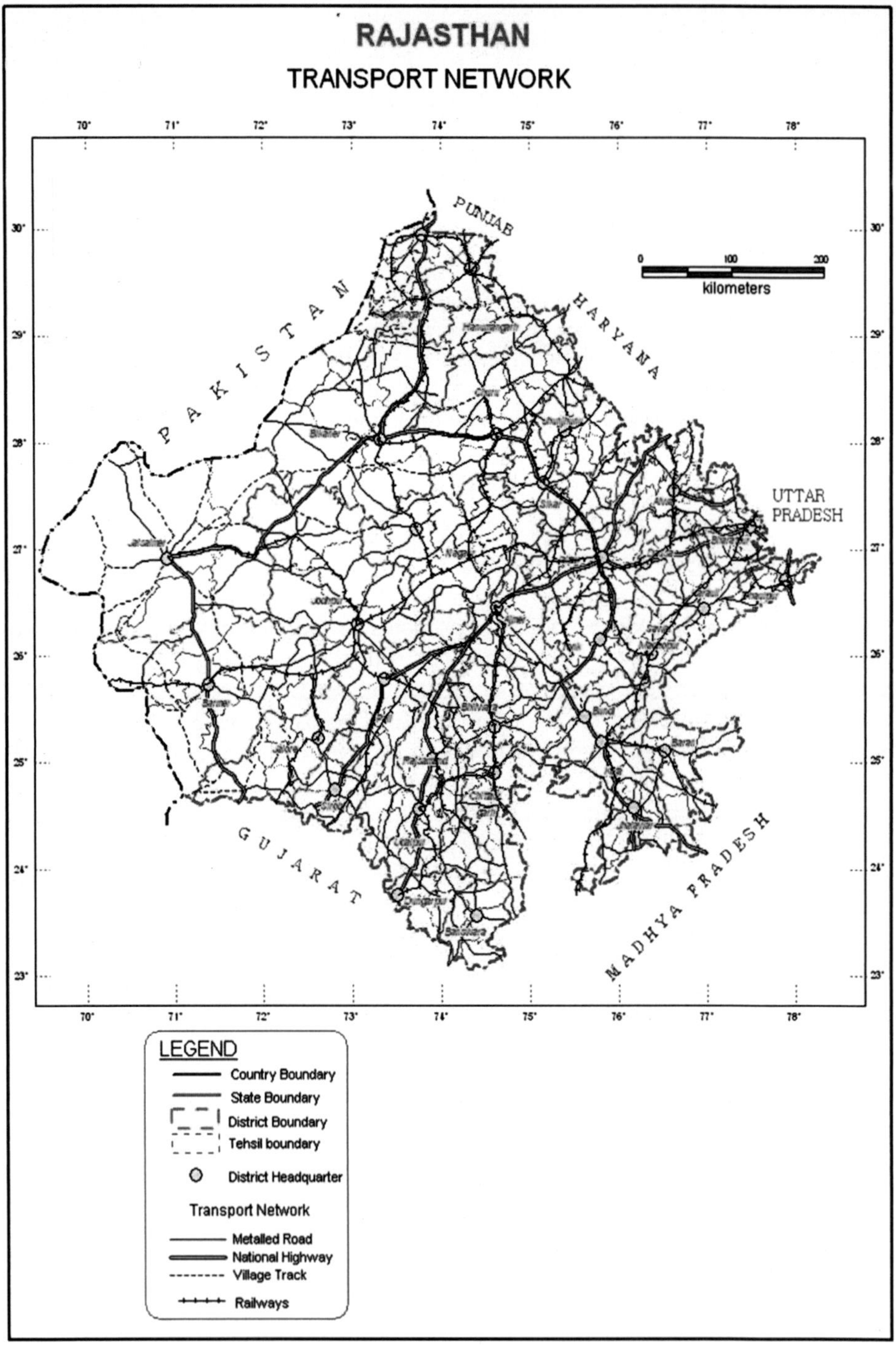

Figure 10.1 Transport Network Map (Source: RSRTC)

Table 10.1. Road Length in Rajasthan

S. No	Districts	Type of Roads					Total
		National Highway (N.H.)	Painted (B. T.)	Metaled (W.B.M.)	Graveled (G.R.)	Fair weather and dressed up tracks (F.R.)	
1	Ajmer	242	2490	81	31		2844
2	Alwar	35	3344	105	156		3640
3	Banswara		1476	110	121		1707
4	Baran	232	886	52	315	11	1496
5	Barmer	189	3998	238	793		5218
6	Bharatpur	71	1869	53	6		1999
7	Bhilwara	129	2551	357	838	13	3888
8	Bikaner	325	2908	87	316	10	3646
9	Bundi	107	1121	95	187		1510
10	Chittorgarh	202	2299	178	162		2841
11	Churu	239	2513	203	145		3100
12	Dausa	88	1148	83	273		1592
13	Dholpur	28	978	83	49		1138
14	Dungarpur	34	1265	156	63		1518
15	Ganganagar	124	3101	11	92		3328
16	Hanumangarh		2460	23	44		2527
17	Jaipur	401	3455	164	44	42	4106
18	Jaisalmer	196	2887	42	731		3856
19	Jalore	41	2153	180	268	18	2660
20	Jhalawar	126	1002	116	144		1388
21	Jhunjhunu		2069	267	140		2476
22	Jodhpur	218	4490	226	1004		5938
23	Karauli	4	1227	164	105	2	1502
24	Kota	120	1299	78	139	2	1638
25	Nagaur	330	3444	533	731		5038
26	Pali	218	2959	440	1028		4645
27	Rajsamand	155	1283	214	231		1883
28	Sawai Madhopur		976	86	149		1211
29	Sikar	191	1842	168	308		2509
30	Sirohi	143	1131	78	97		1449
31	Tonk	109	1343	37	185	8	1682
32	Udaipur	250	2929	197	106	7	3489

Source: PWD

Table 10.2 Road Transport Route Nationalized

S. No	Depot (Districts)	No. of Routes	Route Distance (Km.)	Average Route Distance (Km.)
1	Ajmer	67	8520	127
A	Ajaymeru	71	8896	125
B	Beawar	78	11645	149
2	Alwar	78	8985	115
A	Matsya Nagar	53	5675	107
B	Tizara	33	4630	140
3	Banswara	50	10874	217
4	Barmer	62	12844	207
5	Bharatpur	51	9656	189
A	Lohagarh	61	8796	144
6	Bhilwara	51	8452	166
7	Bikaner	42	12285	292
8	Bundi	62	11765	190
9	Baran	29	5160	178
10	Chittorgarh	93	15722	148
11	Churu	12	2411	201
A	Sardarsahar	44	8980	204
12	Dausa	77	7953	103
13	Dholpur	39	7132	183
14	Dungarpur	70	12322	176
15	Ganganagar	79	15826	200
16	Hanumangarh	66	13408	203
A	Anoopgarh	56	13001	321
17	Jaipur	67	19775	295
A	Delux	34	13534	398
B	Jhalana Dunagri	74	3810	51
C	Sanganer	70	2429	35
D	VaishaliNagar	71	7697	108
E	Vidhyadhar Nagar	90	5038	56
F	Kotputli	38	5094	134
18	Karauli			
A	Hindaun	45	7054	157
19	Jaisalmer			
20	Jalor	41	10634	259
21	Jhalawar	60	9872	165
22	Jhunjunu	39	6724	172
A	Khetri	45	7835	174
23	Jodhpur	64	15808	247
A	Phalodi	28	7193	257
24	Kota	65	13751	212

S. No	Depot (Districts)	No. of Routes	Route Distance (Km.)	Average Route Distance (Km.)
25	Nagaur	41	9475	231
26	Pali	50	6685	134
A	Falna	41	7494	183
27	Rajsamand			
28	Sawai Madhopur	33	5283	160
29	Sikar	103	16251	158
30	Sirohi	40	7307	183
A	Abu Road	43	7982	186
B	Rewdar	43	9360	218
31	Tonk	37	9530	258
32	Udaipur	87	17927	206

Source: PWD

Table 10.3 Villages Connected with Roads

S.N	District	Villages connected with Road				Villages not connected with Painted Road
		Painted	Metalled	Gravelled	Total	
1	Ajmer	568	104	154	826	159
2	Alwar	1020	43	155	1218	728
3	Banswara	596	52	146	794	637
4	Baran	329	23	97	449	621
5	Barmer	570	29	385	984	641
6	Bharatpur	640	27	75	742	603
7	Bhilwara	527	66	325	918	647
8	Bikaner	364	7	54	425	155
9	Bundi	299	19	55	373	453
10	Chittorgarh	669	83	234	986	1186
11	Churu	518	33	19	570	356
12	Dausa	348	20	72	440	569
13	Dholpur	254	20	51	325	226
14	Dungarpur	419	63	69	551	295
15	Ganganagar	1364	4	17	1385	1523
16	Hanumangarh	785	4	1	790	740
17	Jaipur	1105	64	117	1286	845
18	Jaisalmer	208	3	80	291	227
19	Jalore	405	50	141	596	69
20	Jhalawar	298	32	70	400	1048
21	Jhunjunu	479	37	91	607	217
22	Jodhpur	648	29	129	806	54
23	Karauli	335	56	19	410	331
24	Kota	331	22	40	393	418
25	Nagaur	730	92	186	1008	366
26	Pali	571	60	155	786	118

S.N	District	Villages connected with Road				Villages not connected with Painted Road
		Painted	Metalled	Gravelled	Total	
27	Rajsamand	383	55	129	567	306
28	Sawai Madhopur	245	22	32	299	424
29	Sikar	467	100	90	657	274
30	Sirohi	237	19	71	327	119
31	Tonk	304	21	66	391	628
32	Udaipur	750	87	256	1093	1213

Source: PWD

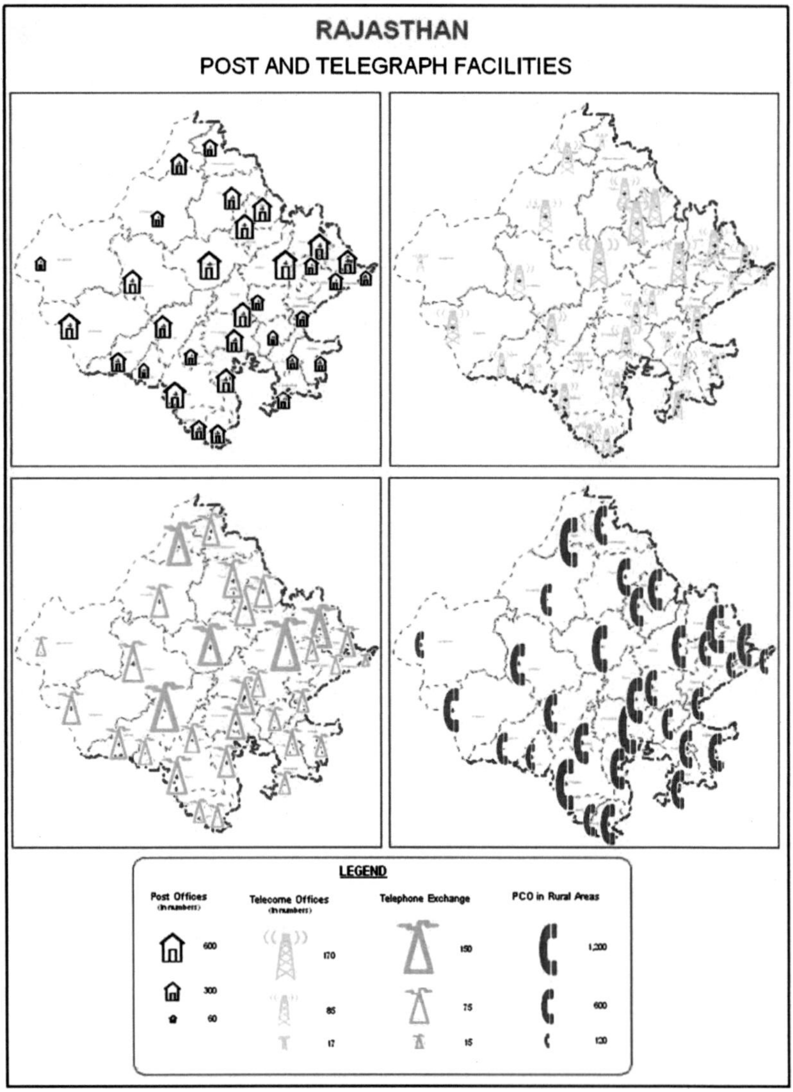

Figure 10.2. Post and Telegraph

10.2 RAILWAYS

Railways, the artery of the economic system, have a supra-regional feature of integrating various States in India. Development of railway system is shaped by interplay of factors such as laying of new lines, conversion of meter gauge into broad gauge, renewal and replacement of tracks and electrification of lines. The railways are functioning as a commercial undertaking as well as public utility service. Though the National Transport Policy Committee suggested a model split in which the share of long distance freight for railways should be 72 per cent and the rest for road transport system. Total rail track length is of 5917 km which is about 9.42% of the all India rail track length. 17.29 According to the statistics Railway in Rajasthan has a density of 17.29 km per 1000 sq.km. of the geographical area of the state.

10.3 AIRWAYS

At present 5 airports at district head quarters of Jaipur, Jodhpur, Jaisalmer Kota and Udaipur districts are in operation and are linked to almost all the metro cities of India. The Sanganer Airport in Jaipur has got recognition as International Airport and is going to play a main role in the state's economy in coming years. Twelve Airstrips and 128 Helipads are already available in various districts in the state. Seven new airstrips have also been developed under relief works recently. Indian Airlines along with other private airlines operate their flights in Rajasthan. They have regular flights to and fro from Rajasthan. An Air-Cargo complex is located at Sanganer near Jaipur. Cargo is custom-cleared at the complex and is transshipped to Delhi by air. Bonded trucking facility is available, to move cargo by road from the Air-Cargo complex to International Air-Cargo Terminal at New Delhi.

10.4 PORTS

Ports on the western coast of India are easily accessible from Rajasthan. These include ports at Bombay/Nhavashera. Gandhidham/Kandla, Jamnagar and Bhav-nagar. Cargo from Rajasthan moves, mostly by road to these ports.

10.5 INLAND CONTAINER DEPOT

ICDs are located at Jaipur and Jodhpur and proposed at Udaipur, Kota, Bhilwara and Bhiwadi. Services provided at wet ports are available at the ICDs including custom-clearance and warehousing facility. Bulk-cargo moves from the ICDs to the ports on the western coast of India.

10.6 TELECOMMUNICATION

Telecommunication facilities in Rajasthan have been improving rapidly of late. As on September 30, 2004, there are 2341 telephone exchanges in the State with 2216751 installed telephone instruments. There are 53879 STD PCOs and 136 cities covers with GSM mobiles. The overall telephone density as per new census is 269 per thousand populations in Rajasthan. The State has inter city (both national and international) direct dialing (STD/ISD) facilities at all district centers and most other major towns. Facsimile facilities can be availed of at all these towns. Telex facilities are available at all major towns. Jaipur is one of the locations included in the Department of Telecommunications recent tender for paging services.

10.7 POST AND TELEGRAPHS

Total numbers of post office in urban areas are 929 while in rural the number is 9359. They are catering about 21.70 lakh persons of the state with 0.30 numbers per 10 sq. km. density in the state of Rajasthan. From the Table 10.4 and Figure 10.2 it is clear that Jaipur being the capital of the state has the highest number of Post Offices, Telegraph offices, Telephone exchange but in PCO category Bhilwara leads the tally.

Table 10.4. Post, Telegraph and Telephone Offices

S.No	District	Post office	Telegraph Offices	Telephone Exchanges	Public call Offices (Rural)
1	Ajmer	430	61	84	989
2	Alwar	485	82	102	1073
3	Banswara	277	58	34	815
28	Baran	197	53	33	724
4	Barmer	479	104	68	896
5	Bharatpur	415	79	57	880
6	Bhilwara	393	95	75	1180
7	Bikaner	222	97	69	537
8	Bundi	176	31	38	512
9	Chittorgarh	403	61	61	780
10	Churu	397	89	85	682
29	Dausa	244	19	41	638
11	Dholpur	188	31	14	340
12	Dungarpur	298	54	36	532
13	Ganganagar	328	93	105	1118
31	Hanumangarh	249	22	62	760
14	Jaipur	600	143	150	859
15	Jaisalmer	154	27	27	374
16	Jalore	265	60	62	513
17	Jhalawar	250	69	33	702
18	Jhunjunu	414	106	70	839
19	Jodhpur	400	86	93	822
32	Karauli	247	27	26	373
20	Kota	196	59	45	699
21	Nagaur	542	165	101	1026
22	Pali	378	95	133	772
30	Rajsamand	210	33	50	790
23	Sawai Madhopur	214	71	34	551
24	Sikar	470	153	93	773
25	Sirohi	180	40	49	378
26	Tonk	221	62	47	664
27	Udaipur	494	85	82	1136

Source: Post, Telegraph and Telephone Offices.

Chapter - 11

CASE STUDIES

During the study period number of GIS based studies has been accomplished both in Rural and Urban areas scattered all over the Rajasthan. These studies have been focused on the burning issues in the context of Rajasthan. One of study deals with the prime concern of rainwater harvesting in the Jhalawar area, second in the desert area of Bikaner district and third in Udaipur district. Another study focus mineral locations and status of mining activities and mineworkers for the entire state has been done. A case study utilizing GIS tools for Traditional Sex Workers located in the three districts (Jaipur, Tonk, and Alwar) of Rajasthan has also been carried out. A study concentrating in identification of water harvesting structures in Alwar district of Rajasthan has been carried out with GPS using GIS technology. During these studies findings based on Analysis and thereafter conclusions have been drawn to facilitate the planner in decision-making processes. In all the above-mentioned studies GIS has played the major and important role in not only in analysis but also in presentation part.

CASE STUDY - 1

11.1 GIS BASED ECONOMIC DEVELOPMENT LEVELS ASSESSMENT STUDY OF OSIAN – MATHANIA TRACT IN JODHPUR DISTRICT

Agricultural planning needs the scientific analysis of the basic resources. It also requires the correlation of these resources with the socio-economic conditions of the rural population. This can be scientifically done by natural resources survey and analysis of the agro-geography of the region. Agro-geography deals with the regional position of agriculture of a particular area. It deals on one head, the particular system and the current position of agricultural resources and potential of the area and its prospect and future development on the other. To this end, it is necessary in the first instance to work out the levels of agro-economic development of the area. This can be worked out on the basis of the efficiency of the data of land utilisation, crop production and the rural occupational structure. These aspects of development are in turn susceptible of further analysis. The arid and semi-arid zones, where agricultural prospects are yet to be assessed, such a study will be of primary need to locate the potential and problem areas in relation to agricultural development. Therefore, a study was undertaken in an area in close proximity to Jodhpur where the rocky (south) and sandy (north) desertic conditions are met with.

11.1.1 Study Area Location

The area studied is located between 26°45' N to 26°28' N and 73°00' E to 72°45' E. From the administrative point of view the entire area belongs to the Jodhpur district of Rajasthan with 10 villages in the south (rocky desert) and 12 villages in the north (sandy tract) of Osian tehsil. Prior to reorganization of States, the area under consideration belonged to Jodhpur tehsil. This is the area where rocky desert of the south (Mathania) area gradually merges in the sandy desert of Osian area in the north. The study region covers a total area of 71810.9 ha or about 718 sq. km. (Fig. 11.1.1 and 11.1.2)

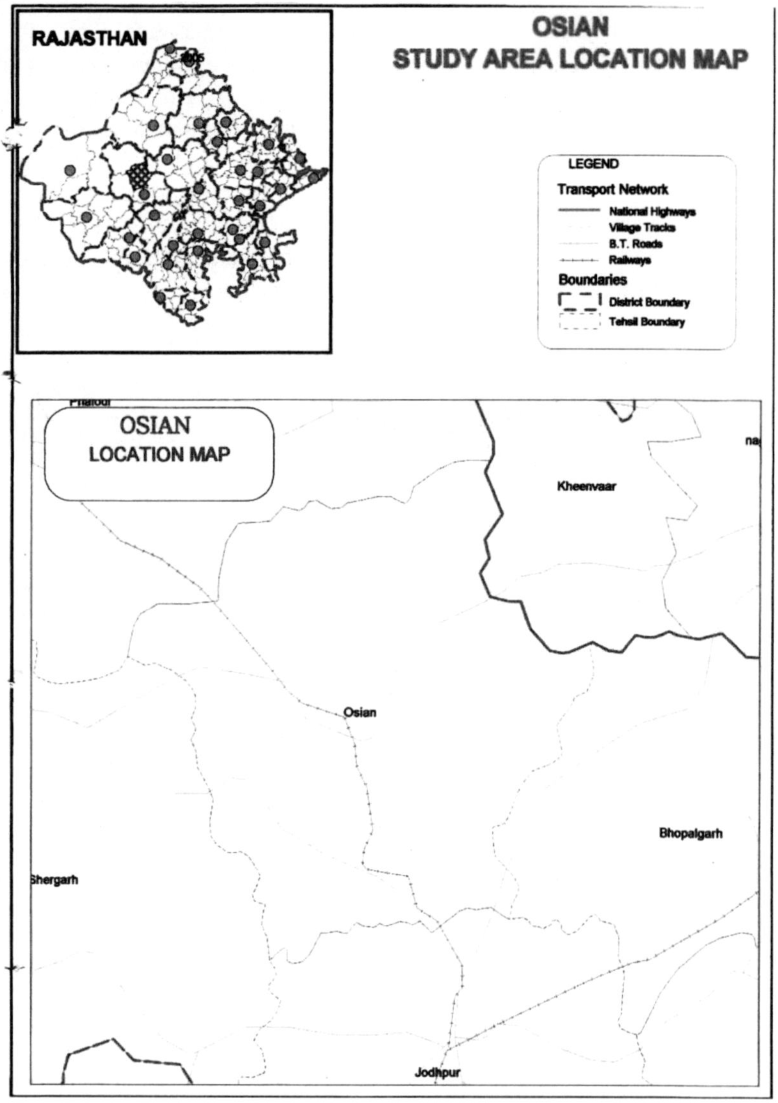

Figure 11.1.1 Study Area Location map

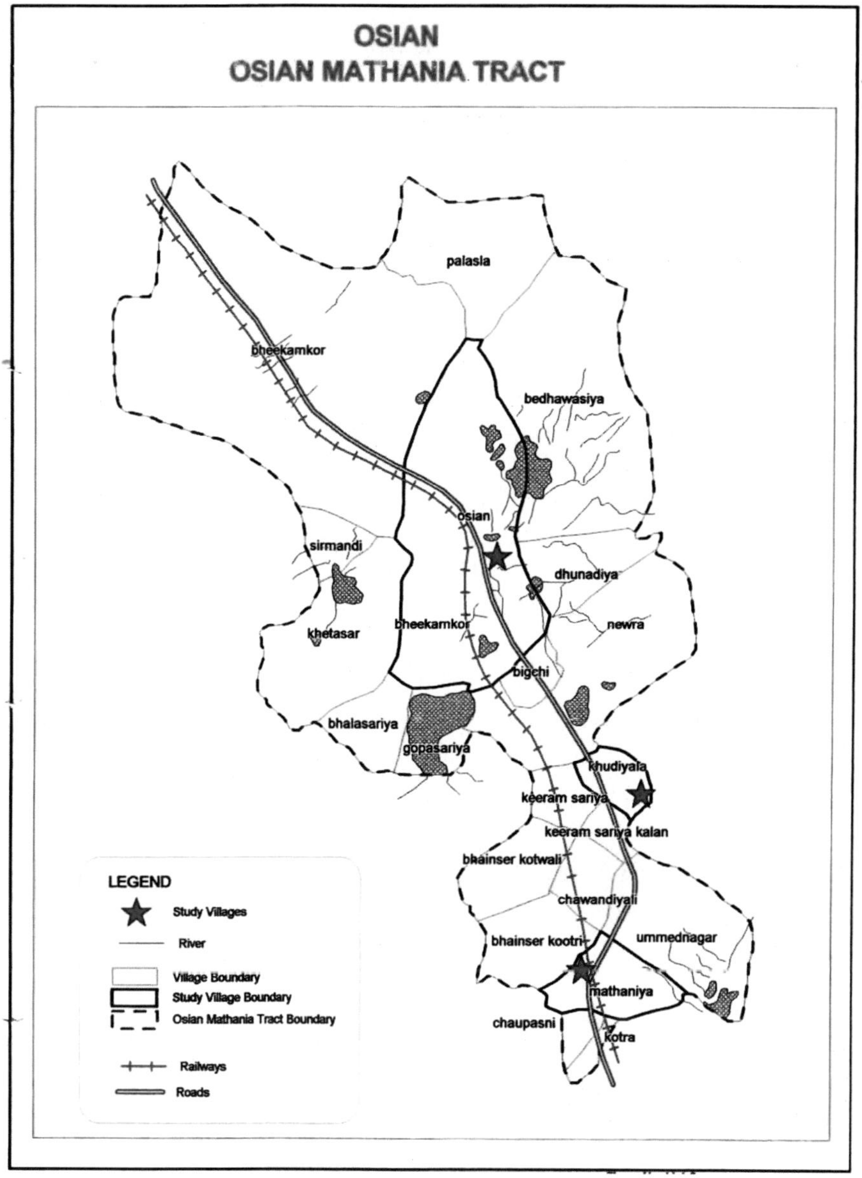

Figure 11.1.2 Village Map

11.1.2 Study Area

The Jodhpur-Phalodi road and railway line run in North-South direction almost through the center of the region. The distance of Jodhpur from the area is 30 kms in the south and 63 kms in the north. Jodhpur to Mathania represents the features of a rocky desert. Upto Mathania a series of rhyolite and sandstone hills are met with and landscape is often gravelly and rocky. From Mathania the sandy plain appears and dunes become the most important aspect of the landscape from Osian. The desert represents 330 meters facet. The area is mostly devoid of any surface drainage making the region a typical arid one from the geomorphic point of view. The water table is low, the static water level varies between 180 m to 240 m. Geographically the tract is well within the boundary of the Great Indian desert. Actually this comes within the present eastern boundary of the desert. The entire tract receives an annual rainfall of about 400 mm. Based upon Koppen (1931) the region is an arid one as the aridity index of the tract is about 48.5 which is the total annual rainfall i.e., 40 cm. If we examine Gorezynski's theory (1945) of aridity factor or Gaussen's (1952) xerothermic index figures read 30 and 300 respectively which are indices of true arid climate. Besides, based on the line drawn according to Thornthwaite's (1948) evapo-transpiration ratio indicates that the annual precipitation here is less than total annual evapotranspiration. Hence climatically the region is essentially an arid one. This climate analysis is further supplemented by the following geomorphic considerations:

- The rivers are misfit and soon die out of a process of alluvial suffocation. There are no permanent flowing rivers but intermittent streams. The drainage is essentially internal.
- There is no network of integrated drainage system and no controlling level to which the landscape is lowered. Instead the rivers deposit their load and sink into alluvium covered ground.
- All the debris resulting out of gullying and dissection of mountains are accumulated in the intermont basins.
- Alluvial sand becomes laterally confluent and continuously slopping 'bajadas' are formed along mountain fronts.

These physiographic parameters are indices of an arid climate. Hence climatically and geomorphologically the tract under study is an arid one.

11.1.3 Methodology

The study was based on field observation and collection of secondary data. Since economic development is complex phenomenon the reliability of the secondary data were verified by detailed field study of three case study villages. The selection of villages has been based on stratified sampling taking biophysical resource units as the basis. Accordingly Mathania (rocky desert), Osian (sandy desert) and Khudiyala (Transitional between rocky and sandy deserts) are selected for this investigation. While selecting the villages the secondary data collected were also consulted. Mathania and Osian have some urban character although functionally both of them are rural according while Khudiyala is a typical rural area of the terrain. The levels of agro-demographic development in the different biophysical units had been worked out on the basis of the efficiency of rural occupational structure and rural agricultural and industrial production of the Study villages selected. Accordingly 19 indicators were selected for deducting synthetic or synoptic index to work out and map the agro-demographic development. These consist of Structural indicators relating to rural population; Occupational indicators relating to rural working population and rural economy and Productivity indicators to per capita production of food crops. The data on structural and occupational indicators have been compiled from the publications of Census of India and that of productivity indicators from field study. The indicators selected are:

(A) Structural indicators (in percentage)

1. Rural population to total population; 2. Cultivators to total population; 3. Agricultural laborers to total population; 4. Density of agricultural lands; 5. Intensity of cropping; 6. Percentage of irrigated land to total cropped land.

(B) Occupational indicators (in % to total workers)

1. Cultivators to total workers; 2. Agricultural laborers to total workers; 3. Rural workers in livestock, forestry, fishing and hunting; 4. Rural workers in mining and quarrying; 5. Rural workers in household industry; 6. Rural workers to other than household industry.

(C) Productivity indicators

1. Per capita agricultural land; 2. Per capita production of cereals; 3. Per capita production of pulses; 4. Yield of principal crops; 5. Per capita income; 6. gross value of agricultural output; 7. Per capita agricultural land to total agricultural population. Some indicators were also considered on broad levels. These are Banking and Cooperative facility; marketing facility; accessibility of the villages by roads and railways; distribution of agro-based industry. Accordingly separate maps were prepared to depict the regional pattern of agro-economic conditions. All these maps were further cartographically analysed to prepare the synoptic map of levels of agro-economic development. This was done mainly by following the methods evolved by Mitra (1968) and Schwartsberg (1969). The synthetic indices are worked out after calculating the village-wise data of the above indicators (Study villages only). A 'score' for each indicator was arrived by standardizing the data tabulated. This was done by assigning zero to the least favored region with respect to each variable and a score of 100 to the most favored one. Intermediate scores were given to each intermediate region according to the arithmetic distance of its non-standardised scores for the two extremes. This can assign comparable sets of scores or index numbers for each variables considered for mapping (Schwartzburg 1969). The synthetic index for each region was then calculated by adding the scores of all the indicators. Accordingly the synoptic maps the levels of agro-economic development and planning units were mapped.

11.1.4 Physical Settings and Attributes of Terrain

11.1.4.1 *Climate*

The climate of Osian-Mathania tract can be classified as arid to extreme arid. The aridity index for the district varies from 70 to 87 and increases from south east to

North West. Here the evaporation is more than seven times the precipitation. In summer the day temperature rises considerably high with frequent sand-storms. The winters are extremely cold.

11.1.4.1.1 *Rainfall*

The rainfall data are available for Osian with effect from 1957. The number of rainy days based on data for the period 1957-2000 are shown in Table 11.1.2. A rainy day for this purpose was considered as a day receiving a rainfall of 2.5 mm or more.

Table 11.1.1: Number of Rainy days in Osian

Station	Month	Number of rainy days	Annual total
Osian	January	0.3	
	February	0.3	
	March	0.3	
	April	0.1	
	May	0.4	
	June	1.6	
	July	5.0	
	August	4.3	
	September	2.6	
	October	0.1	
	November	0	
	December	0.3	15.3

It is noticed that the bulk of the rainfall is received here during the period from June to September. The maximum rainfall as well as number of rainy days occur during August followed by the months of July, September and June respectively. The highest rainfall in respect of Osian occurs in July. Normal date of on set of monsoon in Jodhpur district is last week of July. A few pre-monsoon shower occurs in the month of June. The contribution of winter rainfall (December to February) as well as rain fall during June to September expressed as percentage of annual rainfall are as follows in respect of Osian.

Table 11.1.2 Rainfall Percentages

Station	Winter (December to February)	June to September
Osian	3	94

The contribution of winter rainfall is only 3 to 4 per cent of annual rainfall.

Table 11.1.3: Mean annual and extreme rainfall in Osian and its variability

Station	Normal annual rainfall (mm)	Standard deviation	Coefficient of variation	Extremes of annual highest			Rainfall (mm) lowest		
				Amount	Year	% of normal	Amount	Year	% of normal
Osian	251.8	87.8	35	383.1	1959	152	98.6	1963	39

The standard deviation value in respect of Osian is rather small which may be due to very less number of years of data used in compilation. Defining a high surplus rainfall year (flood year) as a year receiving a rainfall of 150 per cent or more of the normal annual value and a large deficit rainfall (drought year) as a year receiving a rainfall of 50 per cent or less of the normal annual value, the following are the years of floods and droughts at Osian. In Osian the number of high surplus rainfall years are equal to large deficit rainfall year.

11.1.4.1.1.1 *Rainfall Distribution:*

The period in which weekly mean rainfall is 7 mm or more covers the following periods at Osian. Osian – 25th to 37th week (18th June to 16th September). There are only 4 weeks at Osian when average of total rainfall for the weeks exceed 21 mm (i.e. at the rate of 3 mm/day). They cover mostly periods of July and August. The highest mean rainfall is recorded in 34th to 35th weeks (covering second half of August) indicating susceptibility of that stations for getting very heavy rainfall during this period. There is another maximum during second fortnight of July also.

11.1.4.1.1.2 *Percentage Probabilities of Weekly Rainfall*

In view of high variability from year to year, mean values for different weeks would not provide complete information. Accordingly weekly rainfall frequency was prepared and is presented in Table 11.1.4. Though the mean weekly rainfall generally exceeds 7 mm in Jodhpur district for the period from mid June to mid September (25th week of 37th week), the period when probability of obtaining measurable rainfall exceeds 50% is much less than this period as is evident from the data presented below. Actual mean percentage probabilities of obtaining measurable rainfall during three periods are also indicated.

Table 11.1.4. Mean Percentage Probabilities

Station	Duration when percentage probability of measurable rainfall exceeds 50 percent	Mean percentage probability of the period
Osian	26th week to 36th week with a week's break (25th June to 9th September) (10 weeks)	57

Thus the duration of period when probability of occurrence of measurable rainfall exceeds 50 per cent covers only 7 to 11 weeks. The mean percentage probability of the same is 67 for eastern portion of the district and 50 to 57 for the western portion. The median values for each week are much smaller than the mean values showing extreme skew ness in their distribution. It is the few spells of large intensities in a few years that increase the mean values of the respective weeks. Further for the monsoon weeks the percentage frequency does not diminish for increasing intensities.

Table 11.1.5. Percentage frequencies of weekly rainfall data (in mm) of Osian station.

Standard Week no.	Nil rainfall	0.1 to 5.0	5.1 to 10.0	10.1 to 15.0	15.1 to 20.0	20.1 to 30.0	30.1 to 40.0	40.1 to 50.0	50.1
1.	85.7	0	7.1	0	7.1	0	0	0	0
2.	92.8	0	7.1	0	0	0	0	0	0
3.	85.7	7.1	7.1	0	0	0	0	0	0
4.	0	0	0	0	0	0	0	0	0
5.	92.8	0.0	0	7.1	0	0	0	0	0
6.	92.8	7.1	0	0	0	0	0	0	0
7.	92.8	7.1	0	0	0	0	0	0	0
8.	0	0	0	0	0	0	0	0	0
9.	92.8	0	0	0	0	7.1	0	0	0
10.	0	0	0	0	0	0	0	0	0
11.	0	0	0	0	0	0	0	0	0
12.	85.7	14.3	0	0	0	0	0	0	0
13.	85.7	0	7.1	0	0	7.1	0	0	0
14.	0	0	0	0	0	0	0	0	0
15.	0	0	0	0	0	0	0	0	0
16.	0	0	0	0	0	0	0	0	0
17.	0	0	0	0	0	0	0	0	0
18.	92.8	0	7.1	0	0	0	0	0	0
19.	85.7	14.3	0	0	0	0	0	0	0
20.	85.7	14.3	0	0	0	0	0	0	0
21.	92.8	0	7.1	0	0	0	0	0	0
22.	92.8	0	0	0	0	7.1	0	0	0
23.	78.6	7.1	14.3	0	0	0	0	0	0
24.	71.4	0	7.1	7.1	7.1	7.1	0	0	0
25.	57.1	7.1	0	0	14.3	0	21.4	0	0
26.	50.0	21.4	7.1	7.1	0	7.1	7.1	0	0
27.	28.6	7.1	7.1	7.1	7.1	14.3	7.1	7.1	14.3
28.	42.8	7.1	21.4	0	7.1	7.1	0	0	14.3
29.	21.4	14.3	0	0	7.1	14.3	0	7.1	7.1

Standard Week no.	Nil rainfall	0.1 to 5.0	5.1 to 10.0	10.1 to 15.0	15.1 to 20.0	20.1 to 30.0	30.1 to 40.0	40.1 to 50.0	50.1
30.	50.0	14.3	0	0	0	0	14.3	7.1	14.3
31.	42.8	21.4	7.1	0	0	0	14.3	0	14.3
32.	57.1	0	7.1	7.1	7.1	14.3	0	0	7.1
33.	42.8	0	14.3	14.3	0	14.3	7.1	0	7.1
34.	50.0	0	7.1	0	0	0	14.3	14.3	14.3
35.	50.0	7.1	0	7.1	7.1	7.1	21.4	0	0
36.	50.0	7.1	7.1	0	0	7.1	7.1	7.1	14.3
37.	71.4	0	0	0	7.1	0	0	7.1	14.3
38.	85.7	7.1	7.1	0	0	0	0	0	0
39.	85.7	0	0	0	0	0	7.1	0	7.1
40.	0	0	0	0	0	0	0	0	0
41.	0	0	0	0	0	0	0	0	0
42.	0	0	0	0	0	0	0	0	0
43.	0	0	0	0	0	0	0	0	0
44.	0	0	0	0	0	0	0	0	0
45.	0	0	0	0	0	0	0	0	0
46.	0	0	0	0	0	0	0	0	0
47.	92.8	0	0	0	7.1	0	0	0	0
48.	92.8	0	0	0	0	7.1	0	0	0
49.	0	0	0	0	0	0	0	0	0
50.	0	0	0	0	0	0	0	0	0
51.	0	0	0	0	0	0	0	0	0
52.	85.7	14.3	0	0	0	0	0	0	0

The weeks in which the highest probability of measurable rainfall occurs at Osian stations is as follows. The same occurs in August.

Table 11.1.6. Highest Probability of Measurable Rainfall

Osian	1	29th week	1	79%

From the above data, it is seen that it is the few spells of good rainfall intensities in some weeks which would increase the soil moisture storage which is later utilized by the vegetation in the subsequent weeks.

11.1.4.1.2 *Seasons:*

The weather is strongly seasonal and the year can be divided into 4 distinct seasons like:

Winter	December – February
Hot weather period	March – June
Monsoon	July – September
Post-monsoon	October and November

A general description of weather over the district and the synoptic features is given below season wise.

11.1.4.1.2.1 *Winter*

During winter, this part forms part of the subtropical high pressure belt. With the result, clear skies, fine weather low humidity and large diurnal variation of temperatures and northeasterly winds characterize the climate of the tract during the season. But during the occasional passage of western disturbances which are either low pressure areas or waves with associated upper air troughs in the westerly and which usually move from west to east across northwest India. During winter, cloudiness occurs in the tract. Further, on occasions when there is a formation of secondary or western disturbance over Rajasthan, there can be occurrence of rain also. The contribution of such winter rainfall is however not much (only 3%). The humidity is very low during winter. The lowest vapour pressure values in the year are recorded in the winter months only. During the season, the wind directions are steady north easterly. Wind speed in Osian region is very high throughout the year explaining to a certain extent, a large number of barren sand dunes in those localities and the oscillations in their periodic movements. Though winter rainfall contributes only 3 to 5% of annual rainfall, the rainfall values in respect of rare winters are very high as indicated in the total values in the wettest month and the heaviest rainfall in 24 hours. These usually occur when induced 'low' from over Rajasthan in association with moving western disturbances, resulting in large influx of moist air from the Arabian sea.

11.1.4.1.2.2 *Hot Weather*

The temperature begins to increase in the tract from the second fortnight of March. However, high temperature commences to occur only from April. The hot weather season is one of continuous increase of temperature and decrease of barometric pressure over the tract. During summer there is a steady northward transference of region of greatest heat in the Indian continent. Simultaneously the equatorial low pressure belt is also shifted north-wards. The high pressure system over north-west India during winter gradually breaks down and gives away to a low pressure area with a trough extending upto Chota Nagpur. The desert in north-west India becomes the location of greatest heat in the country. May and June are the hottest months of the year. During these months dry and hot dust raising winds, popularly known as 'Loo' and dust storms (andhi) occur very frequently. The minimum temperatures in Osian are lower on account of the higher nocturnal cooling provided by the desert sands around Osian. Though the normal monthly value at these stations during the hottest months of May and July are only 40°C occur often causing heat waves. There is an abrupt fall in the normal maximum temperature in the beginning of June as well as in the first week of July. The former is probably associated with the general decrease of temperature and increase of humidity pattern in north India occurring in association with the onset of the S.W. monsoon in the country in Kerala. The latter coincides with the period of the normal onset of the south-west monsoon in Jodhpur region. The highest daily normal is 42.9°C for 26th May while the five day mean value for the period 26th to 30th May works out to be 42.5°C. The fluctuation of daily means around five day mean pattern is not prominent. Mean daily maxim temperatures exceeding 40°C occur during the period from 26th April to 22nd June.

During 25 years under study, the numbers of heat waves that occurred in the months of April, May, June and July are 21, 14, 10 and 29 respectively indicating probabilities of occurrence of 0.84, 0.56, 0.40 and 1.16 heat waves for year on the average. Of the above mentioned heat waves, 3 in April, 1 in May and 4 in July became wholly or partly severe during their duration. The rest of the heat waves were of moderate intensities. The comparatively larger occurrence of heat waves in the month of July is of significance from the agricultural point of view since crops would have been sown and young seedlings emerged are likely to be damaged by these heat waves. Since the actual normal values themselves are very high (40°-43°C) during May and June and there are comparatively less occasions of heat waves when individual values exceed the normal by 4°C or more. The relative humidity values in the hot weather period especially during March to May are lower than in the winter. However, if the vapour pressure values are compared, they clearly indicate that the actual humidity is higher in summer than in winter. The lower relative humidity values in summer are mainly due to higher temperatures occurring in those seasons. Particularly humidity increases by the third week of May. The Month of June experiences a high vapour pressure of 26 mb though rain does not occur much in the district during that month due to the fact that the normal onset of monsoon takes place only by the last week of June. A significant change in wind regime takes place. Light and variable winds of cold seasons are gradually replaced by strong and steady winds from south-westerly direction. June is the windiest month of the year. High extreme values of rainfall intensities for the individual months as well as for 24 hrs. in a few rare summers also indicate clearly the erratic nature of rainfall in Osian.

11.1.4.1.2.3 *Monsoon Season*

The normal date of onset of south-west monsoon over this tract is the first week of July. But the dates of commencement fluctuate widely with the onset of monsoon, the weather cools down providing a welcome relief after the prolonged heat spells during summer season. The mean vapour pressure in tract is 28 to 29 mb in July and August and 24 to 26 mb in September and thus the humidity over the district is generally as high as the neighbouring area of east Rajasthan and west Madhya Pradesh (average vapour pressure for Kotah with 842 mm annual rainfall is 28 mb and for Bhopal with 1209 mm rainfall the value is 27 mb and so on) getting much higher rainfall. Inspite of such high humidities in the surface air as well as in the upper air, the rain is scanty over the district mainly due to the existence of anticyclonic circulation with subsiding dry air in the upper levels about 2 km during monsoon season which is the main cause for the dissipation of the clouds formed by the ascending moist monsoon air. It is only when westward moving monsoon depressions reach west Madhya Pradesh and east Rajasthan and bring in a large volume of deep moist air, the above divergence layer is penetrated and well distributed rain occurs over the tract. Occasional movement of these depressions through west Rajasthan causes floods as shown by high rainfall intensity figures indicated in Tables. Large scale dust movement in atmosphere over the arid zone of north-west India due to wind erosion during summer months is also listed recently as a possible cause of the occurrence of considerable divergence over the area and associated lack of rainfall. Thus establishment of good and dense pasture over the district by proper management and by controlled grazing would decrease considerably the soil erosion hazards in the district.

The major rainfall in the tract occurs during monsoon season. The period July to September accounts for 78% of annual rainfall. Due to occurrence of pre-monsoon thunder showers and occurrence of monsoon rain in case of years of early onset of monsoon/June also accounts for 10% in this tract. The important features of the rainfall during the season in the tract is high variability from year to year. This tract is susceptible to occurrence of very high rainfall intensities in rare years. Such high intensities cause flood havoc in this desert tract. Wind speeds continue to be very high and steady south westerly during July to September also, broken only when active monsoon conditions persist in the area when light north easterly winds blow. This tract receives the strongest wind regime in Jodhpur district.

11.1.4.1.2.4 *Post Monsoon Season*

The monsoon withdraws from this tract by the second or third week of September. A second hot season prevails over till the end of October. The temperature curve shows a secondary peak during October. The humidity values decrease considerably. The winds are light and variable during the season and the least mean monthly wind speed in the year occur during the season.

11.1.4.2 *Geology*

The rocks present a range from Early Palaeozoic to Recent ages. These are mainly under volcanics and plutonics of Malani igneous suite, sandstone and limestones of upper Vindhyan system and alluvium of Quaternary system.

11.1.4.2.1 *Geologic Units and Their Properties*

11.1.4.2.1.1 Alluvium

The alluvium comprises of wind blown sand kankar, silty clays and sand with occasional gravel. It is in general a heterogeneous mixture with tongues and lenses of granular zones within clayey matter. It is mainly formed of transported fluviatile matter with basal parts and weathered bed rocks in situ. It generally lies above the zone of saturation, except when thick and water table being shallow. Wells located in this horizon have low to moderate yield of water and the quality of water is relatively inferior due to poor permeability of the formation and subsurface drainage. The alluvium is underlain in general by Vindhyan sandstone over most part of the area and by the Malani suite of igneous rocks in the south eastern and south western parts of the tract. In some well sections the alluvium has attained a thickness of about 18.00 meters and overlies the Malani volcanics.

11.1.4.2.1.2 Bhander Limestone

The limestones are exposed as low raised stony waste hummocks in the extreme northern part of the tract. These are dolomite and cherty in nature. These are grey, fine grained, break with smooth conchoidal fractures and often show vague stratification which can be inferred from the general lie of the cherts. When broken the limitation often gives an odour of hydrocarbons. These generally strike N.S. and dip at low angles 4° – 8° towards east. It is often concealed under the wind blown sands. The lime stones have not been tapped by wells except one, which yields moderate supply of water through solution channels.

11.1.4.2.1.3 Bhander Sandstone

These are exposed over large part of the tract and forms flat topped hills and plateau in the region. These are fine to coarse grained, at times hard and compact, ferruginous and occasionally gritty sandstones. Few bands of conglomerates have also been observed in the bore holes sections. It indicates shallow deposition characters like current bedding and ripple marks. Veins of gypsum are found intercalated with sandstones indicating arid climate during sedimentation. The shale becomes frequently intercalated with sandstone, particularly at depth. The sand stones generally strike east-west to NE-SW with very low dips towards 2° to 5° north and south. The two distinct varieties of sandstone can be classified with respect to their water bearing properties. One is gritty porous, semi-consolidated, thin bedded ferruginous sandstone, a potential source of ground water. The other is consolidated, hard sands-tone, mostly serving as confirming bed when occurring below the zone of saturation. The sandstone achieves varying thickness between 50-160 meters as these have been deposited on uneven floor of volcanic. The well constructed in sandstone, yield moderate to high discharge of good quality.

11.1.4.2.1.4 Malani Volcanic

The Malani volcanics are basically felsic lavas and pyroclastic mainly rhyolites of porophyritic texture. The phenocrysts of orthoclase and occasionally of Quartz are mainly observed in the ground mass of the porphyries. The volcanics in general are impervious and hard and little water moves through joints, fractures and other partings. These partings have tendency of decreasing at depth restricting the deeper percolation of water. Few wells have been constructed in the volcanics which yield meager discharge.

11.1.4.2.1.5 Jalore granites

The Jalore granites which probably represent an early phase of the same magomatic (of Malani volcanics) are fine to coarse grained, non-foliated, non-prophyritic/plutonics of pink and grey in colour. With few isolated surface exposure, it is clearly exposed in the well section of village Gangani (26°30' : 73°13'), where it is represented as coarse grained hornblende and biotite granite. The water is mainly held and transmitted through the joints, sheeting partings and weathered mantles. The wells tapping weathered sequence yield higher discharge as compared to those sunk in compact and hard granites. In one of the well section, near village Gangani (adjoining to granites), Epidiorite was observed, along with the formation of secondary minerals like Zeolite and calcite.

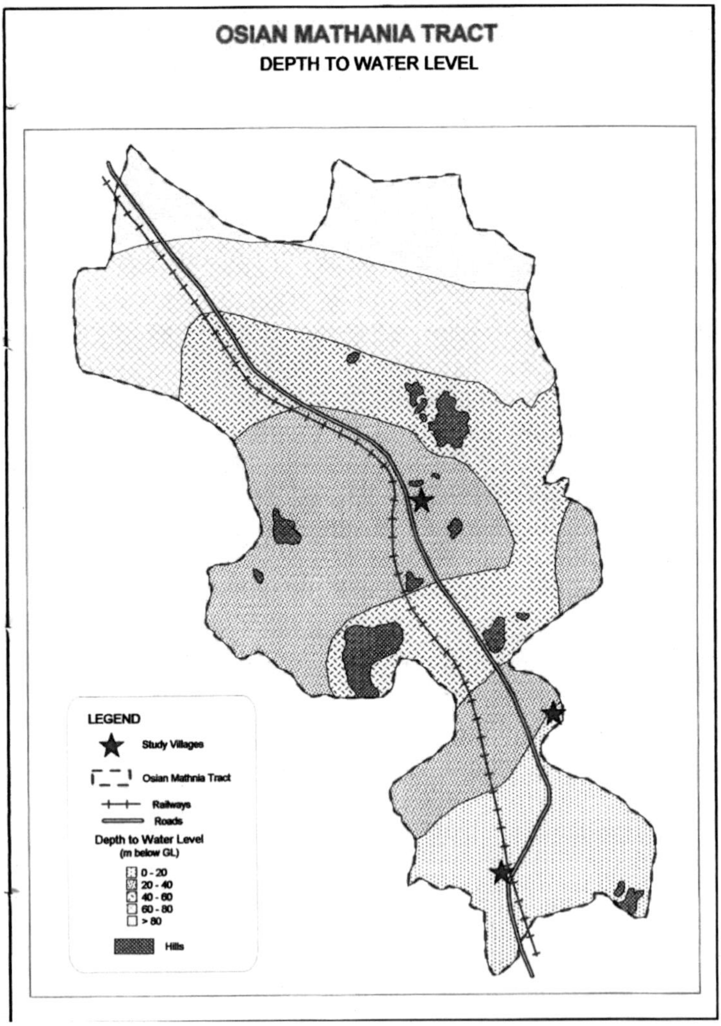

Figure 11.1.3 Depth to water.

11.1.4.3 *Groundwater Potential*

11.1.4.3.1 *Groundwater Occurrence*

Report of Ground Water Board Rajasthan reveals that the principal source of ground water in Osian tract is sandstone of Vindhyan system. The yield of wells tapping sandstone is highly variable, as its hydrologic parameters are many and independent of each other. The ground water occurs in the pore spaces of gritty, semi-consolidated sandstone and in the joints, bedding planes of free stone. Occasionally solution cavities have also been located while drilling. The ground water generally occurs under water table conditions except in localized places, where it occurs under low hydrostatic pressure, due to overlying compact impervious sandstones and argillaceous sandstone.

The rocks of the Malani suite of igneous rock have very low permeability and the wells tapping water from these are less productive except in cases where the wells are having sufficient weathered mantle. Occurrence of ground water in alluvium and limestone has been observed in a few wells only. These formations in general are relatively less permeable and the wells are less productive. The depth to water table in the tract varies within a wide range and it lies at less than 20 meters in the southern part and as deep as146 meters below the land surface in the north western part of the tract. (Fig. 11.1.3).

Range of depth to water table in the wells inventoried in the tract during the course of reconnaissance is tabled below:

Table 11.1.7: Depth to water below ground level (bgl)

Depth to water below land surface in meters	No. of wells measured in the range	Percentage of total wells inventoried	Remarks
20	37	41	76% of the wells inventoried in the tract have water levels with 60 metres from the land surface.
20 – 40	25	27.5	
40 – 60	7	7.5	
60 – 80	5	5.5	
80 – 100	13	14.3	
100 – 120	2	2.2	
120 – 140	1	1.1	
140 – 160	1	1.1	

11.1.4.3.2 *Groundwater Recharge*

The ground water in the tract is mainly recharged through precipitation within the tract and the adjoining areas. A part of this precipitation is lost due to evaporation during rainfall and a part is retained by the soil, which is lost by later on evapotranspiration. Since there is no well developed drainage, the surface runoff is negligible and significant part joins the ground water storage. The amount and intensity of precipitation, composition and texture of the soil, the nature of the exposed and underlying rocks, the vegetation, the geomorphic features of the region, and the depth of the zone of saturation are amongst the important factors which control the ground water recharge in area through precipitation. Due to continuous drought years, there had been limited recharge of ground water in the area. This has resulted in the general lowering of water levels in the areas which are being intensively exploited and less increase in ground water storage, where the ground water exploitation is limited.

11.1.4.3.3 *Ground Water Extraction*

The ground water in the area is mainly extracted by transpiration, subsurface outflow and by wells. The transpiration is maximum where the deserted flora is thickly grown and water table is relatively not deep. The ground water replenishes the capillary fringe which mainly supplies the water for plant growth. The sub-surface outflow is noticed in the southern and south-eastern parts of the tract as is evident by the general direction of the ground water flow. The main source of ground water discharge is from 1520 wells, which are being operated for irrigation and domestic purposes. The pumping from the irrigation wells is maximum in the southern parts, particularly in Rampur-Mathania basin and Teori basin and has been estimated to be about 25 million cubic meters annually. The water level fluctuations in the wells have been observed to varying within 15 meters and is due to changes in balance between ground water recharge and discharge.

11.1.4.3.4 *Agricultural Quality*

The water used for irrigation has profound effect on the nature and properties of the soil and influences crop growth to a great extent. These effects are caused by the chemical constituents present in the water. Thus, making it essential to study its chemical analysis with due consideration to the following factors:

1. Chemical composition of irrigation water.
2. Nature of the soil irrigated.
3. Nature of the crops to be grown.
4. Climatic conditions like rainfalls, evaporation and drainage etc.

Number of standards have been proposed by various workers in different places for the assessment of quality of water. The one which is discussed here based on salinity and sodium absorption ratio. According to the classification nearly 50% of the water covering nearly $^2/_3$ rd area of the tract (shown in enclosed map showing agricultural quality) fall either in $C_1 - S_1$ or $C_3 - S_1$ class map. There is no alkalinity hazard and salinity hazard is not much to induce salinization of soils. A little excess of water at the time of irrigation shall wash out the excess salt. Nearly 20% waters fall in $C_3 - S_2$ or $C_4 - S_1$ class. These waters definitely have higher alkalinity and may present appreciable sodium hazard on heavy soils, but can be readily used with organic matters and gypsum and on light soils as are available in this tract. The rest waters fall in $C_4 - S_2$, $C_5 - S_3$, $C_5 - S_4$, $C_6 - S_4$, class and as such deserve the specific attention due to their both higher salinity and alkalinity. They belong to doubtful to unsuitable class, but in actual practice it is observed that crops at medium to high tolerance can be grown without reduction in the yield, particularly on light and permeable soil with special management practices. Adequate amount of water will have to be supplied, so as to leach out the salts.

11.1.4.3.5 *Present Development of Groundwater*

Groundwater for irrigation and domestic purposes is chiefly obtained from 1,520 wells in operation in the tract. The wells are being pumped to their maximum and new wells are under construction every year. These are large diameter open dug wells; a good number of them depend by boring and about 30 tube wells. About 25% of the wells are fitted with pumping sets in the area and majority of them being operated by electric power. Wherever tube-wells have been sunk and electric power is not available, generators on individual tube well have been provided.

11.1.4.3.6 *Groundwater Yield*

The yield of wells in the area varies from 0.10 to 37.0 Lakh liters per day. Open dug wells which have been bored by 50 to 100 meters and tube wells sunk in semi-consolidated gritty porous sandstones of Vindhyan age, particularly in the south-eastern part of the block have been observed to yield maximum discharge in the region. Wells from other formations like Malani suite of igneous rocks in weathered portions, alluvium and limestone of Vindhyan age yield meager to poor quantity of water. Table 11.1.8 gives the water yielding capacities of different aquifers in the tract.

Table 11.1.8: Water yielding capacities of different lithological formations.

S. No.	Water bearing formations	No. of wells observed	Maximum observed capacity in lpd	Minimum observed capacity in lpd	Average capacity in lpd
1.	Alluvium	3	4.50	0.15	0.20
2.	Sandstone	86	37.00	0.10	0.43
3.	Limestone	1	0.36	0.36	0.36
4.	Rhyolite porphyry	2	1.50	0.60	1.00
5.	Granite	2	0.20	0.20	0.20

11.1.4.3.7 *Groundwater Utilisation*

The agricultural statistics of the tract indicates that only 4.7% of the cultivable land in the tract is being irrigated by wells and a vast cultivable land is lying without irrigation due to lack of irrigational facilities. Such a poor development of groundwater and consequently agricultural development of the tract especially in the northern and north western part of the tract is due to the following reasons:

1. The deep water levels in the north and north-western parts of the block and consequently expensive pumping for water.
2. The deep water level, pumping cost by diesel division pumps high, which make agriculture uneconomical.

11.1.4.3.8 *Groundwater Potential Zones*

The zone marked "A" in the map, mostly around Mathania; comprises of semi-consolidated, gritty, ferruginous sandstones with occasional conglomeratic horizon. The depth of water in this zone ranges between 10 to 20 meters and the existing bored wells yield high discharge. Few wells, which have encountered water bearing cavities at depth, indicate exceptionally high discharge to the extent of 1.5 Lakh liters per hour. The majority of the existing wells in this zone have not been deepened so as to lap the entire thickness of the aquifer with the result that the yield is limited. The discharge of such wells can be increased considerably provided these are deepened by boring for another depth of 50 to 70 meters. This zone is also suitable for construction of shallow tube-wells to a depth of about 100 meters. The discharge from the bored wells and tube wells is expected to vary between 25,000 to 50,000 liters per hour. The quality of water is extremely superior for domestic and irrigation purposes. The present annual pumpage from this zone is to the extent of 25 million cubic meters, which is definitely much more than the anticipated recharge. The post-monsoon inventory has also indicated the lowering of water levels as compared to the pre-monsoon levels. It is, therefore, proposed that the zone marked as "A" may not be exploited further, which otherwise is likely to bring serious depleting conditions. However, the electrification programme of this area can be implemented, which will make the irrigation more economical and convenient.

Another potential zone marked as "B" in the plate lies in the north of zone "A". The water table is relatively high and lies at a depth of 15 to 25 meters below ground level. The existing wells, yield low discharge, but this can be increased by further boring to the extent of 100 meters. The dug-cum-bore wells and tube-wells sunk to an average depth of 150 meters are expected to yield 20,000 to 35,000 liters per hour. The quantity of water is suitable for domestic and irrigation purposes. The remaining area in zone "C" particularly in the north has been observed to be not so productive and therefore, have limited scope for exploitation for agricultural purposes. The deep water levels ranging from 90 to 146 meters and very low permeability are the main handicaps for its development for irrigation. This area can, however, be exploited for drinking and livestock requirement, by way of tube-wells which will yield sufficient potable water.

11.1.4.4 *Geomorphology*

The dominant theme in this account is the close dependence of the landforms on Aeolian action. A secondary theme is the influence of fluvial action in sculpturing the landform. Similar in importance is the influence of geological structure on landform, which forms the base and which has already been described earlier. The area represents an intermont basin landscape towards the north round Osian and south around Mathania with the land towards the center from all around. The central part represents a piedmont surface at an altitude between 300 m to 260 m (W-N.W. to S.S.E.). The piedmont surface is often interrupted by isolated hills of Malini volcanic and Vindhyan sandstone highly weathered and sand hills. Constant deposition of fresh sands are well evident throughout the area. The region slopes in general from north (360 m) to south 260 m. Drainage lines are ill-defined and internal network of integrated drainage are noticed only along the hills, towards the north, round Osian. Each drainage line has its own local base level. In all the cases valley floors are considerably wide. The adjustment of the drainage to the structure can hardly be observed due to original topography being mostly filled up by porous alluvium and in blown sands. All the rivers in the north and south likewise, are lost in the sandy deposits of central piedmont plain, or mostly in the detritus brought down by them during the period of 'cloud bursts'.

The present topography is marked by steep sided residual mountains and hills on an extensive and very nearly level plains (300 to 350 m summit). The hills are too high to be mere monadnocks on a peneplane surface. The steep sides and abrupt transitional are due to the process of weathering. The steep gradient persists to the bareness of every slope. This cannot be explained due to 'normal cycle'. With a constant slope of the hill sides, it appears that the nearly level extensive plain is due to back weathering of the mountains. The back weathering has caused the 'slope retreat' of the hills which has resulted in the replacement of the hills by gently sloping piedmont formed progressively at the base of the back-worn slopes, diversified in the north and south by isolated residual hills. The landscape has thus a characteristic resemblance with 'Isenberg' type of topography essentially produced by mechanical weathering and wind erosion. There are enough evidences to show, as accounted below, that there was a long period of aridity, before the present relief had been attained, which characteristically modified the form of relief features. Following observations may be recorded:

(i) Along the hill front of Osian and Mathania, due to sudden change of slope under the continued process of weathering of bare hills, all slopes, remain steep. This is kept mainly due to wind abrasion at the base. The slope, once determined, remains constant.

(ii) Towards the north east, near Gausainji Ka Bhakar and further north dunes are formed on structural terraces.

(iii) Continued wind action has eroded the softer materials in Mathania and Osian hills.

(iv) The highly weathered and disintegrated rocks deposited at the lower reach of the drainage line show the effects of a long period of aridity and mechanical weathering.

(v) In north, on the south western edge of the hills area number of isolated irregular sand hills or dunes. The sands are carried by the prevailing wind and brought down from the high relief on the isolated irregular hills, where they played round in the form of dunes.

11.1.4.4.1 *Geomorphogeny*

It is evident that the present relief form is mostly due to the relative importance of wind over water action. The arid conditions in north west India were established long after the retreat of the last glacial phase and only in recent period. Heron (1938) examined three peneplane surfaces in Rajasthan and according to his finding the area was probably peneplaned during the mesozoic. The peneplane has since been wrapped so that the position near Osian is 350 m and Mathania and south 260 m. It might be that the area was again uplifted during Tertiary. The water cut valleys with overlapping spurs in south of Osian show the past fluvial action even at the beginning of quaternary era. But the absence of such features on the central part on the piedmont surface is a challenge to the relative importance of humid cycle to form the present relief. Due to the retreat of the last phase of ice sheets these might have occurred an enstatic change of base level that caused the climatic accident in the subtropical desert belt of the globe. The resultant earth movement might have produced fracture to give intermont basin as marked in the central part of the tract under study, that initiated the arid cycle. In the hills and uplands south and around Osian Wadis have markedly developed due to cutting in of gullies by intermittent streams. Narrow rock piedmonts in front of the hills are evident. These are clear evidences to show that some relief features were formed in the youthful stage of an arid cycle.

The retreating of slope and their remnants, flanked by outlying hills, rising abruptly from broadening piedmonts, which slope down the plains, in evident in Osian and Gosainji Ka Bhakar area – a feature characteristic of the maturity of an arid cycle. The present landscape with piedmont and narrow pediment along the sloping hill-side and isolated hills of Malani volcanic due to the back weathering of mountains took place much later only when the uplands were lowered with the decreased rainfall. The general lowering continued due to deflation and abrasion. This occurred in the stage of late maturity or early youth. Wind erosion continued to lower the surface towards the water table – local base level of an arid cycle. It might have been that a rise of water table caused the termination of the arid cycle. It seems that the arid cycle continued to early late stage or at least to late maturity. The present land surface of the plain land is a depositional one – sand deposits which often led to the formation of dunes. In the recent and sub recent depositional land the history of landscape prior to quaternary cannot be traced due to the thickness of land and other superficial deposits. The pattern of ill-defined drainage lines indicate dissection even in the very recent periods before the climate change to desert conditions. The evidences of the cyclic land surfaces are not clear, but the contour indicates that the desert falls in N.W. direction. It is possible, therefore, that the original topography represents pleistocene peneplane surface.

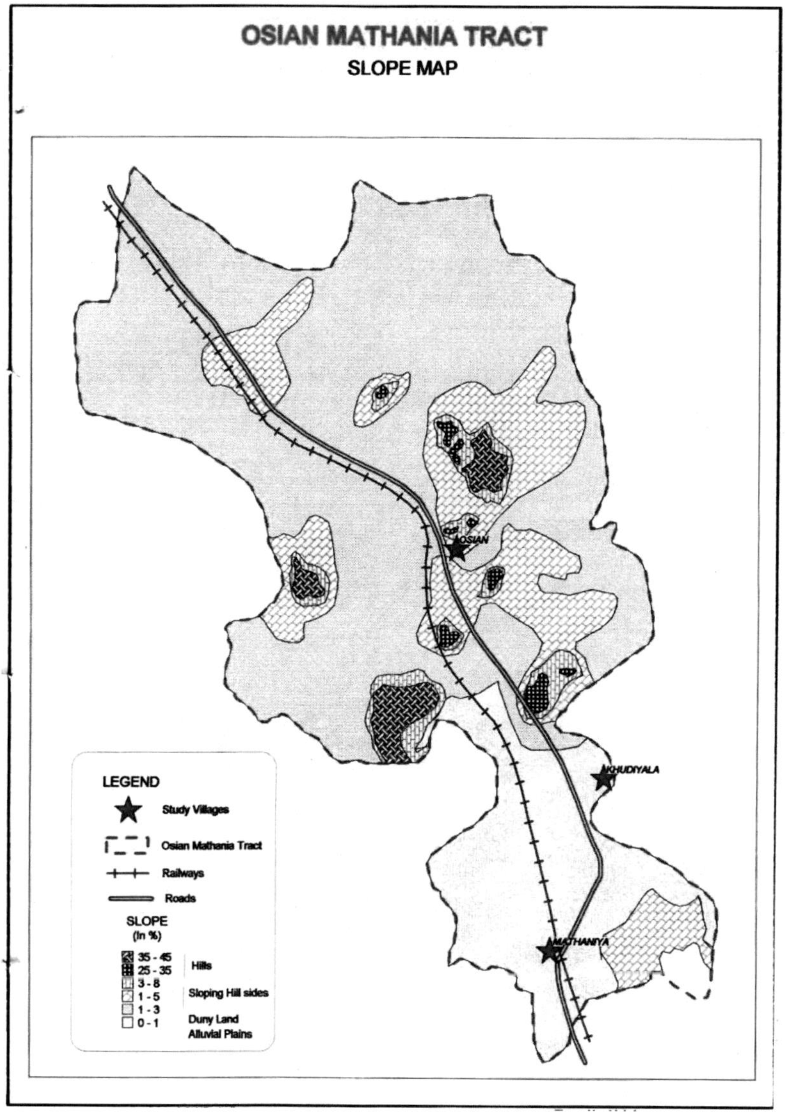

Figure 11.1.4 Slope

11.1.4.4.2 *Slope and Erosion*

The region exhibits wide variation in slope condition (Figure 11.1.4). The same is described as below:

- In the hilly areas the slope is high. In the sandstone hills % of slope varies from 25% to 40% and in some areas in the north it often exceeds 40 to 45 per cent. The rhyolite hills exhibit 20 to 40 per cent slope. Along the front of the hills the slope is 3 to 8 per cent in the zone of pedimentation. The bajadas have slope varying between 1 to 5 per cent. The slope is 0 to 1 per cent in the piedmont tract and in the alluvial tract of the plains of the intermittent streams.

- The gullying action has increased the variation of slope % along the hill fronts.
- Gullies and rill erosion dominate the hills and pediment areas. The degree of erosion is very high affecting 60 to 75 per cent of the surface cover. Towards the north of Mathania, along the hilly terrains of the Osian and the entire tract belonging to Osian and north is affected by wind erosion which covers 70 per cent and above of the surface cover. The degree of erosion varies from slight in the south to high in the north (Fig. 11.1.5).

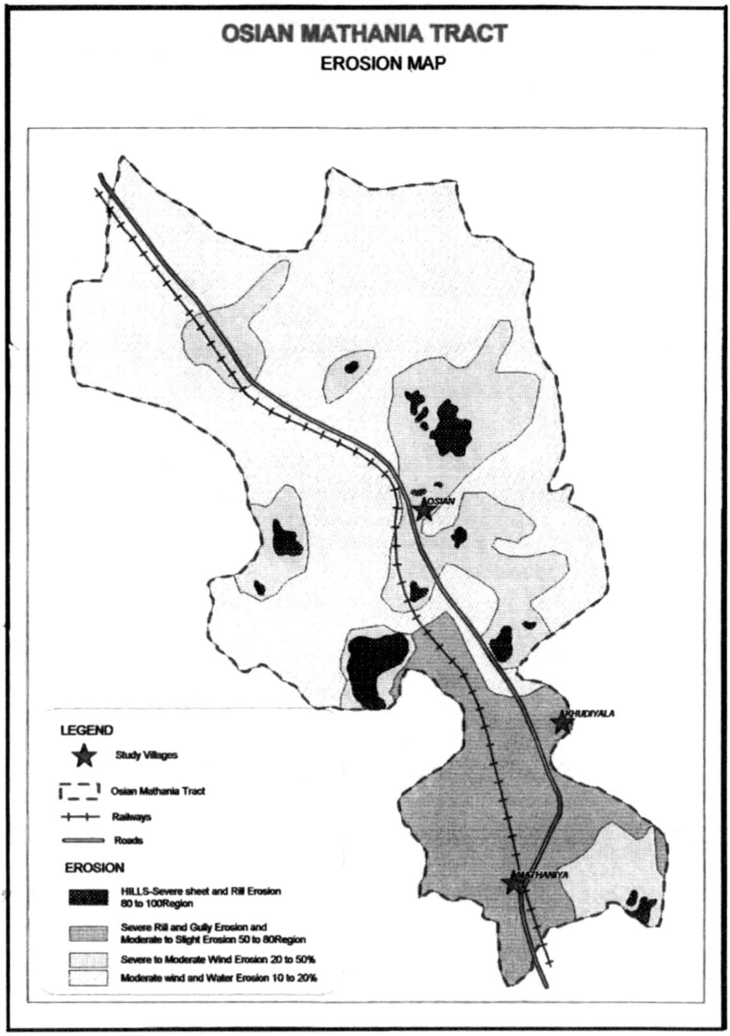

Figure 11.1.5 Erosion

The extent of geomorphic units are shown in Fig. 11.1.6 and their respective areas.

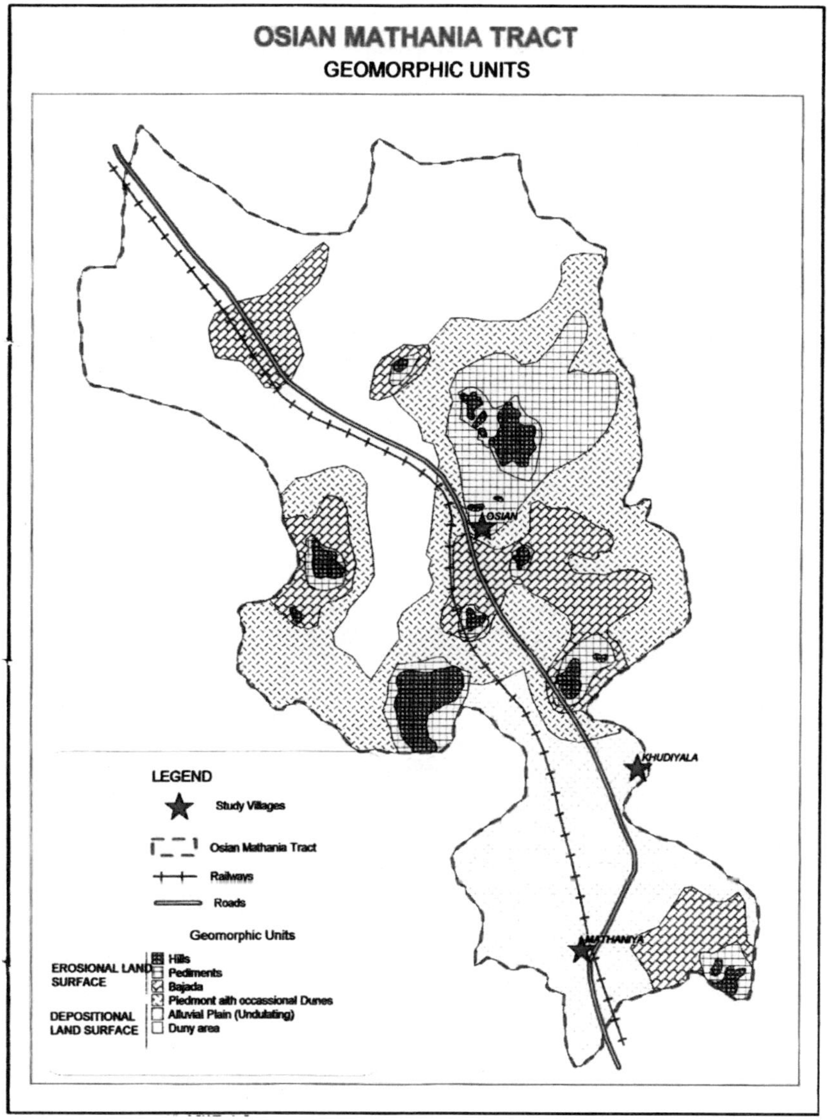

Figure 11.1.6 Geo units

From the foregoing account, it is possible to classify the region into following geomorphic divisions (Fig. 11.1.7)

1. The undulating terrain of Mathania in the south. This extends from Mathania in the south to Krimriya in the north – a desert steppe.
2. Krimiriya to south of Bigchi – the piedmont.

3. Bigchi to Osian: Transitional between rocky to sandy desert. Deposition of sand and formation of dunes along the mountain fronts are common.
4. Osian: Bhikamkor area: Sandy desert and duny country.

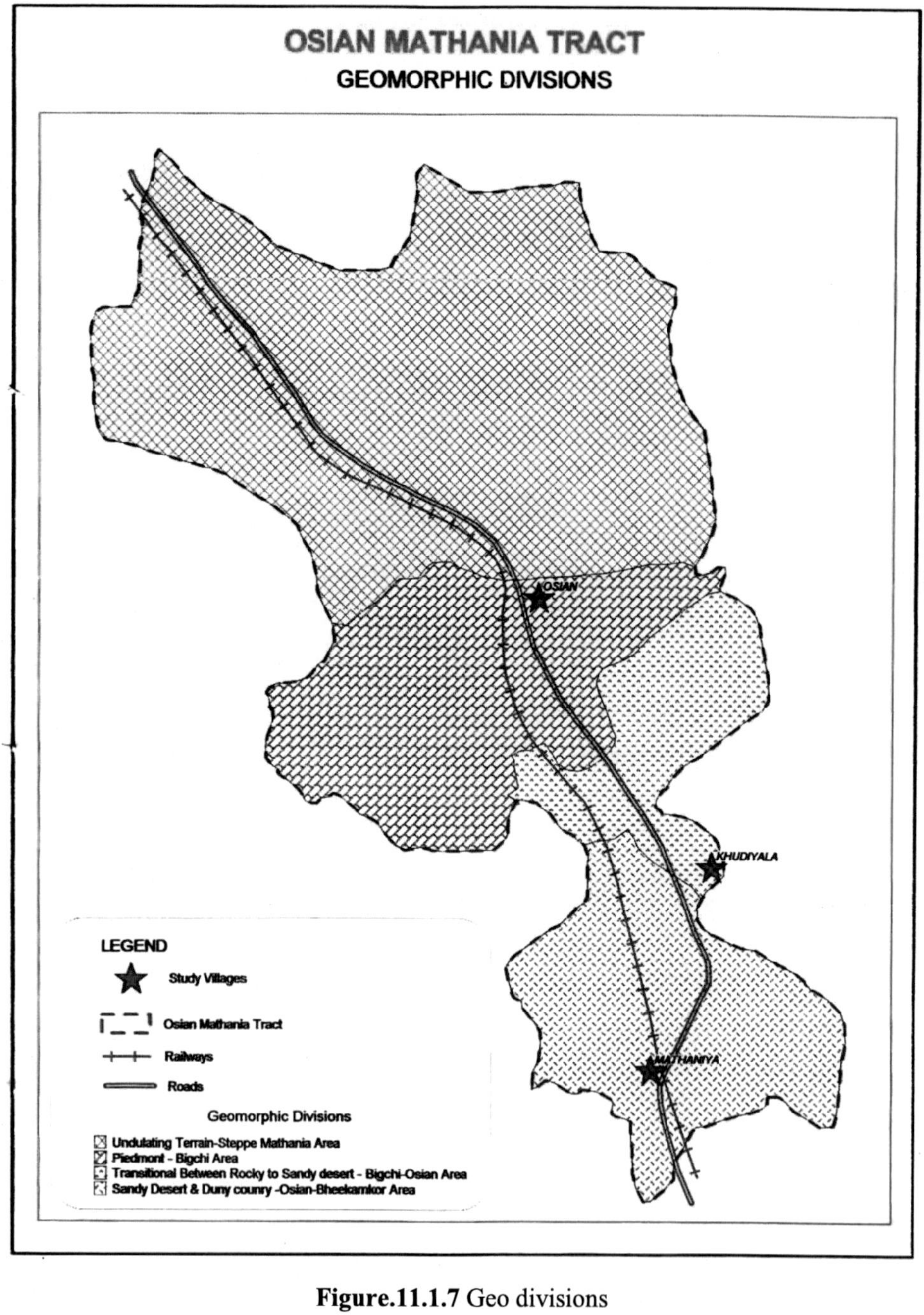

Figure.11.1.7 Geo divisions

11.1.4.4.3 *Land Type*

Based on the investigations and data obtained it is possible to classify the area into three land types – a concept which is based on the homogeneity of the land form, soil and vegetation (Fig. 11.1.8).

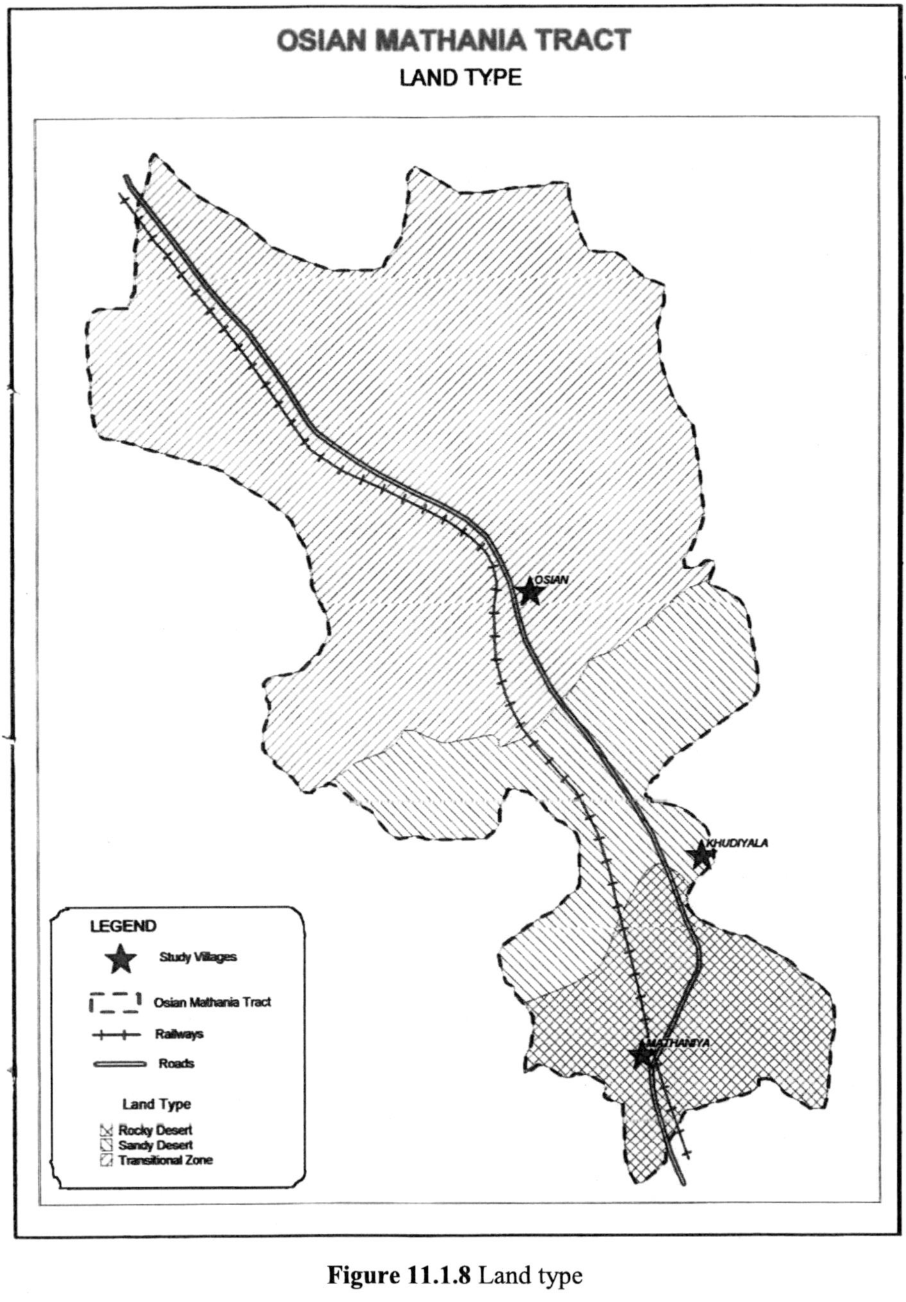

Figure 11.1.8 Land type

These are: 1. Rocky desert in the south round Mathania and Umednagar; 2. Sandy desert in the north in Osian (central) and Bikhampur area; 3. The central region from Umednagar to Bigchi – the zone transitional between rocky and sandy desert. This is the region where the rocky desert gradually merges into the sandy desert. The sandy desertic conditions, however, predominates.

11.1.5 Soils

Soil cover in Mathania-Osian tract shows considerable heterogeneity because of complex history of landscape evolution (Figure 11.1.9) Light textured old aggraded alluvial plains further transformed under the highly arid phases of Late Quaternary have soils with a moderately differentiated solum. The associated dunes are almost devoid of any significant soil forming features. Soils developed from weathering products of Mathania have their own peculiar characteristics. A brief account of these soils is as follows:

11.1.5.1 *Soils of light texture (Brown to light brownish grey soils*

These soils are deep to very deep with a light yellowish brown to pale brown (10 YR $^{6}/_{4}$-$^{6}/_{3}$ D), wind sorted sandy, slightly to moderately hummocky surface, a pale brown to yellowish brown (10 YR $^{6}/_{3}$-$^{5}/_{4}$ D), loamy fine sand, very weakly blocky, slightly to moderately calcareous subsoil underlain by a weakly to moderately developed zone of calcium carbonate concretions at depth of 90 cm or more. Most of the field boundaries in the cultivated land have fence line hummocks and besides the field proper also generally shows slight to moderate hummockiness. The thickness of wind sorted sand varies at most places 10 to 30 cm. *Present land use:* Almost the entire area of this series is under cultivation with bajra, moong, moth, gaur and till and main crops. *Extent and distribution:* These soils occupy a large area in the west, south-east of Osian. The important village are as Bedhwasiya, Kharda, Mewasa west of Tinwri, Khari Khurd, Bigchi, Mandia, Khudiala, Basani, Neora, Dunana Bhawad and west of Umednagar.

11.1.5.2 *Brown to pale brown dune soils*

The dune soils are very deep, loose, pale brown (10 YR $^{6}/_{3}$ D), slightly calcareous sand with low water salinity capacity. **Range in characteristics :** The colour of sand is dominantly pale brown (10 YR $^{6}/_{3}$ D) some time light yellowish brown (10 YR $^{6}/_{4}$ D). The soil masses is loose, structure-less, single grained with a clay content of 4.9 to 5.1% and that of silt from 0.8 to 1.7%. Calcium carbonate content is 0.2 to 0.7 sometime ranges 2.8 to 4%. Moisture eq. 3.2 to 4.7 and WHC 23.2 to 27.1%. *Present landuse:* The bulk of the area of this series in Osian tehsil is under cultivation in good rainfall area. In the low rainfall area the dunes are oultivated occasionally. Crops are taken once in 3 to 5 years and rest of the time the lands are kept fallow and used as open grazing lands. *Extent and distribution:* This is distributed in west, east and south east of Osian proper.

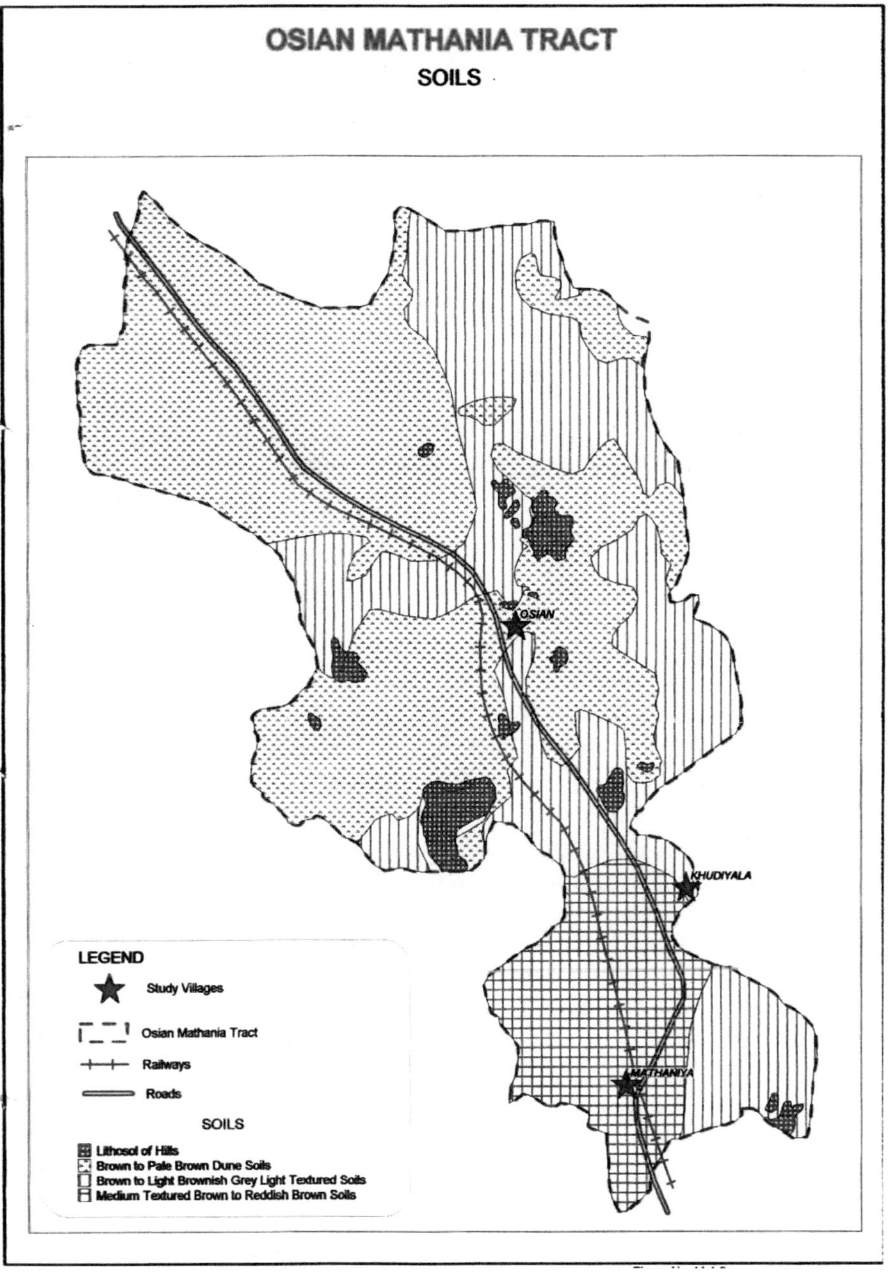

Figure 11.1.9 Soils

11.1.5.3 *Medium textured brown to reddish brown soil*

These soils are characterized by a reddish brown (7.5 YR $^{5}/_{4.5}$ YR $^{4}/_{4}$). Generally loamy sand surface and sandy loam subsoil followed by a hard weathered micaceous sandstone at depths of mostly 30 to 50 cms. The soils and substrata are non-calcareous. There are frequent exposures of a sandstone and the area occupied by these varies from 20-40%. Although soil depth is a limitation, the area occupied by these soils has been found as good aquifers and irrigated farming is practiced extensively. Sandstone exposures occur as small pockets of 1 to 5 ha within the larger area of moderately deep soils. *Range in characteristics:* Texture of the surface soil is invariably loamy sand and that of subsoil sandy loam occasionally loamy sand with a range of 10.3 to 18.4% clay and 5.9 to 7.4% silt. Colour of subsoil is 7.5 YR 5/4 to 5 YR 4/3 in dry state. Moisture equivalent values are mostly between 8.4 to 11.2% in the subsoil and WHC ranges from 27.0 to 35.7%. *Present landuse:* Except rocky and rock outcrop area, almost the entire area is under cultivation and a sizeable part of it mostly in Mathania, Rampura, Twiri is under irrigation with high value cash crops. Under unirrigated condition Bajra and Guar are the common crops. *Extent and distribution:* These soils are distributed between the Umednagar and Mathania. The important villages are Rampura, Binzwaria, Jur, Guidya, east of Tinwri, Rajasani and east of Umednagar.

11.1.6 Vegetation

The sparse and highly variable precipitation, extreme variation of diurnal and annual temperature and high evaporation prevalent in the tract, are also intrinsically reflected in the vegetation types. The physiognomic aspect of the desert is, therefore, poor, scanty xerophytic vegetation, widely dispersed leaving large areas bare. The formation of vegetation in various habitats are enumerated below:

Mixed xenomorphic woodland: This formation occurs on flat older alluvial plains and lower piedmont plains with deep soil deposition such as near Umednagar, Jud, Kudiyala, Neura. The natural vegetation is highly modified because of biotic action and cultivation although in general the climatic climax is *Prosopis cineraria* which is distributed throughout the plains of wester Rajasthan. Depending on the degree of degradation the following plant communities are met with in this formation. *Prosopis cineraria-Zizyphus nummularia-Capparis deciduas* community: This is the most prevalent type on a large part of the flat plains. Because of yearly cutting, *Zizyphus nummularia* does not show any visible impact on the cultivated field whereas this species plays an important role in low-rainfall zones where as a forage (Pala) it is considered as the 'alfalfa' of the desert. Saxena (1978) reported that in such situations, the density of *Prosopis cineraria* varied from 15 to 20 plants/ha, the height from 10-14 m and diameter at breast height 50-80 cm. In village grazing lands or Orans, which are protected more by religious beliefs, the plant density is invariably high ranging from 150-390/ha.

Depending upon the degree of anthropogenic influence and grazing pressure of livestock the species in the plant community undergo considerable changes leading to the establishment of a spectrum of recombined plant communities as follows: 1. *Capparis deciduas-P. cineraria-Z. nummularia;* 2. *Prosopis cineraria-Capparis deciduas;* 3. *Prosopis cineraria-Zizyphus nummularia;* 4. *Capparis decidua;* 5. *Zizyphus nummularia- Capparis decidua* and 6. *Zizyphus nummularia*. The inclusion or exclusion of any one species in the community has immense bioedaphic significance of indicator value. In the absence of all the above species an extremely degraded stage is visible in many parts where *Tephrosia purpurea, Crotalaria burhia* and *Indigofera cordifolia* predominate.

The common shrub associates in all the above communities include *Calotropis procera.* (Aak), *Balanites aegyptiaca* (Hingoto) and *Acacia jacquemontii* (Bawli). Among the common herbs and grasses, mention may be made of *Aerva persica, Tephrosia purpurea, Crotalaria burhia, Convolvulus microphyllus, Heliotropium subulatum, Pulicaria wightiana, Arnebia hispidissima, Celosia argentea, Eleusine compressa, Dactyloctenium sindicm, Eragrostis cilliaris* and *Cenchrus biflorus.* *Salvadora oleoides-Prosopis cineraria* community: This community is generally met with in the lower piedmont plains having very deep sand deposition such as in the neighbourhood of Mathania. *Salvadora oleoides* occurs both in trees and shrub form. Other associated trees and shrubs are *Maytenus emarginatus, Acacia Senegal, Z. nummularia, C. deciduas, Euphorbia caducifolia, Grewia tenax* and *Balanites aegyptiaca.* The plant density is reportedly (Saxena, 1978) ranging from 180-215 plants/ha with 8-10 per cent crown cover. *Lithophytic scrub desert:* This formation occurs on eroded rocky surfaces, gravelly plains of sandstone west of Mathania on the route to Teori, at Bhainsar Kutai and on sides of road from Neura to Osian. These serve as upland link to the adjoining flat plains. The area is generally covered with gravel and rock fragments of various sizes and are subject to water erosion. The soil is invariably very shallow such a habitat supports a scrub vegetation of the following type.

Cappris deciduas community: It is most common on the eroded rocky surface and piedmont plains. It is an open community with low density 5-25 plants/ha and very poor growth. Occasional trees of *Acacia senegal* and *Prosopis cineraria* 2-3 m in height (i.e. stunted) also occurs in this community.

Associated shrubs, herbs and grasses are given below:

Shrubs: *Leptadenia pyrotechnica, Calcotropis procera, Aerva persica, Sericostoma pauciflorus.*

Herbs: *Tephrosia purpurea, Crotalaria burhia, Bonamia latifolia, Orygia decumbens, Cleome papillosa, Cleome brachycarpa, Boerhavia diffusa* and *Indigofera cordifolia.*

Grasses: *Dactyloctenium sindicum, Oropetium thomaeum, Eleusine compressa, Eragrostis ciliaris, Tragus biflorus* and *Aristida hirtigluma.*

Psammophytic scrub desert: As the name implies this formation has its genesis on the sand dunes which become a prominent feature from Osian and beyond such as west of Osian enroute to Khetasar. The size of dunes vary from 9 to 30 m in height. Five types of sand dunes namely parabolic, longitudinal, transverse, barchan and shrub coppice dunes have been recognized in the tract (Saxena and Singh, 1976). The formation of barchan dunes on flat land as also stabilized parabolic dunes was observed near settlements in Osian perhaps because of cultivation of dunes for kharif crops right from lower slopes of the flanks to the crest. The vegetation on the dune is scanty with a limited number of species. The botanical composition of the stabilized dunes varies in different parts of the dune as will be described below:

(i) Dune crest: *Panicum turgidum, Cymbopogon jwarancusa, Panicum antidotale, Aerva pseudotomentosa, Citrullus colocynthis, Farsetia hamiltoni, Crotalaria burhia* and *Cenchrus biflorus.*

(ii) *Leeward side: Calotropis procera, Aerva pseudotomentosa.*

(iii) Dune flanks upper slope: *Clerodendron phlomoides, Acacia jacquemontii, Maytenus emarginatus, Calotropis procera, Cyperus bulbosus, Citrullus colocynthis, Crotalaria burhia.*

(iv) Dune bases and lower slopes: *Prosopis cineraria, Tecomella undulata, Leptadenia pyrotechnica, Lycium barbarum, Aerva persica, Indigofera cordifolia, Lasiurus sindicus, Cymbopogon jwarancusa, Cenchrus biflorus* and *Aristida funiculate.*

Thus one of the most significant points of dune vegetation is the complete absence of *Calligonum polygonoides* in the dunes near Osian which is otherwise a prominent feature in most of the dunes in western Rajasthan.

11.1.7 Occupational Pattern and Classification of Land

The Rajasthan desert is basically agricultural and rural and this is equally true of Osian region which is located in the heart of the desert. Out of the total worker population, about 87% (1991 Census) is engaged in agriculture and allied activities with animal husbandry as an important occupation. Large scale industries are absent. An inadequate net work combined with low educational standard (literacy 20.09%), further add to the industrial backwardness. The main features of occupation in Osian tehsil are given in Table 11.1.9. In a population of 2,56,542 the total number of workers are 31%. These are engaged in various activities as follows:

Table 11.1.9: Main Feature of Occupation in Osian Tehsil

S. N.	Activities	Percentage
1	Main Workers	100.00%
2	Cultivators	77.22%
3	Agricultural Laborers	9.33%
4	Livestock, Forestry, Fishing, Hunting, Plantation Orchards and Allied Activists	0.49%
5	Mining and Quarrying	0.45%
6	Manufacturing, Processing Servicing and Repairs in Household industries	1.60%
7	Manufacturing, Processing Servicing and Repairs in Other than Household industries	0.94%
8	Female in Construction	0.83%
9	Commerce and Trade	1.59%
10	Transport, Storage and Communication	0.37%
11	Other Services	7.18%

Source: Field Survey

11.1.7.1 *Factors in landuse pattern*

We have already seen earlier how the topography of the tract varies from rocky desert in the south to the sandy desert in the north and a transitional zone in between. Such a topography with sandy soil along with low precipitation, leads to water scarcity and arid conditions not conducive for stable agriculture. The landuse pattern (Table 11.1.10) of the region is obviously dictated according to the climatic and environmental conditions of the region.

Table 11.1.10: Occupational set up in Osian tehsil

District/ Tehsil/ Urban area/ Town	(I-IV) Total workers		I Cultivators		II Agricultural labourers		III Livestock, Forestry, Fishing, Hunting, Plantations, Orchards and allied activities		IV Mining and quarrying	
	Male	Female	Male	Female	Male	Female	Male	Female	Male	Female
Osian	53259	17907	47894	14915	4642	2950	359	41	364	1

(V-IX) Total workers		V Manufacturing and servicing (a) House-hold Industries		V Manufacturing and servicing (b) Other-than house-hold industry		VI Construction		VII Trade and commerce		VIII Transport, Storage and commu-nication		IX Other services		X Non-worker	
M	F	M	F	M	F	M	F	M	F	M	F	M	F	M	F
9782	392	1186	114	737	25	667	7	1242	55	299	0	5651	191	70388	89280

T = Total Workers (I – IX) Male Female
63041 18299

It shows that out of a total area of 349956 sq.km. in Osian tehsil, the forest is negligible, unirrigated area is 79.7%, irrigated area 4.1%, culturable waste 5.5% and land not available for cultivation 10.4% (Table 11.11). The low rainfall which causes deficiency of moisture in the soil for which reason much of the land is left fallow.

Table 11.1.11. Land utilization pattern (Area in Hectare.)

Tehsil	**Area (Hectare) of Village**	**Total Forest Area**	**Total Irrigated Land**	**Total Un-irrigated Land**	**Culturable Waste Land**	**Area not available for Cultivation**
Osian	349956	749	14408	278912	19411	36477
%	100.0%	0.2%	4.1%	79.7%	5.5%	10.4%

11.1.8 Case Study of Three Villages

The three villages studied in details are Mathania (Population 9592), Khudiyala (Population 1063) and Osian (Population 9317) all located in the Osian tehsil.

Table 11.1.12: Occupational Structure in Study Villages (1991)

S. No.	Category		OSIAN	KHUDIYALA	MATHANIYA
1	Total Population		9317	1063	9592
		M	23.47%	25.49%	26.09%
2	Main Workers	F	1.38%	12.70%	5.34%
		Total	24.86%	38.19%	31.43%
		M	19.86%	60.84%	24.58%
3	Cultivators	F	2.42%	32.27%	6.27%
		Total	22.28%	93.10%	30.85%
		M	1.99%	0.00%	24.48%
4	Agricultural Laborers	F	0.09%	0.49%	9.92%
		Total	2.07%	0.49%	34.39%
		M	0.91%	0.00%	0.20%
5	Livestock, Forestry, Fishing, Hunting, Plantation Orchards and Allied Activists	F	0.13%	0.00%	0.00%
		Total	1.04%	0.00%	0.20%
		M	1.17%	0.25%	0.00%
6	Mining and Quarrying	F	0.00%	0.00%	0.00%
		Total	1.17%	0.25%	0.00%
		M	3.93%	0.00%	6.27%
7	Manufacturing, Processing Servicing and Repairs in Household industries	F	0.39%	0.00%	0.30%
		Total	4.32%	0.00%	6.57%
		M	8.25%	0.49%	4.08%
8	Manufacturing, Processing Servicing and Repairs in Other than Household industries	F	0.04%	0.49%	0.13%
		Total	8.29%	0.99%	4.21%
		M	4.27%	0.25%	2.65%
9	Female in Construction	F	0.00%	0.00%	0.07%
		Total	4.27%	0.25%	2.72%
		M	16.67%	1.23%	5.31%
10	Commerce and Trade	F	0.78%	0.00%	0.03%
		Total	17.44%	1.23%	5.34%
		M	1.60%	0.49%	3.75%
11	Transport, Storage and Communication	F	0.00%	0.00%	0.00%
		Total	1.60%	0.49%	3.75%
		M	35.79%	3.20%	11.71%
12	Other Services	F	1.73%	0.00%	0.27%
		Total	37.52%	3.20%	11.97%

11.1.8.1 *Occupational structure of the three villages*

The data on the occupational structure in the three villages are given in Table 11.1.12.

It will be seen that out of the total population, the percentage of total workers in Mathania is 31.43, that of Khudiyala 38.19 and that of Osian 24.86. Of these, 65.5% are engaged in Agriculture and allied activities in Mathania, 93.59% in Khudiyala and 24.35% in Osian. Khudiyala being typically rural accounts for wholly agricultural occupation.

Table 11.1.13: Occupational structure in study villages:

Name of village	Area of village in Ha	House-holds	Occu-pied houses	Popu-lation	Male	Fe-male	Cultivators I	II	III	IV	Produ-ction other than culti-vation	Comm-erce	Trans-port	Other serv-ices
1. Year 1951														
Khudiyala	1154	54	54	253	136	117	110	106	-	-	1	-	-	36
Mathania	2368	635	653	3412	1791	1621	161	1040	159	48	550	639	51	764
Osian	8828	661	661	4100	2166	1934	196	70	-	-	-	-	-	2730
2. Year 1991														
Khudiyala	1171.18	166	166	1063	564	499	378	2	0	1	5	5	2	13
Mathania	2559.82	1493	1492	9592	5127	4465	930	1037	6	0	407	161	113	361
Osian	3883.34	1619	1618	9377	5275	4102	516	48	24	27	391	404	37	869

It is significant to observe that highest number of nonworkers occur in Osian (75.14%) followed by Mathania (68.57%) and least in Khudiyala (61.81%). The population and occupational structure of three villages in the years 1951 is given in Table 24. It will be seen that the population which was 253 in 1951 increased to 1063 in 1991 in Khudiyala village. Likewise in Mathania the population in 1951 was 3412 which increased to 9592 in 1991. The corresponding figure in Osian was 4100 in 1951 which increased to 9377 in 1991. Thus in all the villages the population seems to have doubled in the course of 20 years. The occupational structure also shows a very distinct trend. In the case of Khudiyala the occupation has been entirely agricultural activities and there was practically none in commerce and transport occupation in 1951. Even in 1991 the picture had not substantially changed in Khudiyala. From the Table 11.1.13 it is magnitude of changes in occupational structure in the three villages is quite pronounced over the period of 20 years and there is distinct trend for emphasis on agriculture as an occupation which seems to give a better subsistence economy to the rural population. The data on schedule caste and schedule tribes are given in Table 11.1.14.

Table 11.1.14: Percentage of schedule caste and schedule tribes

Name of caste	Mathania	Khudiyala	Osian
1. Year 1951			
Schedule caste	16.10%	12.65%	11.54%
Schedule tribe	0.12%	9.88%	3.56%
2. Year 1991			
Schedule caste	20.27%	23.05%	17.21%
Schedule tribe	0.06%	0.00%	4.12%

It will be seen that highest number of schedule caste are present in Mathania (20.27%) followed by Khudiyala (23.05%) and Osian (17.21%). The schedule tribe however seems to have moved from typical rural village of Khudiyala (0%) in last 40 year followed by Mathania (0.06%) and then Osian (4.12%).

11.1.8.2 Literacy

The data on literacy is given in Table 11.1.15 below for the years 1951-1991.

It will be seen that the literacy percentage in 1951 was 11 per cent in Mathania compared with 46.73 per cent in 1991 for male population. The literacy in female population in this villages was lower i.e. 0.5 per cent in 1951 and now 16.19 per cent in 1991. The picture is still poor in Khudiyala. In this village literacy percentage in the male was 0.8 per cent in 1951 which increased to 37.23 per cent in 1991. In the female population, however, the literacy remained very poor with nil percentage in 1951 and only 2.20 percent still after 40 years. It is interesting to point out that the rapid improvement in the literacy in Osian village is remarkable. The literacy percentage in male population in 1951 was 0.63 per cent which showed a phenomenal increase to 60.68 per cent in 1991. In female population the percentage was 0.04 in 1951 which increased to 24.43 per cent in 1991. In general the literacy percentage in the female was lower in this village also.

Table 11.1.15: Literacy in Study Villages

Name of village	Literacy in %					
	Mathania		Khudiyala		Osian	
	Year					
Sex	1951	1991	1951	1991	1951	1991
Male	11.0	46.73	0.8	37.23	0.63	60.68
Female	0.5	16.19	0.0	2.20	0.04	24.43

11.1.8.3 *Basic Amenities*

The phenomenal rise in the literacy percentage in Osian may perhaps be ascribed to the better educational amenities available in Osian village which has the maximum number of 5 primary schools and 2 high schools unlike Mathania which have only 3 primary school and 2 middle and 2 high school and Khudiyala which has only one primary school with no middle or high school.

Table 11.1.16: The Amenities available in Study Villages (Year 1991)

S. No.	Facilities	OSIAN	KHUDIYALA	MATHANIA
A	Education			
1	No. of Primary Schools	5	1	3
2	No. of Middle Schools	4	0	2
3	No. of High Schools	2	0	2
4	No. of PU Colleges	2	0	0
5	No. of Adult Literacy Centers	0	0	1
6	No. of Other Schools	0	0	1
B	Health Care			
1	No. of Maternity Centers	0	0	1
2	No. of Primary Health Centers	1	0	1
3	No. of Health Centers	1	0	0
4	No. of Primary Health Sub Centre	1	0	0
5	No. of Dispensary Centers	2	0	1
6	No. of Family Planning Centre	1	0	1
C	Drinking Water			
1	Tap	1	0	0
2	Well	1	0	1
3	Tank	1	0	1
4	Tube-well	1	1	1
5	Hand-pump	1	1	0
D	Communication			
1	Post Office	1	1	0
2	Telegraph Office	0	0	0
3	Post and Telegraph Office	1	0	1
4	Phone Connections	1	0	0

Of the three villages, Osian has the best medical facility having two dispensaries, one PHC and a family planning centre. Mathania comes second with one dispensary but Khudiyala has no medical facility worth the name. Post and telegraph facility is present in both Osian and Mathania but a post office now exist in Khudiyala. Osian is covered with Tap connection for drinking water. While Khudiyala is covered only with H.P. for drinking water. In the Mathania and Khudiyala villages the staple food of the people is both wheat and bajra, but in Osian in the north it is only bajra because no rabi crop is grown in Osian.

11.1.9 Level of Economic Development and Agro-Demographic Structure

An attempt has been made in this study to analyse the agro demographic and occupation structure of the region in order to bring out the level of economic development in the different land type units earlier worked out. Thus a regional analysis of the entire area has been worked out with a synoptic or synthetic map showing the pattern of the level of economic development.

11.1.9.1 *Concept of Economic Indicator*

The levels of economic development and agro-economic and occupational structure of different land units have been worked out on the basis of efficiency of rural occupational structure and rural agricultural and industrial production. But these aspects of development are in turn susceptible of further analysis. Accordingly 14 indicators are selected for deducting synthetic index to work out and map the level of economic development and occupational structure of the region. These consist of four *structural indicators* relating to working population; nine *occupational indicators* relating to working population and rural economy (the region has no urban center) and three *productivity indicators* to per capita production of agricultural crops. Data relating to rural industries are not, however, available. The data on structural and occupational indicators are compiled from the publications of Census of India (Chandar Shekhar 1972; Verma 1973) and that of productivity indicators from the revenue records of the Government of Rajasthan (1971).

All these indicators were digitiesed in Geographical Information System and analysed to prepare thematic map to depict level of economic development. This was done mainly by following the methods evolved by Mitra (1968); and Schwartzberg, J.E. (1969). The synthetic indices for agro-demographic developments and agro-demographic regions are worked out after calculating the village wise data of the above indicators. A 'score' for each indicator is arrived at by standardizing the data tabulated. This is done by assigning zero to the least favored village with respect to each variable and a score of 100 to the most favored one. This can assign comparable sets of scores or index number for each variables considered for mapping (Schwartzberg 1969). The synthetic index for each land type is then calculated by adding the scores of all the indicators and by taking their average values.

Table 11.1.17: Agrodemographic indicators affecting economic development – Structural indicators

Land Type	Study village	Total workers to total population		Cultivators to total population		Agricultural labour to total population	
Rocky desert	Mathania	31.43%	**50.70**	9.70%	**86.15**	10.81%	**100.00**
Transitional between rocky and sandy desert	Khudiyala	38.19%	**100**	35.56%	**100**	0.19%	**0**
Sandy desert	Osian	24.86%	**0**	5.54%	**0**	0.52%	**3.18**

11.1.9.2 *Level of Economic development*

Following the score obtained from the data of the Study villages a map showing the level of economic development has been prepared. Tables 11.1.17, 11.1.18, and 11.1.19 indicate the condition and score of different agro-demographic indicators affecting the levels of economic development. The average score of each unit showing levels of economic development is indicated in Table 11.1.20 (Fig. 11.1.10).

Table 11.1.18. Agro-demographic indicators affecting economic development – Occupational indicators

Land Type	Study villages	I		II		III		IV		V		VI	
		Cultivators		Agricultural labour		Livestock forestry, fishing etc		Mining and quarrying		Manufacturing servicing		Construction	
		%	Score	%	Score	%	Score	%	Score	%	Score	%	Score
Rocky desert	Mathania	30.85	**87.91**	34.39	**100**	0.20	**19.20**	0.00	**0**	10.78	**84.27**	2.72	**61.40**
Transitional between rocky and sandy desert	Khudiyala	93.10	**100**	0.49	**0**	0.00	**0**	0.25	**21.13**	0.99	**0**	0.25	**0**
Sandy desert	Osian	22.28	**0**	2.07	**4.89**	1.04	**100**	1.17	**100**	12.61	**100**	4.27	**100**

Land Type	Study villages	VII		VIII		IX	
		Trade and commerce		Transport and communication		Other	
		%	Score	%	Score	%	Score
Rocky desert	Mathania	5.34%	**25.34**	3.75%	**100.00**	11.97%	**25.56**
Transitional between rocky and sandy desert	Khudiyala	1.23%	**0**	0.49%	**0**	3.20%	**0**
Sandy desert	Osian	17.44%	**100**	1.60%	**51.39**	37.52%	**100**

Table 11.1.19: Agrodemographic indicators affecting economic development – Productivity indicators

Land Type	Study villages	Cereals and pulses		Commercial crops – Guar	
		Production per head in kg	Score	Yield in kg/ha	Score
Rocky desert	Mathania	318.6	94.3	600	100
Transitional between rocky and sandy desert	Khudiyala	333.4	100	400	0
Sandy desert	Osian	73.3	0	500	50

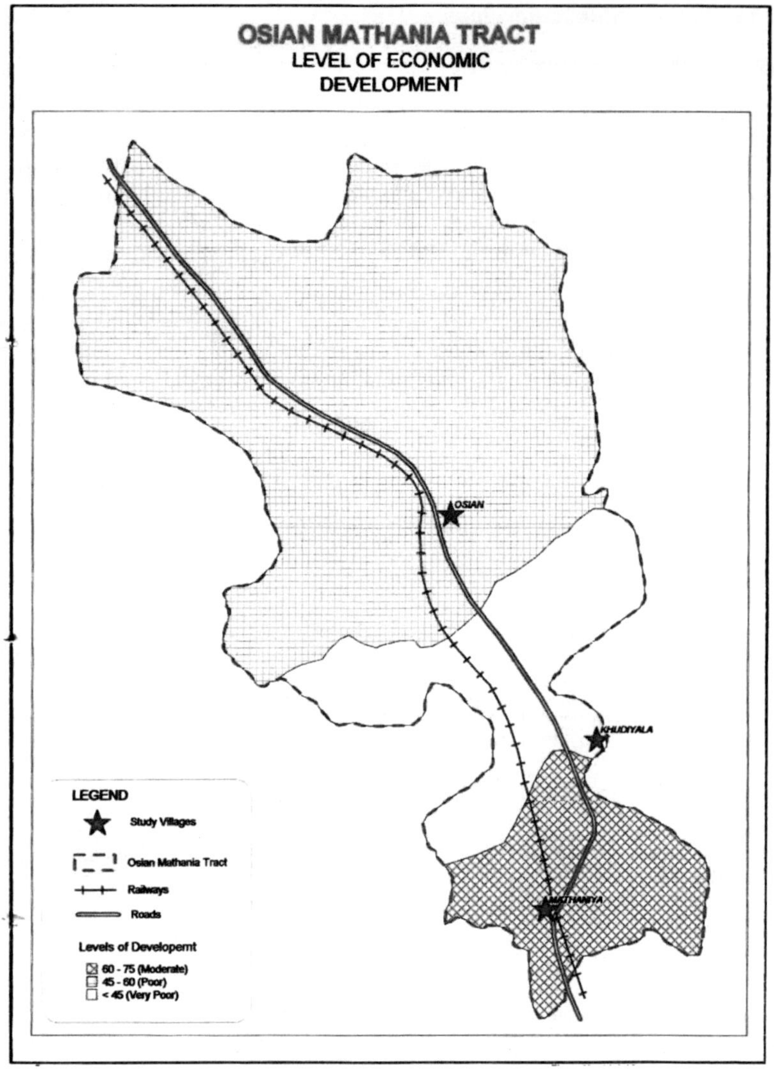

Figure 11.1.10 Level of economic development

Table 11.1.20: Levels of Economic Development

Land Unit	Study villages	Score			
		Structural indicator	**Occupational indicator**	**Production indicator**	**Levels of development**
Rocky desert	**Mathania**	236.85	503.68	194.3	**62.32 (moderate)**
Transitional between rocky and sandy desert	**Khudiyala**	200.00	121.13	100	**28.08 (very poor)**
Sandy desert	**Osian**	3.18	656.27	50	**47.30 (poor)**

The tables clearly indicate that there is a distinct relation between the land types and levels of economic development. Thus the development of transitional unit is very poor. Secondary and tertiary occupation is practically absent. The sandy desert is also poorly developed. Cultivation is the main occupation. The rocky desert is moderately developed with cultivation and secondary activities like mining or quarrying being the main occupational structure.

■■■■

CASE STUDY-2

11.2. IMPACT ASSESSMENT STUDY OF RAINWATER HARVESTING/ RECHARGING STRUCTURES IN BIKANER DISTRICT

As a result of complex physiographic conditions, drought conditions are endemic in the state of Rajasthan especially in western region. These conditions recur in varying intensity, year after year. Owing to its adverse impact the people many a times' experiences extreme distress conditions. Livelihoods of large proportion of the agrarian community get affected due to crop loss and water scarcity. This is despite the fact that this section of society, since time in memorial has learnt to cope with recurrent droughts in the state. In year 2003 the state felt the rigors of drought. The impact on already despaired communities was severe in magnitude, as the monsoons had failed for the fifth consecutive year. This could be visualized from the fact that 41000 villages of all 32 districts in Rajasthan were declared drought affected. Looking at the austerity of drought, a study was carried out to map and assess the benefits from the water harvesting activities.

Based on the initial community based assessment data, activities that would support in long term drought proofing were considered for the GIS based mapping study. As a matter of fact, the study has mapped water harvesting/recharging structure in selected villages of three blocks in the Bikaner district. The villages identified for the study are shown in Figure 11.2.1. Considering the fact that implementation of WHRS activities on large scale would generate an impact on the existing surface and ground water availability, which eventually improve the drought coping capacities of the community, it was felt prudent to assess direct and indirect benefits of the activities. Study thus was focused to map the study area using GIS' and assess cost benefit (direct and indirect benefits) of water harvesting/recharging activities implemented in the study villages.

11.2.1 Objectives of the study

Overall, the studies aims at assessing cost benefit of the rainwater harvesting works implemented in the selected study villages using GIS technology.

Following specific tasks were considered to keep the study focus;

- Preparation of base Map of Study area using GIS technology.
- Data collection on social composition and infrastructure facilities existing in the study villages.
- Identification of existing harvesting/recharging structures and estimation of their storage capacity.

- Identification of existing water sources, their average water level (seasonal), number and type of community served with in the recharging zone of each structure.
- Benefit estimation.
- Increment in water storage capacity of the WHS.
- Number of water sources to be impacted and increment in water level (seasonal) to be expected and type of community to be benefited within the recharging zone.
- Infrastructure improvement in selected villages and benefit to the community.
- Monetary value of the accrued benefits.

11.2.2 Methodology

11.2.2.1 *Location of Study Area*

Bikaner district is located between 27°11' to 29°03' North latitude and 71°54' to 74°12' East longitude in the northwestern part of Rajasthan covering a geographical area of about 27244 sq. km. Administratively, Bikaner district is divided into four tehsil - (1) Bikaner, (2) Lunkaransar, (3) Kolayat and (4) Nokha. It is bounded by Ganganagar district on the north, partly by Jaisalmer and Pakistan on the west, Churu and Hanumangarh district on the east, north-east, Nagaur and Jodhpur districts on the south and south-east respectively. The total population of the district is about 1211140 among which 729998 are rural and 481142 is urban population.

11.2.2.2 *Study Sample Size*

Study has been focused in 100 villages where water harvesting/ recharging related activities exist. These villages are scattered over three tehsils – Lunkaransar, Kolayat and Bikaner Rural of the Bikaner district. The cost-benefit analysis of WIIS activities is based on the findings from 100 villages. The locations of these villages are shown in Figure 11.2.2.

11.2.2.3 *Data Collection*

Application of Geographical Information System requires data accuracy. Since this study also utilised GIS for analysis and drawing conclusion, collection of accurate information was emphasized from different sources. To ensure this data was collected from secondary as well as primary sources. The following gives the details of the information that was collected from these sources.

11.2.2.3.1 *Secondary Sources*

Existing literature already available with state and Central Government departments were collected for review and carry out analysis. Type of information retrieved from different sources based on the checklist is reflected in the following:

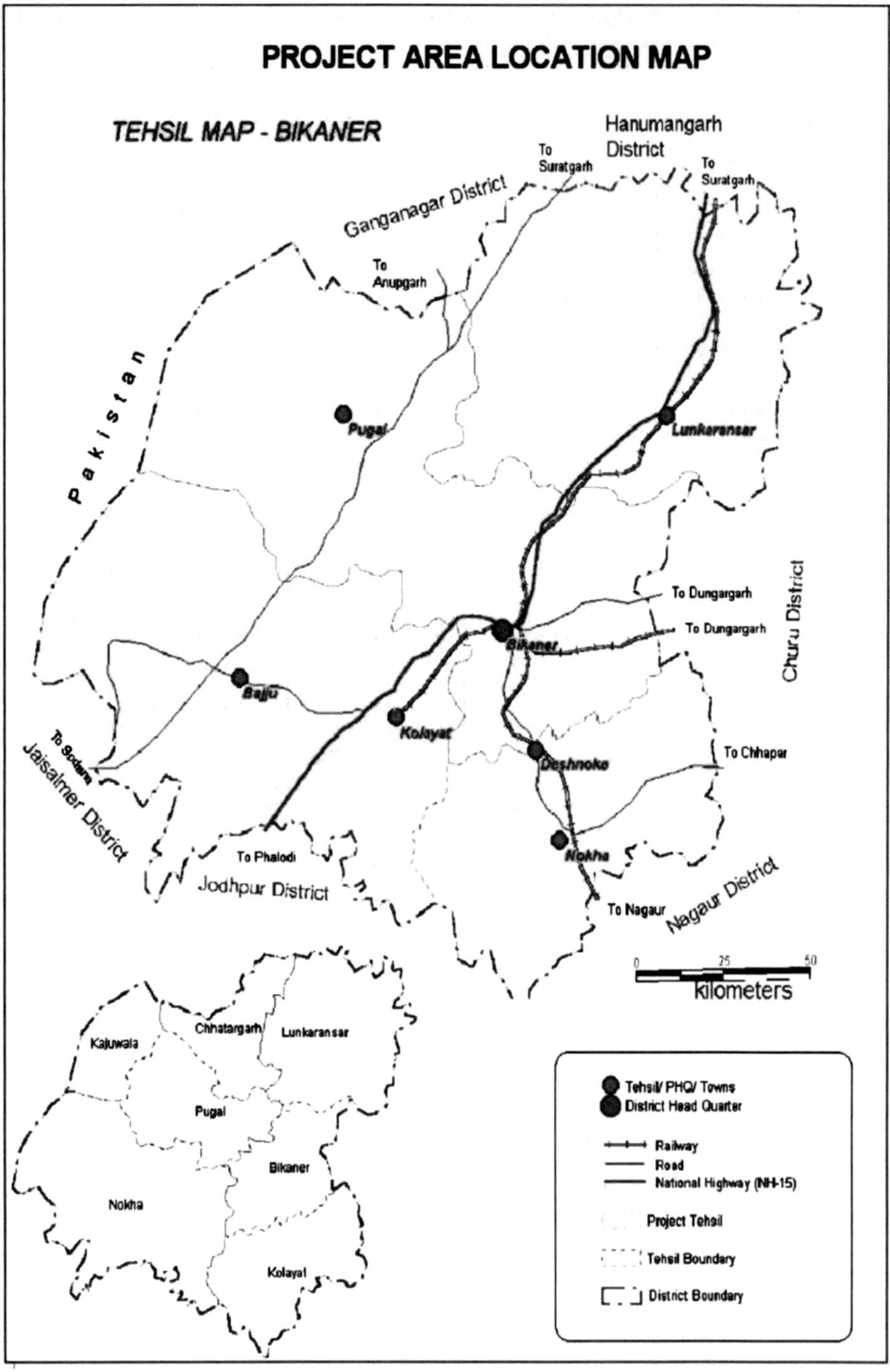

Figure 11.2.1. Project Area Location Map (Source: Census of India)

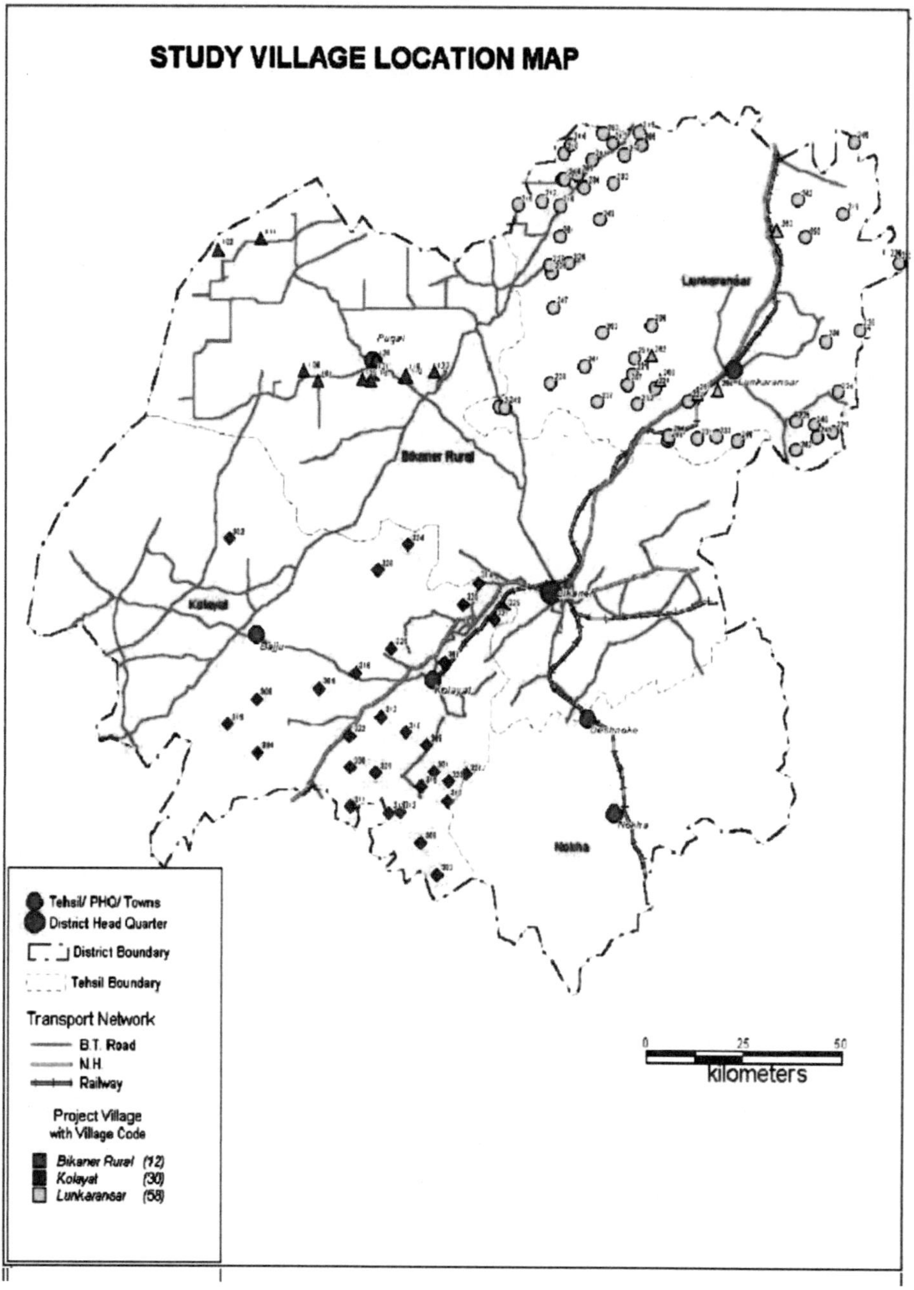

Figure 11.2.2. Study village Location Map (Source: Census of India)

Secondary Data and its Sources

Source:	Information/ Data
Base map of Bikaner District and Geographical Topo sheets (G.T. sheets) from Survey of India, Jaipur	• Village location • Site Location • Transport Network • Existing water bodies
Census map and report from Census of India, Jaipur	• Village population • Social composition • Demographic details
Annual Reports from Groundwater Board, PHED, Indira Gandhi Nahar Pariyozana (IGNP), Meteorological Department based at Bikaner.	• Rainfall and PET • Soil characteristics • Geological, Hydrological and hydro-geological settings • Command area plan • Groundwater quality and quantity • Agricultural production

11.2.2.3.2 *Primary Sources*

Focused group Discussion

Since the general information such as social composition and natural features of the village was required it was felt appropriate to conduct FGDs with the community members in each of the study village. These FGDs were conducted using schedules to collect information regarding the water harvesting structures, its storage capacity and recharge potential and dependent community within the recharging zone of the structure. Also social composition was understood during the discussion to supplement the secondary data.

Field survey and observation

Global positioning system (GPS) was applied to triangulate the exact location of the villages represented in the base maps. It was felt necessary to use this system to ensure data accuracy as most of the study villages falling under the restricted area have been represented symbolically in the base maps. Also GPS was used to locate the sites and trace out the transportation and canal network present in the study villages, which are not marked in the base maps. During the survey of the study villages, data related to the quality of the catchment, soil and its structure necessary for GIS based mapping was also collected using schedules.

11.2.2.4 *Process of Data Collection*

Before initiation of fieldwork preliminary meetings with Local organisations were organised to discuss the objectives, scope of work, methodology and timeline of the study. The preliminary information about the study and related components of the study was discussed with Local organisations at Bikaner. Based on the discussions, checklist and schedules for data collection from secondary and primary resources respectively were developed. Since Local NGOs and government departments were the implementing agencies for the WHS activities in the study villages, a formal meeting with these organisations at their respective offices in blocks were organised time to time to discuss the study design and solicit their support in data collection. Information/data already available with them were initially collected and reviewed. As required, focused group discussions were conducted with the help of support staff of these NGOs. Field GPS survey was conducted to obtain GIS location with the active involvement and participation of the NGO representatives in their respective study areas.

11.2.3 Data Processing and Analysis

The collected information both from primary and secondary were fed in the computer. This includes text data in tabular form and the vectorisation of maps updated through GPS. The fed information was analysed for its accuracy. The data in tabular form related was linked to the base map using MAP INFO software. The processed data was interpolated and mapped to generate thematic mappings. Through the developed thematic maps existing status of ground water in terms of its quality and quantity was assessed. To assess the impact of each of the water harvesting structure data related catchment area, soil characteristics, Climatological parameters and groundwater quality, quantity and fluctuations were drawn from the thematic maps.

11.2.4 Direct Benefits

Immediate response to water harvesting/ recharging activities implemented has been considered to draw direct benefits. In this category incremental availability of water was converted in to LPCD requirement to assess the drinking water benefits of the raised assets. Monetary value of increased crop and milk production with improvement in water availability has also been estimated. Each site has been discussed in details individually and a comparison note has been prepared for all sites in the end.

11.2.5 Indirect Benefits

Induced ground water recharge in the down stream sources and impact on its quality has been assessed to draw indirect benefits. The estimation was done after deducting evapotranspiration (specified for the area) losses and drinking water draft from the total runoff harvested in the structure. The fluctuations in the water level at the down stream sources have been calculated on the basis of seepage rate, distance of the water source from the structure and the actual water depth in the source.

11.2.6 Measures of Quality Control

The precision achieved in the analysis primarily depends upon the correctness of the database used in the analysis. Hence to optimize it, the fed information/ data was screened and checked thoroughly for its correctness in each stage of the study. The first level of check was got introduced by way of involvement of single person-having experience in the similar field and well versed with the study. The chances of error in the study were also curtailed by the fact that only single person without any assistant carried out data collection as well it analysis. The data collected from the government sources was triangulated from primary sources using FGD and field survey. Standard software was used to conduct analysis and arrive at conclusion.

11.2.7 Findings and Analysis

Following section present finding and analysis based on the data collected through FGD, field survey and processed using standard software.

11.2.7.1 *Profile of the Study Area*

The study was carried out in 100 villages spread over tehsil of Lunkaransar, Kolayat and Pugal between February - July 03. As per 2001 census, these tehsil are inhabited by 455391 rural population (27% of the total population of the district) belonging to different castes, predominantly Rajput, Purohit, Jat, Meghwal, Jogi and Muslims. Of these, the last three could easily be classified as vulnerable communities.

11.2.7.1.1 *Social Composition*

Social composition data collected from study villages revealed wide distribution of different castes in all the villages. This includes Meghwal, Jat, Rajput, Brahmin, Sansi, Naik, Muslims etc. The total population comprises of 52.6% of male population and remaining 47.4% of female population. Out of the total population 24.24% belongs to Scheduled caste while only 0.4% belongs to schedule tribes. The remaining 75.36 consist of general cast distributed in the villages with varying compositions of Rajputs, Brahmins, Vaishyas etc.

11.2.7.1.2 *Transportation*

Findings of the GPS survey as shown in figure 11.2.3 reveals that about 50% of the villages are covered with B.T. Road and W.B.M. Rest of the 40% is having Kacha village track connected with W.B.M. or B.T. Road with less than 1km distance. There are 10% of villages, which are situated in the remote areas where there is no facility except village track for the connectivity. As far as other transportation systems are concerned, only 5% of the villages are connected by rail.

11.2.7.1.3 *Drinking Water Sources*

Community meets their drinking water needs from different sources including tap water connected served by PHED. As per the data collected, 35 villages are covered by piped water supply. The villages not covered by this facility, meets its drinking water requirement from open wells, village pond and tube wells. There are about 61 open wells, 71-village pond and 15 tube wells, which are located in the study area including the one covered by tap water supply.

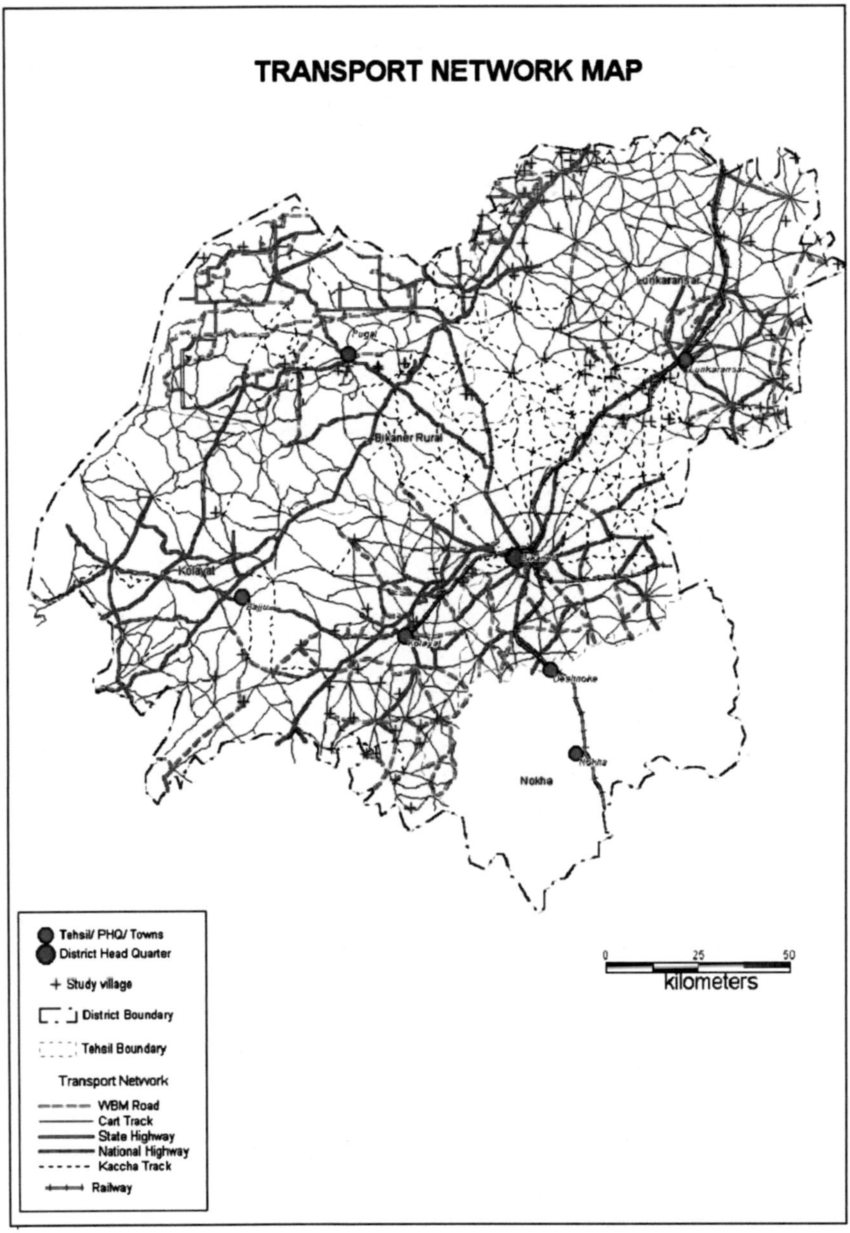

Figure 11.2.3. Transport Network Map (Source : Census of India)

11.2.7.1.4 *Climate*

Rainfall features of the study villages are no different from Bikaner district. The data given below indicate that majority of study villages features mean annual rainfall less than 300 mm (259.60 mm).

Table 11.2.1. Mean Annual Rainfall

S. No.	Mean Annual Rainfall (in mm)	Study Villages	
		Number	%
1	< 300	98	98.00
2	300 - 400	2	2.00

The data outlined below infers high rate of potential Evapotranspiration in the study villages. As per the analysis, it is 6.75 times higher than the annual rainfall, which consequently affects the surface water sustainability.

Table 11.2.2. Mean Annual Evapo-transpiration in Study Area.

S.No.	Mean Annual Potential Evapo-transpiration (in mm)	Study Villages	
		Number	%
1	160 - 170	58	58.00
2	170 - 180	38	37.00
3	180 - 190	4	4.00

11.2.7.1.5 *Soils*

The soil condition widely varies in the study villages. As shown in Table 11.2.3 about 57% villages signify suitable agriculture conditions with occurrence of Grey brown soil. Sandy soil dominates 25% of the study villages, which limits the agriculture production.

Table 11.2.3. Type of Soil

S. No.	Soils	Study Villages	
		Number	%
1	Desert Sand Dunes	25	25.00
2	Grey Brown Soil	56	56.00
3	Calcic Brown Soil	4	4.00
4	Non Calcic Brown Soil	15	15.00

11.2.7.1.6 *Hydrogeology*

Overlay of village location and hydro-geological settings as shown in Figure 3.2 infers that about 56.43% of the villages possess saline groundwater concentration. Similarly, about 67.33% villages are having water table in the range of 40 to 80 m. Depth of more than 80 m has been noticed in 11% of the villages.

Table 11.2.4. Depth to Ground Water

S. No.	Depth to Groundwater	Study Villages	
		Number	%
1	< 40 m	22	22.00
2	40 – 80 m	67	67.00
3	80 – 120 m	7	7.00
4	> 120 m	4	4.00

The ground water data indicates positive development of ground water in villages influenced by IGNP canal. About 9% of study villages are in such locality, which are expected to get maximum benefit with construction/renovation of recharge structures.

Table 11.2.5 : Groundwater Depth Fluctuation

S. No.	Groundwater Depth Fluctuation (1984-97)	Study Villages	
		Number	%
1	2 to 4	74	74.00
2	0 to 2	17	17.00
3	0 to -4	9	9.00

11.2.7.1.7 *Groundwater Yield*

Study indicates that about 35% of villages are situated in the groundwater yield range of 70 – 180 cum/day and 8% in the range of 180 –200 cum/day. Rests of the village are having saline groundwater not suitable for drinking. The summary of study results has been summarized in Table 11.2.6. the ground water potential zones mapped in Figure 11.2.4.

Table 11.2.6. Groundwater Yield

S. No.	Groundwater Yield (in m^3/day)	Aquifer	Study Villages		Tehsil
			Number	%	
1	45 - 90	Older Alluvium	0	0.00	Kolayat
	70 - 180		13	12.87	Lunkaransar
2	180 - 225	Tertiary Sandstone	0	0.00	Bikaner
	180 - 200		8	7.92	Kolayat
	70 - 180		10	9.90	Lunkaransar
3	70-180	Nagaur Sandstone	13	12.87	Kolayat
4	-	Saline Groundwater	57	56.43	All

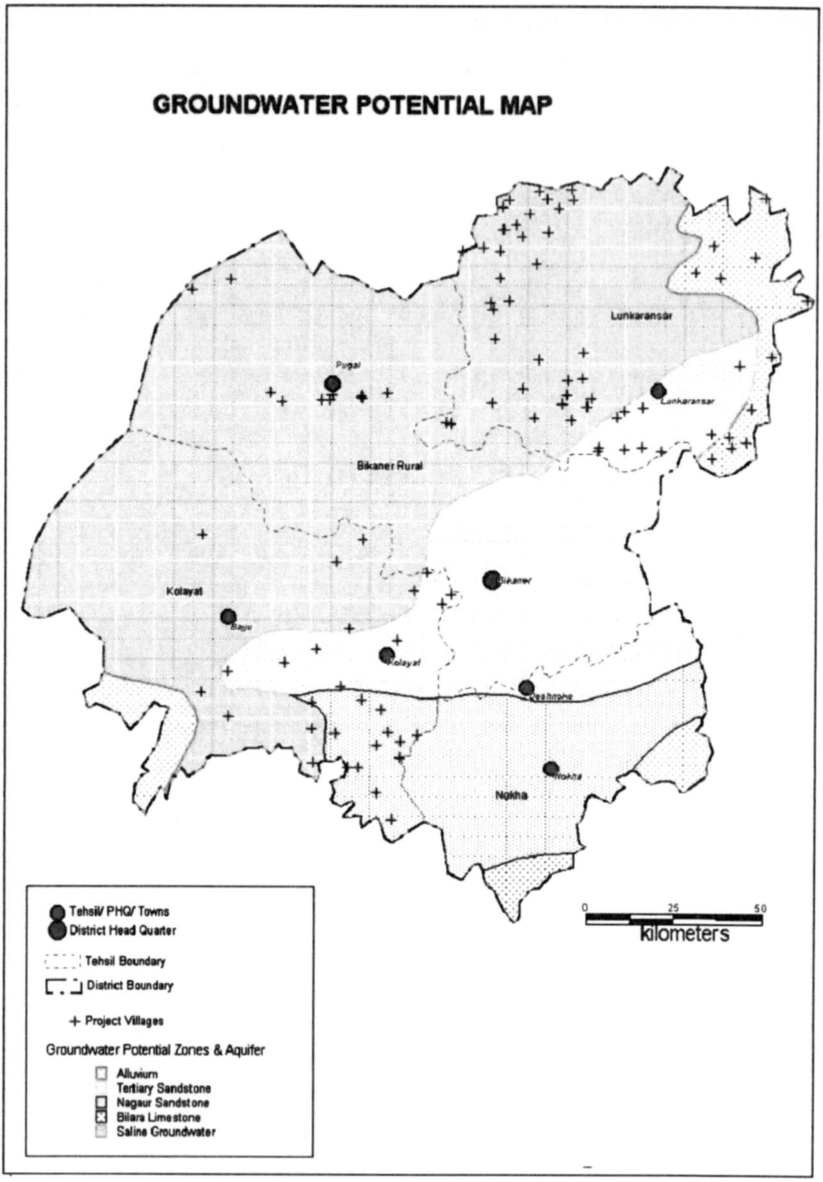

Figure 11.2.4. Groundwater Potential Map (Source: GWD and CGWD)

11.2.7.1.8 *Groundwater Quality*

Due to presence of high salinity, fluoride and nitrate content, 89% of the study villages have un-potable groundwater sources. The areas have been marked in Figure 11.2.5 Consequent to poor quality of ground water, rainwater harvesting thus becomes central to improve the quality of ground water.

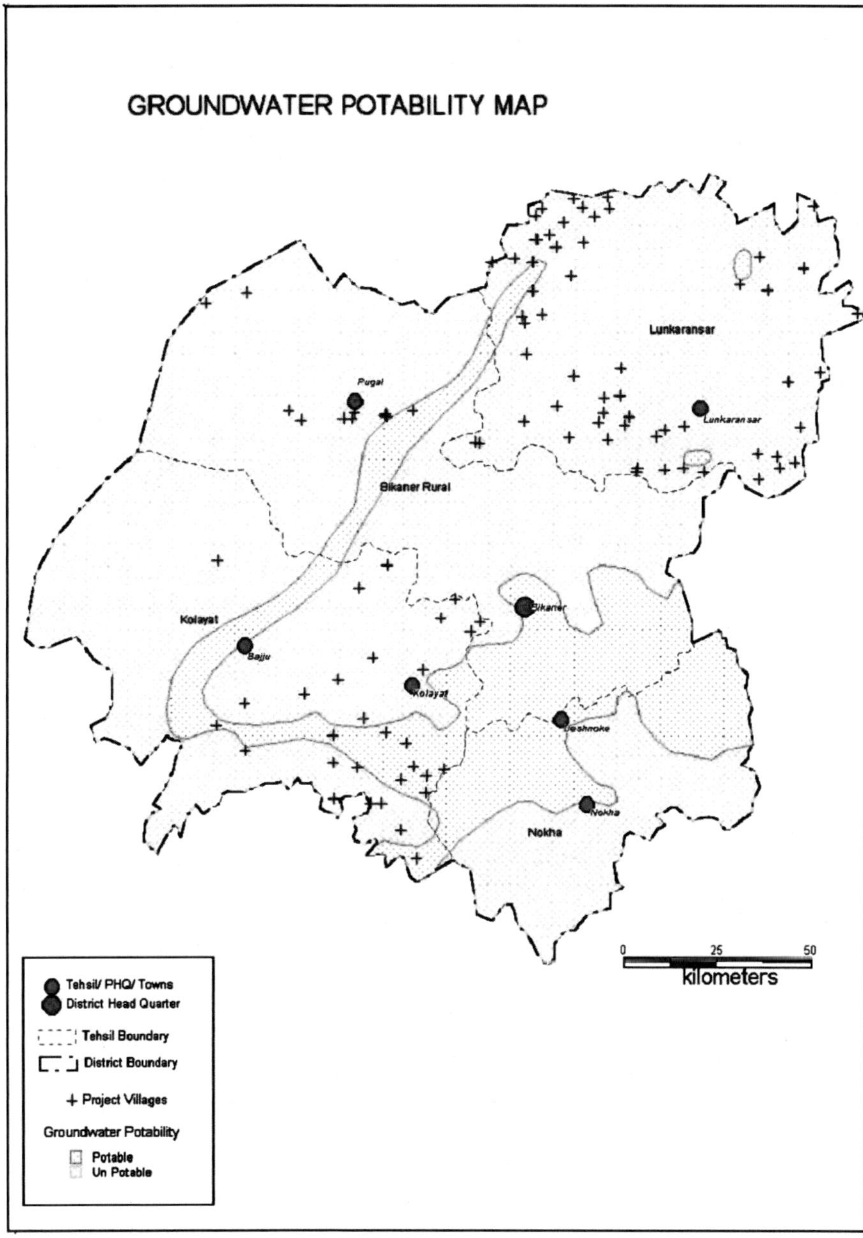

Figure 11.2.5. Groundwater Portability Map (Source: GWD and CGWD)

11.2.7.2 *Water Harvesting/ Recharging Activities in Study Villages*

The split up of 100 study villages covered by the partner NGOs under the study is presented in following table:

Table 11.2.7: Village Distribution

S. No.	Tehsil	Study Villages	
		Number	%
1	Lunkaransar	53	53.00
		5	5.00
2	Bikaner Rural	12	12.00
3	Kolayat	30	30.00
	Total	100	100.00

The information concerning beneficiary coverage retrieved from secondary sources reveals that against the target of 4400, the study has provided relief to 6993 vulnerable households. With in WHRS category different set of activities has been implemented in the study villages. In IGNP command area, Khala-Desilting work with a focus on earth removal was undertaken to remove sand deposition from the watercourses or minor channel. Similarly, suitable sites at selected villages were identified for renovation and construction of old and new ponds respectively. At few sites, earthwork excavation work was also undertaken to remove large amount of sand deposited in school campus and on road obstructing routine work. The following table along with the Figure 11.2.6 represents the activity distribution across the study area.

Table 11.2.8 : Distribution of activities

S. No.	Type of Activities	GIS Study Villages	Tehsil
1	Canal Desilting	18	Lunkaransar
		11	Bikaner Rural
		0	Kolayat
	Total	29	
2	Talai Work	35	Lunkaransar
		0	Bikaner Rural
		30	Kolayat
	Total	65	
3	Canal Desilting + Talai Work	3	Lunkaransar
4	Kachchi Road and Talai Work	1	Lunkaransar
5	Earth Removal School	1	Lunkaransar
6	Earth Removal Road	1	Bikaner Rural
7	Diggi construction	1	Bikaner Rural
	Total Villages	100	3 blocks

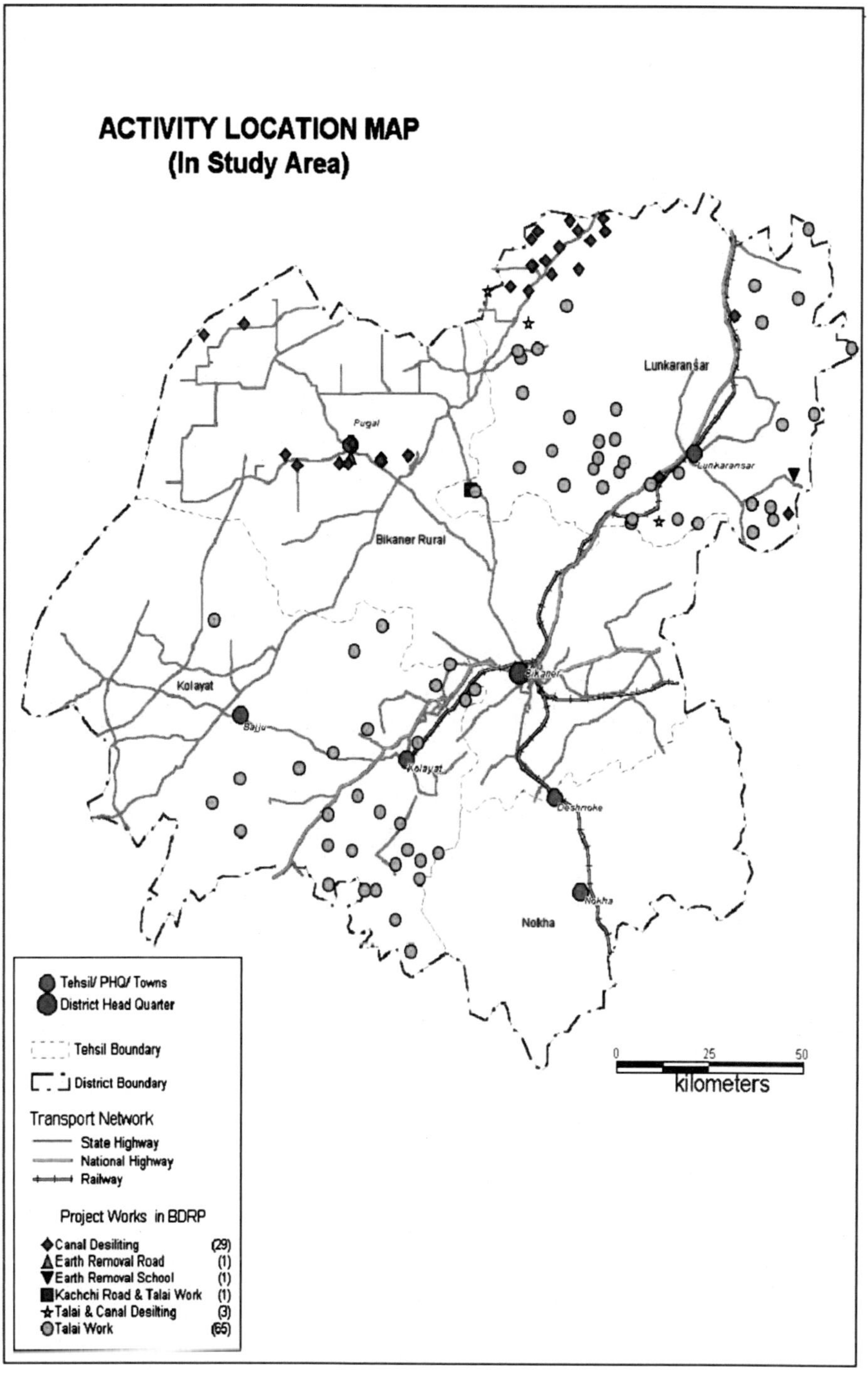

Figure 11.2.6. Activity Location Map (Source: Census of India)

11.2.7.3 *Cost Benefit Estimation*

11.2.7.3.1 *Direct Benefits*

11.2.7.3.1.1 Talai Works

Out of total 67 sites considered, 50 new sites were identified for construction while rest lying defunct in the absence of maintenance was renovated. These categories of works were widely undertaken in Lunkaransar and Kolayat.

a. Input Parameters and Data Processing

- Based on the rainfall and PET series developed by the State Groundwater Department monthly rainfall and PET data has been retrieved and considered for the analysis.
- The catchment area of each Talai was divided in three broad categories, based on the reconnaissance and the FGD with the villagers. These categories include catchment area less than 5.0 ha, more than 5 and less than 10 and more than 10 ha. The catchment was defined in these limits to facilitate the data processing and software operation.
- Data on earthwork removal signifying quantum of work accomplished at each site was collected to estimate the increased storage capacity.
- The soil conditions at each site were identified to arrive at the infiltration rate. The infiltration rate for sand, silt and clayey soil has been defined as $1x10^{-4}$, $1x10^{-5}$ and $1x10^{-6}$ cm/sec.
- Looking at the site conditions a figure of 30% of the average rainfall has been considered as runoff coefficient to calculate runoff volume generated from the catchment area.
- In view of limited storage capacity of the pond/Talai, benefit has been estimated solely for drinking water fulfillment for human and cattle. Considering the fact that villages may have other drinking water sources, a factor varying from 0.1 to 0.5 has been taken to estimate the dependent population on the source. This factor has been verified during the FGD at individual sites before incorporating it in the analysis.
- Since a tractor trolley Tanker of capacity of delivering 4000 liters cost about Rs. 60.00, an average rate of Rs. 15 per thousand liters has been used in the cost analysis.
- It is assumed that 50% of the cattle population getting drinking water stored in the structure will produce 2-liter milk per cattle per day. Market rate of Rs. 10 per liter has been considered for the analysis.
- It is further assumed that the drinking water cost contribute half to the total cost of the milk produced as the fodder is generally available free of cost at the Charnot land. This assumption is on safer side, which can be calibrated with actual data subsequently.

- Drinking water demand for the human being and cattle has been calculate @ 10 lpcd and 15 lpcd respectively. The sum of both the drinking water demand is termed as drinking water draft.
- After setting the above listed assumption and input parameter, the average runoff generated from the catchment was calculated multiplying the average rainfall with catchment area and runoff coefficient month wise.
- Factors like evaporation and other losses have been calculated using available figures of the PET for estimating net month wise drinking water availability.
- The volume of water available for drinking purpose has been multiplied by per unit cost to arrive at the monetary figure. Similarly the duration for which the water will remain available has been considered to estimate the monetary value of the milk produced per cattle for that period.

b. Result Based Finding

Figure 11.2.7 depicts the variable increase in storage capacity of the Talai where the works were executed under the study. It is estimated that the catchment area of the 68 Talai would generate a total runoff of 410817.00 cum. With the execution of work in these Talai, the storage capacity is expected to increase by 319260.50 cum. As per the analysis carried out, a total volume of 213905.00 cum would be available for drinking. For the given human and cattle population this quantity of water is expected to last for 4.19 months. It is estimated that through the Talai works a total direct benefit of Rs. 19035271 would be generated by the study, which is 116.29% of the total amount invested under the activity. Figure 11.2.7 presents the summary of benefit accrual with respect to actual cost in the study villages. The Figure 11.2.8 indicates that more than 48% villages would meet their construction cost in their first year of operation. Data analysis and field observation suggest that water in 5 sites located in Bandhali, Grandhi, Khikhania Patta, Miya Kor and Salasar villages may not last more than a month due to permeable feature of the soil.

11.2.7.3.1.2 Canal Desilting Works

Out of 100 study villages, canal-desilting work or removal of sand from Khala was undertaken in 29 villages. These works were exclusively implemented in Lunkaransar and Pugal blocks.

a. Input Parameters and Data Processing

- Through FGD, data on increase in command area has been collected from the field to estimate the benefit that would accrue in case wheat crop is produced during Rabi and Kharif seasons with irrigation facility from IGNP.
- Based on FGD, average production of 2 and 1.5 tones per hectare has been considered in kharif and Rabi seasons respectively.
- Looking at the market trend an average cost of Rs. 3000.00 per tones has been considered for wheat.

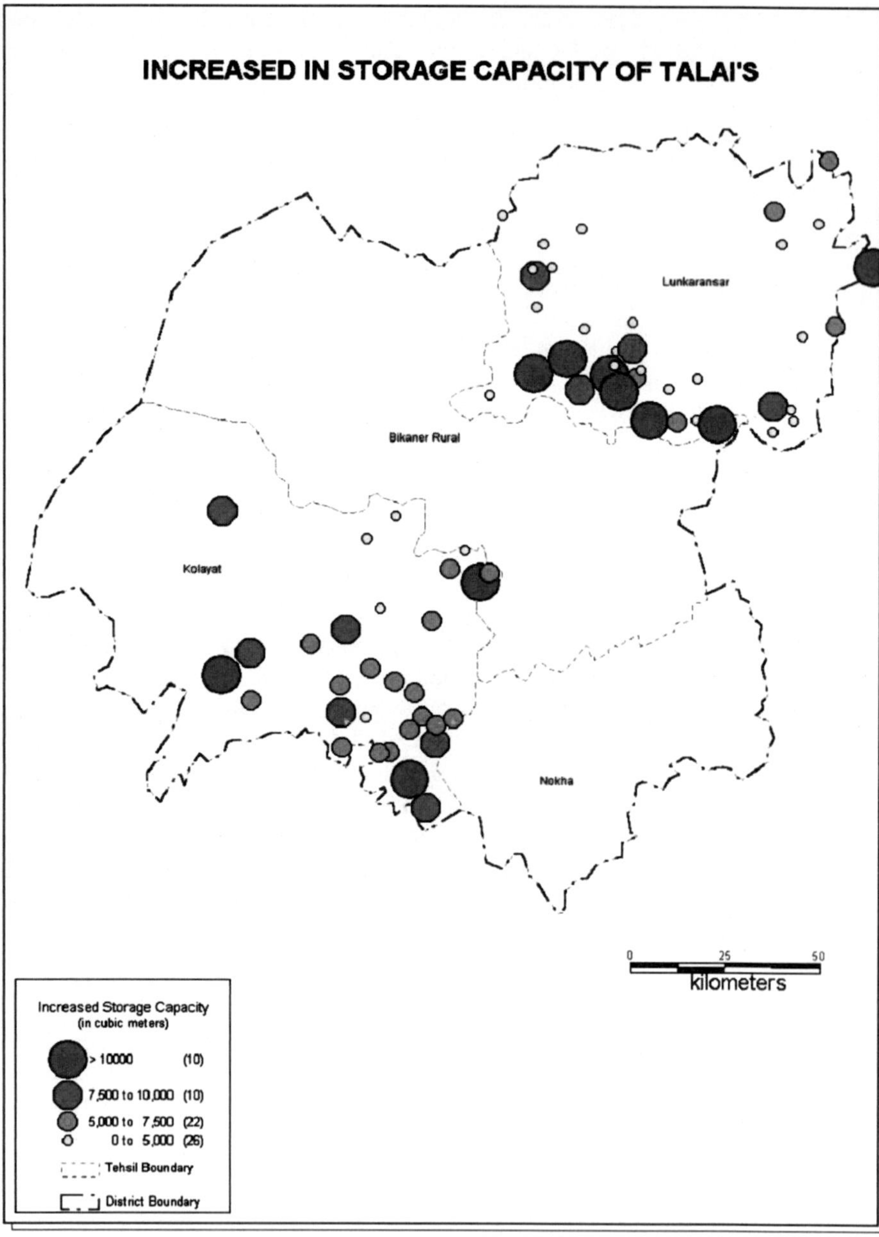

Figure 11.2.7. Increased in storage capacity of Talai's (Source: TALAI's)

- Since, canal water is also used for drinking purpose, its monetary value has also been estimated to assess the benefits. While estimating the value, it is assumed that 10 to 25% of the total population is depended on canal water for meeting their drinking water needs.

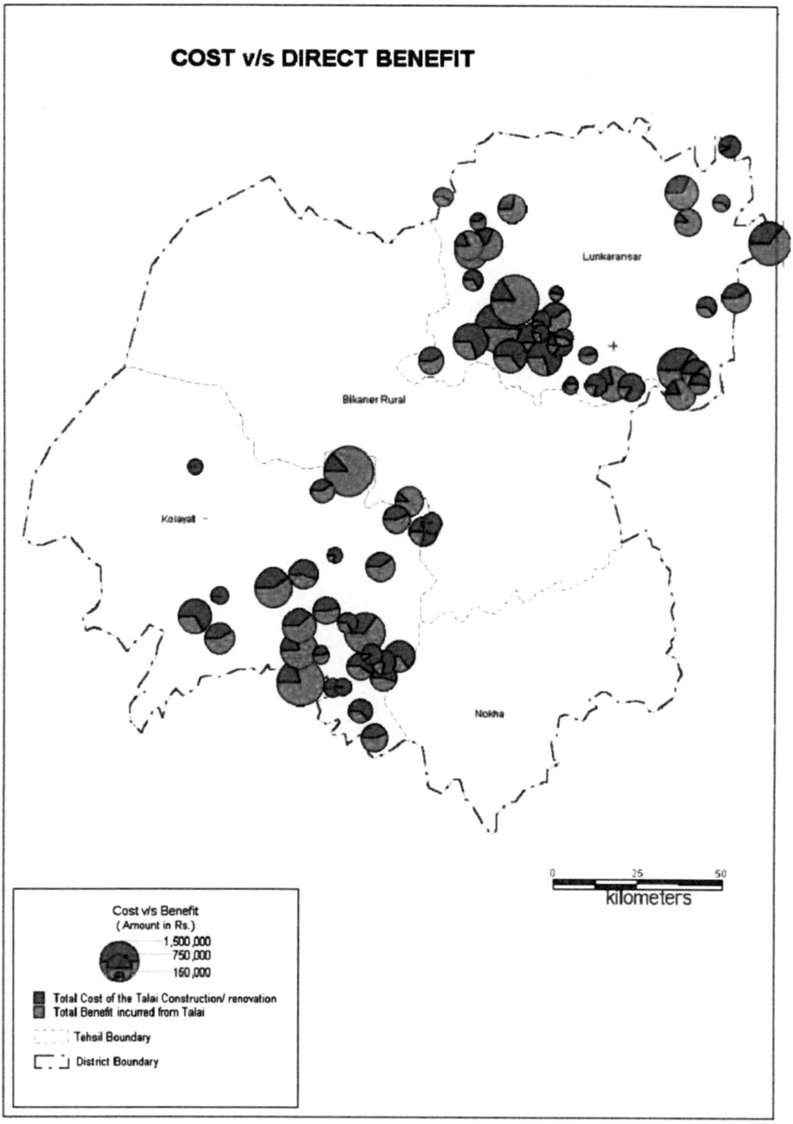

Figure 11.2.8. Cost v/s Direct Benefit (Source: Talai)

b. Result Based Finding

Figure 11.2.9 shows the command area that would increase as a result of canal desilting works undertaken by the study. As per study estimates about 212.50-hectare area that is about 6% of the Total Command Area will increase and would be available for the agriculture as a result of implementation of canal desilting work. As per the analysis, 758.75 tones of food grain will be additionally harvested from increased command area amounting to Rs. 22,76,250.00, which is 42.45% of the total cost invested. Additionally drinking water for whole year will be available to the cultivators and their cattle.

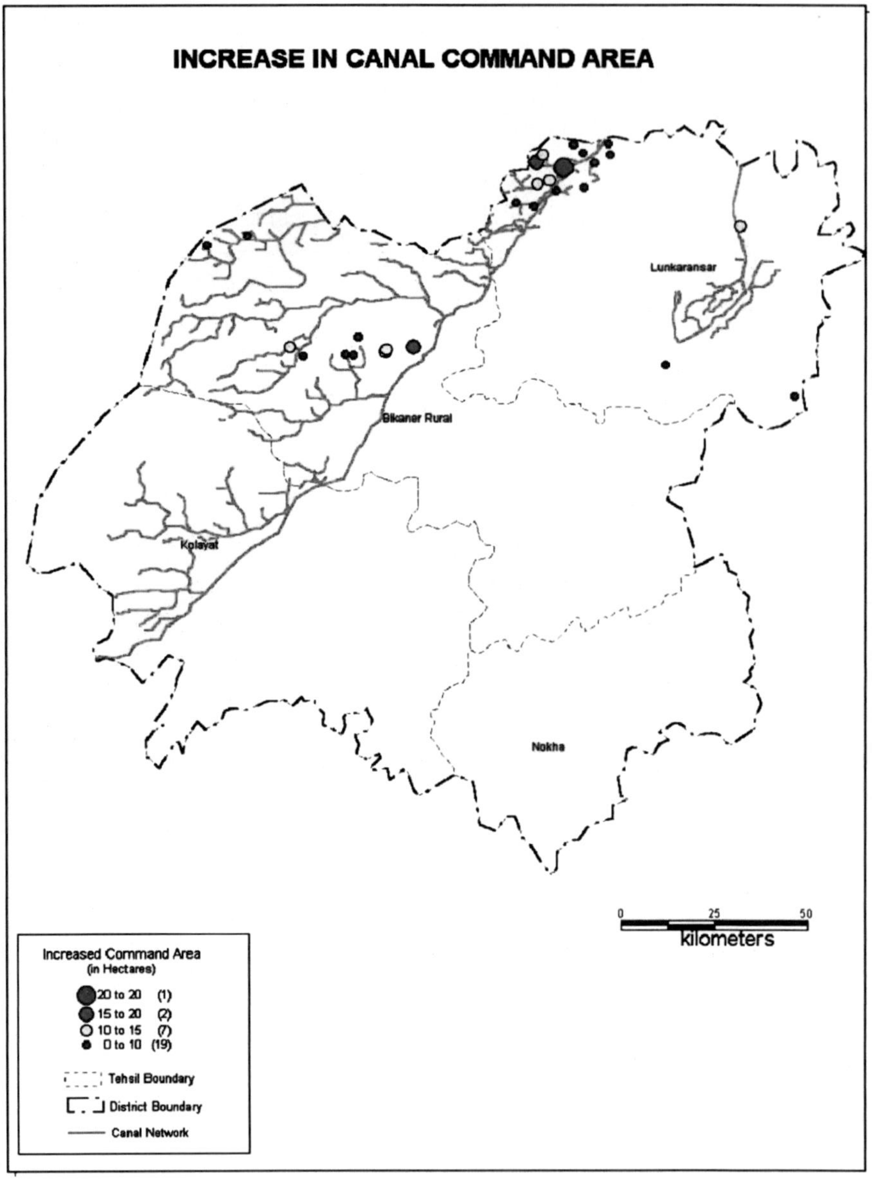

Figure 11.2.9. Increase in General Command Area (Source: CAD)

It is estimated that with implementation of Canal desilting works, a total direct benefit of Rs. 5362100.55 would be generated by the study, which is 68.9% of the total amount invested under the activity. Figure 11.2.10 presents the summary of percentage of benefit accrued in the study villages with respect to its actual cost. The analysis presented indicates that about 90% villages would accrue benefits of more than 50% of their construction cost in their first year of operation.

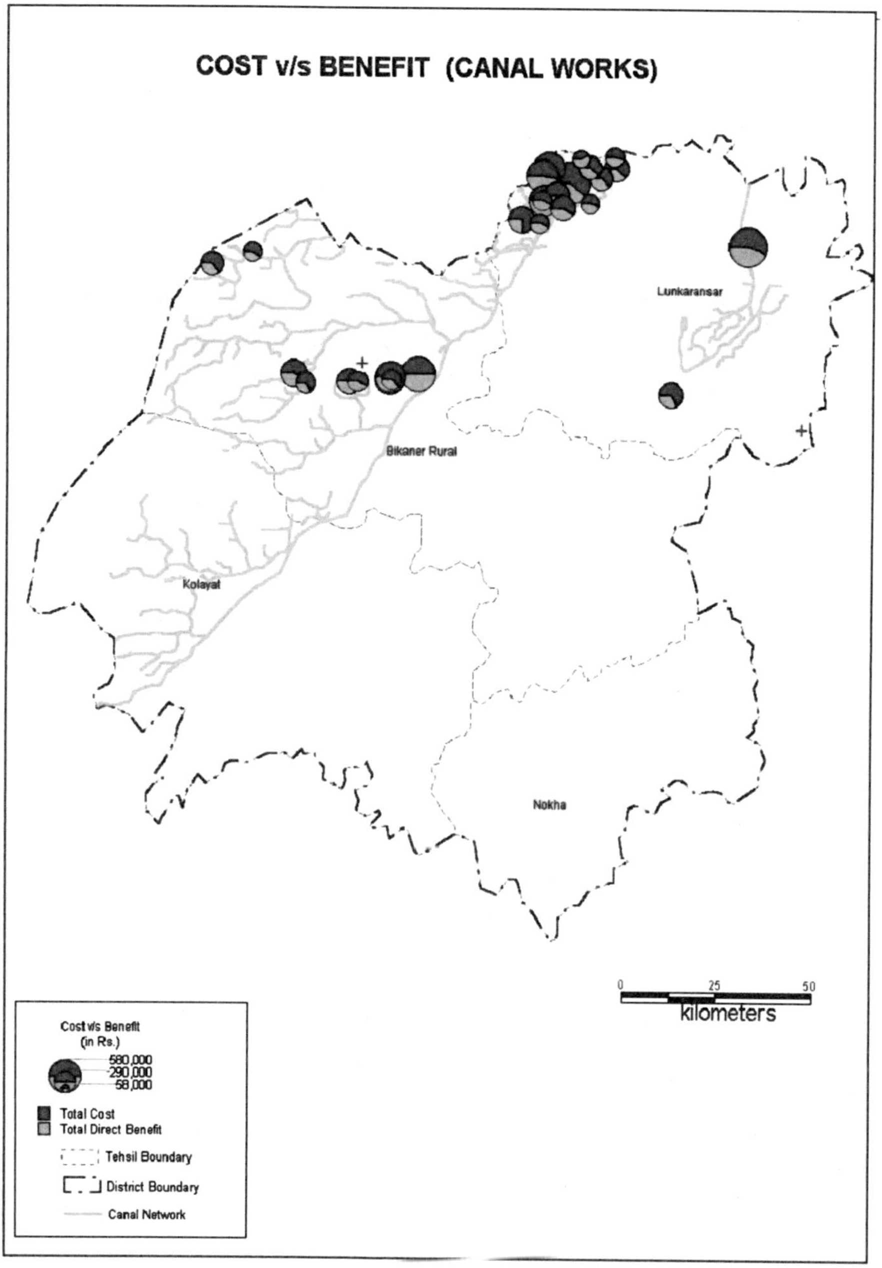

Figure 11.2.10. Cost v/s Benefit (Canal Works) (Source: CAD)

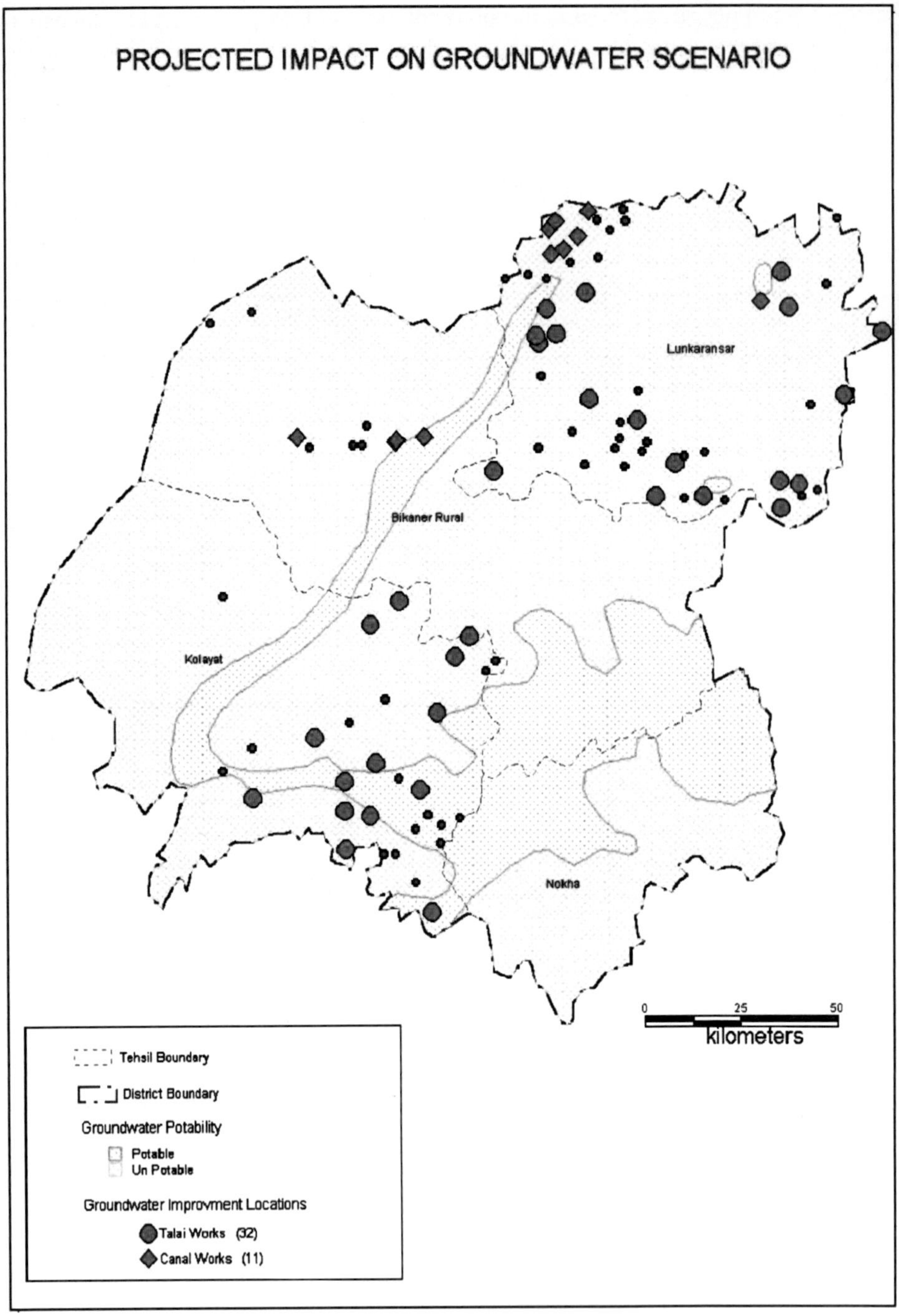

Figure 11.2.11. Project Impact on Ground water Scenario
(Source: CAD, GWD and CGWD)

11.2.7.3.2 *Indirect Benefit*

As per the study estimates, total seepage of 43,713.50 cum, which is 10.64 % of the total runoff and 13.69% of the total storage capacity, would be yielded with the renovation/construction of Talai's. This water will not only increase the ground water availability but would also improve its quality in 61 villages, where un-potable state has been reported. Similar impact would also be visible in 29 villages where canal desilting works has been implemented under the study. Figure 3.10 shows the location where ground water scenario is expected to improve as a result of WHRS works implemented under the study. The incremental availability of water in the structures constructed is estimated to raise the water table of 55 kuis and 15 open wells located in the downstream. As per study analysis it will raise the water table by 1.5 to 2.0 meters or increase the availability for additional 2 months in the sources falling inside the 100-m radius in the down stream. It is expected that increase in moisture content will also improve the fodder availability for the cattle in the surroundings.

■■■■

CASE STUDY - 3

11.3 ASSESSMENT OF POTENTIAL RECHARGE AREAS IN FOUR DISTRICTS OF RAJASTHAN

In the wake of the third consecutive year of drought in the state of Rajasthan, a GIS based assessment study was carried out in select four districts in order to help alleviate the situation. The four districts selected on the basis of severity of drought. Relief interventions in the drought- affected areas must be designed to address the immediate felt and perceived needs of the affected population, with special attention to the differential impact that the situation has on vulnerable populations including the SC/ST and women and girls within the households. However, from past experiences and current analysis of the situation it has emerged that interventions must address the need for household food/cash security, water availability and protection of livestock. Government Organisations as well as NGOs intends that the interventions thus designed should not duplicate any other interventions but rather compliments other initiatives and are based on collaborations with other stakeholders. It is for the purpose of assessing the recharge areas so that benefit of the implementation of the designed interventions could actually percolate to the vulnerable communities.

11.3.1 Objectives

The specific objectives of the study were as follows:

1. GIS based mapping of the existing natural resources in the study districts.
2. Assess the availability of water in the areas using GIS technology.
3. Identification and mapping of the Potential Recharge Zones based on realistic analytical approaches.

11.3.2 Methodology

The aim of the study was to make an assessment of potential areas for rain water harvesting based on GIS technology using available data. Therefore, final results were dependent very much on availability, quality and year of existing data. In order to make optimum use of existing data and to ensure achievement of all objectives the following approach was followed:

(i) *Collection and review of previous works existing data and reports/ atlas:* Atlas/ reports and available data on aquifer characteristics, water levels, ground-water extractions and groundwater quality were collected from various government institutions.

(ii) *Data treatment and storage:* Since data were collected from a wide variety of institutions they had to be checked. All data were computerised and structured per topic. A number of maps were digitized to use computerized data in a Geographic Information System (GIS). This made it possible to make a combined analysis of data from different sources.

(iii) *Data analysis and mapping:* After compilation, data were mapped and processed for final analysis.

11.3.3 Data Collection and Processing

Existing data (Remote sensing data, Watershed Atlas, Groundwater Atlas and data from various state and central govt. departments) in the selected districts were collected, processed and interpreted. Data were classified as much as possible, stored in a data base for analysis, interpolation and mapping. The description covers the following aspects:

(i) Tehsil-wise Geomorphological, Hydrogeological study of the area under consideration. Aquifer geometry. Depth to bedrock, boundary conditions, lateral extension, depth and thickness of alluvium etc.

(ii) Groundwater characteristics. Transmissivity and storage coefficient (if available); or at least, the specific capacity of wells/tube wells, water level data, water level trends and ground water quality

The mapping approach enabled to identify and locate the effect of abstraction on the aquifer, to identify the areas which are potential zones and suitable for water harvesting and clarify the most deteriorated zones of the aquifer with reference to water quality.

The potential areas for recharging were based on approach consist of following criteria:

1. Geo-morphological and hydro-geological aspect.
2. Groundwater fluctuation trend aspect.
3. Ground water quality and irrigation practice aspect.

11.3.4 Resource Mapping of the Study Area

11.3.4.1 *Location of Study Area*

District Jodhpur: Jodhpur is located in the western part of Rajasthan. It stretches between 26°0' and 27°31' North latitude and 72°55' and 73°52' East longitude. The district covers an area of 22850 sq.km. Jodhpur district is bounded on the north by Bikaner, north-west by Jaisalmer districts, on the east by Nagaur and Pali districts, on the south by Pali and Barmer districts and on the west by Jaisalmer district.

District Pali: Pali is located between 24°45' and 26°29' North latitude and 72°47' and 74°18' East longitude, covering an area of about 12,387 sq.km. Pali district is bounded by Jodhpur in north, Jalore in south west, Sirohi in the south, Udaipur and Rajsamand in the south east, Ajmer in north-east and Nagaur district in north.

District Sirohi: The district is situated in the South West of Rajasthan between the parallels of 24°20' and 25°17' North latitude and 72°16' and 73°10' East longitude.

It covers geographical area of about 5136 sq.km. It is bounded on the north by Pali district in the east by Udaipur district in the south by Banaskantha district of Gujarat state on the West and Northwest by Jalore district.

District Udaipur: The district is situated in the Southern part of Rajasthan. It lies between the parallels of 23°45' and 25°10' North latitude and 73°0' and 74°35 East longitude covering geographical area of about 12590 sq.km. It is bounded on the north by Rajsamand district, north-west by Pali district, west partly by Sirohi and Gujarat state, south by Dungarpur and partly by Banswara and east by Chittorgarh district.

11.3.4.2 *Administrative Setup*

District Jodhpur: Administratively Jodhpur district is part of Jodhpur division. This district is divided into 6 tehsils namely : (I) Bhopalgarh (2) Bilara (3) Jodhpur (4) Osian (5) Phalodi and (6) Shergarh. It comprises of 9 Panchayat Samitis namely: (1) Balesar (2) Bap (3) Bhopalgarh (4) Bilara (5) Luni (6) Mandore (7) Osian (8) Phalodi (9) Shergarh. Although some of the blocks names are common to tehsil names, the boundaries vary at number of places.

District Pali: Administratively Pali district is a part of Jodhpur division. This district is divided into 7 tehsils namely: (1) Jaitaran, (2) Raipur, (3) Sojat, (4) Marwar Junction, (5) Bali, (6) Desuri and (7) Pali. The district also divided into 10 panchayat samities namely (1) Jaitaran, (2) Raipur, (3) Rohat, (4) Sojat, (5) Pali, (6) Kharchi, (7) Rani, (8) Sumerpur, (9) Desuri, (10) Bali.

District Sirohi: Administratively Sirohi is a part of Jodhpur division. The district comprises of 5 tehsils, namely:- (1) Sirohi, (2) Shivganj, (3) Revdar, (4) Abu Road and (5) Pindwara. Which are also panchayat samities of the district.

District Udaipur: Administratively Udaipur is a part of Udaipur division. The district is divided in 10 tehsils, namely:- (1) Kotra, (2) Jhadol, (3) Salumber, (4) Kherwara (5) Sarada, (6) Girwa, (7) Gogunda, (8) Dhariwad (9) Mawli and (10) Vallabhnagar. The district also divided into 11 panchayat samities namely (1) Gogunda, (2) Bargaon, (3) Mawli, (4) Kotra, (5) Girwa, (6) Bhinder, (7) Jhadol, (8) Sarada, (9) Salumber, (10) Dhariwad and (11) Kherwara.

11.3.4.3 *Geological Settings*

District Jodhpur: The geological configuration of the district is characterised by the presence of the formation of the Aravallis, the Malani suite and the Vindhyan Super Group. The Malani beds consist of felsite and quartziferous porphyries.The Vindhyans lie uncomformable upon the Malani rhyollite and are much interrupted by alluvium and wind blown sand and consists of scattered flat patches of rocks.

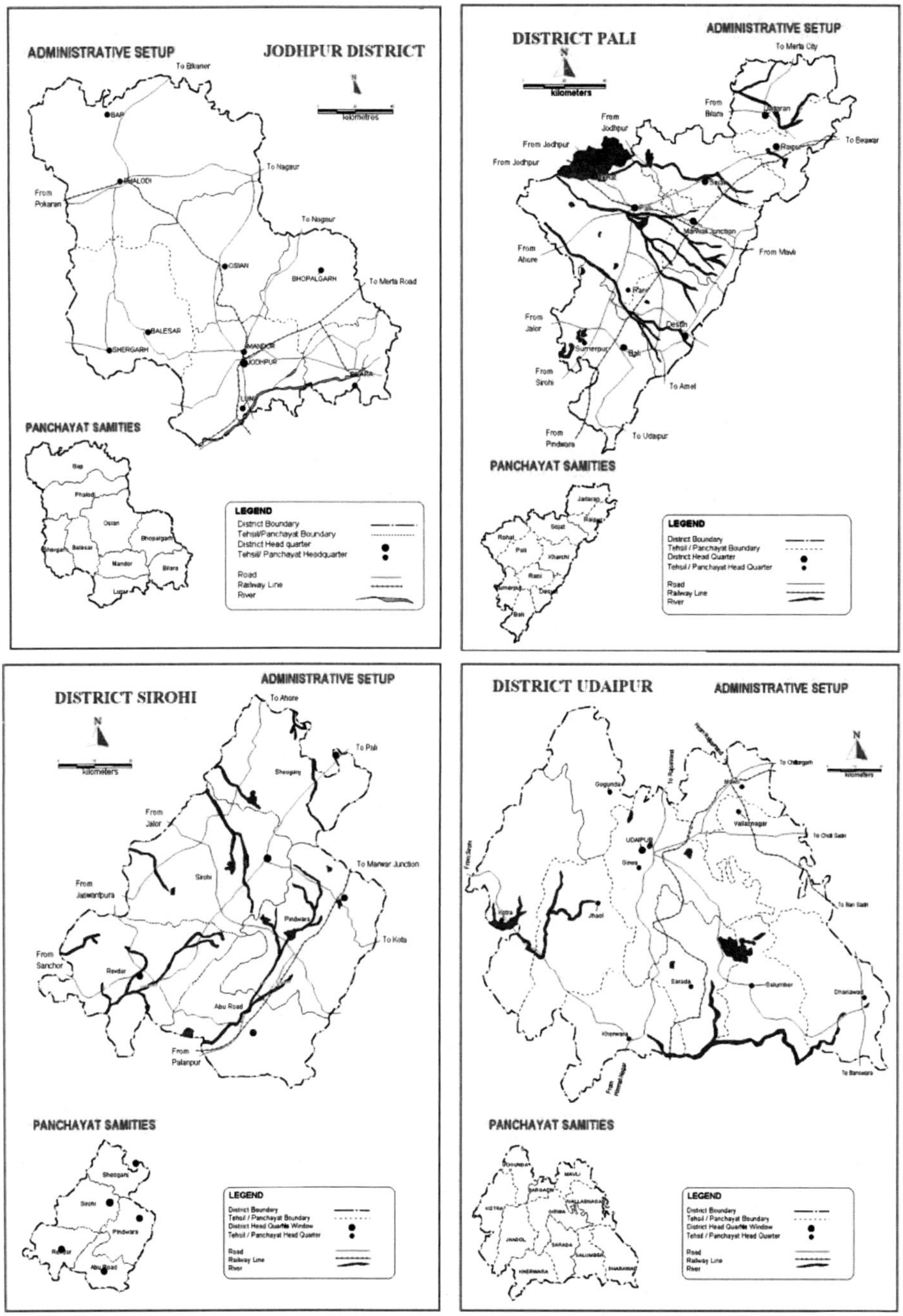

Figure 11.3.1. Administrative setup of the districts (Source: GWD and CGWD)

District Pali: Geological formation of the district is represented by Igneous, Sedimentry and Meta-sedimentary rocks. The Delhi Super Group is represented by

Ajabgarh group exposed near the eastern border of the district and consists of schist, phyllite, marble and basic volcanic and they are intruded by granite and rhyollite, predominant among which is Erinpura granite, covering the south and the south-eastern part of the district. The Jalore type granite is exposed south of the Pali town and is generally pink in colour. The Malani rhyolite (volcanic) covers only small areas restricted to the south-west of Pali. Marwar Super Group occurs in the northern part of the district and are represented by limestone, dolomite, sand stone and shale.

District Sirohi: The Delhi Super Group of rocks occupies the Aravalli ranges in the eastern part of the district and also small chain of hills to the west of Mount Abu. Among the main rock types are phyllite, mica-schist, limestone, marble, calc-silicate, quartzite, etc. these rocks ranges is the age from 1650 to 1400 million year and are intruded by the grey Erinpura and pink Jalore granite. The subsequent Rhyolite are ancient volcanic rocks that erupted during Cretaceous period.

District Udaipur: Aravalli Super Group covers the major parts of the district. Udaipur district is particularly rich in mineral resources, as a large variety of important minerals are found in the district. Important metallic and non-metallic minerals found in the district are ores of copper, lead, zinc and silver, among rock phophate, asbestos, calcite, limestone, bayrites, emerald and marble etc. are important.

11.3.4.4 *Physiography and Soil*

District Jodhpur: Physiographically, the district can be divided into three distinct units viz., alluvial plain, escarpment, and sand dunes. The alluvial plain between Jodhpur and Bilara has gently undulating topography. The plain from Bilara in the east to Shergarh in the west is encirculated by a group of hills, isolated hillocks and ridges separated by alluvium and sand filled valley. The western and north-western parts of the district are characterised by sand dunes. The district has a poor drainage system which is too is internal. The river Luni and its major tributaries, Mithri is the only ephemeral streams. The Luni river enters the district near Jhak and flows in southwesterly direction covering Bilara, Mandore and Luni blocks, leaving the district at Dhundara to enter in the Barmer district.

The north-western part of Jodhpur is a vast sandy area. The dominant soil of dune areas is light pale brown to brown, very deep fine sand to loamy sand. These soils can be classified as Torripsamments great group belonging to Aridizol order. In central part of the district have sandy soil with petrocalcic horizon at depth of 40-80 cm. The horizon is compact and it does not permit entry of water in arid roots. These soils have been classified as solids belong to Aridizol order. The eastern part of district is having dark grayish brown to brown, medium textured and moderately sub angular blocky with sub surface and well developed sub angular blocky structure. Soil profile occurrence of calcic and/or cambic horizon permits them to classify as cambids/calcids great group of Aridizols order. At some places soil salinity problem occurs and these are classified as Aridizol order soils.

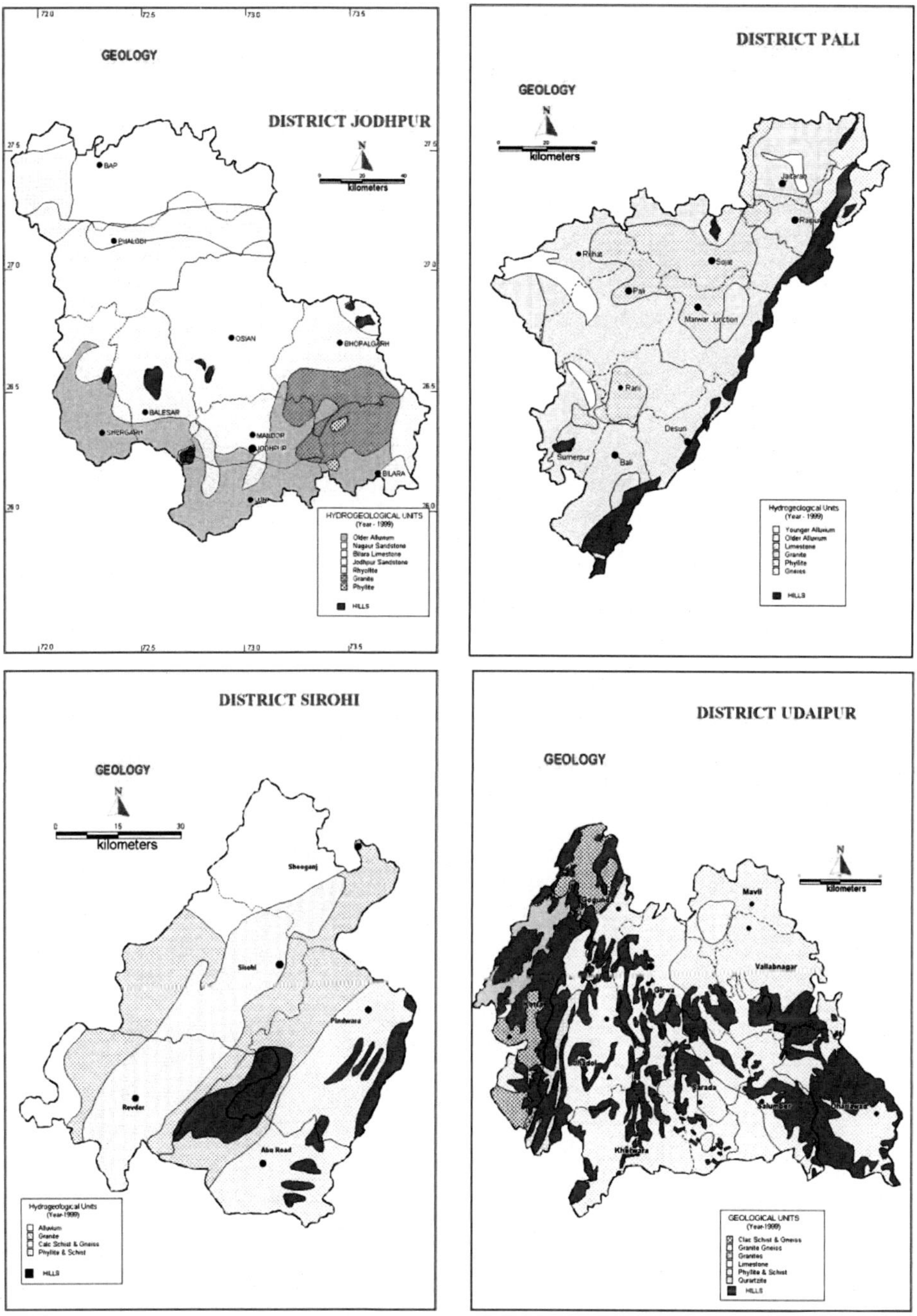

Fig. 11.3.2. Geology (Source: GWD and CGWD)

District Pali: The area of the district may be called sub mountainous and has undulated plains with scattered hills. The district is surrounded by Aravalli range on its south-east. The general slope of the district is from east to west. The texture of the soil is generally sandy loam. The lower level of the sand is made up of rock of calcium carbonate. There is no perennial river in the district. Four tributaries of river Luni viz. Sukhri, Lilri, Bandi and Jawai flow in district. There is no lake or natural spring in the district. Important dam constructed in the district is Jawai Dam in Ball tehsil. The soils of Pali, Desuri and Rani panchayat samities have good permeability and are therefore appropriate for agricultural purposes. On the other hand, the soils of Kharchi, Sojat and Rohat panchayat samities have lesser permeability and are saline in nature.

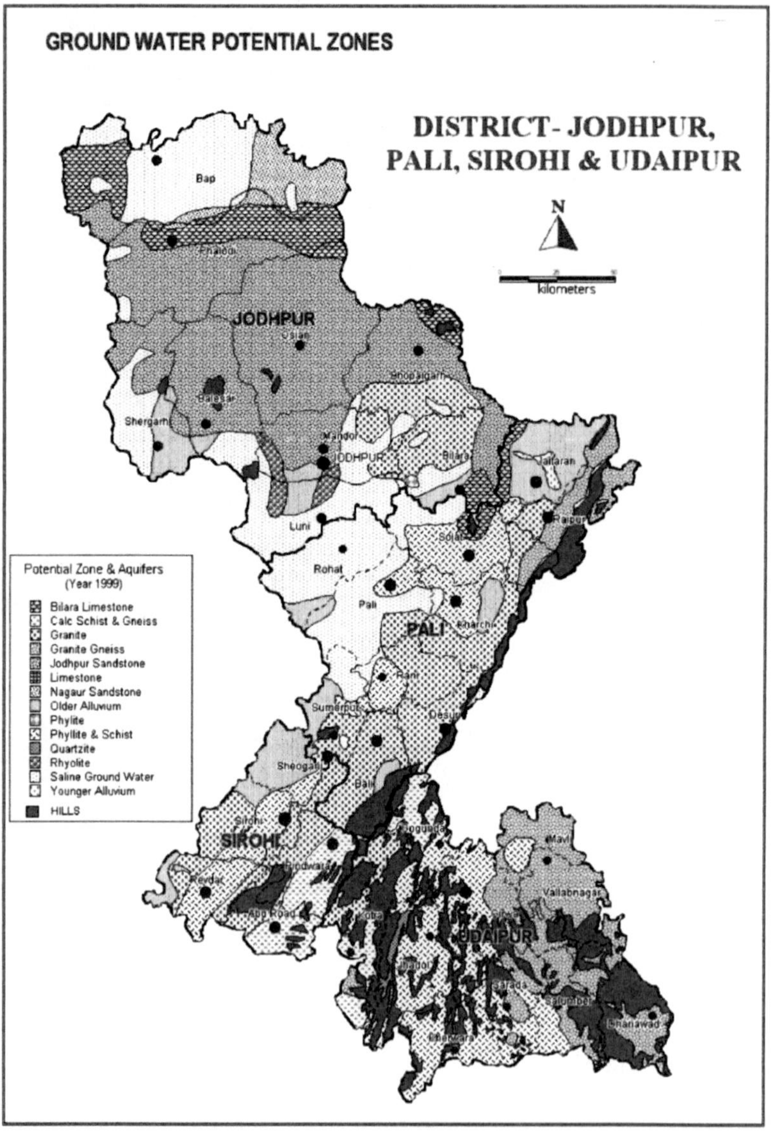

Fig. 11.3.3. Groundwater Potential zones (Source: GWD and CGWD)

The soils occurring in this area is dark grayish brown to brown, medium textured and moderately sub-angular blocky with heavier subsurface and well-developed subangular blocky structure underlained by thick strata of lime coated concretionary zone. The occurrence of calcic and/or cambic horizon permits them to classify as Carnborthids/Calciorthids. Some patches of salinity is observed and they are classified as Salorthids

District Sirohi: A large part of the district is a vast semi desert plain, marked by isolated hills and chains of hillock forming the eastern and south-western extending Aravalli ranges in the east. Abu Sirohi range divides the district into two parts. In the western portion, four group of scattered hills (offshoots of the Aravalli range) an available one each in the tehsil Sirohi and Sheoganj and two in Revdar tehsil. Detached hills of the Aravalli range are situated in the south east of the district, Mount Abu is situated at about 1219 meter above sea level. Another important plateau is Oriya in Abu road tehsil and lies below the main peak of Guru Sikhar which is 1722 meter above the mean sea level. Jawai is the longest and largest river of northwest, which eventually joins Luni river. Other important rivers are Banas, Khari, Sukri, Badi, Kapal Ganga and Krishnawati. The west to southern part of Sirohi district having the soils of the area are generally medium in texture. In the valley part the soils are deep but in the hills it is shallow to moderately deep. Generally the colour of soils are brown to reddish brown. The north to eastern part of Sirohi district, the soils having yellowish brown in colour and light textured. The soils of other region are yellowish brown to grayish in colour and medium to heavy texture. These are rich in nutrients having medium to high fertility status

District Udaipur: The district is encircled by Aravalli ranges from north to south. The northern part of the district consists of generally elevated plateau, while the eastern part has vast stretches of fertile plains. The southern part is covered with rock, hills and dense forest while as the western portion of known as the hilly tracks of Mewar is composed of Aravalli range. Except hard or partially weathered rocks, all types of soil in the district are deep to moderately deep. Clay loam soil is available in tehsil Gogunda, Kotra, Jhadol, Girwa, Mawli and Vallabhnagar while red loam soil is available in Kherwara, Sarada, Salumber and Dhariwad. Generally soils in the western part of the district are stony while yellowish brown soil is met in small portion of eastern and southern part. Som, Jhakham, Wakal, Sabarmati and Berach are main rivers of the district, which are non-perennial and flow during the rainy season only. Soils of the area are medium to heavy textured having brown to grayish brown colour. The fertility of these soils is medium and soils have good water retention capacity. In the foothills soils are shallow to moderately deep. In the valley soil are deep to very deep.

11.3.4.5 *Climate*

District Jodhpur: The climate of the district is characterised by extremes of temperature, fitful and uncertain rainfall and dryness. The winter season which starts in November lasts upto February is followed by summer lasting up to June. Period from July to mid September is monsoon season. There is a good variation in maximum and minimum temperature within the district. In average maximum temperature is about 47°C and minimum is 3°C, during spells of cold waves sometimes it reaches to freezing point (in Phalodi). The average normal annual rainfall is 318.70 mm. The rainfall is very scanty and erratic. Its intensity declines as we move from east to west and south-east to north-west.

District Pali: The climate of the district is dry and has extremes of temperatures. It is very hot during summer and very cold during winter. The district has registered the maximum temperature of 45°C and minimum of 0°C. The average mean temperature is 22.5°C. January is coldest, while May and June are the hottest months. Normal rainfall of the district is 490.4 mm. The average numbers of rainy days are only 22 a year.

District Sirohi: The district has a dry climate with a hot season. Generally cold season starts from December and last till February followed by hot season which continues upto middle of June, while the period from mid June to mid September is of the south-west monsoon, next remaining period till winter is post monsoon season. Maximum, minimum and mean temperature recorded are 37°C, 6°C and 25.5°C respectively. The normal annual rainfall is 638.4mm.

District Udaipur: The district has on the whole moderate and healthy climate without significant seasonal variations, January is the coldest. The maximum, and minimum recorded are 45°C, 03°C respectively. The average annual rainfall is about 624.5 mm. Generally rainfall decreases from the southwest to the northeast, on an average there are 31 rainy days in a year.

11.3.4.6 *Forest*

District Jodhpur: Only 6590 ha. i.e. (0.29%) of the total reported area in the district is under forests. Which mainly on the rocky terrain

District Pali: The forest of the district constitutes about 6.28 per cent of the total area of the district. The forest comes under subsidiary edaphic type of dry tropical forest. These are widely scattered over hillocks and ridges.

District Sirohi: Total 151225 hectares come under forest, which comes to be 30.62 per cent of the total area of the district.

District Udaipur: Major portion is covered with rocks and hills, which are stocked with forest. Forest covers about 240047.86 hectares, which is 23.43 percent of the total area of the district. The forest fall under northern tropical dry deciduous type.

11.3.4.7 *Population*

Table 11.3.1. Tehsil-wise population

S. No.	Tehsil	Total Population - 1991	Male	Female	Schedule Tribe	Schedule Caste
1	Phalodi	161456	84257	77199	5775	23874
2	Osian	306437	160069	146368	10151	50466

S. No.	Tehsil	Total Population - 1991	Male	Female	Schedule Tribe	Schedule Caste
3	Bhopalgarh	156516	81088	75428	1469	29553
4	Shergarh	113079	59679	53400	7665	18855
5	Balesar	127529	66288	61241	4343	21351
6	Bap	103647	54995	48352	3530	16570
7	Mandore	111928	58734	53194	3765	16976
8	Luni	166729	86946	79783	5890	26534
9	Bilara	141630	73774	67856	706	31194

District Pali: The total population of the district is 1486432 out of which 1163085 is rural and 323347 urban. Tehsil-wise population is as under -

Table 11.3.2. Tehsil-wise population

S. No.	Tehsil	Total Population - 1991	Male	Female	Schedule Tribe	Schedule Caste
1	Jaitaran	134851	69221	65630	43	28796
2	Raipur	144710	73820	70890	42	21247
3	Sojat	140478	71229	69249	143	28695
4	Pali	84540	43740	40800	5837	16145
5	Rohat	81290	41822	39468	3827	16367
6	Marwar Junction	164492	83088	81404	4071	31801
7	Desuri	85369	42543	42826	4151	16394
8	Rani Station	94105	46950	47155	3169	17849
9	Bali	149232	75114	74118	37430	25611
10	Sumerpur	115337	58478	56859	14897	20964

District Sirohi: The total population of the district is 654029 among which 526447 are rural and 127582 are urban population. Tehsil-wise population is as under:

Table 11.3.3. Tehsil-wise population

S. No.	Tehsil	Total Population - 1991	Male	Female	Schedule Tribe	Schedule Caste
1	Shivganj	99742	50942	48800	13796	15915
2	Sirohi	133923	67468	66455	6991	23642
3	Pindwara	152236	77548	74688	53069	15770
4	Abu Road	136006	71697	64309	54885	4904
5	Revdar	132122	67862	64260	14890	43517

District Udaipur: The total population of the district is 2066580 among which 1670817 are rural and 395763 are urban population. Tehsil-wise population is as under:

Table 11.3.4. Tehsil-wise population

S.No.	Tehsil	Total Population - 1991	Male	Female	Schedule Tribe	Schedule Caste
1	Mavli	180953	92307	88646	32091	17696
2	Gogunda	122578	62073	60505	51490	10670
3	Kotra	130449	66743	63706	116263	1724
4	Jhadol	147044	75228	71816	104744	4293
5	Girwa	557694	293598	264096	141524	8969
6	Vallabhnagar	195424	99440	95984	37834	15748
7	Dhariwad	164836	84182	80654	126520	8331
8	Salumber	175611	87933	87678	86852	8086
9	Sarada	185497	92935	92562	105116	9048
10	Kherwara	206494	102345	104149	155125	9168

11.3.4.8 *Meteorology*

11.3.4.8.1 *Rainfall*

Table 11.3.5. District wise Rainfall

S. N.	District	Mean Rainfall (mm)												
		Jan.	Feb.	March	April	May	June	July	Aug.	Sept.	Oct.	Nov.	Dec.	Annual
1	Jodhpur	3.9	5.3	3.7	2.8	9.3	30.9	97.5	119.3	44.9	4.8	1.1	1.8	325.3
2	Pali	3.5	4.7	2.1	2.0	10.9	41.6	144.8	182.4	70.9	6.5	1.5	1.3	472.2
3	Sirohi	4.2	5.2	1.7	2.5	12.0	57.7	238.1	226.3	76.3	9.2	3.0	2.2	638.4
4	Udaipur	8.0	3.2	2.9	1.7	9.2	68.0	228.9	204.9	107.9	10.0	4.5	2.0	650.3

11.3.4.8.2 *Evaporation*

Table 11.3.6. District wise Evapo-transpiration

S. No.	District	Mean Annual Potential Evapo-transpiration (cm)
1	Jodhpur	170 - 200
2	Pali	150 - 160
3	Sirohi	140 - 150
4	Udaipur	130 - 140

11.3.4.9 *Aquifer Characteristics*

11.3.4.9.1 *Alluvial deposits*

District Jodhpur: It comprises younger and older alluvium mainly composed of unconsolidated to semi consolidated clays, sand gravel, pebble in varying proportions. Thickness of the lithe unit varies considerably due to undulating bed rock topography. It has been tapped maximum in area south west of Jodhpur around villages Pal and Deli where thickness recorded upto 60 m. These are composed of rounded to sub-rounded poorly sorted gravel, pebble, cobble and boulders of limestone, granite, rhyolite, quartzite and slaty formation. The lithe unit has poor specific yield and quality of ground water is saline; so it has not been included in ground water potential area. Available data indicates that the alluvium thickness forming the water bearing formation has variable depth from 10 m to 90 m.

District Pali: It mainly comprises stream laid deposits consisted of sand, gravel with some silt, clays and at places pebbles and cobbles. The lithe unit normally has thickness upto 50 m but in north eastern part attains thickness of about 100 m. It is heterogenous mixture of semi consolidated to consolidated sediments comprising sand, silt, clay, kankar and silt. Sand is generally medium to coarse grained, sub-rounded, while gravel include sub-angular to sub-rounded quartz grains. Rock fragments in the lithe unit are angular to sub-rounded. Thickness of alluvium varies considerably due to undulating bed rock topography and observed maximum in north eastern area (about 100 m).

District Sirohi: It comprises unconsolidated to semi-consolidated sand, gravel, pebbles, boulders and at places clay in varying proportion. These are fluvial deposits and their thickness in major part of the area varies between 10 to 15 m.

District Udaipur: Valley fill formed by fluvial activities, usually at lower topographical locations, comprising of boulders, cobbles, pebbles, sand, silt and clay. The unit has consolidated sediment deposits. Pediments are broad gently slopping rock flooring, erosional surface of low relief between hill and plain, comprised of varied lithology, criss crossed by fractures and faults. Pediments covered essentially with relatively thicker alluvial, colluvial or weathered materials.

11.3.4.9.2 *Ground Water Flow*

District Jodhpur: The general direction of ground water flow varies significantly. It has been inferred southeast to northwest in the northern part whereas in Phalodi block it become south west to north east. The southern and north western area have NE to SW or NNE to SSW flow. Ground water mounds and troughs having small spread have been observed in Mandore, Osian, Balesar and Bap blocks. In area south west of Phalodi and west of Jodhpur hydraulic gradient has been worked out as 4.0 and 5.0 m/km respectively. It indicates low permeability aquifer underlying the area. Other parts have hydraulic gradient between 1.14 to 1.81 m/km.

District Pali: Ground water flow generally follows surface drainage pattern. In north eastern area, flow has been inferred E to W or SE to NW. Western and southwestern area has E to W and SE to NW flow. Hydraulic gradient is gentle in north eastern and central area (1.3 m/km) and steeper in the southern most. It indicates low permeability aquifers in the southern area.

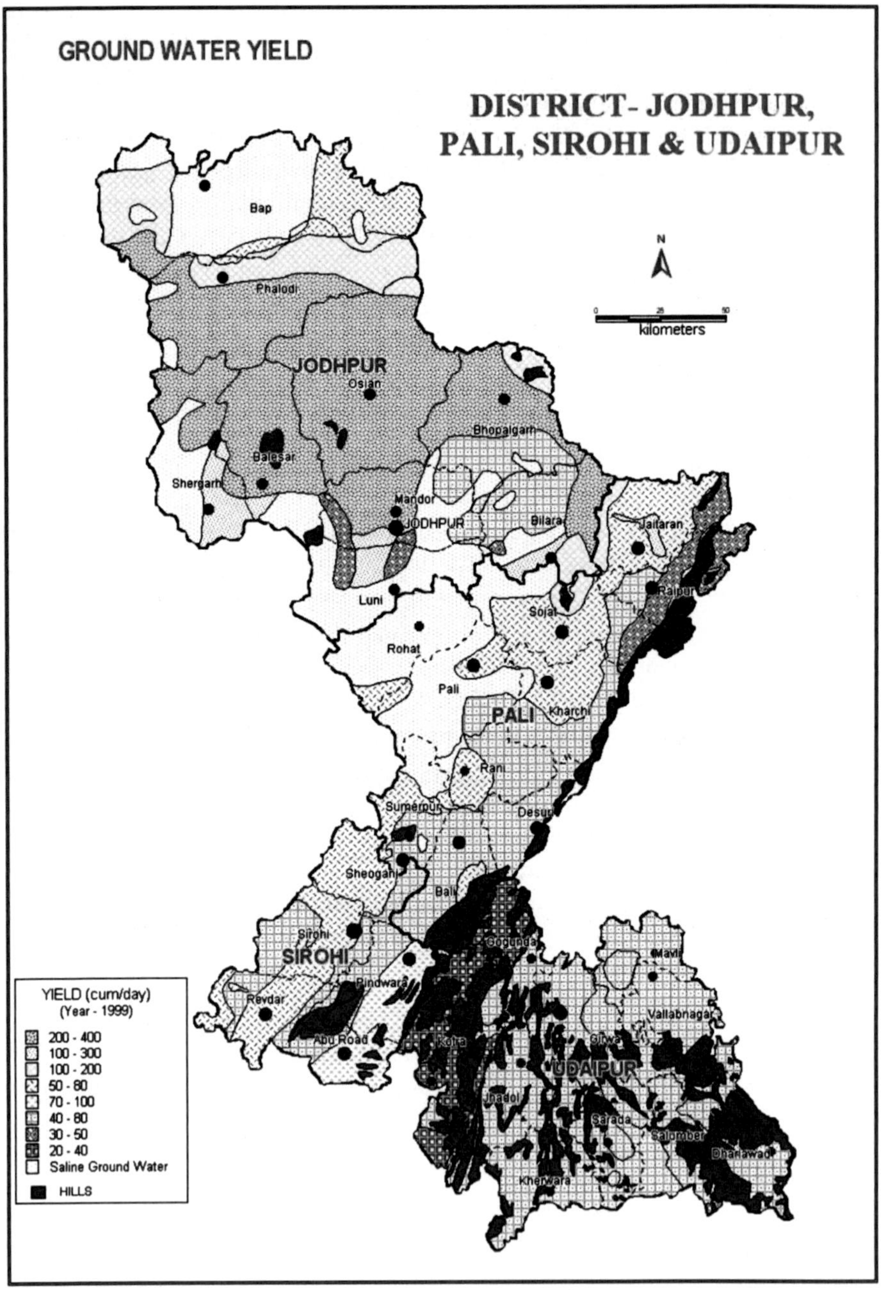

Figure 11.3.4. Groundwater Yield (Source: GWD and CGWD)

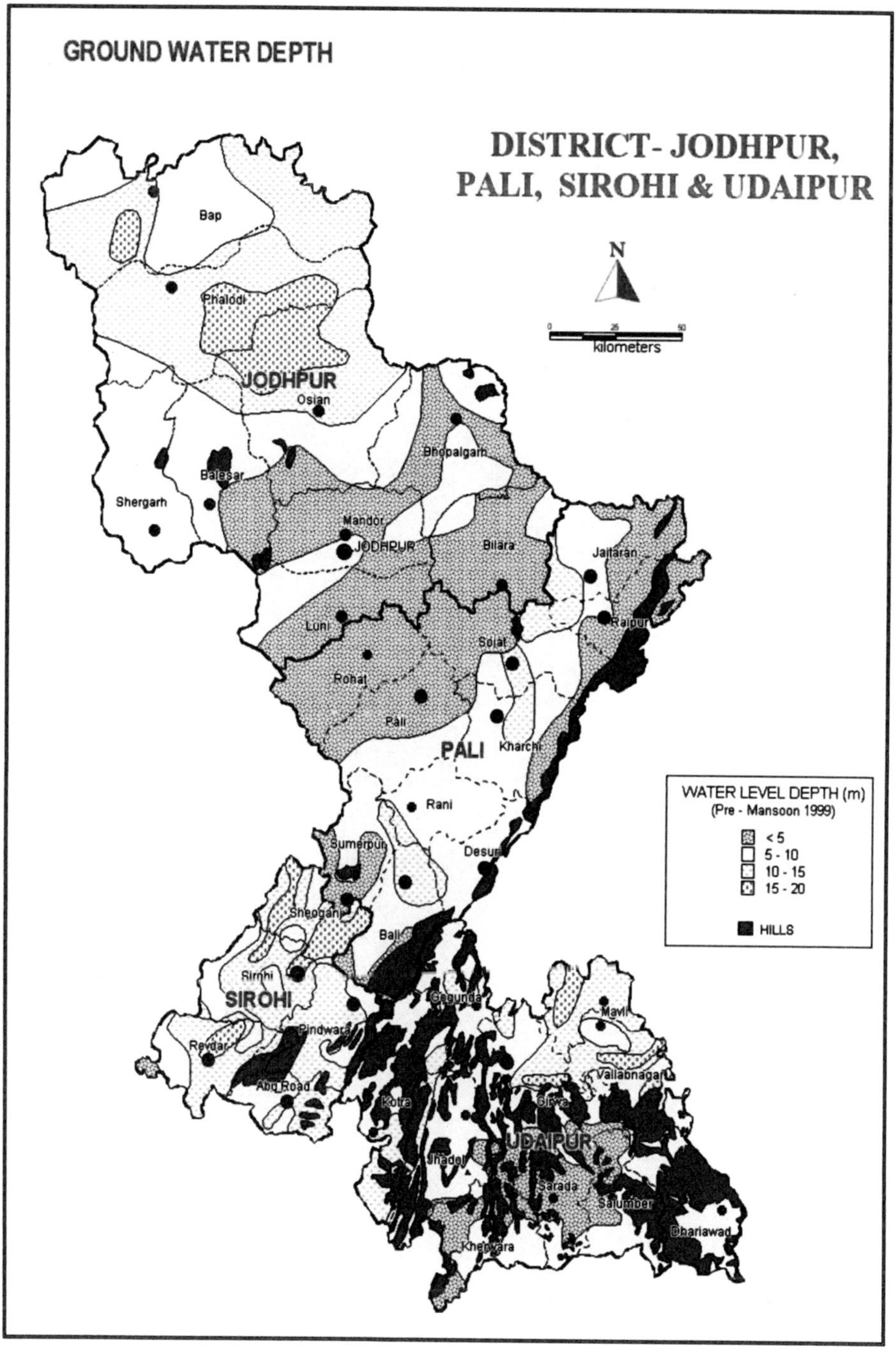

Figure 11.3.5. Groundwater depth (Source: GWD and CGWD)

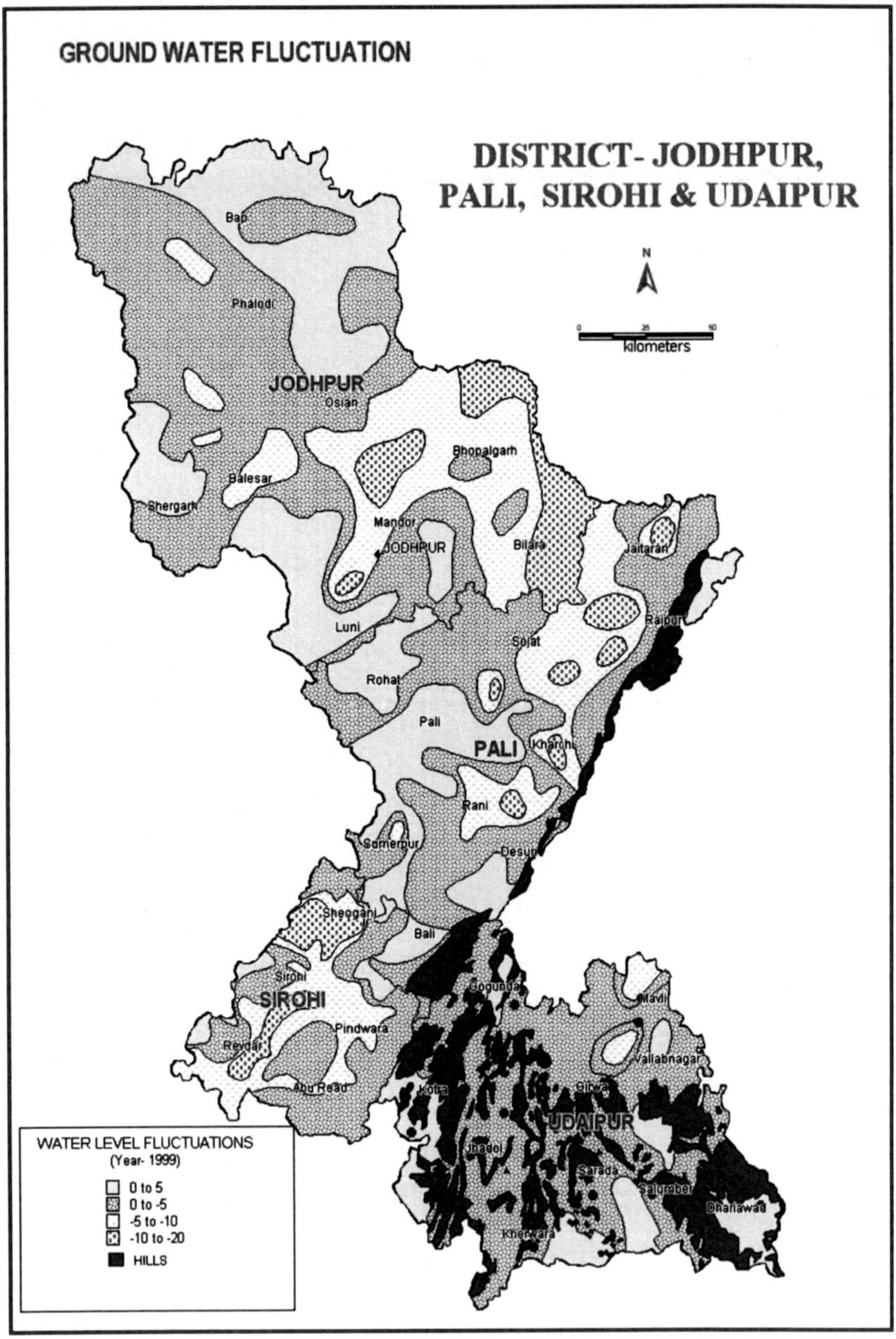

Figure 11.3.6. Groundwater Fluctuation (Source: GWD and CGWD)

District Sirohi: General direction of ground water flow north of Sirohi has been inferred toward SE to NW. In other parts of the district, it is N to S whereas in southern portion it is from NE to SW. Hydraulic gradient is more steep in hilly terrain, in other parts of the area it has been computed as 4.0 m/ Km.

District Udaipur: General direction of ground water flow varies considerably. In Mavli and Bhinder blocks, flow direction varies from NW to SE or W to E, or N to S, in Sarada Salumber and Kherwara blocks, SE to NW or E to W, in Dhariwad block, while in other remaining blocks NE to SW or N to S. Hydraulic gradient varies from 5.00 to 11.40 m/km.

11.3.4.9.3 *Groundwater Extraction*

Table 11.3.7.

District Jodhpur

S. No.	Block	Category
1	Balesar	Safe
2	Bap	Safe
3	Bhopalgarh	Over Exploited
4	Bilara	Safe
5	Luni	Critical
6	Mandor	Safe
7	Osian	Western Part Safeand Other Parts Over Exploited
8	Phalodi	Two Localized Area Over Exploited
9	Shergarh	Safe

District Pali

S. No.	Block	Category
1	Bali	Safe
2	Desuri	Critical
3	Jaitaran	Over Exploited
4	Kharchi	Critical
5	Pali	Safe
6	Raipur	Over Exploited
7	Rani	Safe
8	Rohat	Safe
9	Sojat	Over Exploited
10	Sumerpur	Safe

District Sirohi

S. No.	Block	Category
1	Abu Road	Safe
2	Pindwara	Over Exploited
3	Reodar	Semi Critical
4	Sheoganj	Safe
5	Sirohi	Safe

11.3.4.9.4 *Groundwater Quality*

District Jodhpur: High salinity, alkalinity, fluoride and nitrate in ground water at some places in the district renders it unsuitable for irrigation and drinking use. Due to high sodium content nearly 25 percent ground water in the district have residual sodium carbonate (RSC) above 2.0 meq/L. Similarly, nearly 20.4 per cent groundwater samples are characterised by sodium adsorption ratio (SAR) more than 18. In Bilara, Luni and Mandore blocks, high SAR values are more common and these waters usually have soluble sodium percentage above 70. Continuous use of such waters for irrigation leads to soil alkalinity which may result in low soil permeability. Use of gypsum to overcome hazardous effect of sodic water is therefore suggested for optimum utilisation of such water.

District Pali: The ground water in the district is characterised by low to moderate hardness. 49.2, 31.0 and 19.7 per cent water samples of the district are characterised by hardness in the range 0-300, 300-600 and above 600 mg/L as calcium carbonate. The hardness of water is comparatively more in northern part of the district covering Rohat, Pali and Sojat blocks. 45%, 46% and 30% ground waters in these blocks have hardness above 600 mg/L respectively. Rani block in south, on the other hand, has modest values for hardness in ground water, 83% well waters in the block are characterised by hardness below 300 mg/L. The maximum hardness in ground water is observed at Kharchigaon in Kharchi block (2350 mg/L) whereas the lowest value is seen at Takhatgarh in Sumerpur block (85 mg/L). Ground water is the major source of irrigation in the district. As per the Anonymous water quality criterion for rating the irrigation waters, 77.3 per cent ground waters having salinity below 6000 μ/cm are suitable for irrigation on the sandy loam soils of the district for all type of crops. The next 5.7 per cent waters of the salinity range 6000-8000 μ/cm could be used for irrigating 'salt-tolerant crops including wheat and mustard. The rest 17.1 per cent ground waters having salinity above 8000 μ/cm, however, unsuitable for irrigation owing to high salinity. These waters are mostly found in Rohat and Pali blocks. It is further observed that the saline waters have high percentage of sodium (>70%) whereas the fresh waters generally have residual sodium carbonate (RSC) more than 2.0 meq/L. The two factors adversely effect the soil properties by reducing the permeability when such water is used for irrigation, high sodium waters generally occur in Jaitaran, Kharchi, Pali, Raipur, Rohat and Sumerpur blocks whereas high RSC waters are seen in Ball, Desuri, Jaitaran and Kharchi blocks. An integrated map for drinking water quality based upon the maximum permissible limits for salinity, nitrate and fluoride content shows area in central and western part of the district have ground water of unsuitable quality owing to one or more constituent above the maximum permissible limits.

District Sirohi: The ground water in the district is characterised by low to medium range of hardness usually less than 600 mg/L. The per cent frequency of water samples falling in the total hardness (TH) range of 0-300 and 301-600 mg/L is found as 42.5% and 46.9% respectively. Only 10.6 per cent ground water in the district show hardness value above 600 mg/L. This is due to low ratio of calcium and magnesium ions in ground water. The maximum TH is seen as 1920 mg/L in village Warain of Pindwara block. Blockwise, the ground water in Pindwara is inferior to ground water in other blocks in respect of hardness values. In this block 18.2% samples show hardness value above 600 mg/L as compared to only 5.2 to 14.3% of such waters occurring in other blocks. Owing to low salinity, the ground water in the district is suitable for agriculture use. However, due to relative abundance of sodium and bicarbonate ions, the ground water is marginally alkaline in character. It is observed that nearly 14.3% samples have soluble sodium percentage above 70 while 18.8% samples show residual sodium carbonate (RSC) above 2.0 meq/L. The percentage of such waters is even more in Shivganj and Sirohi blocks where 38.1% and 20% samples respectively have sodium percentage above 70. Similarly, 28.6 and 20% samples in the two blocks show RSC above 2.0 meq/L respectively. The Bithura village in Shivganj block is characterised by highest RSC value of i4.4 me/L. An integrated map for drinking water quality based upon ICMR standards shows considerable area non-conforming to the standards. It is observed that only 37.8% ground water in the district have all the three constituents within the maximum permissible limits. In rest, one or more constituent is above the prescribed limits. Salinity i, not a major factor problem for rejection of ground water for drinking use, but high fluoride affects the water quality most. Due to granite rocks forming major hydrogeology of the district, the ground water is often rich in fluoride content.

District Udaipur: 96.8% of the ground water having EC within the limit of 4000 μS/cm is suitable for irrigation as far as salinity criteria is concerned to the soils of the district of course, the depth to water should not be below 1.5 metres at any time of the year. Water is free from sodium hazard as 98% ground water have sodium within 70% of total cations. Residual sodium carbonate above 2.0 meq/L which may turn soil hard and difficult for germination of seeds popularly known as alkali hazard is found only in 9.2% of ground water of the district. The potability map of the district indicates that the most part of the district have potable ground water. Ground water unsuitable for drinking is observed in two big patches, one in north-eastern part covering part of Mavli and Bhinder blocks, on account of salinity and another in southern part covering part of Salumber and Sarada blocks having high fluoride content. Small patches showing ground water unsuitable for drinking are also seen in block Girwa, Mavli and Bhinder blocks.

11.3.4.10 *Irrigation Sources*

District Jodhpur: Jodhpur district has total geographical area i.e. 22850 sq.km. The area under irrigation is 1388.17 sq.km. which is about 6.07 per cent of the total geographical area of the district. The main source of irrigation in the district is wells and few darns also contribute for it. Among the tanks suitable for irrigation purposes, Soorpura tank, Pichiyak (Jaswant Sagar) and Golejor bundh are worth mentioning.

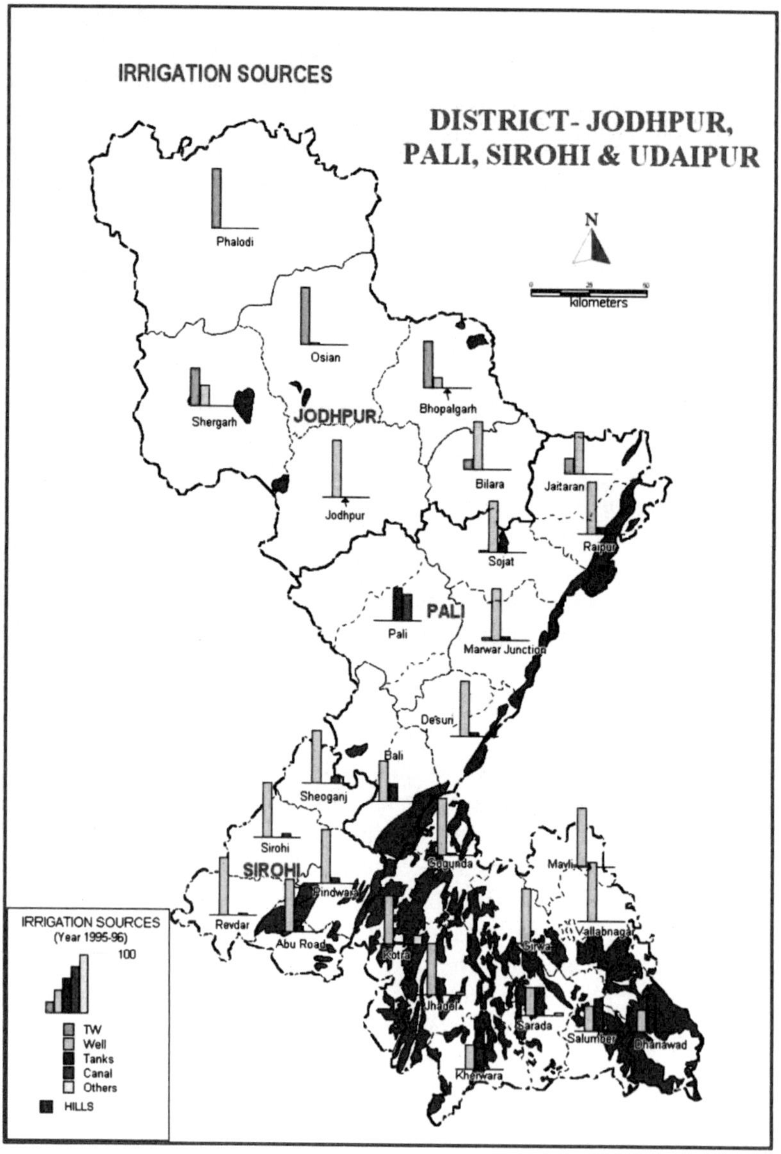

Figure 11.3.7. Irrigation Sources (Source: GWD and CGWD)

District Pali: Pali district has total geographical area 12,387 sq.km. The area under irrigation is 2824.02 sq.km. which is about 22.79 per cent of the total geographical area of the district. The main source of irrigation in the district is wells, which constitute 75 per cent of total irrigated area followed by pond tanks 20 per cent, tube wells 5 per cent. In all; district has 92 darns out of which three darns, one each at Gajni, Potalia and Chipatia are broken. Total catchment area of all the dams is 238150.14 acres. The capacity of 89 dams was 20197.3 million cubic feet in 1989-90. The largest dam in the district is Jawai dam (in Bali tehsil) whose storage capacity is 7887.5 million cubic feet and covers an area of 102315 acres of cultivable command area.

District Sirohi : Sirohi district is having total geographical area 5136 sq.km. The area under irrigation is 1152.38 sq.km. which is about 22.43 per cent of the total area. Main source of irrigation in the district is wells. Area irrigated by wells is 82 per cent, ponds and canals 14 per cent, Canals 4 per cent to total irrigated area. During pre independence, there were 8 dams in the district with an irrigation capacity of only 3289 acres of land exists. After independence various minor and medium irrigation projects were taken up in the district and about 33 darns were got completed. Thus total number of 41 dams with their irrigation capacity of 42,255 acres of land were available in the district, during the year 1988-89.

District Udaipur : Udaipur district has total geographical area 12590 sq.km. The area under irrigation is 874.27 sq.km. which is about 7.71 per cent of the total geographical area of the district. Main source of irrigation in the district is wells. Followed by tanks/ponds. Prominent rivers of the district are Wakal, Berach, Som, Jhakham, Mahi and Gomti. Important water bodies of the district are Fateh Sagar, Swaroop Sagar, Pichhola, Bari Talab, Udaisagar and Jaisamand. Important watersheds /river basins of the district are Som, Jhakham, Berach, Sai, Sukri, Vakal and Sabarmati.

11.3.5 Findings

The maps prepared from the collected data clearly indicates the overall scenario of the depleting status both quantity and quality of the existing ground water. The mean rainfall in not equal to the mean evapotranspiration even in the monsoon season (rainy season) in most of the parts of the area study area. Rainfall in the area is not even equal to the mean rainfall since last three years causing the worst effect in the ground water resources as this is the only recharging source to the ground water. Day by day increasing agricultural and domestic draft of the ground water has leaded the situation to the alarming position. Most of the areas are over exploited and depth of the water level gone down to significant deep. Now the time has come to make long term planning so that the ground water is recharged using rainwater harvesting.

11.3.5.1 *Hydro-geological Approach*

"Good radiators are the best absorbers" This natural characteristics is also applicable to the geological formations i.e., good water bearing strata/ formation yields better discharge/ yield of the water and in turns provide best suitable sites for the recharge. The Figure 11.3.8 indicates the suitable areas where the yield of the water is greater and thus indicates good site for carrying out water harvesting works.

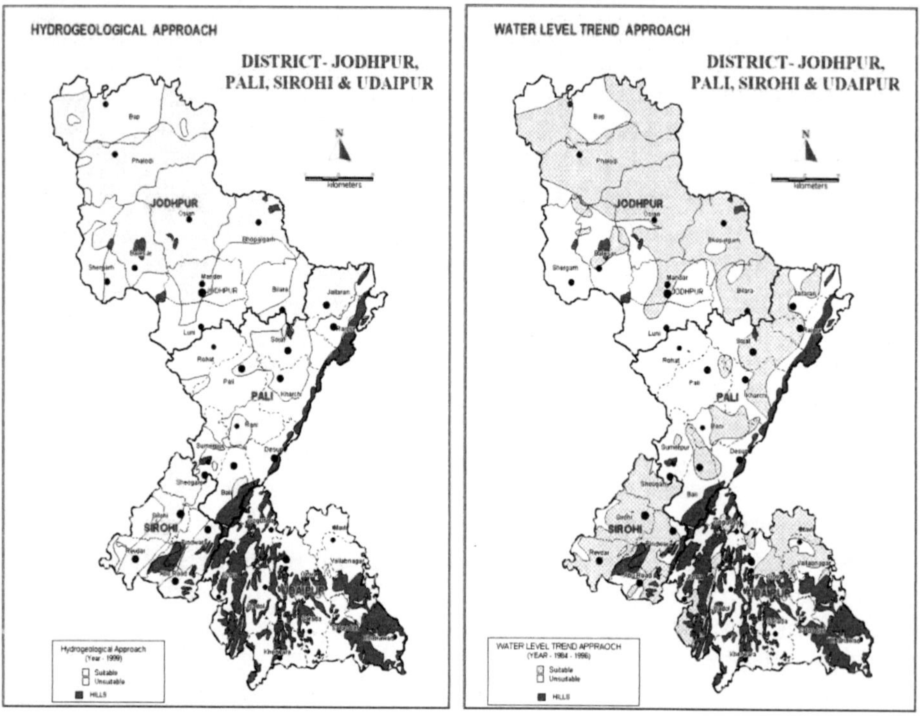

Figure 11.3.8 and 8A. Hydrogeological and Water Level Approach (Source: GWD and CGWD)

11.3.5.2 *Groundwater Trends Approach*

When the map of water level depth is superimposed on change in water level the resultant map indicate the area which is affected the most and thus needing the water conservation works. The Figure 11.3.8A indicates the suitable areas where the fluctuation is much thus indicates suitable site for carrying out water harvesting works.

11.3.5.3 *Groundwater Quality Approach*

The maps of Salinity, Fluorides and Nitrates have been superimposed to find out the area suitable from the quality parameter point of view. The ground water quality differs from place to place and also with different depth at a particular place. The fresh rain water when percolates downwards improves the quality of the ground water which decreases as time passes on thus leads to the need of harvesting rain water as much as possible. The Figure 11.3.9 indicates the suitable areas where the water is potable and thus indicates suitable site for carrying out rain water harvesting works.

11.3.5.4 *Selection of Approach*

Selecting an area suitable for water harvesting depends upon many factors e.g., Land use patterns - agricultural or drinking requirement (Giving weight to quality app-roach). One single approach can not serve the purpose and thus combination of such approaches has to be selected depending upon the topography, land-use / land-cover of the area.

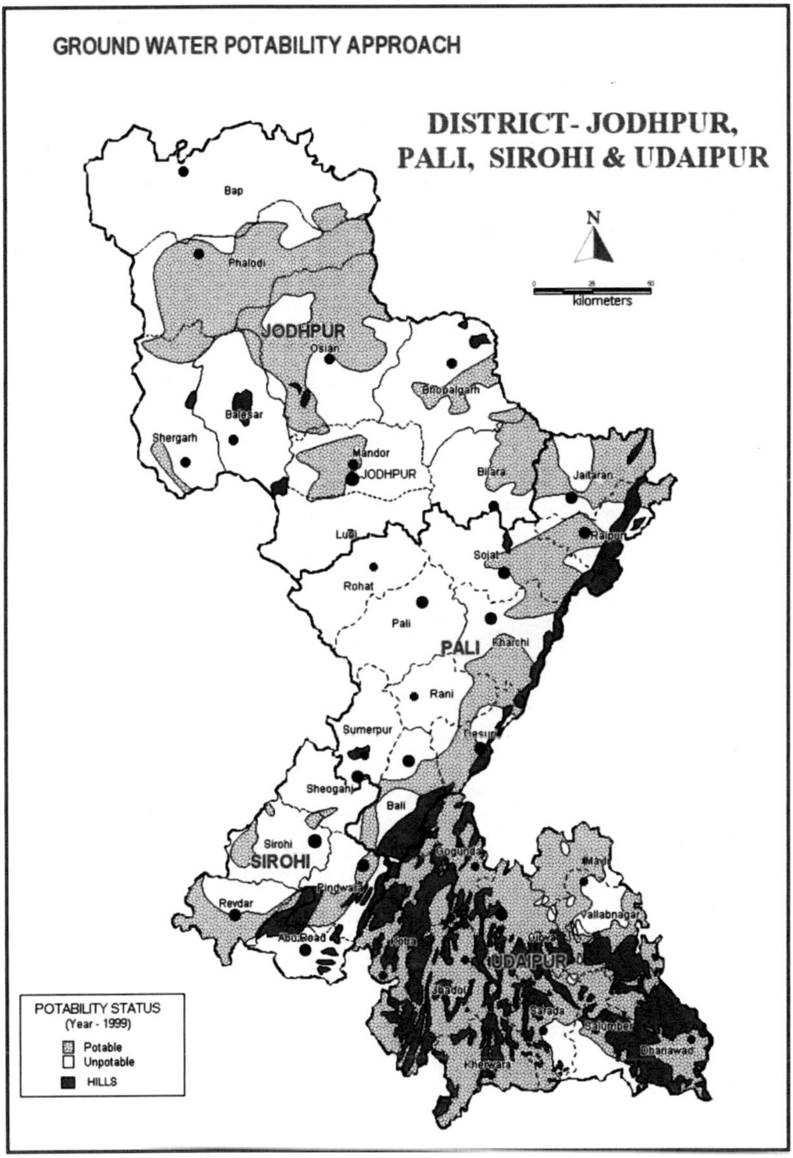

Figure 11.3.9. Groundwater Potability Approach (Source: GWD and CGWD)

11.3.6 Conclusion

11.3.6.1 *Rain Water Harvesting Areas*

The area selection approach should also include some practical aspect by field visits and discussion with various government departments already working in the area. As the prime objective of the study is to find out the potential areas with respect to the Groundwater and to carry out labour oriented Water Harvesting works so as to provide the employment thus wages to the drought affected people, other aspects such as agriculture production, domestic water consumption etc. has to be given some sort of priority.

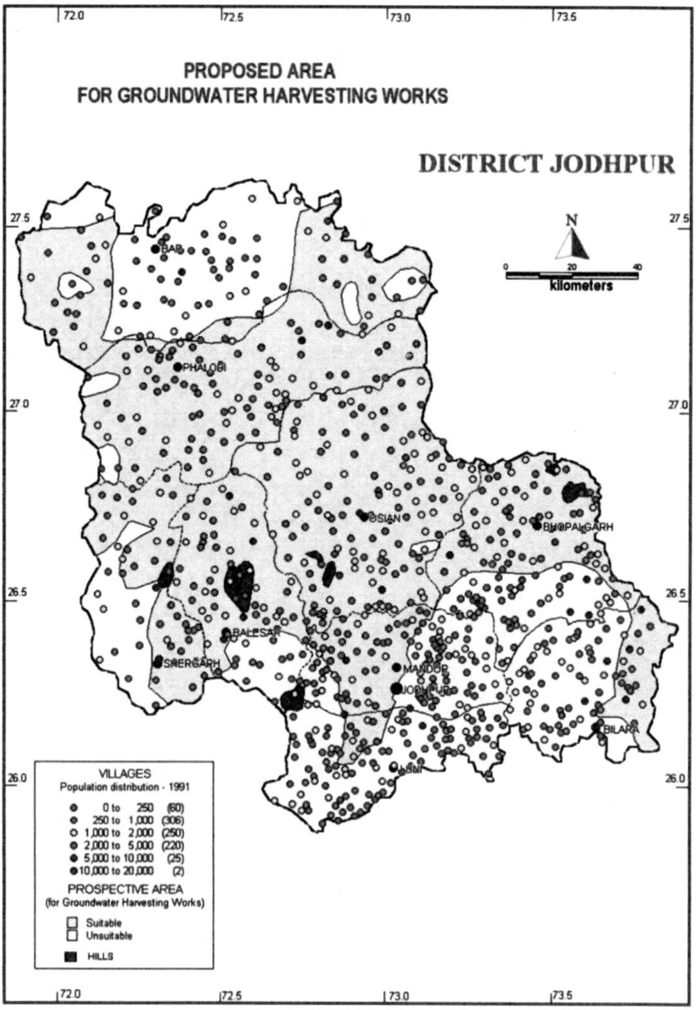

Figure 11.3.10. Proposed Area for Groundwater Harvesting Works (Source: GWD and CGWD)

District Jodhpur: The Prospective Figure 11.3.10. indicates the proposed area demarcation for groundwater harvesting works.

(i) Bap block: Out of 89 villages 37 villages fall in the proposed area covering east and western parts of the block and 49868 out of 101508 persons of the block.

(ii) Phalodi block: Out of 92 villages 88 villages fall in the proposed area covering almost entire area of the block and also covering population 160197 out of 161907 persons of the block

(iii) Shergarh block: Out of 49 villages 36 villages fall in the proposed area covering south-east, east and northeastern part of the block and also covering population 178276 out of 111305 persons of the block

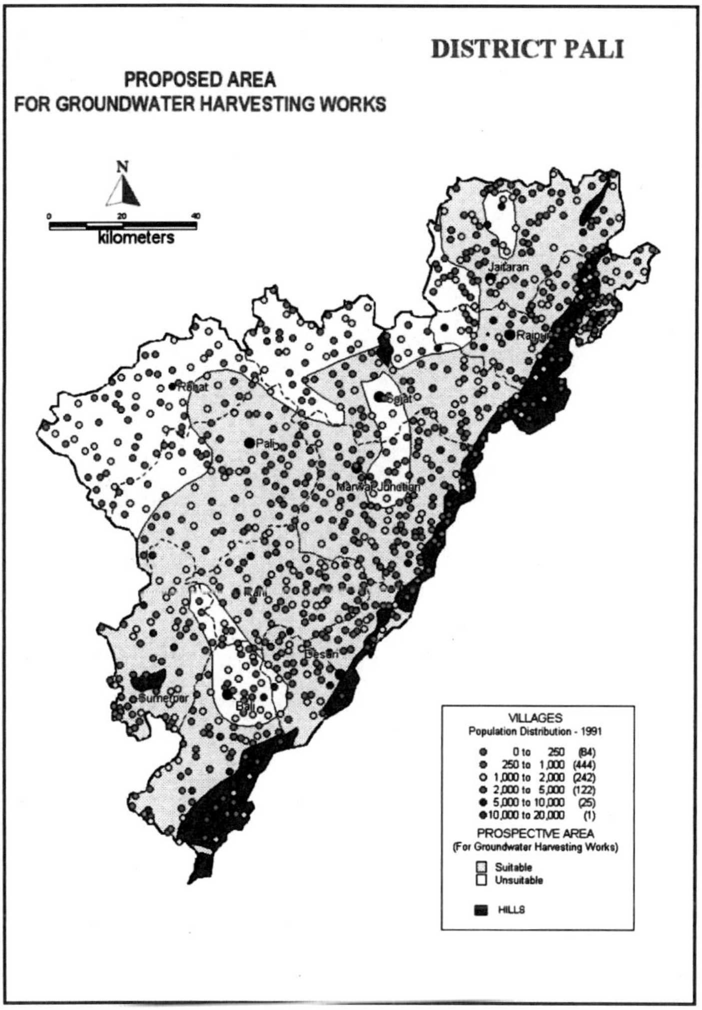

Figure 11.3.11. Proposed Area for Groundwater Harvesting Works (Source: GWD and CGWD)

District Pali: The Prospective Figure 11.3.11. indicates the proposed area demarcation for groundwater harvesting works.

(i) Bali block: Out of 90 villages 65 villages fall in the proposed area covering entire block except northern part and 102060 out of 152523 persons of the block.

(ii) Rani block: Out of 82 villages 78 villages fall in the proposed area covering almost entire area of the block and also covering population 104433 out of 110528 persons of the block

(iii) Desuri block: Out of 69 villages 66 villages fall in the proposed area covering almost entire block and also covering population 61579 out of 69402 persons of the block

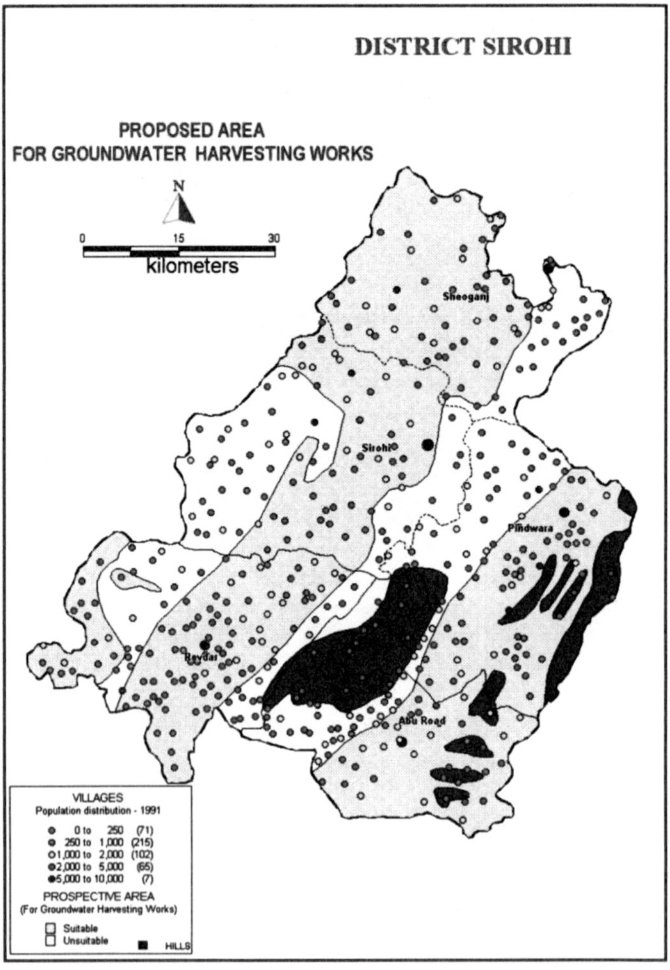

Figure 11.3.12. Proposed area for Groundwater Harvesting Works (Source: GWD and CGWD)

District Sirohi: The Prospective Figure 11.3.12. indicates the proposed area demarcation for groundwater harvesting works.

(i) Abu Road block: Out of 81 villages 50 villages fall in the proposed area covering western part of the block and 46516 out of 81611 persons of the block.

(ii) Reodar block: Out of 126 villages 95 villages fall in the proposed area covering western area of the block and also covering population 94577 out of 130791 persons of the block

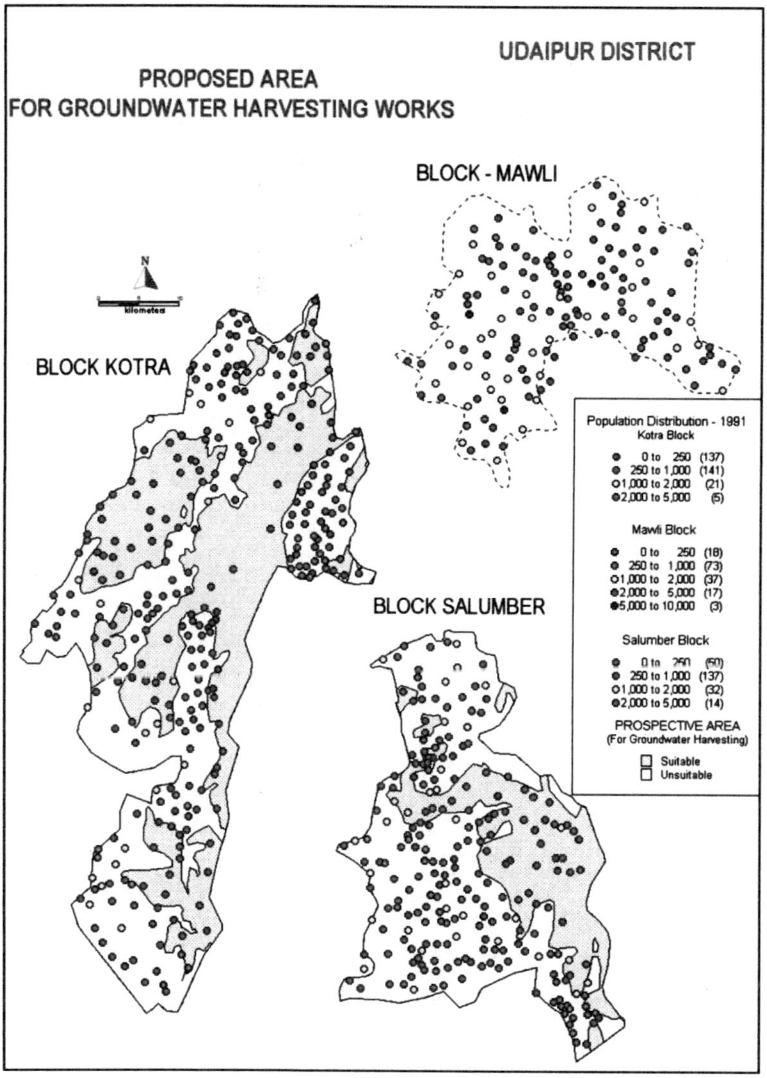

Figure 11.3.13. Proposed Area for Groundwater harvesting Works (Source: GWD and CGWD)

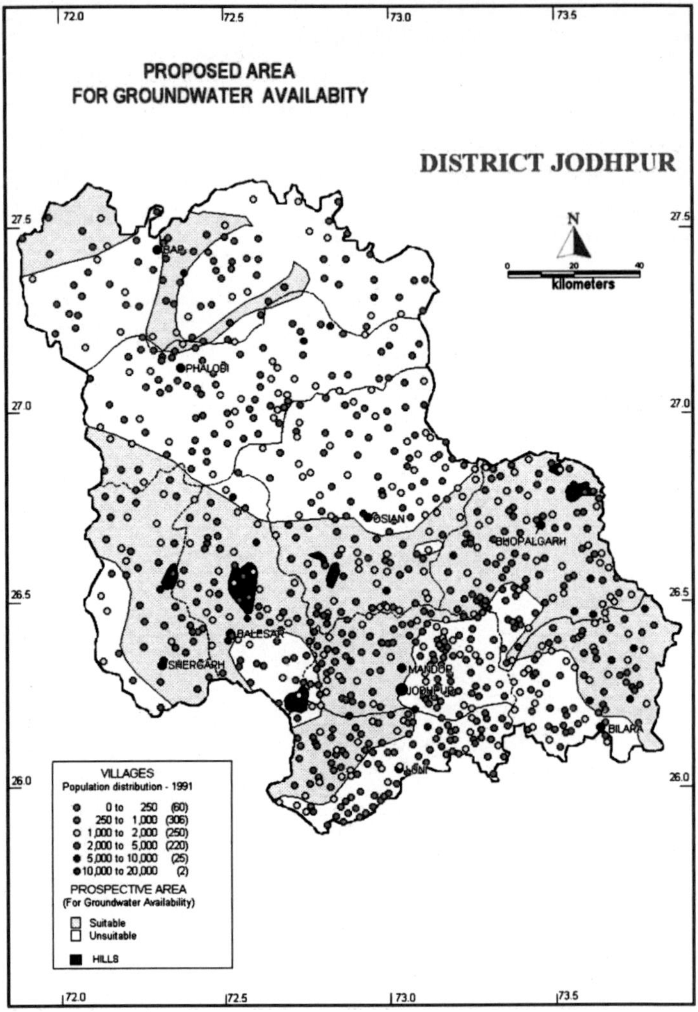

Figure 11.3.14. Proposed area for Groundwater Availability (Source: GWD and CGWD)

District Udaipur: The Prospective Figure 11.3.12. indicates the proposed area demarcation for groundwater harvesting works.

(i) Kotra block: Out of 304 villages 100 villages fall in the proposed area covering scattered area of entire block except western part and 31272 out of 130359 persons of the block.

(ii) Salumber block: Out of 233 villages 58 villages fall in the proposed area covering almost eastern and middle area of the block and also covering population 34688 out of 161720 persons of the block

(iii) Mawli block: Out of 148 villages 164371 population ther is no village which fall in the proposed area as the entire block have no potential area as far as selected approach is concerned.

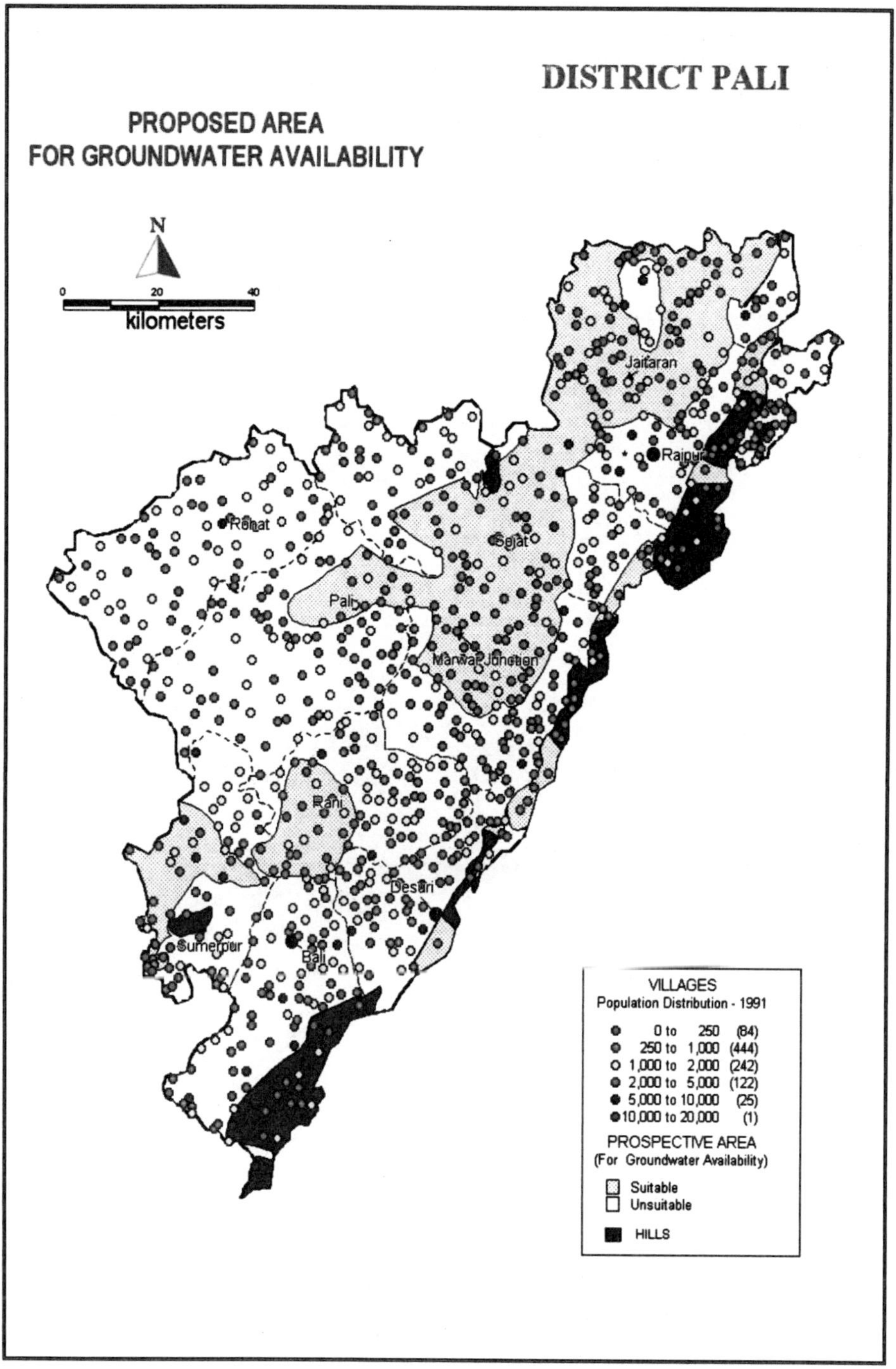

Figure 11.3.15. Proposed area for groundwater availability (Source: GWD and CGWD)

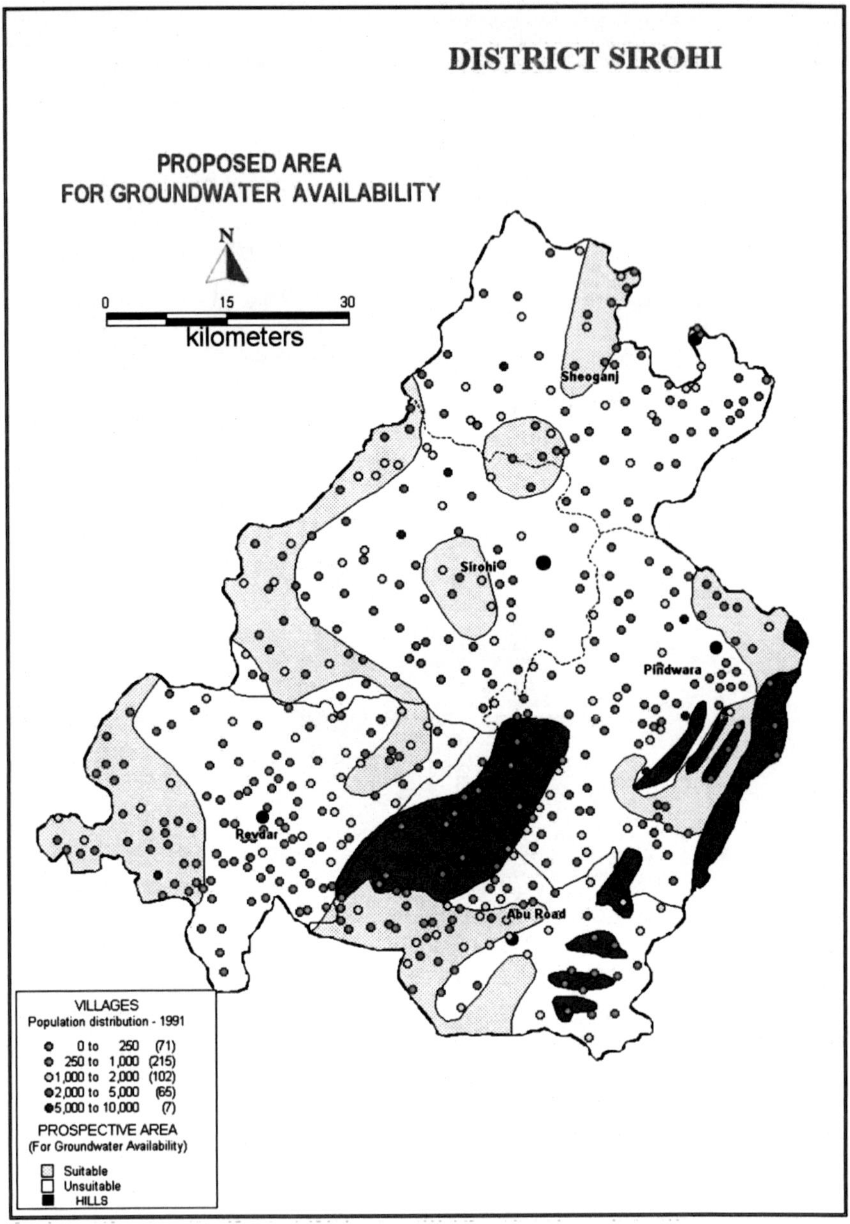

Figure 11.3.16. Proposed area for groundwater availability (Source: GWD and CGWD)

11.3.6.2 Water Availability Areas

The area selection approach should also include some practical aspect by field visits and discussion with various government departments already working in the area. As the objective of the study is also to find out the potential areas with respect to the Groundwater availability from the secondary data available, to find out the possibility of constructing wells in the project area so as to provide the employment thus wages to the drought affected people and other aspects such as agriculture production, domestic water consumption etc.

District Jodhpur: The Prospective Figure 11.3.14. indicates the proposed area demarcation for groundwater harvesting works.

(i) Bap block: Out of 89 villages 25 villages fall in the proposed area covering east and northwest and central parts of the block and 28381 out of 101508 persons of the block.

(ii) Phalodi block: Out of 92 villages 9 villages fall in the proposed area covering southwestern area of the block and also covering population 11805 out of 161907 persons of the block

(iii) Shergarh block: Out of 49 villages 42 villages fall in the proposed area covering entire block except southwestern part of the block and also covering population 93033 out of 111305 persons of the block

District Pali: The Prospective Figure 11.3.15. indicates the proposed area demarcation for groundwater harvesting works.

(i) Bali block: Out of 90 villages 4 villages fall in the proposed area covers northern part of the block and 5432 out of 152523 persons of the block.

(ii) Rani block: Out of 82 villages 20 villages fall in the proposed area covering southwestern area of the block and also covering population 29453 out of 110528 persons of the block

(iii) Desuri block: Out of 69 villages 3 villages fall in the proposed area covering some part in the east of the block and also covering population 1134 out of 69402 persons of the block

District Sirohi: The Prospective Figure 11.3.16. indicates the proposed area demarcation for groundwater harvesting works.

(i) Abu Road block: Out of 81 villages 32 villages fall in the proposed area covering entire block except northern part and 40665 out of 81611 persons of the block.

(ii) Reodar block: Out of 126 villages 37 villages fall in the proposed area covering almost entire area of the block and also covering population 43857 out of 130791 persons of the block

District Udaipur: The Prospective Figure 11.3.17 indicates the proposed area demarcation for groundwater harvesting works.

(i.) Kotra block: Out of 304 villages 223 villages fall in the proposed area covering scattered area of entire block except western part and 86712 out of 130359 persons of the block.

(ii.) Salumber block: Out of 233 villages 222 villages fall in the proposed area covering almost eastern and middle area of the block and also covering population 148373 out of 161720 persons of the block.

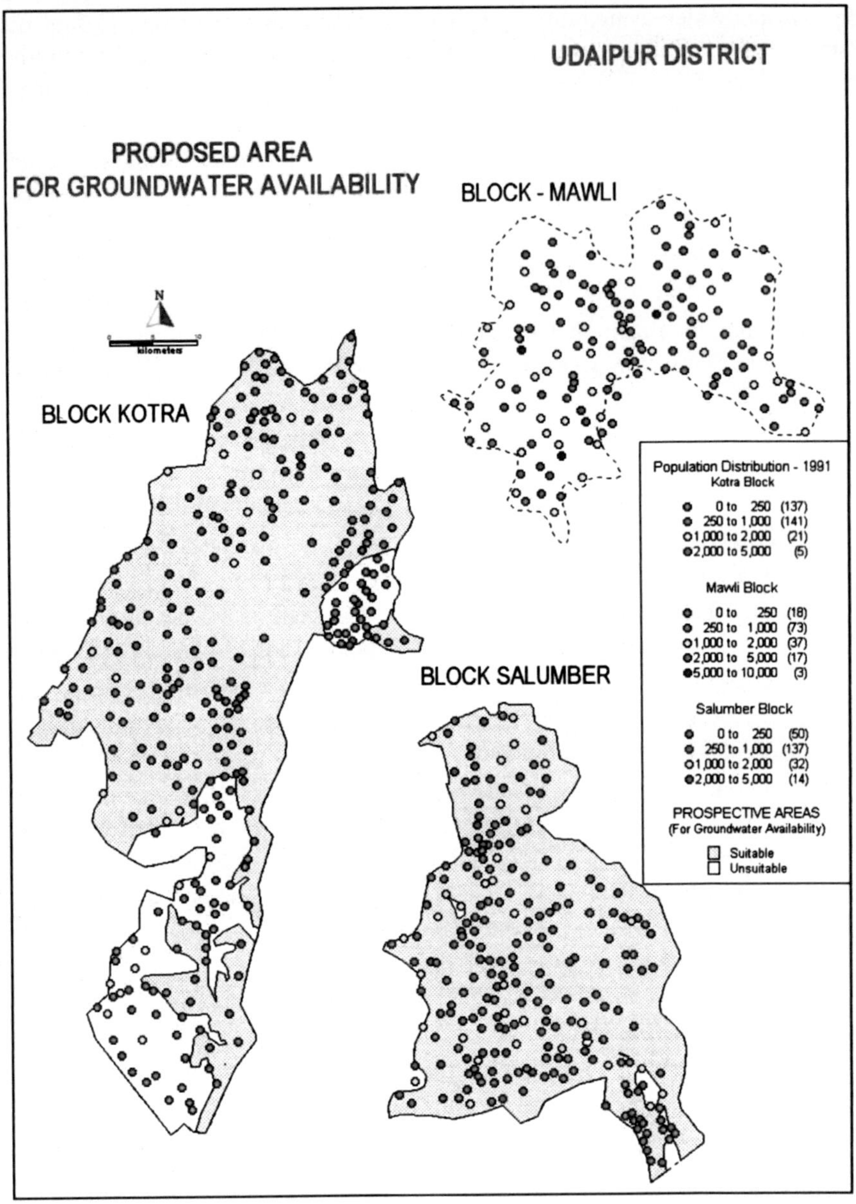

Figure 11.3.17. Proposed area for groundwater availability (Source: GWD and CGWD)

(iii.) Mawli block: Out of 148 villages 164371 population there is no village which fall in the proposed area as the entire block have no potential area as far as selected approach is concerned.

CASE STUDY - 4

11.4 MAPPING OF HIGH RISK BEHAVIOUR POPULATION GROUPS IN RAJASTHAN

The core feature of India's response to HIV/AIDS pandemic has been through targeted interventions among high-risk behaviour and other specific population groups. It is based on the assumption that certain groups of people, by virtue of their behavioral attributes as well as environmental conditions of living, are prone to contact HIV infection more quickly and spread the disease in a very short time. It is believed that direct interventions among these groups will help in containing further spread of HIV, and reducing transmission from population with high prevalence to population with low prevalence. Several population groups at high risk for HIV/AIDS have been identified in various studies carried out by governmental as well as non governmental organisations. These include:

- Commercial sex workers
- Clients of commercial sex workers
- Truckers
- Transport workers and their partners
- Sexually Transmitted Disease (STD) clinic attendees,
- Migrant workers
- Men who have sex with men (MSM)
- Street children
- Intravenous Drug Users (IDUs)

This list of population groups at high risk for HIV/AIDS, however, may not be exhaustive. There may be several other population groups vulnerable to HIV/AIDS. It is believed that factors depicting high-risk behaviour and vulnerability to HIV are present in every society. It is their intensity and combinations, which make population groups particularly vulnerable. Hence, it is desirable to view vulnerability from the perspective of behavioural attributes as well as environmental conditions of living. These include; *demographic, socio-cultural and economic characteristics, infrastructure and other developmental factors*. The challenge is therefore, to identify the presence of these factors in different combinations and intensity in various geographical areas and societies, to determine how far their combinations and intensity are responsible for making population groups vulnerable to HIV/AIDS. Also, for planning intervention programmes for prevention and control of HIV/AIDS, it is imperative to have an adequate understanding of presence of high-risk population groups in different geographical locations. Such an understanding is desirable from two perspectives; for controlling the spread of epidemic with-in the high-risk behaviour group, and for reducing transmission of HIV from high-risk to low risk groups.

The basic information desired for this purpose includes;

- What are high-risk behaviour population groups?
- Where are they located?
- What is their numerical presence in an area?

Mapping is the most widely used method for obtaining this information. It enumerates and creates an inventory of population groups at risk for HIV/AIDS in an area. Mapping exercise provides information on these parameters in the form of a map or database. It has a potential utility for the donors, policy makers, programme managers and social analysts to plan the resource allocation and assess the progress of intervention programmes.

11.4.1 Objectives

The study was designed with following specific objectives as follows:

- To Identify the (intermediate) determinants of high-risk behavior with respect to HIV/AIDS. (These include variables directly or indirectly influencing the high-risk behavior such as truck drivers' population, tourist centers, industrial and migrant workers, inmates in jails, poverty etc.).
- To prepare Geographical Information System (GIS) maps of the state on the basis of intermediate determinants of high-risk behavior and identify the locations (rural/urban) where high-risk behaviour population groups are likely to be found.
- To prepare vulnerability Index (for HIV/AIDS) on the basis of determinants of high risk behaviour and rank the districts of the state on the basis of this Index.

11.4.2 Process

The study was based on the analysis of secondary data drawn from various sources viz., census, national population surveys, health surveys, official statistics etc. These data were collected from different government departments, research and census reports. As the district is the smallest administrative unit for which relevant data are available, it has been taken as the unit of analysis in the study. All the efforts were made to select variables depicting presence of high-risk behaviour and associated characteristics. Appropriate statistical techniques were used to determine the direction and magnitude of their relationship with HIV/AIDS. Since the variables were different in shape, size and nature, different statistical procedures were applied to make them comparable and bring them at a uniform level, in order to develop a composite index.

11.4.2.1 *Selection of Variables*

The first step in the process of developing vulnerability index was the selection of appropriate parameters depicting high-risk behaviours and associated environmental conditions. Many of the key socio-cultural, economic, and demographic and other environmental conditions, under which HIV transmission occurs and spreads, are not sufficiently known or understood. On the basis of the existing researches, the following 25 variables short-listed for the study are given below. The reference year is shown in the bracket.

Demographic Variables

- Percentage of Urban Population (2001)
- Sex-ratio (2001)
- Percentage of scheduled castes to total population (1991)
- Percentage of scheduled tribes to total population (1991)
- Proportion of total in-migrants to urban areas (1991)
- Female work participation (2001)

Economic Development Variables

- Proportion of road length of National Highways passing through the
- District (1998)
- Index of Infrastructure development (1994-95)
- Human poverty index (1997)

Health Variables

- Health Index (1997-98)
- Proportion of AIDS cases registered (2002)
- Prevalence of STI among men (1999-2000)
- Ratio of TB cases reported in the district (2001)

Social Development Variables

- Proportion of urban population living in slums (2001)
- Human development Index (1997)
- Literacy rate of males (2001)
- Literacy rate of females (2001)
- Percentage of boys attending higher (IX-XII) education classes (1997-98)
- Percentage of girls attending higher (IX-XII) education classes (1997-98)

Socio-Cultural and other Variables

- Proportion of rural traditional sex workers (2002)
- Ratio of taxis registered in the district (2001)
- Ratio of trucks registered in the district (2001)

- Ratio of inflow of domestic tourists to the district (2001)
- Ratio of inflow of international tourists to the district (2001)
- Proportion of inmates in jails (2002)

11.4.2.2 *Sources of Data*

The major sources of data were;

- Department of Statistics
- Census of India
- Department of Urban Development
- Department of Tourism
- Department of Social Welfare
- Department of Tribal Welfare
- Department of Education and Higher Education
- Department of Medical and Health
- Department of transport
- Reports of various studies in Rajasthan
- Reports of surveys- e.g.; RCH (RHS)
- Human Development Report
- Other documents providing district level data

11.4.3 Development of Vulnerability Index

Several methods were in used for developing composite indices. The basic tenet of developing such indices was to get a comprehensive picture of the particular characteristics on the basis of relevant variables. The procedure was very simple. The relevant variables or the variability present in variables was pooled to construct the desired index. However, before pooling, the variables were subjected to specific statistical procedures/methods to bring them at an equal level and make them comparable. These processes normally involved standardising and assigning appropriate weights to the variables.

For the purpose of developing vulnerability index for HIV/AIDS, following alternative methods were used:

1. Method of Difference
2. Method of Standard Deviation
3. Factor Analysis
4. Geographical Information System (GIS)
5. Vulnerability Index Specific to Environmental Conditions
6. Vulnerability Index Specific Components of Targeted Interventions

11.4.3.1. *Method of Difference*

This method of standardization, termed as the *"Method of Difference",* is extensively used by UNDP for constructing Human Development Index (HDI), Gender Development Index (GDI), Gender Empowerment Measures (GEM) etc. Due to its simplicity, this method has become an important tool for developing composite indices for different purposes. For developing vulnerability index for HIV/AIDS by method of difference, rotated data (where all the variables are made uni-directional) were used. This process of standardisation took care of variations present in the range of different variables. It was noted that the variables considered for analysis had different magnitude and range. For instance, birth rate of the districts varied between 30 and 40 per 1000 population, whereas death rates ranged from 8 to 14 per 1000 population. When these rates were merged to calculate the composite index, difference in their magnitude was likely to distort the results. The method of difference, overcame this problem, by standardizing all the variables on the basis of difference between their extreme values. Difference of actual and minimum values of the variables across all observations was divided by the difference between its maximum and minimum values to get the standard value. The procedure employed was as follows.

$$\text{Standard value of the } i^{th} \text{ observation of (p) variable} = 100 * \frac{V_{pi} - V_{pmin}}{(V_{pmax} - V_{pmin})}$$

Where:

V_{pi} = Value of $(i)^{th}$ observation of p variable

$V_{p\,min}$ = Minimum value of p variable

$V_{p\,max}$ = Maximum value of p variable

The standard value of series ranged from 0 to 100 in this exercise. The district with lowest score got a value of 0 while the district with highest score got a standard value of 100.

The standardized value for each of the variables was assigned weights on the basis of coefficient of the correlation between the variable and the proxy variable for prevalence of HIV (percent of males reporting symptoms of STIs). The standardised value of every variable was multiplied by the corresponding coefficient to get the weighted standard values.

The weighted standard values for all variables were added to get the composite index value, which was termed as vulnerability index for HIV/AIDS. The district with the lowest index value was expected to be least vulnerable to HIV, while the district with maximum index value, most vulnerable. Further, the districts were ranked on the basis of their index values.

11.4.3.2 *Method of Standard Deviation*

The method of standard deviation was used by the National Institute of Health and Family Welfare, New Delhi for prioritizing the districts on the basis of reproductive and child health needs (Alag et al., 2001). This prioritisation was aimed at assisting the planners for appropriate resource allocation on the basis of the felt needs of the programme. Keeping in view the robustness and replicability of this method for getting a consolidated picture of the existing situation, it was employed for computing the vulnerability index for HIV/AIDS. In this method mean and standard deviation of the variables were used to get the standardized values.

The standard value was computed by taking the difference of mean value of the series from the individual value and dividing it by standard deviation of the series.

$$\text{Standard value of ith observation of p variable} = \frac{(V_{pi} - \overline{X}_p)}{SD_p}$$

Where:

V_{pi} = Value of (i) th observation of p variable

$\overline{X}_p$ = Mean value of p variable

SD_p = Standard deviation of p variable

Every standard value was assigned weights on the basis of its correlation with the proxy of prevalence of HI/AIDS.

The weighted standard values for all the variables were added to get the composite vulnerability index.

11.4.3.3 *Factor Analysis*

This factor analysis could also be used to standardize the variables on the basis of factor scores. These factor scores could be calculated from the factor loadings and other parametric estimates in rotated factor structures. A principal component factor analysis was employed to arrive at the communalities and number of factors to be retained. Varimax rotation method was tried out to arrive at simple factor structure.

11.4.3.4 *Geographical Information System (GIS)*

GIS mapping is an alternative way to depict the level of vulnerability to HIV/AIDS. This technique highlights the regional variations present in the vulnerability index on a geographical map, with varied colours, patterns and graphics. The level of vulnerability can be identified by the thickness of a bar or special colour or specific pattern. For using this technique, the actual values of variables were assigned weights and pooled to arrive at a composite value. The pooled averages were plotted on a geographical map with the help of a special GIS mapping software. The different colours and patterns were used to depict different levels of vulnerability. The district-wise differentials in vulnerability were also depicted through vertical bars.

The GIS mapping technique was also used to present the variation in level of vulnerability calculated by different methods.

11.4.3.5 *Vulnerability Index specific to environmental conditions*

It is assumed that the vulnerability to HIV/AIDS is a multi-dimensional phenomenon and needs to be viewed in the context of behavioral attributes of the population as well as the external environmental conditions to which the population is exposed. The intensity and combinations of these factors make the population particularly vulnerable to HIV/AIDS, in the sense, that some of the combinations may create conditions conducive to promote high-risk behaviour and spread of HIV epidemic. To understand the vulnerability in context of different components of the external environment, relevant variables were grouped into five clusters representing different environmental conditions. The ideology behind developing this index was to assess the vulnerability of the districts under different environmental conditions. For computing the vulnerability index for clusters, the variables of the clusters were standardized using the method of standard deviation. The standard values were weighted with the coefficient of the correlation between the variable and the proxy variable. Composite index for each cluster was computed by summing up all the weighted standard values.

11.4.3.6 *Vulnerability Index Specific to Targeted Interventions*

One of the core features of India's response to HIV/AIDS pandemic is targeted intervention among high-risk behaviour and other specific population groups. Keeping the relevance of these interventions in view, specific vulnerability index were computed on the basis of the size of high-risk behaviour population groups in a district. The rationale behind this method is to assess the specific vulnerability with respect to particular component of high-risk behaviour (such as, presence of traditional sex-workers, length of national highways in the district, reported cases of AIDS, etc). It could also be utilized to prioritize the district for implementing targeted intervention programmes, particularly, in absence of complete information on the size of specific high-risk group population in an area. In this method variable depicting particular risk behaviour was used as weight (wherever necessary with an appropriate controlling/ general variable) for the index value. The scores obtained were considered as specific vulnerability index values with respect to particular risk behaviour variable.

11.4.4 Findings

The vulnerability indices obtained via different methods are discussed independently. Attempts are also made to compare these findings.

11.4.4.1. Method of Difference

In this method the vulnerability index is computed by standardizing the variables using *method of difference*. The standardized values of variables were assigned weights equivalent to their correlation coefficient with proxy variable, and pooled to develop the composite vulnerability index. The ranking of districts on the basis of this method is presented in Table 11.4.1.

Table 11.4.1. Ranking of Districts on the Basis of Vulnerability Index by Method of Difference

District	Rank	District	Rank
Jaipur	1	Kota	17
Sawai Madhopur	2	Rajsamand	18
Bharatpur	3	Ganganagar	19
Udaipur	4	Bhilwara	20
Tonk	5	Barmer	21
Dungarpur	6	Bikaner	22
Karauli	7	Baran	23
Chittorgarh	8	Sirohi	24
Banswara	9	Nagaur	25
Ajmer	10	Hanumangarh	26
Alwar	11	Pali	27
Jhalawar	12	Sikar	28
Jodhpur	13	Jaisalmer	29
Dausa	14	Jalore	30
Bundi	15	Churu	31
Dholpur	16	Jhunjhunu	32

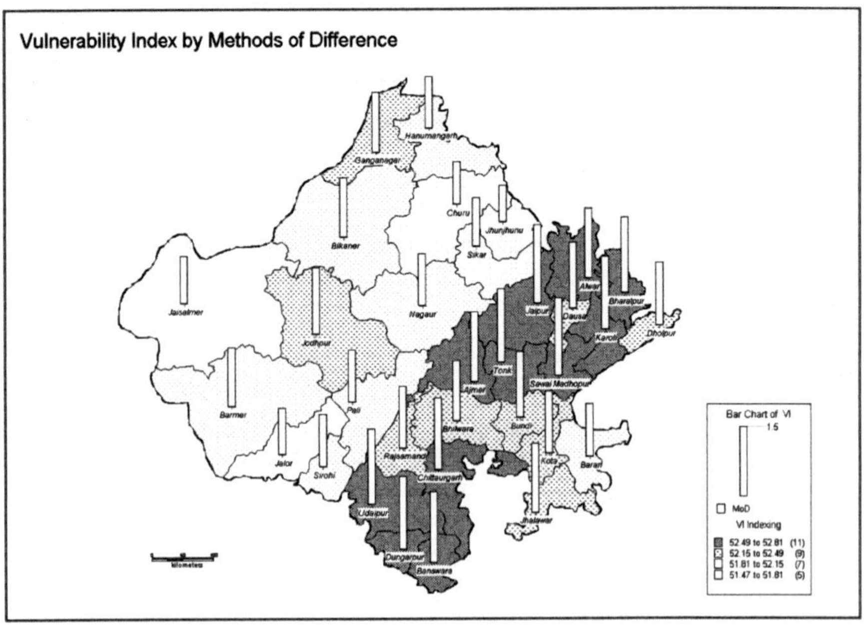

Figure 11.4.1. Vulnerability Index by Methods of Difference (Source: Table)

Figure 11.4.1 indicates that the districts of *Jaipur, Sawai Madhopur, Bharatpur, Udaipur and Tonk,* are more vulnerable to HIV. The districts ranking low in vulnerability are; *Jhunjhunu, Churu, Jalore, Jaisalmer and Sikar.*

11.4.4.2. *Method of Standard Deviation*

In the second method, the vulnerability index is computed by the *method of standard deviation*. The standardized values of the variables were assigned weights equivalent to their correlation coefficient with proxy variable, and pooled to develop composite vulnerability index. The ranking of districts by this index is presented in Table 11.4.2.

Table 11.4.2: Ranking of Districts on the Basis of Vulnerability Index by Method of Standard Deviation

District	Rank	District	Rank
Jaipur	1	Kota	17
S. Madhopur	2	Dholpur	18
Bharatpur	3	Rajsamand	19
Udaipur	4	Bhilwara	20
Tonk	5	Bikaner	21
Karauli	6	Barmer	22
Alwar	7	Hanumangarh	23
Chittorgarh	8	Sirohi	24
Dungarpur	9	Nagaur	25
Ajmer	10	Baran	26
Baswara	11	Pali	27
Jhalawar	12	Sikar	28
Jodhpur	13	Jalore	29
Dausa	14	Jaisalmer	30
Bundi	15	Churu	31
Ganganagar	16	Jhunjhunu	32

Figure 11.4.2 shows that the districts of *Jaipur, Sawai Madhopur, Bharatpur, Udaipur* and *Tonk* are more vulnerable districts. The districts low in rank are, *Jhunjhunu, Churu, Jaisalmer, Jalore* and *Sikar.* A comparison between the vulnerability indices by methods I and II indicates that 90 per cent of more vulnerable districts and 100 per cent of the less vulnerable districts are common, although the ranks of the districts were a little different in the two methods.

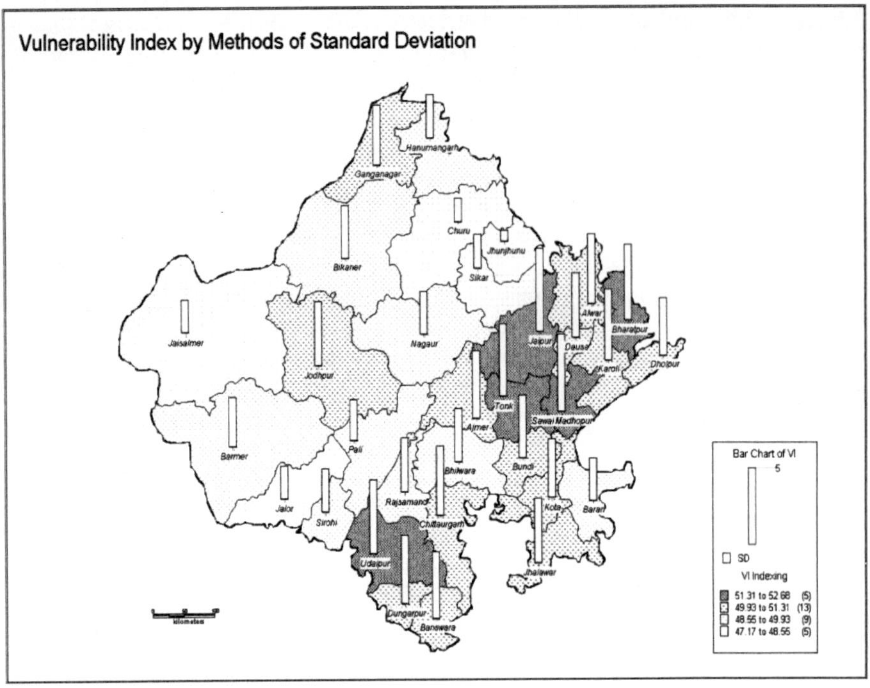

Figure 11.4.2. Vulnerability Index by Methods of Standard Deviation (Source: Table)

11.4.4.3: *Factor Analysis*

A principal component factor analysis with varimax rotations was carried out to extract the main factors. This exercise had thrown up seven factors explaining 81.8 percent variability present in the variables. The last three factors had the significant loadings of 2-3 variables only. Therefore, for the purpose of analysis first four factors were taken into consideration. These factors explained two thirds (67.57 percent) of the variability present in the variables.

The **first factor** had a significant loading on 17 variables. These included; level of urbanisation, literacy and schooling level, infrastructure and other development indicators, poverty, mobility of national and international tourists and length of national highways passing through the districts. Broadly, these variables reflected the "*level of development*". Therefore, we have termed this factor as development factor.

The **second factor** had the significant loadings on the following variables; Human development index, prevalence of STI among men, sex ratio, scheduled tribes' and scheduled castes' population, migration etc.

These variables reflected *"demographic and social characteristics*" of the districts, and were named as such.

The ***third factor*** had a significant loading on only two variables, male and female literacy levels. Similarly, the fourth factor had significant loadings of the health index and size of traditional sex workers population. Due to the loadings of fewer variables in the third and fourth factors, these had become less relevant to the analysis. Further, due to smaller loadings on the first two factors also, the exercise for calculating factor scores did not yield the consistent results. Therefore, this method was not employed to calculate the vulnerability index.

11.4.4.4 *Geographical Information System (GIS)*

To rank the districts on the basis of their vulnerability to HIV/AIDS by GIS mapping procedure, non-standardised values of variables were used. These values were weighted with the coefficient of their correlation with the proxy variable, and pooled to draw the maps. The results are presented in Figures 11.4.3 (A and B). These results closely corroborate with the ranking of districts on the basis of method 11.4.4.1 and 11.4.4.2.

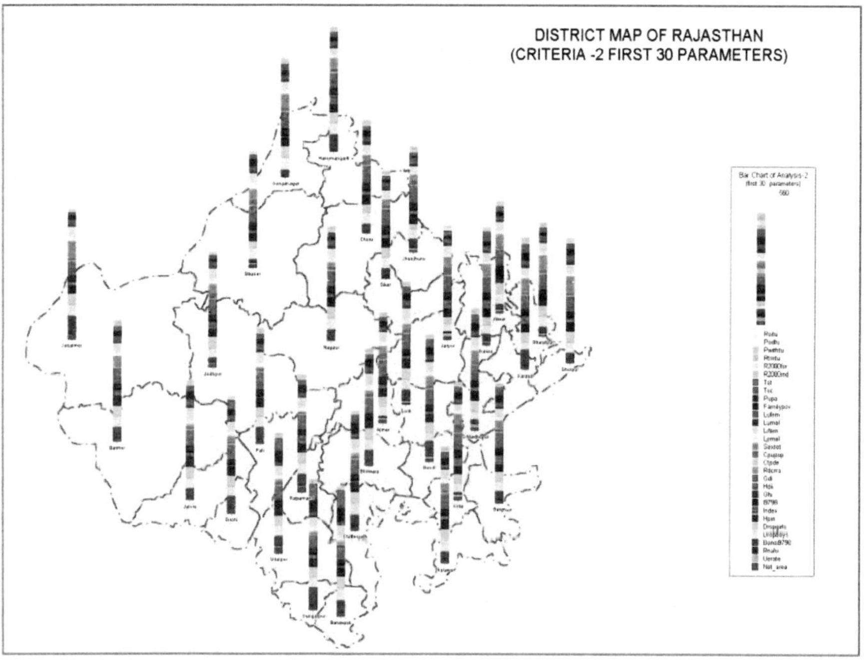

Figure 11.4.3 (A). District map of Rajasthan (Criteria-2 first 30 parameters) (Source: Table)

11.4.4.5 *Vulnerability Index Specific to Environmental Conditions*

To understand the vulnerability to HIV/AIDS under varying environmental conditions, the relevant factors were grouped in five clusters namely;

1. Demographic Variables
2. Economic Development Variables
3. Health Variables

4. Social Development Variables
5. Socio-cultural and other variables

The vulnerability index for each of these groups of variables was calculated by the *method of standard deviation* for standardising the variables and assigning weights according to the correlation coefficient of the variables with the proxy variable. This vulnerability index is termed as the V*ulnerability Index Specific to Environmental Conditions.*

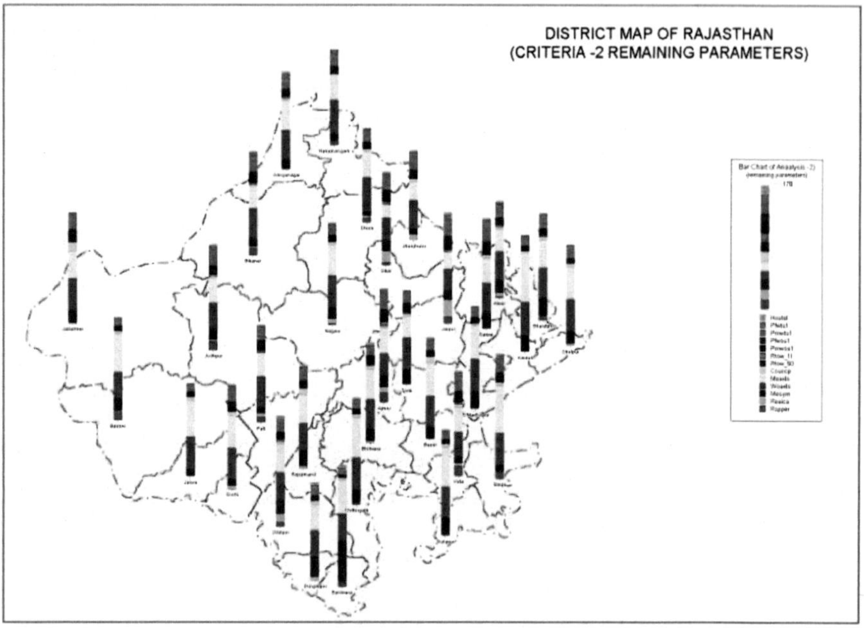

Figure 11.4.3 (B). District Map of Rajasthan (Criteria -2, Remaining Parameters) (Source : Table)

11.4.4.5.1 *Demographic Variables*

Table 11.4.3. presents the ranking of the districts on the basis of vulnerability index specific to demographic variables.

Table 11.4.3: Ranking of Districts on The Basis of Vulnerability Index Specific to Demographic Variables.

District	Rank	District	Rank
Banswara	1	Hanumangarh	17
Ganganagar	2	Bhilwara	18
Kota	3	Jalore	19
Sawai Madhopur	4	Udaipur	20
Karauli	5	Jhalawar	21

Dungarpur	6	Bharatpur	22
Bundi	7	Bikaner	23
Chittorgarh	8	Dholpur	24
Dausa	9	Pali	25
Baran	10	Jodhpur	26
Tonk	11	Barmer	27
Churu	12	Nagaur	28
Sirohi	13	Jhunjhunu	29
Alwar	14	Sikar	30
Jaipur	15	Jaisalmer	31
Ajmer	16	Rajsamand	32

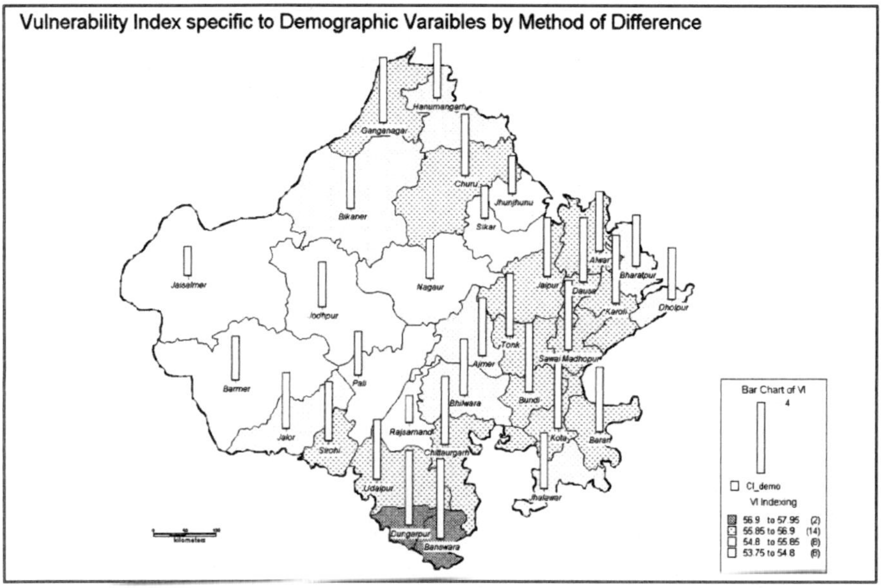

Figure 11.4.4 (A). Vulnerability Index specific to Demographic Arabibles by Method of Different (Source Table)

Figure 11.4.4 (A and B) shows that *Banswara, Ganganagar, Kota, S. Madhopur and Karauli* are more vulnerable to HIV/AIDS with respect to demographic parameters. On the other hand, the districts of *Rajsamand, Jaisalmer, Sikar, Jhunjhunu and Nagaur* occupies lower rank on this vulnerability index. This ranking could be explained in the context of influence of the individual demographic characteristics. Interestingly, both the categories of districts, more as well as less developed (demographically), have scored higher ranks. For example, Banswara, S. Madhopur and Karauli are more vulnerable due to the higher proportion of SC and ST population (symbolizing low level of development). Whereas, Kota and Ganganagar are more vulnerable due to higher level of urbanization and in-migration respectively. It means that the presence of different demographic characteristics adds to the vulnerability of population to HIV/AIDS in different ways.

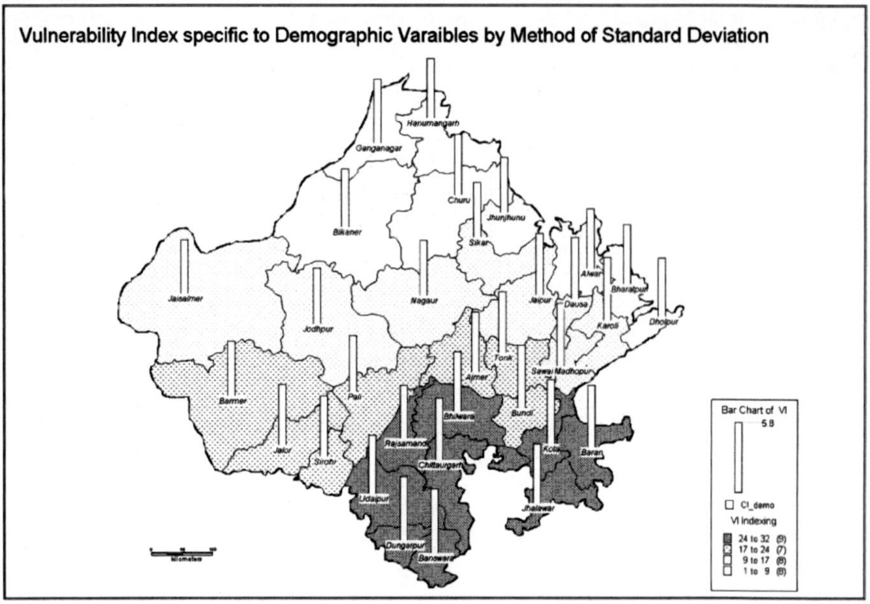

Figure 11.4.4 (B) Vulnerability Index specific to Demographic Arabibles by Method of Standard Deviation (Source Table)

Table 11.4.4: Ranking of Districts on the Basis of Vulnerability Index Specific to Economic Development Variables.

District	Rank	District	Rank
Barmer	1	Bikaner	17
Jalore	2	Bundi	18
Rajsamand	3	Banswara	19
Sirohi	4	Dungarpur	20
Jaisalmer	5	Chittorgarh	21
Karauli	6	Bharatpur	22
Dausa	7	Baran	23
Churu	8	Bhilwara	24
Tonk	9	Alwar	25
Dholpur	10	Jhunjhunu	26
Jhalawar	11	Ganganagar	27
Nagaur	12	Jaipur	28
S. Madhopur	13	Hanumangarh	29
Sikar	14	Jodhpur	30
Udaipur	15	Ajmer	31
Pali	16	Kota	32

11.4.4.5.2 *Economic Development Variables*

The variables considered to construct the vulnerability index specific to economic development, included; human poverty index, infrastructure development index and the proportion of length of national highway passing through the district. The ranking of districts on the basis of vulnerability index specific to economic development variables is shown in Table 11.4.4.

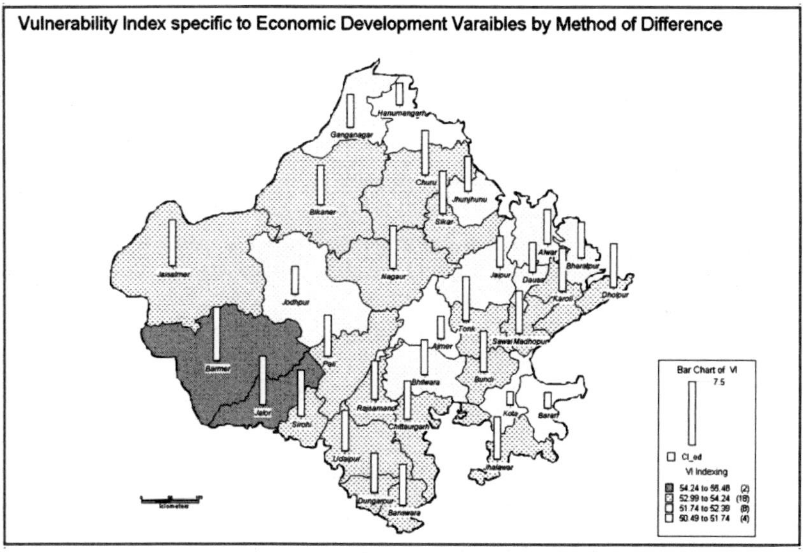

Figure 11.4.5 (A). Vulnerability Index specific to Economic Development Variables by Method of Different (Source Table)

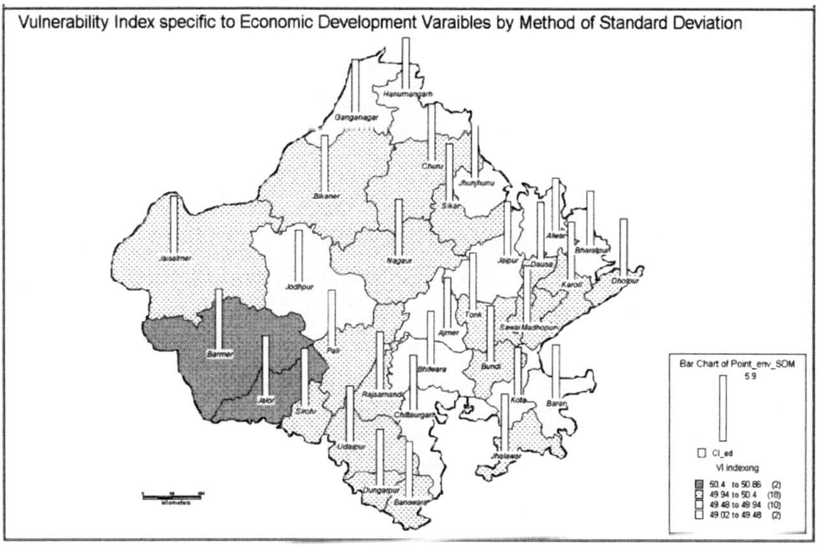

Figure 11.4.5 (B). Vulnerability Index specific to Economic Development Variables by Method of Standard Deviation (Source Table)

Figure 11.4.5 (A and B) shows that the districts of *Barmer, Jalore, Rajsamand, Sirohi and Jaisalmer* are more vulnerable to HIV/AIDS. On the other hand, the districts of *Kota, Ajmer, Jodhpur, Hanumangarh and Jaipur* are less vulnerable. The districts in former category are economically less developed, whereas in later category, relatively more developed. The later districts had shown a higher vulnerability when viewed from a composite index (Table 11.4.1 and 2). It would mean that the economic development might precipitate or impede vulnerability to HIV directly under varying conditions due to the presence of different combinations of variables.

11.4.4.5.3. *Health Variables*

The variables considered for developing this index were; health index, number of reported cases of AIDS, prevalence of STI among men and number of reported cases of TB. The ranking of the districts on the basis of vulnerability index specific to health variables is shown in Table 11.4.5.

Table 11.4.5: Ranking of Districts on the Basis of Vulnerability Index Specific to Health Variables.

District	Rank	District	Rank
Bharatpur	1	Jaipur	17
Dungarpur	2	Barmer	18
Karauli	3	Nagaur	19
Sawai Madhopur	4	Sikar	20
Jhalawar	5	Jaisalmer	21
Tonk	6	Jodhpur	22
Udaipur	7	Sirohi	23
Rajsamand	8	Pali	24
Chittorgarh	9	Kota	25
Dausa	10	Baran	26
Dholpur	11	Ganganagar	27
Banswara	12	Jalore	28
Bhilwara	13	Jhunjhunu	29
Ajmer	14	Bikaner	30
Alwar	15	Hanumangarh	31
Bundi	16	Churu	32

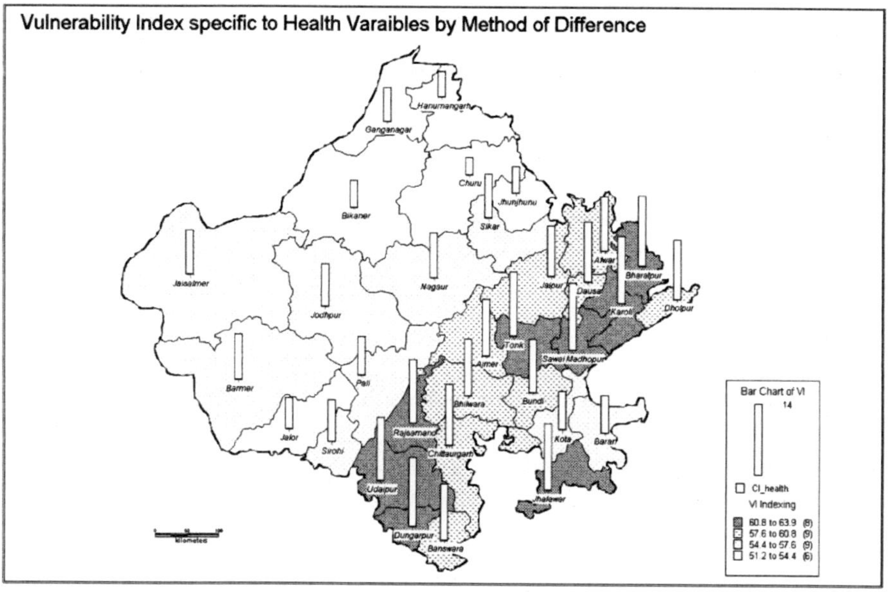

Figure 11.4.6 (A). Vulnerability Index specific to Health Variables by Method of Different (Source Table)

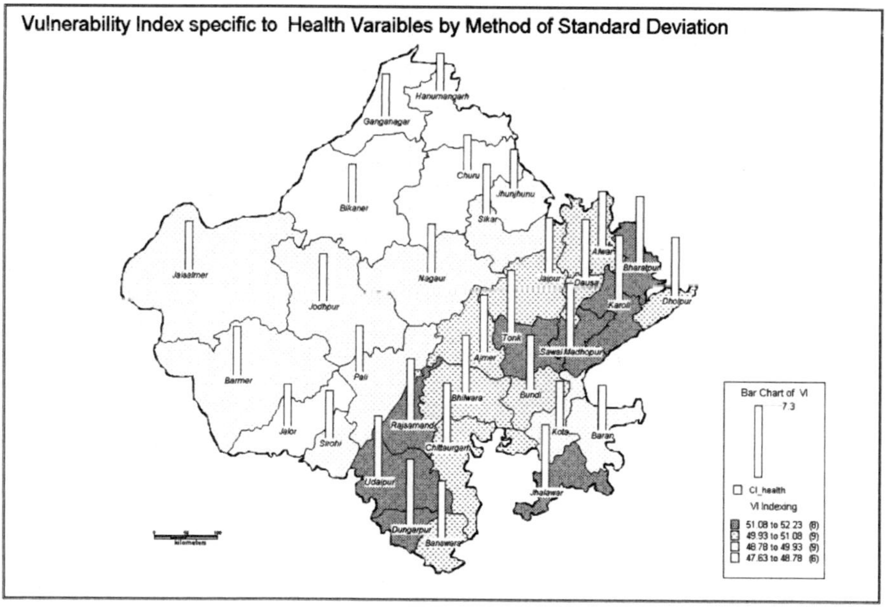

Figure 11.4.6 (B). Vulnerability Index specific to Health Variables by Method of Standard Deviation (Source Table)

Figure 11.4.6 (A and B) indicates *that Bharatpur, Dungarpur, Karauli, S. Madhopur and Jhalawar* are more vulnerable to HIV/AIDS in the context of variables specific to health. *Churu, Hanumangarh, Bikaner, Jhunjhunu and Jalore* had emerged as less vulnerable districts on this index. The former category of districts has a relatively lower rank on health index and a larger proportion of STI positive people. The districts in the later category occupied an intermediate position with respect to these variables, but at the same time, had a lower number of reported cases of AIDS and TB. Jaipur Sikar and Udaipur, though had a higher number of reported cases of AIDS and TB, but had a higher rank on health index. That is why, on the vulnerability index specific to health variables, these districts had occupied intermediate positions.

11.4.4.5.4. *Social Development Variables*

The variables considered for constructing this index included; literacy level of males and females, proportion of boys and girls in higher education classes, proportion of people living in slums and human development index. Table 11.4.6 shows the ranking of districts on the basis of vulnerability index specific to social development variables.

Table 11.4.6: Ranking of Districts on the Basis of Vulnerability Index specific to Social Development Variables.

District	Rank	District	Rank
Kota	1	Churu	17
Bikaner	2	S. Madhopur	18
Jodhpur	3	Bharatpur	19
Hanumangarh	4	Udaipur	20
Ganganagar	5	Sikar	21
Baran	6	Jaisalmer	22
Ajmer	7	Tonk	23
Jaipur	8	Bhilwara	24
Alwar	9	Barmer	25
Jhunjhunu	10	Sirohi	26
Dausa	11	Rajsamand	27
Pali	12	Karauli	28
Bundi	13	Jalore	29
Chittorgarh	14	Jhalawar	30
Nagaur	15	Dungarpur	31
Banswara	16	Dholpur	32

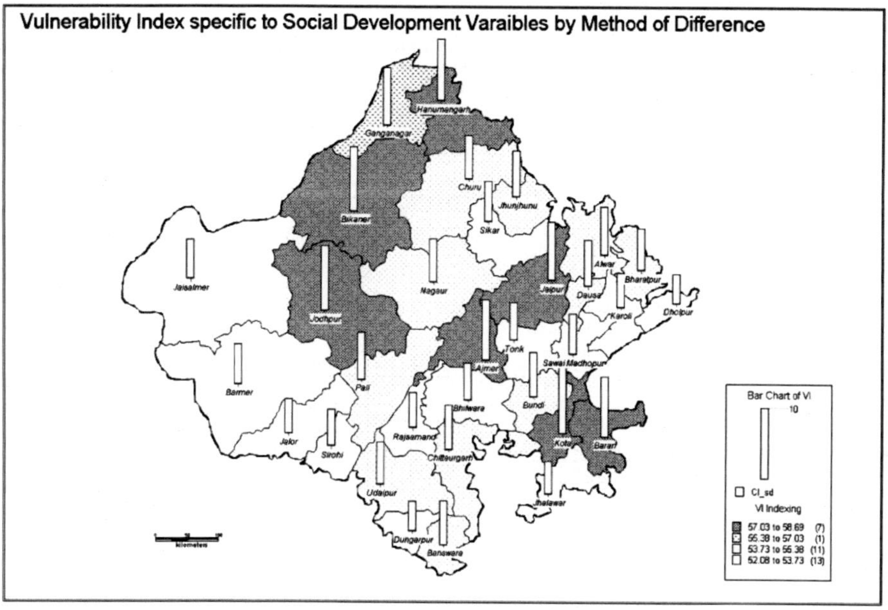

Figure 11.4.7 (A). Vulnerability Index specific to Social Development Variables by Method of Difference (Source Table)

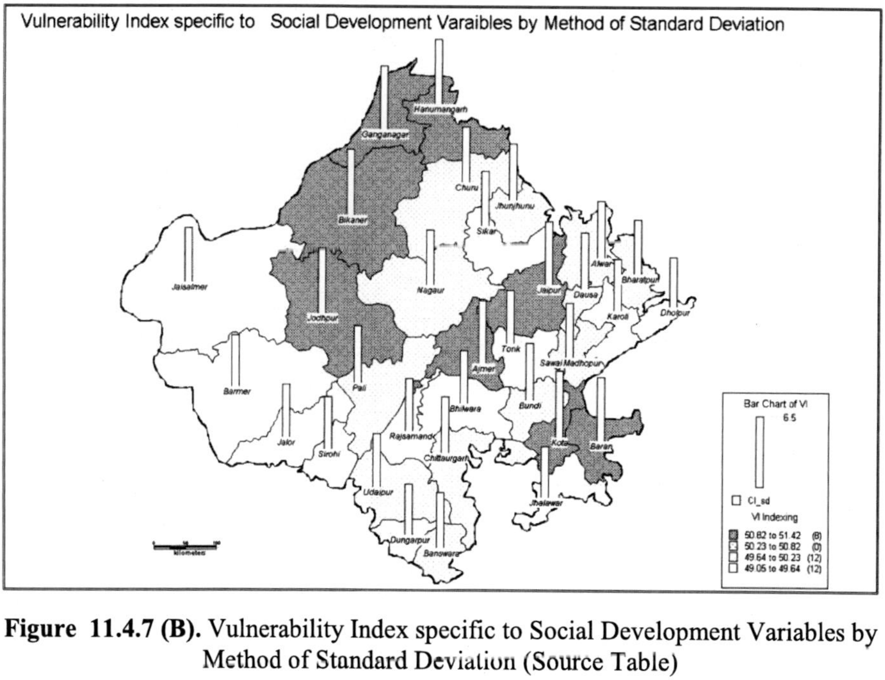

Figure 11.4.7 (B). Vulnerability Index specific to Social Development Variables by Method of Standard Deviation (Source Table)

Figure 11.4.7 (A and B) shows that the socially more developed districts like, *Kota, Bikaner, Jodhpur, Hanumangarh and Ganganagar* are more vulnerable to

HIV/AIDS as compared to the other districts in the state. The districts scoring low on this index are; *Dholpur, Dungarpur, Jhalawar, Jalore and Karauli*. Analysis of the variables used to construct this index suggests that both negative and positive effects of variables had made a district vulnerable to HIV/AIDS. For example, Kota, Bikaner and Jodhpur are more vulnerable due to higher proportion of slum population. Ganganagar and Hanumangarh had higher vulnerability since they scored low on human development index. Districts low on vulnerability index have had a smaller proportion of slum population and had an intermediate position on human development index. The literacy levels of men and women and schooling of adolescent boys and girls have had marginal influence on the ranking.

11.4.4.5.5. *Socio-cultural and Other Variables*

In this category, variables other than demographic, economic, health and social development are included. Broadly, this body of variable reflects on culture and traditions, law and order situation, tourism, mobility etc. The ranking of the districts on the basis of vulnerability index specific to socio-cultural and other variables is shown in Table 11.4.7

Table 11.4.7: Ranking of the Districts on the Basis of Vulnerability Index Specific to Socio-Cultural and Other Variables.

District	Rank	District	Rank
Jaipur	1	Jhunjhunu	17
Jodhpur	2	Bhilwara	18
Alwar	3	Rajsamand	19
Udaipur	4	Barmer	20
Ganganagar	5	Nagaur	21
Hanumangarh	6	Sikar	22
Bikaner	7	Sirohi	23
Kota	8	Churu	24
Bharatpur	9	Bundi	25
Ajmer	10	Dausa	26
Tonk	11	Jhalawar	27
Dholpur	12	Banswara	28
Jalore	13	Baran	29
Chittorgarh	14	Dungarpur	30
S. Madhopur	15	Jaisalmer	31
Pali	16	Karauli	32

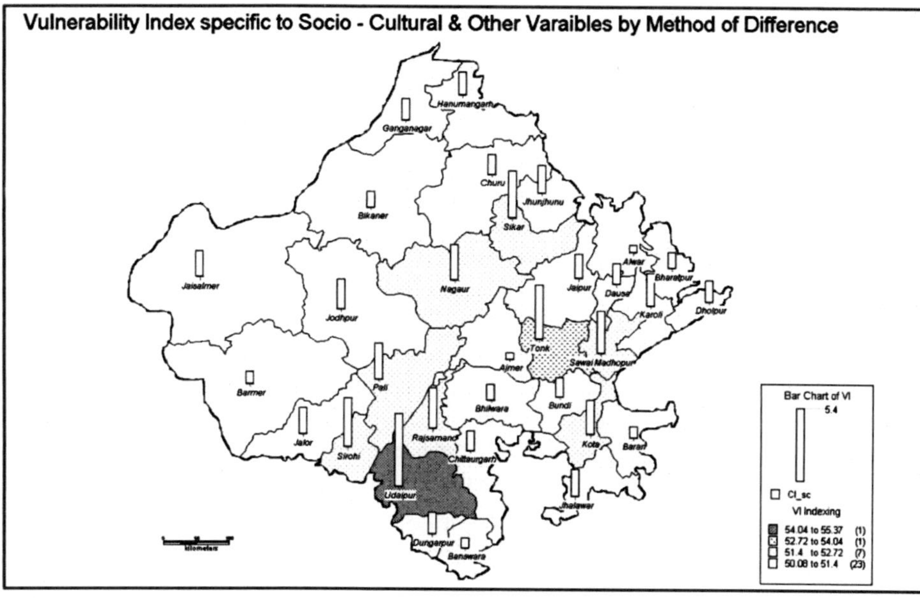

Figure 11.4.8 (A). Vulnerability Index specific to Socio-Cultural & Other Variables by Method of Difference (Source Table)

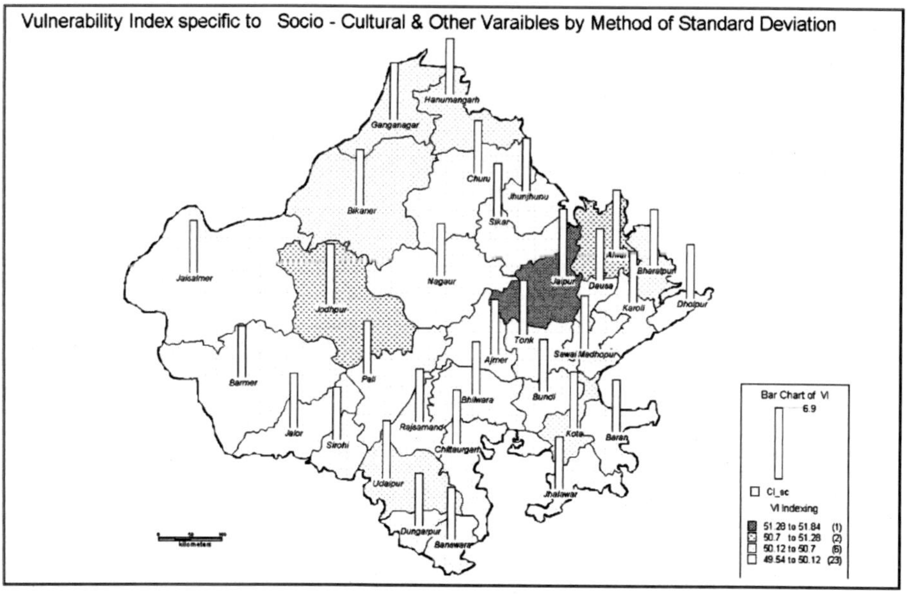

Figure 11.4.8 (B). Vulnerability Index specific to Socio-Cultural & Other Variables by Method of Standard Deviation (Source Table)

As expected, (Figure 11.4.6 A and B) *Jaipur, Jodhpur, Udaipur, Alwar and Ganganagar* have emerged as more vulnerable districts on this index. These districts not only have a higher presence of traditional sex-workers, but a higher proportion of inmates in jails, inflow of domestic and international tourists and movement of trucks and taxis on their highways. *Karauli, Jaisalmer, Dungarpur, Baran and Banswara* have scored low due to probably lower values of these variables.

11.4.4.6. *Vulnerability Index Specific to Targeted Interventions*

As stated earlier, the purpose behind developing specific vulnerability index was to assess the vulnerability of populations with respect to particular attributes of high risk behaviour (such as, presence of traditional sex-workers, jail inmates, reported cases of TB and AIDS) etc. It could be useful for prioritising districts for implementing targeted interventions. The method of standard deviation was used for standardizing the variables. Standardised values were given weights according to correlation coefficient with proxy variable. The pooled values were further weighted by multiplying with the value of variable representing the high-risk behaviour group to get the vulnerability index specific to that particular group.

Specific index was calculated for the following variables;

- Rural traditional sex workers
- Intensity national highway
- Reported cases of AIDS.
- Number of inmates in Jail
- Reported cases of Tuberculosis (TB)

11.4.4.6.1. *Rural Traditional Sex Workers*

It has been noted that in several tribes in Rajasthan commercial sex activities by the daughters is socially accepted. These practices have a historical past. During the days of princely states, women of these castes were traditional entertainers in the courts. After princely states were dissolved these women lost the source of livelihood for themselves and their families. Probably under distress they have adopted the oldest profession in the world. Further some of these castes were nomadic and some were criminal castes also. With the passage of time they have lost their traditional source of livelihood and allowed their women particularly daughters, to support the families through commercial sex activities. For the purpose of this analysis data on traditional sex workers were taken from a study commissioned by I-CHAP. In this study the rural populations traditionally engaged in commercial sex activities were mapped. Table 11.4.9 shows the ranking of the districts on the basis of the vulnerability index specific to the presence of the rural traditional sex-workers.

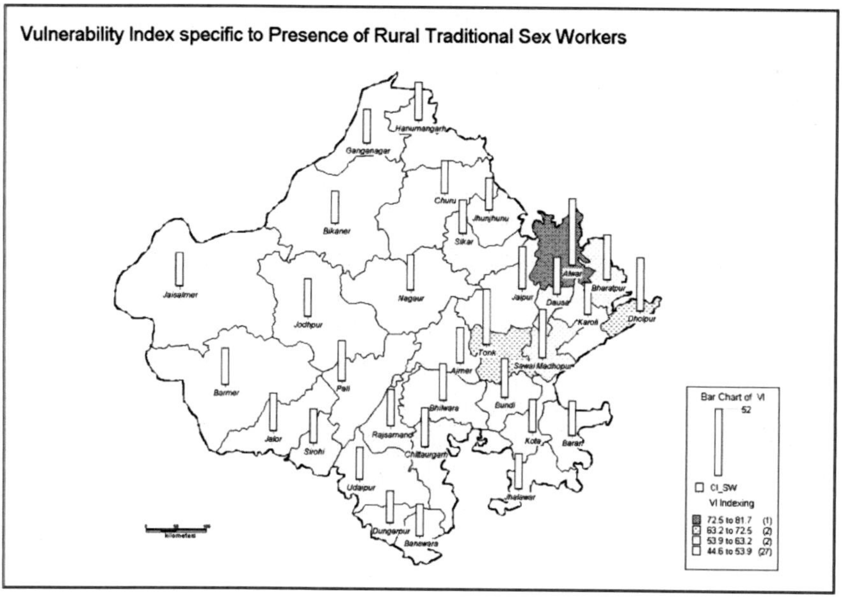

Figure 11.4.9. Vulnerability Index specific to Presence of Rural Traditional Sex Workers (Source Table)

Table 11.4.9: Ranking of Districts on the Basis of Vulnerability Index Specific to Presence of Rural Traditional Sex-Workers.

District	Rank	District	Rank
Alwar	1	Ajmer	17
Tonk	2	Jhalawar	18
Dholpur	3	Nagaur	19
S. Madhopur	4	Baran	20
Bharatpur	5	Ganganagar	21
Jaipur	6	Sirohi	22
Pali	7	Jaisalmer	23
Bundi	8	Sikar	24
Chittorgarh	9	Jhunjhunu	25
Jalore	10	Churu	26
Jodhpur	11	Bikaner	27
Hanumangarh	12	Kota	28
Dausa	13	Banswara	29
Barmer	14	Dungarpur	30
Bhilwara	15	Karauli	31
Rajsamand	16	Udaipur	32

Figure 11.4.9 shows that the districts of *Alwar, Tonk, Dholpur, S. Madhopur, Bharatpur and Jaipur* are more vulnerable to HIV/AIDS due to the presence of traditional sex-workers. The districts ranking low in this list are; *Udaipur, Karauli, Dungarpur, Banswara and Kota.*

11.4.4.6.2. *Intensity of National Highway*

Length of national highways passing through a district is an important indicator to understand the intensity of vehicular movement particularly of trucks, on its roads. As truck drivers are known to practice high risk behaviour, this variable may indirectly reflect the vulnerability of the district due to mobility of truckers in its area. The Table 11.4.10 shows the ranking of the districts on the basis of the vulnerability index specific to the length of national highways in the district.

Table 11.4.10: Ranking of Districts on the Basis of Vulnerability Index Specific to the Length of National Highway in the District.

District	*Rank*	District	Rank
Jaipur	1	Dausa	17
Bikaner	2	Bharatpur	18
Churu	3	Bundi	19
Jaisalmer	4	Jalore	20
Barmer	5	Alwar	21
Pali	6	Dungarpur	22
Rajsamand	7	Dholpur	23
Sikar	8	Jhunjhunu	24
Ajmer	9	Baran	25
Ganganagar	10	Nagaur	26
Sirohi	11	Hanumangarh	27
Tonk	12	Karauli	28
Jodhpur	13	Bhilwara	29
Udaipur	14	Bansawara	30
Jhalawar	15	Chittorgarh	31
Kota	16	S. Madhopur	32

As evident from Figure 11.4.10 Jaipur, *Bikaner, Churu, Jaisalmer and Barmer* are more vulnerable (to HIV/AIDS) due to the longer stretch of national highways passing through their area.

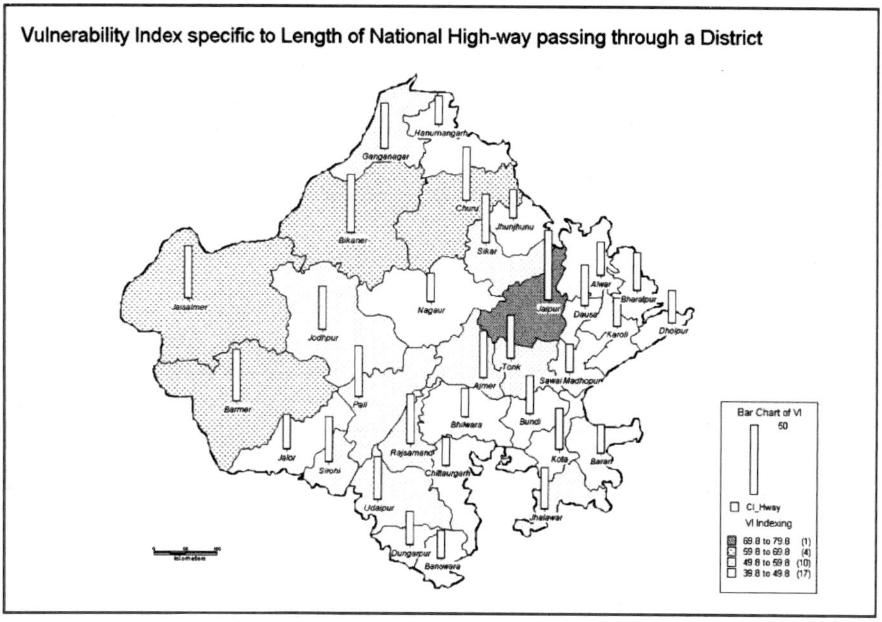

Figure 11.4.10. Vulnerability Index specific to Length of National High-way passing through a District (Source Table)

11.4.4.6.3 *Reported Cases of AIDS*

In spite of the fact that reporting of AIDS cases is highly incomplete, whatever data are available in this respect, reflect the trends over time. The trend could be in terms of increase in the number of AIDS cases or better reporting of the AIDS cases. The vulnerability index specific to reported cases of AIDS should be viewed in these contexts. Table 11.4.11 shows the ranking of the districts on the basis of the vulnerability index specific to reported cases of AIDS.

Table 11.4.11: Ranking of the Districts on the Basis of Vulnerable Index Specific to Reported Cases of Aids.

District	Rank	District	Rank
Jaipur	1	Jodhpur	17
Sikar	2	S. Madhopur	18
Udaipur	3	Bundi	19
Ajmer	4	Banswara	20
Jhunjhunu	5	Chittorgarh	21
Nagaur	6	Karauli	22
Sirohi	7	Jalore	23
Pali	8	Hanumangarh	24

Alwar	9	Bharatpur	25
Dungarpur	10	Baran	26
Kota	11	Tonk	27
Rajsamand	12	Jaisalmer	28
Churu	13	Barmer	29
Bikaner	14	Ganganagar	30
Bhilwara	15	Dausa	31
Dholpur	16	Jhalawar	32

In the Figure 11.4.11 the districts of Jaipur, Sikar, Udaipur, Ajmer and Jhunjhunu have emerged as more vulnerable to HIV/AIDS due to the high number of reported cases of AIDS.

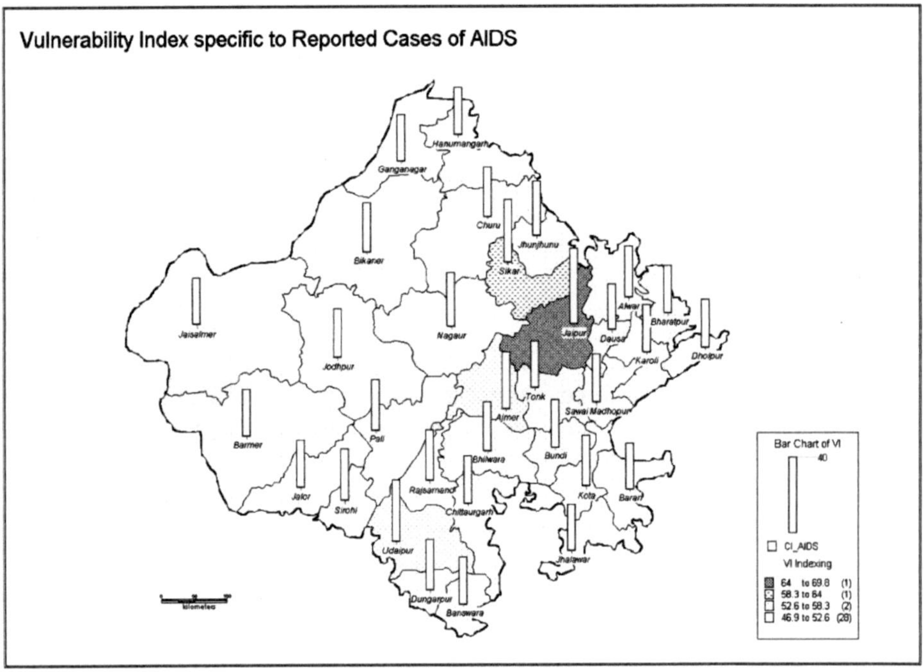

Figure 11.4.11. Vulnerability Index specific to Reported Cases of AIDS

11.4.6.6.4 *Number of Inmates in Jails*

It is believed that the practice of men having sex with men (MSM) is generally higher among the jail inmates due to obvious reasons. In this respect the districts having crowded or larger number of inmates in jails are expected to be more vulnerable to HIV/AIDS. The ranking of the districts on the basis of vulnerability index specific to number of inmates in Jail is shown in Table 11.4.12.

Table 11.4.12: Ranking of the Districts on the basis of Vulnerable Index Specific to Number of Inmates in Jails.

District	Rank	District	Rank
Jaipur	1	Nagaur	17
Kota	2	Banswara	18
Jodhpur	3	Dholpur	19
Udaipur	4	Jhunjhunu	20
Bharatpur	5	Churu	21
Bikaner	6	Karauli	22
Ajmer	7	Pali	23
Alwar	8	Sikar	24
Ganganagar	9	Sirohi	25
Hanumangarh	10	Jalore	26
Chittorgarh	11	Dausa	27
Baran	12	Barmer	28
Jhalawar	13	Dungarpur	29
Bhilwara	14	Rajsamand	30
Bundi	15	S. Madhopur	31
Tonk	16	Jaisalmer	32

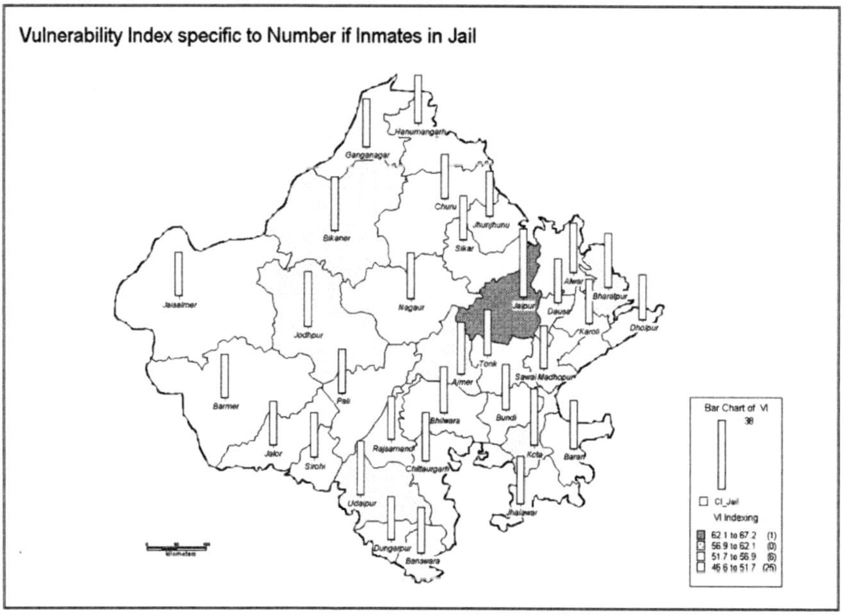

Figure 11.4.12. Vulnerability Index specific to Number if Inmates in Jail

The Figure 11.4.12 shows that districts of *Jaipur, Kota, Jodhpur, Udaipur and Bharatpur* are more vulnerable to HIV/AIDS due to larger number of inmates in jails.

11.4.4.6.5 *Number of Reported Cases of TB*

T.B is one of the largest reported opportunistic infection of HIV/AIDS in India. It is assumed that districts having higher number of cases of TB are likely to have larger number of people infected with HIV/AIDS.

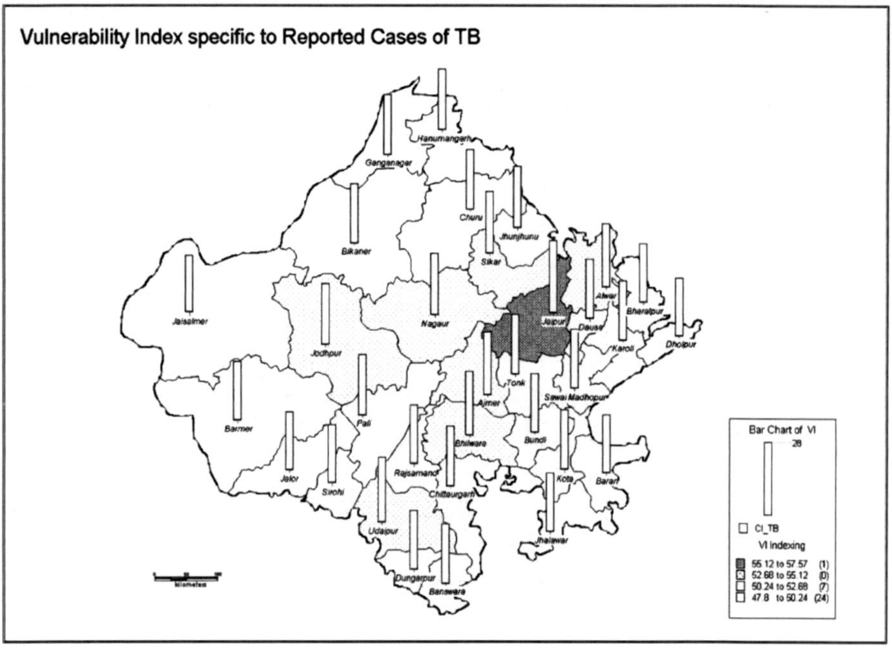

Figure 11.4.13. Vulnerability Index specific to Reported Cases of TB

Table 11.4.13 shows the ranking of the districts on the basis of vulnerability index specific to reported cases of TB.

Table 11.4.13: Ranking of the Districts on the Basis of Vulnerable Index Specific to Reported Cases of TB

District	Rank	District	Rank
Jaipur	1	Tonk	17
Udaipur	2	Kota	18
Bhilwara	3	Dausa	19
Alwar	4	Bharatpur	20
Ajmer	5	Bikaner	21

Nagaur	6	Dungarpur	22
Jodhpur	7	Rajsamand	23
Sikar	8	Sawai Madhopur	24
Jhunjhunu	9	Barmer	25
Ganganagar	10	Jalore	26
Chittorgarh	11	Bundi	27
Pali	12	Jhalawar	28
Hanumangarh	13	Baran	29
Churu	14	Dholpur	30
Banswara	15	Sirohi	31
Karauli	16	Jaisalmer	32

In Figure 11.4.13 districts of *Jaipur, Udaipur, Bhilwara, Ajmer and Alwar have emerged as* more vulnerable to HIV/AIDS due to the presence of higher number of reported cases of TB.

CASE STUDY - 5

11.5 GIS BASED MAPPING OF MINES, MINEWORKERS AND HIGH RISK BEHAVIOUR GROUPS IN RAJASTHAN

HIV AIDS is the most severe and rapidly spreading disease in the world. It has no physical boundaries and has covered almost entire world particularly the developing countries. The number of infected cases and those having AIDS has increased manifold in the last one decade. Scientists and researchers are doing their best to find out its cure but so far prevention is the only protection against this disease. One category that has been identified to be vulnerable recently and is not being addressed so far in Rajasthan is **mineworkers**. Due to their poor and usually unsatisfactory living conditions, they become prone to many kinds of diseases. Most of the mineworkers live without their families near the work places and hence their sexual needs also remain unmet. To meet their sexual needs, they often get involved in sexual activities with CSW or in multiple partner sex. This makes them vulnerable to the risk of acquiring HIV. Therefore, in order to address the sexual health needs of mineworkers it becomes necessary to identify the mines where substantial numbers of mine workers are present. As per the available information there exits more than 9000 registered mines of various kinds employing to more than 5.00 Lacs peoples as mine workers in the state of Rajasthan. To identify the types, locations and number of mineworkers engaged in these mines and gather relevant information as evidence to plan the program, a GIS based mapping and assessment study from primary as well as secondary resources was carried out.

11.5.1 Objective

The main objectives of the study were:

- Generation of comprehensive and districts wise database of mining locations for the state of Rajasthan (including types, size, locations and estimated numbers of mine workers).
- GIS based district-wise mapping of mining areas and it's locations in the state of Rajasthan. Also, to map the High Risk Behaviour Groups (HRBGs) co-located close to /around the mining areas.

11.5.2 Methodology

Since the assignment required deeper understanding of mining which includes identification of geographical areas, types and persons engaged in mining industry, the study used both secondary as well as primary sources to generate information for further analysis. The details of the data collected from different sources based on the checklist developed are discussed as following:

11.5.2.1 *Study Area*

The entire Rajasthan state having 32 districts was considered as study area for the mining study.

11.5.2.2 *Data Collection*

Since mining is the major revenue generating industry in the state of Rajasthan, state government keeps control over the mining activities through various government departments e.g., Rajasthan State Mines and Minerals Development Corporation (RSMMDC), Rajasthan state Mines and Geology Department etc. To develop a quick and comprehensive database on mining, information from secondary as well as primary sources have been collected. The information and its source are as given below:

11.5.2.2.1 *Secondary Sources*

As state government departments control mining, hence, for the study purpose existing literature already available with these departments were collected for review and further analysis. The type of information retrieved from different sources based on the checklist is reflected in the following table:

Source	Information/ Data
Survey of India, Jaipur	• Base map of Rajasthan
Geological Survey of India, Jaipur	• Locations of Mining areas • Type of Mining areas • Mining Potentials • Geological Formations
Rajasthan State Mines and Mineral Corporations, Jaipur & Mines and Geology Department, Udaipur. Registrar, Mining, Ajmer	• District-wise listing of mining areas • District wise listing of type of mines • District-wise and Mineral-wise listing of employment in mines • District-wise and Mineral-wise total leased area for mining • District-wise sale value of minerals • District-wise and Mineral-wise revenue generation.
Statistical Department, Jaipur	• District-wise and Mineral-wise revenue generation.
Mining associations, Jaipur, Bikaner, Udaipur, Jodhpur, Kota.	• District-wise listing of mining areas • District-wise listing of type of mines • District-wise and Mineral-wise listing of employment in mines
Internet and Websites	• Mining areas and type of mines • Case studies

11.5.2.2.2 *Primary Sources*

As discussed earlier, collection of primary data was considered necessary for the study to understand the existing status of mining activities in Rajasthan. Focused group discussions with head of the departments and associations were conducted to fulfill the purpose. Field visits were primarily conducted to triangulate the data collected from secondary sources. The data collected from primary sources was based on the checklist as outlined below:

Method:	**Information/ Data**
A. Focus Group Discussion (FGD)	
• Assistant Manager P and A, Rajasthan State Mines and Mineral Corporations, Jaipur • Director, Mines and Safety departments, Government of India under Labor Ministry, Ajmer. • Registrar, Mines and Engineering Department, Ajmer • In charge, Statistical Division, Mines and Engineering Department, Ajmer • Assistant Director, Rajasthan Mines and Geology Department, Udaipur • Assistant Director, Data Wing, Geological Survey of India, Jaipur	• Status of mining in Rajasthan • Geographical locations of mines • Type of mines and its working status. • Number of mineworkers and their Status

11.5.2.3 *Process of Data Collection*

The preliminary information about the RTFSW and related components of the study were collected from the concerned government and non-government organisations at Jaipur. Based on the information, checklist for data collection was designed for data collection. Since State government departments are the controlling and monitoring agency for the mining activities, a formal meeting with concern officers in those departments were organised at their respective offices to discuss the study design and solicit their support in data collection. Information/ data available with them were initially collected and reviewed. As required, focused group discussions were conducted with the head of the institution/ departments to collect primary data. Literature available with these departments on the concerned issue was also reviewed and relevant data retrieved from them.

11.5.2.4 *Data Processing and Analysis*

The collected information vis-à-vis mining locations, types and mineworkers were fed in to the computer. Base map was developed demarcating state, district boundaries and transport network using GIS software. Same software was used to check the adequacy of the data collected.

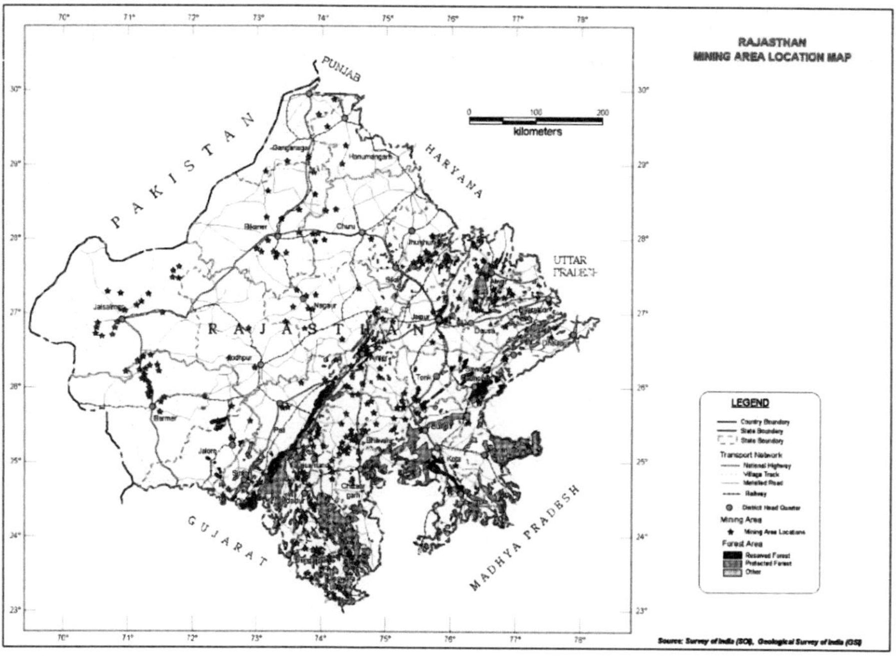

Figure 11.5.1

The database thus generated was sorted district wise and mineral wise to generate district level information. The results generated were superimposed on the base map using GIS software to visualize the current status of the mining activities in Rajasthan. The details of the parameters assessed, which were later superimposed on base figure to demarcate the status of the mining activities, are as outlined below:

- Locations of existing mining area.
- Number of mines
- Type of mines.
- Number of Persons Employed in mining activities.
- Mineral wise Revenue generation from mines
- District wise RTFSW

11.5.3 Findings

11.5.3.1 *Mining Locations*

As stated in the previous chapter Rajasthan is rich in mineral resources and are scattered over the entire geographical area of the state. Figure 11.5.1 gives a geographical view of mining location in the state of Rajasthan. From the figure it is clear that most of the mines are located near the national highways, state highways and railway route.

11.5.3.2 *Mine Numbers*

Table 11.5.1.

S. No.	District	Leases No.'s	Area Hectares
1	Ajmer	397	8963.19
2	Alwar	394	11010.24
3	Banswara	130	1337.09
4	Baran	26	26.37
5	Barmer	209	5284.08
6	Bharatpur	308	6673.93
7	Bhilwara	255	8484.54
8	Bikaner	110	14325.70
9	Bundi	992	6747.59
10	Chittorgarh	74	7188.72
11	Churu	153	118.02
12	Dausa	71	1520.28
13	Dholpur	72	2033.04
14	Dungarpur	216	4661.57
15	Hanumangarh	13	9474.88
16	Jaisalmer	283	4099.16
17	Jalore	372	1925.50
18	Jaipur	792	9449.21
19	Jhalawar	113	557.30
20	Jhunjhunu	152	2467.64
21	Jodhpur	340	2409.51
22	Karauli	252	13560.46
23	Kota	132	3777.69
24	Nagaur	318	15353.61
25	Pali	245	6944.42
26	Rajsamand	1474	7557.93
27	Sawai Madhopur	83	746.14
28	Sikar	138	5676.03
29	Sirohi	178	1903.86
30	Sri Ganganagar	13	2098.87
31	Tonk	107	1459.18
32	Udaipur	691	15200.05
	Total	9103	183035.8

There are about 9103 registered lease mines in the state of Rajasthan. Table 11.5.1 gives district wise details of the mine numbers in the state of Rajasthan. Same data is reflected in Figure 11.5.2 presenting geographical view of the mines. From the data it is clear that Rajsamand district leads the state with 1474 i.e., 16% of the total registered mines followed by Bundi (11%). Sri Ganganagar and Hanumangarh are the lowest having 13 number each. Although Nagaur district is having 318 (3.5%) of total mines but geographically covers the largest leased mining area of 15354 hectares (8.4%) of the total mining area). Udaipur contains the second largest mining area of 15200 hectares. Baran contributes the lowest with 26.37 hectares. According to Mines and Geology Department out of total 9103 mines 7841 (86%) falls in the Minor mineral category and rest 1262 (14%) falls in Major mineral category.

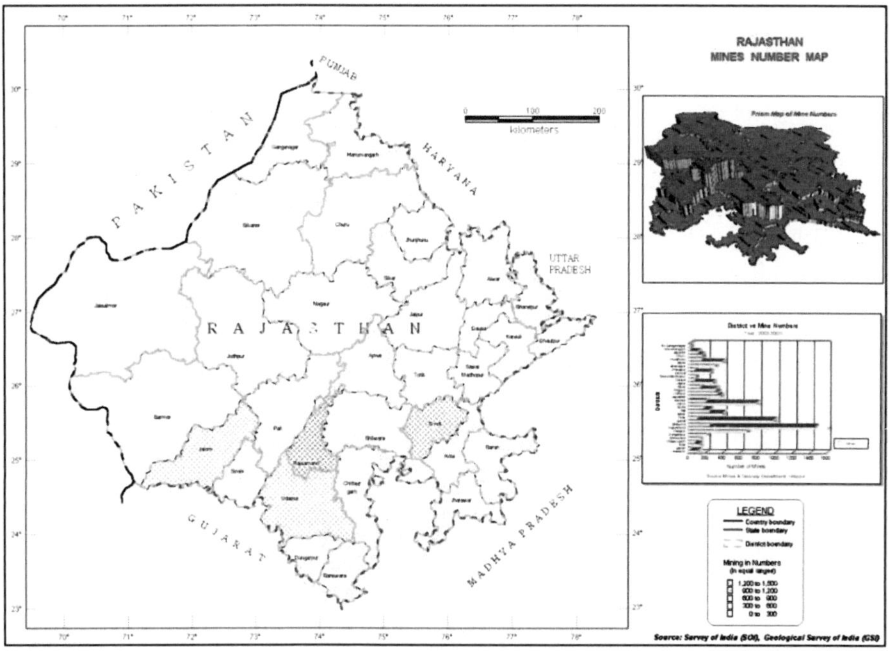

Figure 11.5.2.

11.5.3.3 *Mineral Type*

Table 11.5.2

S.No.	Type	Leases	Area	Employment	Revenue
1	Copper	5	898.55	1406	21444.56
2	Iron Ore	17	1139.65	53	121.46
3	Lead Concentrate	0	0.00	0	0
4	Lead Zinc (Rom)	5	3058.69	3052	850467.45
5	Zinc Concentrate	0	0.00	0	0
6	Manganese	1	72.27	40	116
7	Asbestos	28	2160.83	121	868.8

8	Ball Clay	74	7024.54	393	17642.18
9	Barytes	4	91.58	7	30
10	Calcite	22	1371.09	712	5646.09
11	China Clay	56	5229.51	631	8507.4
12	Dolomite	10	814.35	382	6938.7
13	Felspar	179	5850.70	1221	7200.06
14	Fire Clay	10	311.04	41	22.17
15	Flourite	14	1755.71	134	543.42
16	Graphite	2	382.70	0	134
17	Gypsum	51	22176.72	609	143338.32
18	Jasper	16	664.54	61	138.83
19	Laterite	1	291.40	0	0
20	Limestone	38	21161.08	11778	936742.8
21	Ochres	45	3064.25	261	9810.25
22	Pyrophyllite	5	193.69	219	202.77
23	Quartz	195	7645.80	1606	8604.88
24	Rock Phosphate	9	3299.86	1200	211457.19
25	Selenite	4	1062.45	26	1218.88
26	Silica Sand	80	13408.95	847	3765.39
27	Siliceous Earth	14	553.74	68	252.36
28	Soapstone	311	24045.89	4826	49272.23
29	Wollastonite	6	319.15	823	12586.47
30	Lignite	3	4518.89	152	23949.3
31	Emerald Crude	5	230.63	0	0.55
32	Garnet (Abr. and Crude)	16	568.42	302	296.09
33	Kyanite	1	5.00	0	38.75
34	Magnesite	4	43.37	14	76.68
35	Mica	24	1147.92	86	285.55
36	Sillimanite	1	50.50	0	0
37	Sulphur Ore (Pyrite)	1	200.00	0	80.96
38	Vermiculate	2	29.52	8	11.5
39	Red Oxide	1	38.43	0	7.7
40	Shale	1	5.00	0	0.35
41	Garnet Gem	0	0.00	150	0
42	Gold	0	0.00	3272	1311.872
43	Epidote	1	5.00	0	0
	TOTAL	1262	134891.41	34501	2323132

There are about 43 minerals in Major category and 22 minerals in Minor mineral category found in the state of Rajasthan. Category wise mineral type, lease numbers, areas and revenue generated are given Table 11.5.2 & 11.5.3 and Figure 11.5.3.

Table 11.5.3.

S. No.	Type	Leases	Area	Employment	Revenue
1	Bentonite	18	210.84	165	2711.46
2	Brick Earth	21	581.25	31889	48377.95
3	Chert	8	6.21	20	84.83
4	Chips Powder	50	35.76	611	8005.98
5	Fuller's Earth	29	1356.81	185	1018.77
6	Granite	404	588.92	2213	17726.76
7	Kankar - Bajri	14	44.34	33777	141916.38
8	Lime Kankar	0	0.00	0	17.09
9	Limestone (Burning)	385	14513.11	13351	142558.69
10	Limestone (Dimnl.)	187	2818.65	41642	130391.81
11	Marble	2362	3237.34	52243	657442.25
12	Masonry Stone	2469	5913.60	141046	157868.13
13	Mill Stone	7	1076.84	48	92.07
14	Murram	0	0.00	2120	866.76
15	Patti Katla	60	358.60	2024	2250.23
16	Quartzite	5	4.25	10	27.97
17	Rhyolite	93	125.00	520	2395.36
18	Salt Petre	3	0.00	3	273.00
19	Sandstone	1305	16804.20	115781	319080.50
20	Serpentine	386	416.14	5817	106808.06
21	Shale	2	1.86	4	32.66
22	Slate Stone	33	50.67	63	864.12
	Total	7841	48144.39	443532	1740810.83

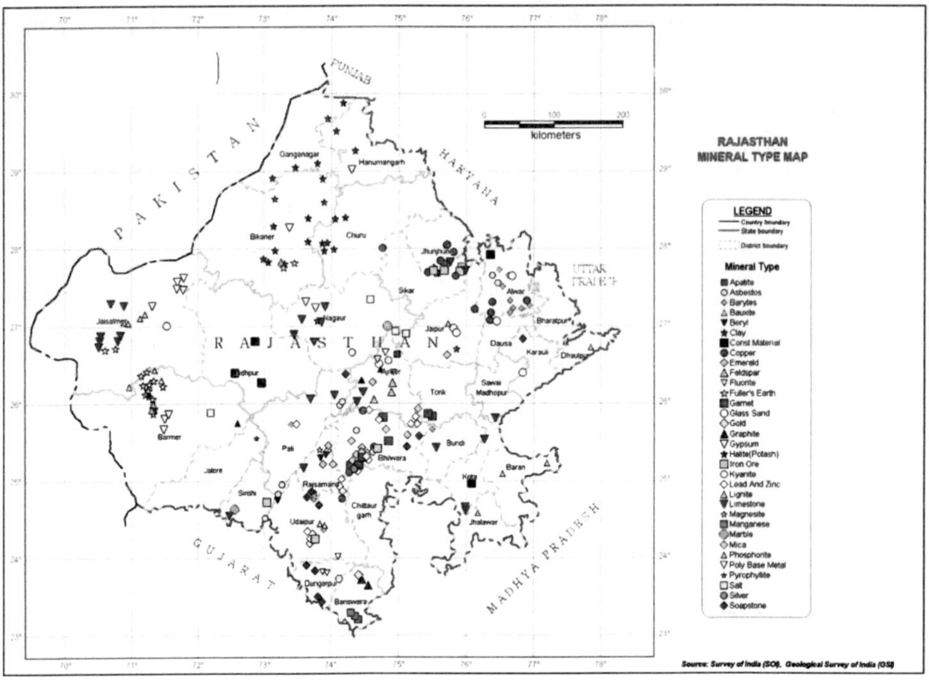

Figure 11.5.3.

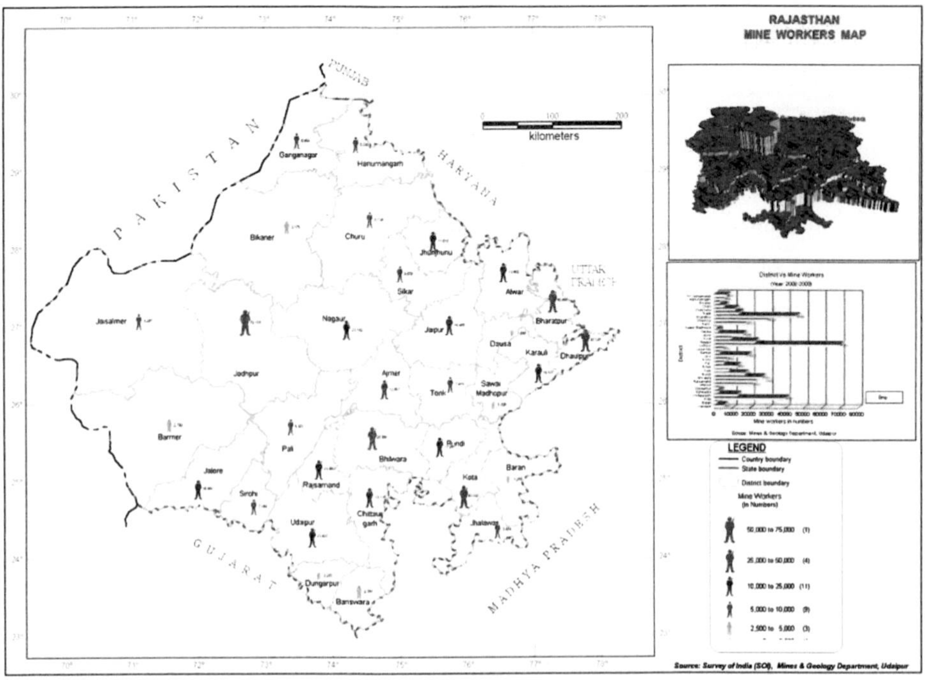

Figure 11.5.4.

11.5.3.3.1 *Major Minerals*

In Rajasthan there are 43 varieties of minerals in the category of Major minerals, which are mined at different places in Rajasthan as reported by M&GD, Udaipur. Soapstone (311) leads the tally followed by Quartz (195) in terms of lease numbers in this category. Soapstone mineral also leads the state in term of largest geographical lease (mine) area of 24046 hectares, which is about 18% of the total major mineral leased areas. Gypsum stands at second place with 16% of total major mineral leased area. In terms of Revenue, Limestone is the largest revenue-generating major mineral, contributing Rs. 93 Crores (about 40% of total revenue from major minerals) in the year 2002-2003. Lead Zinc (Rom) stands second making 37% contribution in revenue generation.

Ajmer leads the tally with maximum 196 (i.e., 15.5%) of total leased mines while Bhilwara leads in revenue generation contributing up to 30% of the total revenue generated from major mineral category. Major mineral category constitutes 14% of total registered leased mine generating 56.5% of the total revenue from all type of mines.

11.5.3.3.2 *Minor Minerals*

In Minor minerals, Masonry stone mineral is having the largest number of mines (2469 i.e., 31.5%) followed by Marble (2362 i.e., 30%) in the state Rajasthan. Sandstone mineral leads the state in term of largest geographical lease area of 16804.2 hectares, which is about 35% of the total minor mineral leased area. In terms of Revenue, Marble is the largest revenue-generating mineral in the state of Rajasthan, contributing Rs. 65 Crores (about 38% of total revenue from minor minerals) in the year 2002-2003. Bhilwara leads the tally with maximum 1405 (i.e., 18%) of total leased mines while Rajsamand leads in revenue generation contributing up to 16% of the total revenue generated from major mineral category. In Minor mineral category registered leased mine are 86% of the total leased mines generating 43.5% of the total revenue from all type of mines.

11.5.3.4 *Mine Workers*

Table 11.5.4 and Figure 11.5.4 depicts district wise the number of workers employed in registered mines in Rajasthan. The figures in the table indicates that maximum 6766 (about 19.6%) mine workers are employed in Jaisalmer in major leased mines while 70093 (about 16%) mine workers are employed Jodhpur district in Minor mine category. And collectively, Jodhpur leads the tally with highest mine workers number (70154), which is about 15% of the total mineworker population.

Table 11.5.4.

S.No.	District	Leases			Employment			Revenue		
		Major	Minor	Total	Major	Minor	Total	Major	Minor	Total
		(Numbers)			(Numbers)			(Rs. X1000)		
	Ajmer	196	201	397	1847	10960	12807	68261.42	100202.59	168464.01
2	Alwar	16	378	394	34	13868	13902	231.380	58052.940	58284.32
3	Banswara	6	124	130	40	2540	2580	633	3170.000	3803
4	Baran	0	26	26	0	1265	1265	0.000	1622.53	1622.53
5	Barmer	15	194	209	202	2588	2790	27730.210	7768.150	35498.36
6	Bharatpur	19	289	308	153	45713	45866	69.710	45106.190	45175.9
7	Bhilwara	152	103	255	1574	25419	26993	666970.42	107588.55	774558.97
8	Bikaner	88	22	110	446	2729	3175	40493.000	9692.840	50185.84
9	Bundi	8	984	992	404	16877	17281	25622.460	76338.070	101960.53
10	Chittorgarh	50	24	74	950	12168	13118	340127.27	15549.470	355676.74
11	Churu	0	153	153	0	5738	5738	586.000	13219.000	13805
12	Dausa	21	50	71	149	1747	1896	1544.490	3925.700	5470.19
13	Dholpur	0	72	72	0	30128	30128	0.000	32701.970	32701.97
14	Dungarpur	72	144	216	682	1575	2257	5186.23	15483.860	20670.09
15	Hanumangarh	11	2	13	28	5517	5545	17555.580	9398.140	26953.72
16	Jaipur	82	710	792	553	18440	18993	11000.73	96030.300	107031.03
17	Jaisalmer	18	265	283	6766	9700	16466	176266.45	13866.00	190132.45
18	Jalore	6	366	372	210	5071	5281	1210.000	20563.000	21773
19	Jhalawar	1	112	113	0	5454	5454	2.800	10924.470	10927.27
20	Jhunjhunu	28	124	152	4767	6848	11615	23119.142	13061.350	36180.492
21	Jodhpur	16	324	340	61	70093	70154	138.830	140408.90	140547.73
22	Karauli	42	210	252	350	17757	18107	1683.830	33489.080	35172.91
23	Kota	2	130	132	2280	37752	40032	66718.160	116889.86	183608.05
24	Nagaur	13	305	318	144	23048	23192	16260.660	221605.24	237865.9
25	Pali	37	208	245	2144	3179	5323	114816.46	30777.490	145593.95
26	Rajsamand	69	1405	1474	1971	21512	23483	2765	270651.18	273416.18
27	Sawai Madhopur	6	77	83	42	966	1008	247.23	2979.110	3226.34
28	Sikar	45	93	138	223	9455	9678	1187.95	29288.790	30476.74
29	Sirohi	10	168	178	1374	5820	7194	248743.10	27994.820	276737.92
30	Sri Ganganagar	13	0	13	44	6915	6959	12089.040	11276.860	23365.9
31	Tonk	39	68	107	1015	6901	7916	1033.92	36509.320	37543.24
32	Udaipur	181	510	691	6048	15789	21837	347790.41	132284.29	480074.71
	TOTAL	1262	7841	9103	34501	443532	478033	2220085	1708420	3928505

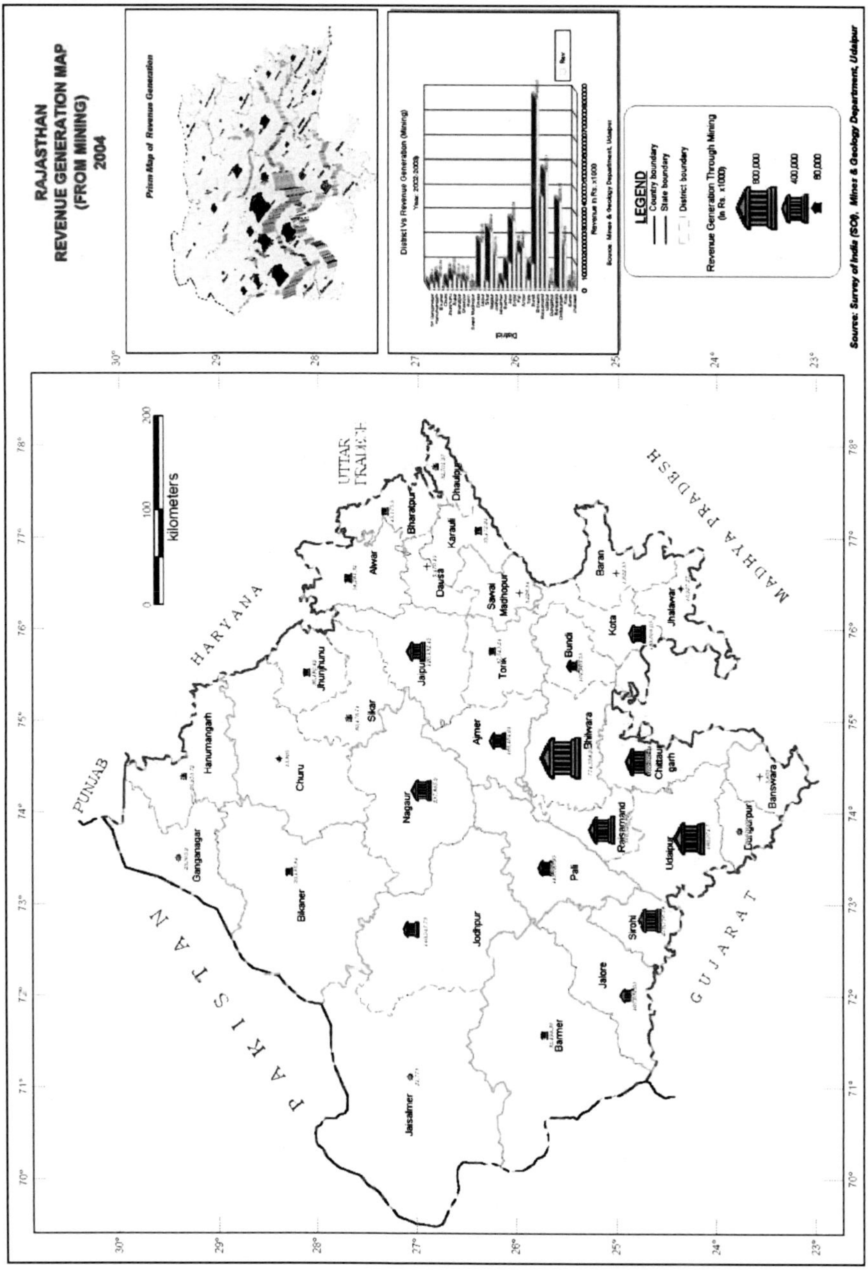

Figure 11.5.5.

In Major mineral category, Limestone mineral employs maximum about 34% of mine workers generating 40% of the total major mineral revenue. Similarly in Minor mineral category Masonry Stone leads the tally employing 32% mine workers but generating 9% of the total minor category revenue.

11.5.3.5 *High Risk Behavior Groups (HRBGs) in Rajasthan*

11.5.3.5.1 *Rural Traditional Female Sex Workers (RTFSW)*

According to collected information from government organization and non-government organisations the numbers of Rural Traditional Female Sex Workers (RTFSW) and their household are 5736 and 7570 respectively. They are engaged in high-risk activities.

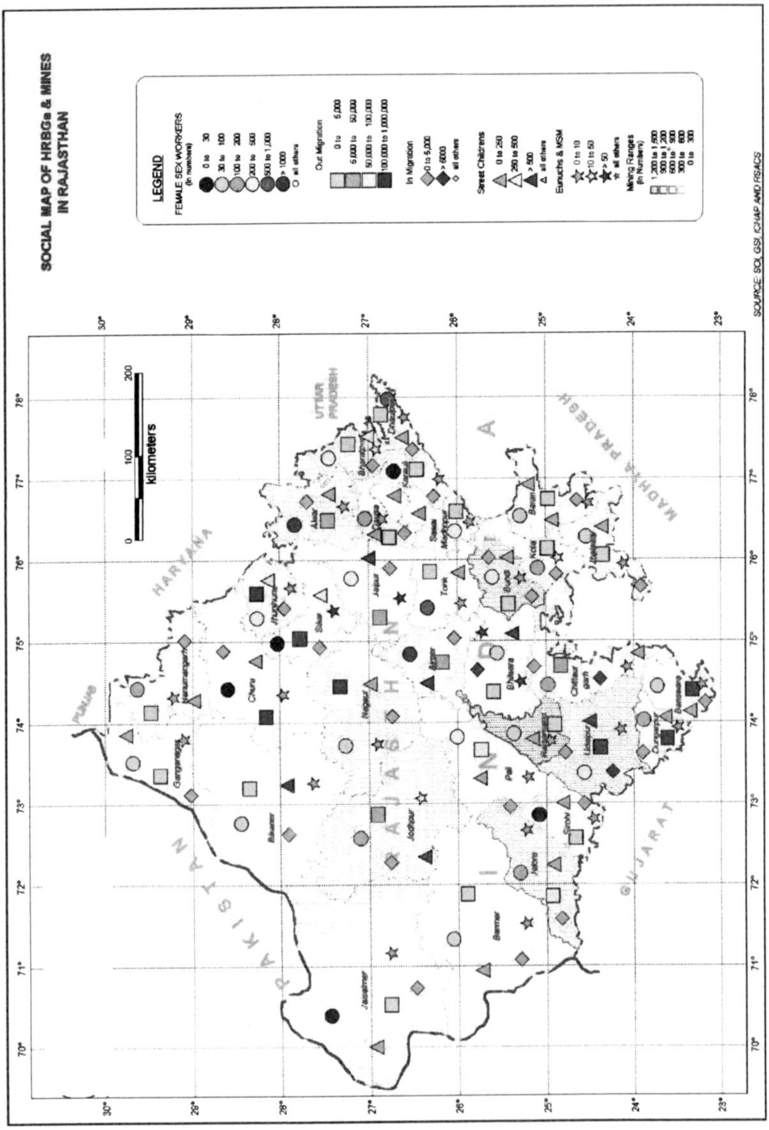

Figure 11.5.6.

According to the statistics Bundi has highest (17%) number of Households and Alwar (18%) has highest number of RTFSW. According to statistics, in Rajasthan RTFSW population is about 1.2% of the total mineworkers population. Tonk district has highest ratio of 9.14% RTFSW per mineworker population followed by Dausa with 6.27%.

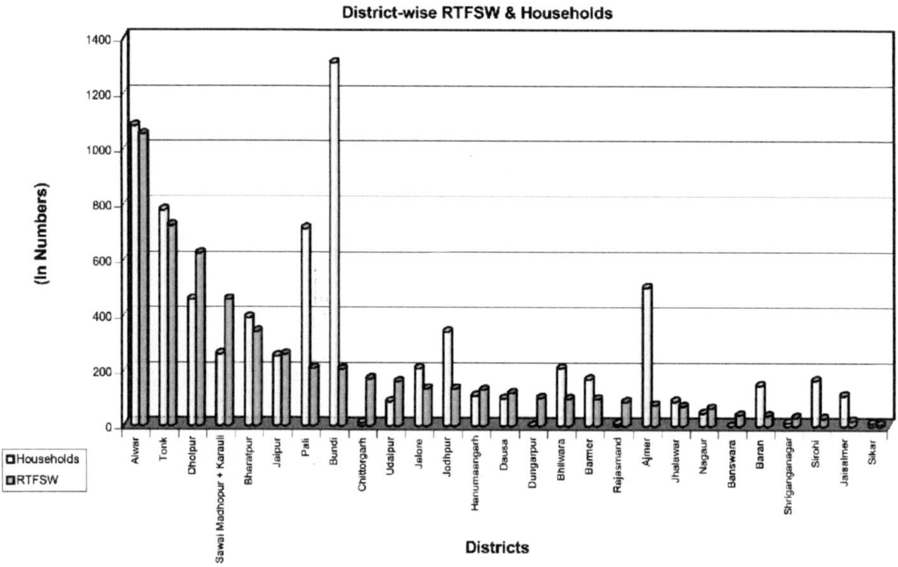

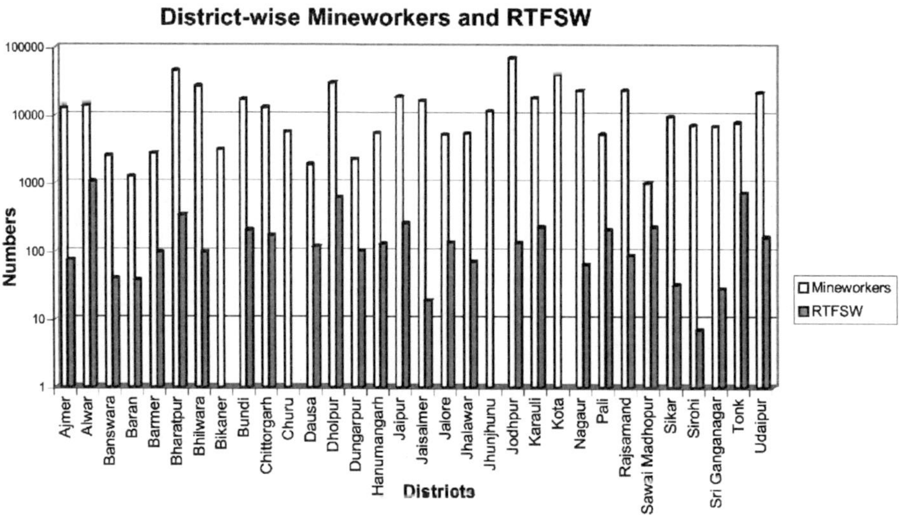

11.5.3.5.2 *Female Sex Workers (FSW)*

According to GOs and NGOs the total numbers of FSW reported by different sources are l as 15,281 in Rajasthan. According to the statistics Ajmer has highest (59.7%) number of FSW followed by Alwar (7.6%).

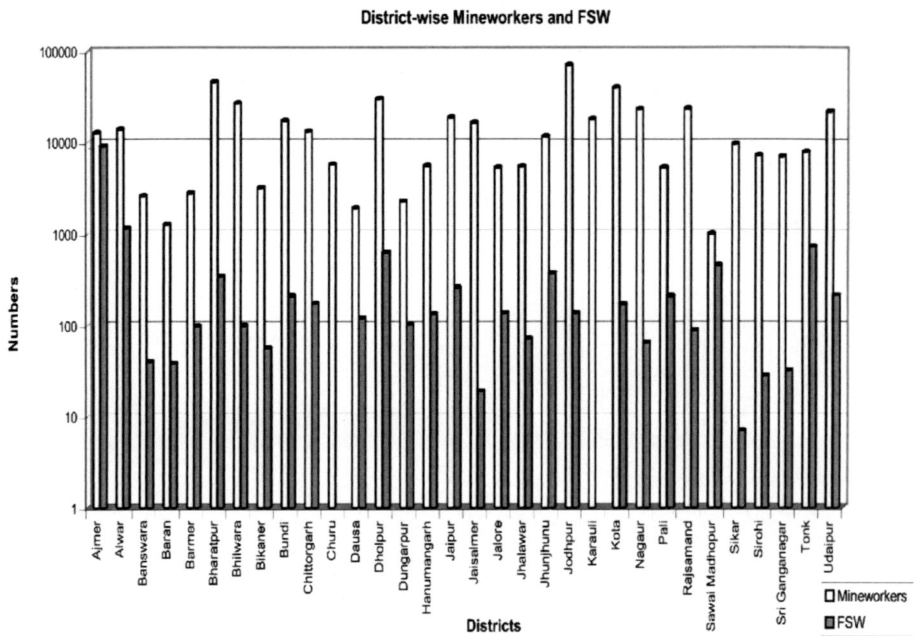

Above Graph shows district wise position of FSW with mineworker population. According to the analysis, in Rajasthan FSW population is about 3.2% of the total mineworkers population. Ajmer district has highest ratio of 71.2% FSW per mineworker population followed by Sawai Madhopur with 45.2%.

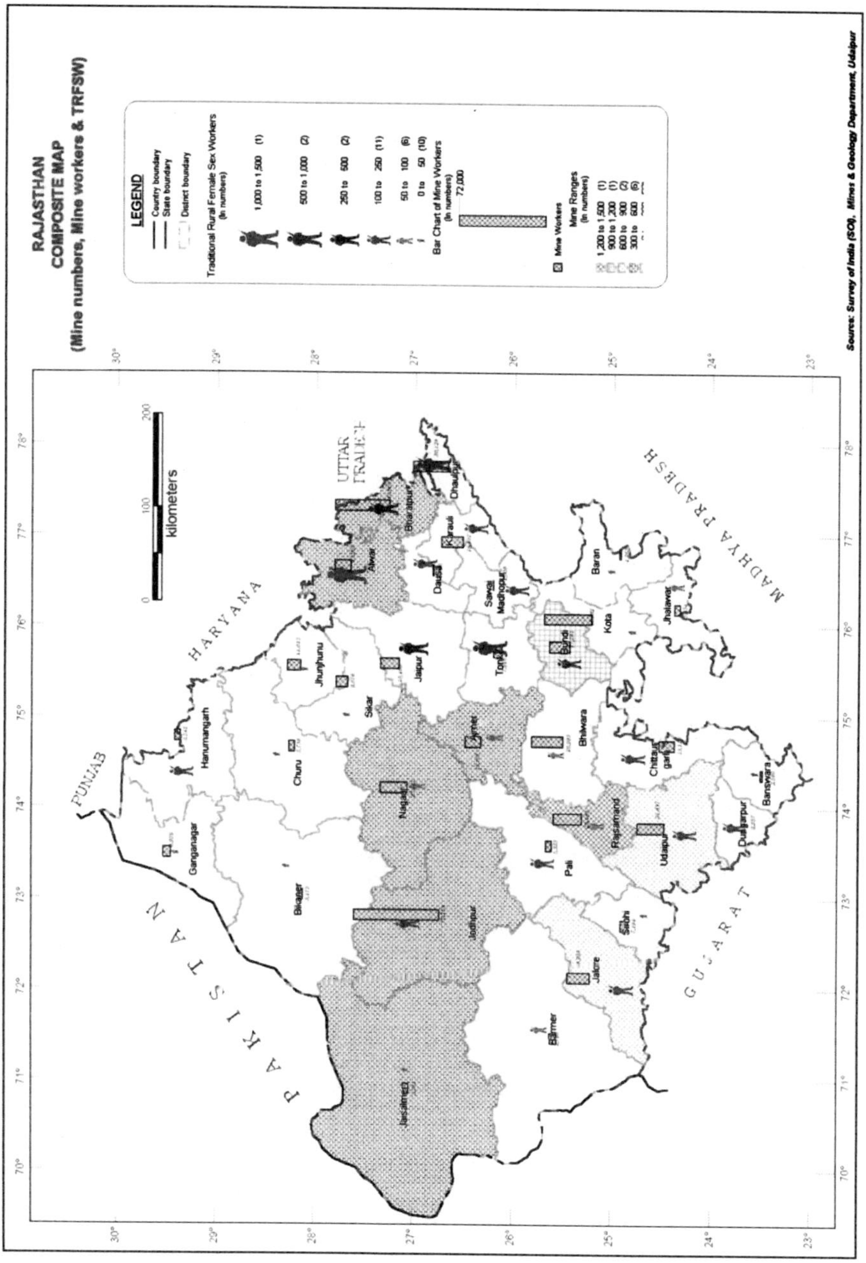

Figure 11.5.7

11.5.3.5.3 Out Migrants

According to GOs and NGOs the total numbers of Out migrants reported by different sources in Rajasthan are 933097.

According to the statistics Udaipur has highest (22.3%) number of out-migrants followed by Dungarpur (17.9%).

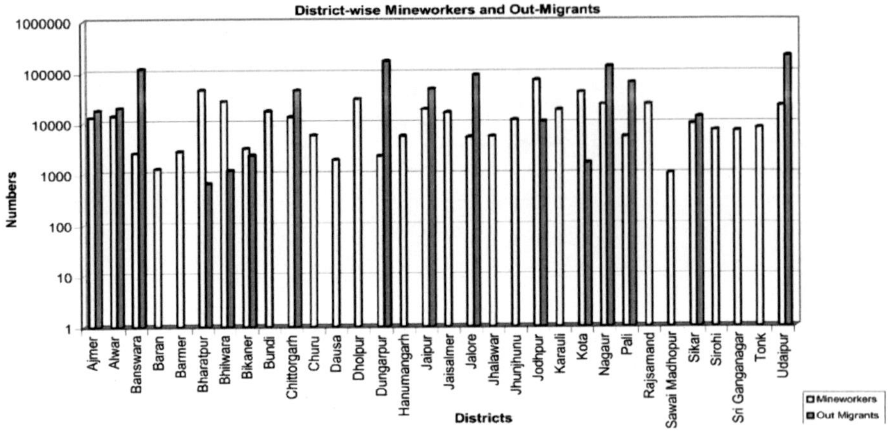

Following Graph shows district wise position of out-migrants with mineworker population. According to the analysis, in Rajasthan Out-migrants population is about 195% of the total mineworkers population. Dungarpur district has highest ratio of 74 out-migrants per mineworker population followed by Banswara with 45.7.

11.5.3.5.4 *In Migrants (Urban)*

The total numbers of In-migrants reported by different sources in Rajasthan are total 19,90,911 reported in Rajasthan. According to the statistics Sikar has highest (42.5%) number of In-migrants followed by Churu (27.7%).

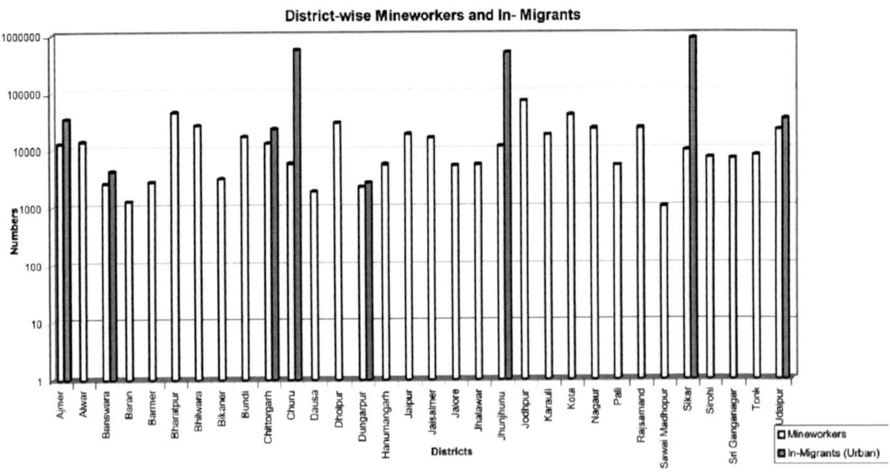

Above Graph shows district wise position of In-migrants with mineworker population. According to the analysis, in Rajasthan In-migrants population is about

416.5% of the total mineworkers population. Churu district has highest ratio of 96.4 In-migrants per mineworker population followed by Sikar with 87.4.

11.5.3.5.5 *Truckers*

According to GOs and NGOs the total numbers of Truckers reported by different sources in Rajasthan are 21,982. According to it Ajmer has highest (56.6%) number of Truckers followed by Jaipur (32.3%). Following Graph shows district wise position of Truckers with mineworker population. According to the analysis, in Rajasthan Truckers population is about 4.6% of the total mineworkers population. Ajmer district has highest ratio of 97.1% Truckers per mineworker population followed by Jaipur with 37.4%.

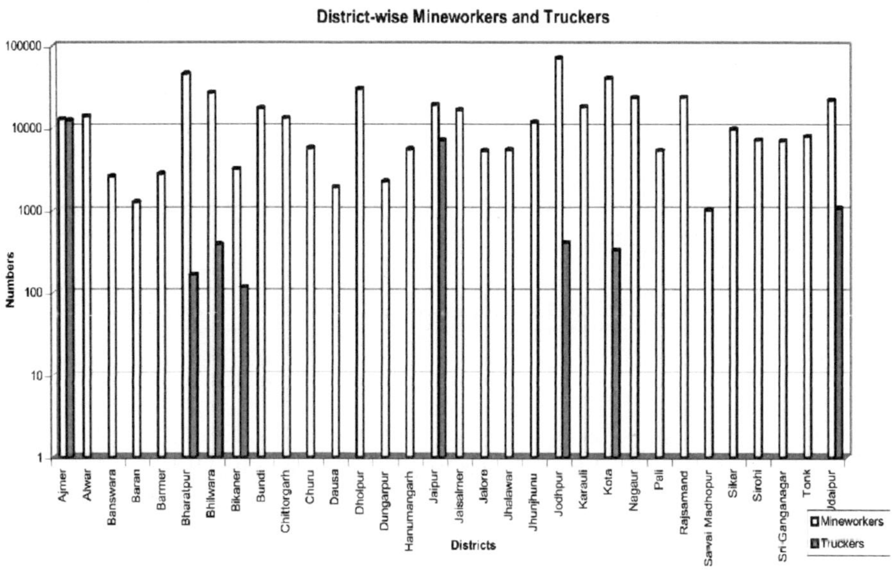

11.5.3.5.6 *Street Children*

According to GOs and NGOs the total numbers of Street Children reported by different sources in Rajasthan are 9002. According to the statistics Jodhpur has highest (32.5%) number of Street Children followed by Udaipur (15.4%). Following Graph shows district wise position of Street Children with mineworker population. According to the analysis, in Rajasthan Street Children population is about 1.88% of the total mineworkers population. Bikaner district has highest ratio of 18.3% Street Children per mineworker population followed by Ajmer with 8.9%.

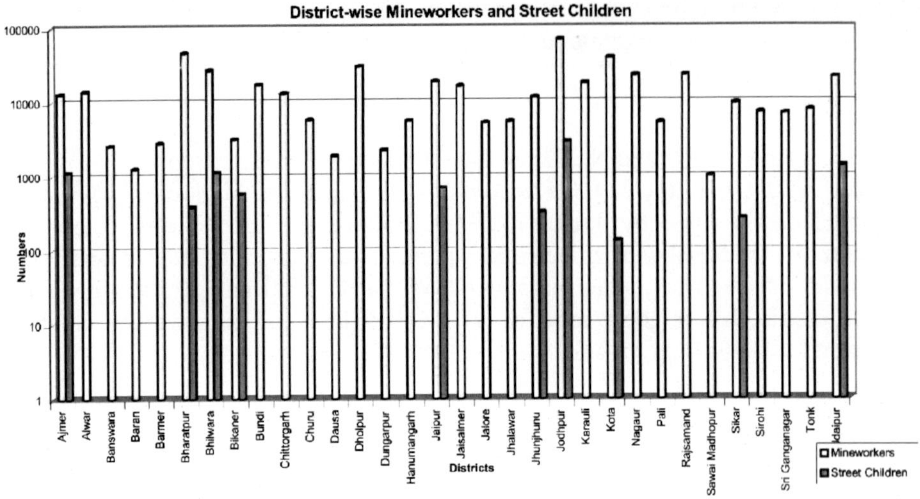

11.5.3.5.7 *Eunuchs*

According to GOs and NGOs the total numbers of Eunuchs reported by different sources there are 725 in Rajasthan.

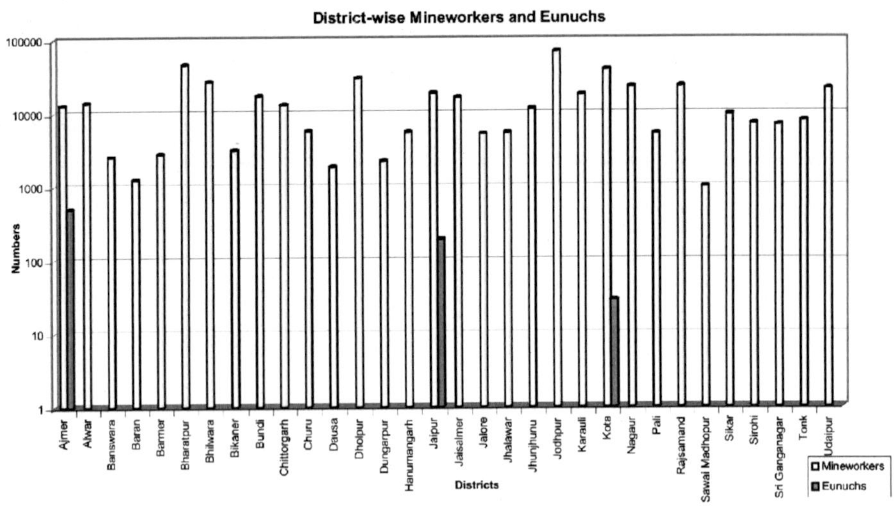

According to the statistics Ajmer has highest (69.2%) number of Eunuchs followed by Jaipur (26.8%). Following Graph shows district wise position of Eunuchs with mineworker population. According to the analysis, in Rajasthan Eunuchs population is about 0.15% of the total mineworkers population. Ajmer district has highest ratio of 3.9% Eunuchs per mineworker population followed by Jaipur with 1.0%.

11.5.3.5.8 *MSM*

According to GOs and NGOs the total numbers of MSM reported are 913 in Rajasthan. According to the statistics Jaipur has highest (43.3%) number of MSM followed by Ajmer (25.4%). Following Graph shows district wise position of MSM with mineworker population. According to the analysis, in Rajasthan MSM population is about 0.19% of the total mine-workers population. Jaipur district has highest ratio of 2.1% MSM per mineworker population followed by Ajmer with 1.8%.

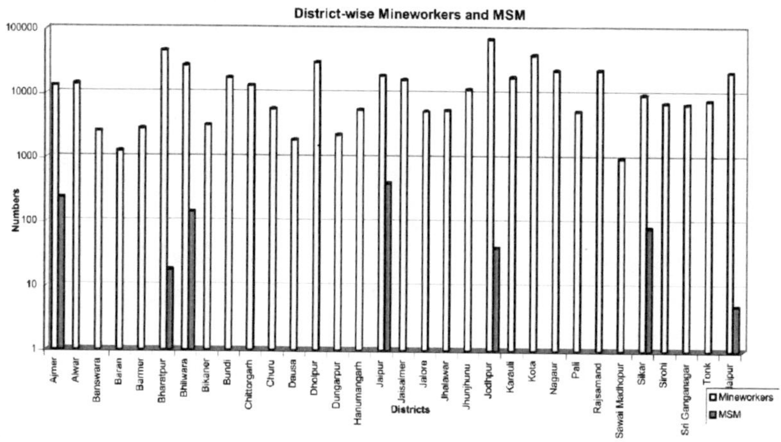

11.5.3.5.9 *IVDUs*

According to GOs and NGOs the total numbers of IVDUs reported by different sources are total 387 in Rajasthan.

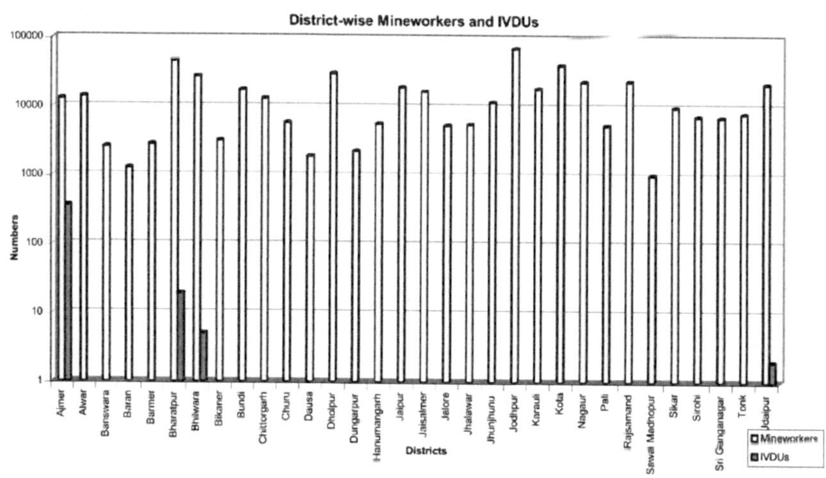

According to the statistics Ajmer has highest (93.3%) number of IVDUs followed by Bharatpur (4.9%). Graph shows district wise position of IVDUs with mineworker population. According to the analysis, in Rajasthan IVDUs population is about 0.08% of the total mineworkers population. Ajmer district has highest ratio of 2.8% IVDUs per mineworker population.

11.5.4 Prioritization of Potential Mining Districts

For further interventions, priority rating was developed based on the data collected during the study. All the districts were sorted in ascending order based on total number of leased mines, Mineworker population and total revenue generation. Priority indexes were assigned based on the sorting criteria. Final collective listing was developed combining the above three indexes. Table 11.5.5 presents priority rating of the districts.

Table 11.5.5 Priority Rating of Districts

Priority Rating	District	Total Leased Mine in No.'s	Mine Workers in No.'s	Total Revenue In Rs '000
1	Rajsamand	1474	23483	273416.18
2	Udaipur	691	21837	480074.71
3	Bhilwara	255	26993	774558.97
4	Jodhpur	340	70154	140547.73
5	Nagaur	318	23192	237865.90
6	Jaipur	792	18993	107031.03
7	Bundi	992	17281	101960.53
8	Bharatpur	308	45866	45175.90
9	Ajmer	397	12807	168464.01
10	Jaisalmer	283	16466	190132.45

CASE STUDY - 6

11.6 BENEFIT ASSESSMENT OF RAIN WATER HARVESTING STRUCTURES IN THREE BLOCKS OF UDAIPUR DISTRICT

To large extent agriculture exist in a climatic environment of uncertainty. This is particularly true in the mid-latitude which experiences a procession of high and lows that move from west to east and gives rise to daily weather changes. A Normal climate, as given by the mean values, does not exist and information concerning the extremes is often more important for planner to identify the cost effective rainwater harvesting activities suitable for their Study areas. In such circumstances it is very important to have the feed back of the activities implemented in the past in that particular geographical area. Hence looking to the potential and magnitude of the issue a study was carried out to assess the benefits from the various type of the rainwater harvesting constructed in the selected villages in the Udaipur district of Rajasthan. Three blocks namely, Girwa, Kotra and Salumbar of Udaipur District were selected covering 7 villages for the assessment study where the activities were mainly the construction of structures like, water harvesting structures for storing runoff, Loose Stone Check Dams, Contours Trenches in pasture lands, excavation of new ponds and renovation of old one, field shoulder bunding and gabion structures. These structures were implemented by NGOs in villages with the local community participation.

11.6.1 Objective

The overall objective of the proposal is to develop GIS based cost benefit analysis of rainwater harvesting construction works based on available data of the villages and to prepare graphical representation/maps of the final analysis.

The specific objectives are as follows:

- Collection of data/information related demographic and water resource conditions from selected three blocks (KOTRA, SALUMBAR and GIRWA).
- Generation of GIS based maps of three blocks to represent pre-monsoon and post monsoon water resource scenario in the three blocks.
- Identification of the location and demarcation of villages selected for Cost Benefit Analysis study using Global Positioning System.
- Field measurement of water level depth and collection of water samples from the selected study villages.
- To carryout COST BENEFIT Analysis for each study village site and prepare a Comparative cost benefit analysis statement using GIS technology.

11.6.2 Methodology

Comprehensive cost benefit analysis study of the area using GIS was carried out in the study area so as to make it **Model study using GIS technology** which could be used as guidelines to replicate in other similar areas.

11.6.2.1 *Location of Study Area*

Study has been focused in three blocks of Udaipur district where water harvesting/recharging works have been implemented by local NGOs. In total seven numbers village were selected such that each individual site represents each category of geological and hydro geological formation that are exposed in the study blocks. Table 11.6.1 indicates the name of village and site along with the actual costing of water harvesting structures. Three blocks of the Udaipur districts were selected (Figure 11.6.2) ensuring proper sample representation of the area as indicated in Table 11.6.1. The aerial distribution of the sites could be seen in the Figure 11.6.1.

Table 11.6.1 Study Villages

S. No.	Code No.	Village	Block	Site	Pop_91	Amount (Rs.)
1	54	Basee Jhuhawat	Salumbar	Joyra naka	2492	84490
2	36	Gaonra Pal	Salumbar	Hingawat Nal	1693	136600
3	57	Basse Singawat	Salumbar	HamaTalav	1477	263600
4	209	Semliya	Griwa	Khedia	300	211914
5	30	Dang	Kotra	Sirwal	208	127170
6	203	Patoo Khera	Griwa	Charnot Land	425	402500
7	57	Basse Singawat	Salumbar	Khirvania ki nal	1477	479356

11.6.2.2 *Data Collection*

The reliability of any Geographical Information System depends upon the accuracy of the database. Since this study also utilised GIS for analysis and drawing conclusion, collection of accurate information was emphasized from different sources. To ensure this data was collected from secondary as well as primary sources. The following gives the details of the information that was collected from these sources.

11.6.2.2.1 *Secondary Sources*

Collection and review of existing data and reports: Reports and available data on aquifer characteristics, water levels, groundwater extractions and groundwater quality were collected from various government institutions and other organisations. To carryout impact assessment study on ground water it is necessary to have the data of pre-monsoon season as it reflects most deteriorated and severe groundwater condition both quality and quantity wise. There are different government department who takes the water sample from each wells and get it analyse in their laboratories. The main departments are: Rajasthan State Groundwater Department, Public Health Engineering Department, Central Groundwater Board etc.

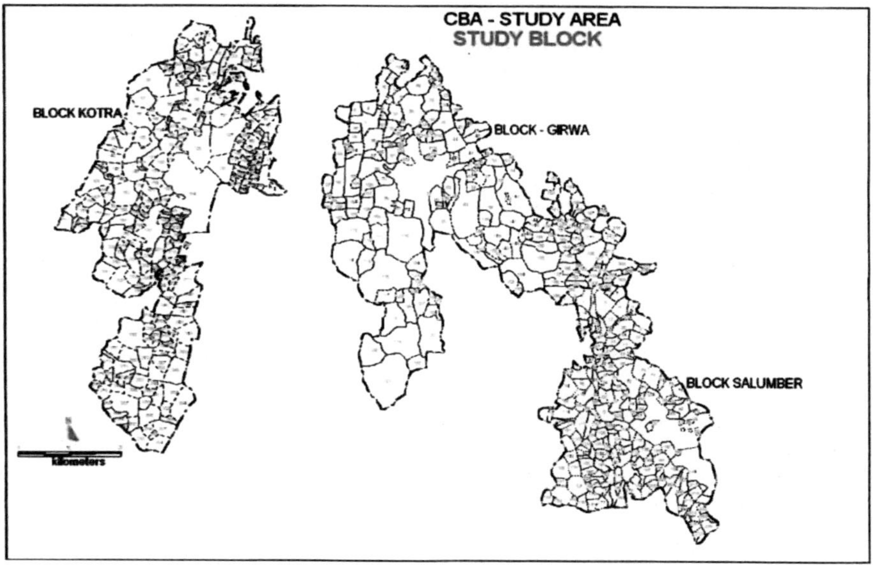

Figure 11.6.1. Location Map of Study Blocks

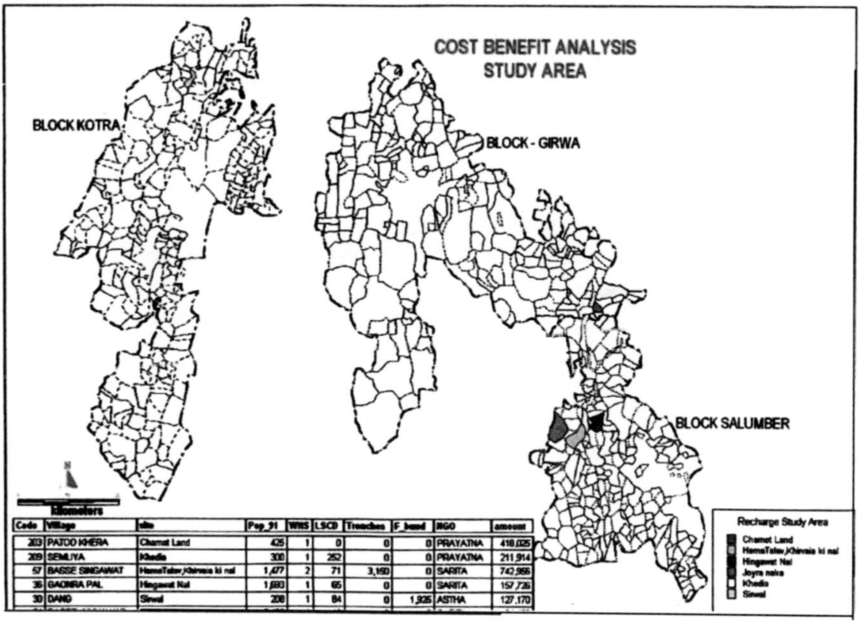

Figure 11.6.2. Location Map of Study Villages

11.6.2.2.2 *Primary Sources*

Field survey and observation: Global Positioning System (GPS) was used to identify the earth coordinates of the study villages. During the field survey of the study villages, data related to the quality of the catchment, soil characteristics, type of structure were also collected. To have the latest information of the groundwater quality and groundwater level depth during the field visit water sample and measurement of the water level depth from the nearest water source like open well/hand pump/ tube well was taken. The sample were got chemically analysed from the state ground water department, Udaipur.

11.6.2.3 *Process of Data Collection*

Before initiation of fieldwork preliminary meetings with Local organisations and government departments were organised to discuss the objectives, scope of work, methodology and timeline of the study. Based on the discussions, checklist and schedule for field visit for data collection from secondary and primary resources respectively was developed. In Udaipur districts many local NGOs and government organisations have implemented water harvesting structures on watershed basis. A formal meeting with some of these organisations at their respective offices in blocks was organised time to time to get their support in data collection and analysis. Information/data already available with them were initially collected and reviewed. As required, focused group discussions were conducted with the help of support staff of these NGOs. Field GPS survey was conducted to obtain GIS location in the respective study areas.

11.6.2.4 *Data Processing and Analysis*

The collected information both from primary and secondary were fed in the computer. This includes text data in tabular form and the vectorization of maps updated through GPS. The fed information was analysed for its accuracy. The data in tabular form related was linked to the base map using MAP INFO software. The processed data was interpolated and mapped to generate thematic mappings. Through the developed thematic maps existing status of ground water in terms of its quality and quantity was assessed. To assess the impact of each of the water harvesting structure data related catchment area, soil characteristics, Climatological parameters and groundwater quality, quantity and fluctuations were drawn from the thematic maps. Cost - benefit analysis was performed based on observed impact of WHS works on drinking water availability, quality, irrigation and thus change agricultural production etc. These results of the analysis were linked to the maps to represent graphical view changes in the overall economics of the people of the villages.

11.6.2.5 *Direct Benefits*

Immediate response to water harvesting/ recharging activities implemented has been considered to draw direct benefits. In this category incremental availability of water was converted in to LPCD requirement to assess the drinking water benefits of the raised assets. Monetary value of increased crop and milk production with improvement in water availability has also been estimated. Each site has been discussed in details individually and a comparison note has been prepared for all sites in the end.

11.6.2.6 *Indirect Benefits*

Induced ground water recharge in the down stream sources and impact on its quality has been assessed to draw indirect benefits. The estimation was done after deducting evapo-transpiration (specified for the area) losses and drinking water draft from the total runoff harvested in the structure. The fluctuations in the water level at the down stream sources have been calculated on the basis of seepage rate, distance of the water source from the structure and the actual water depth in the source.

11.6.3 Findings

11.6.3.1 *Rainfall Distribution*

In Udaipur region Monsoon rains are received from 15th June to August months. And the area receives maximum amounting to more than 90% of the rainfall during this period. Daily rainfall record of the last 15 months has been collected, processed for the analysis.

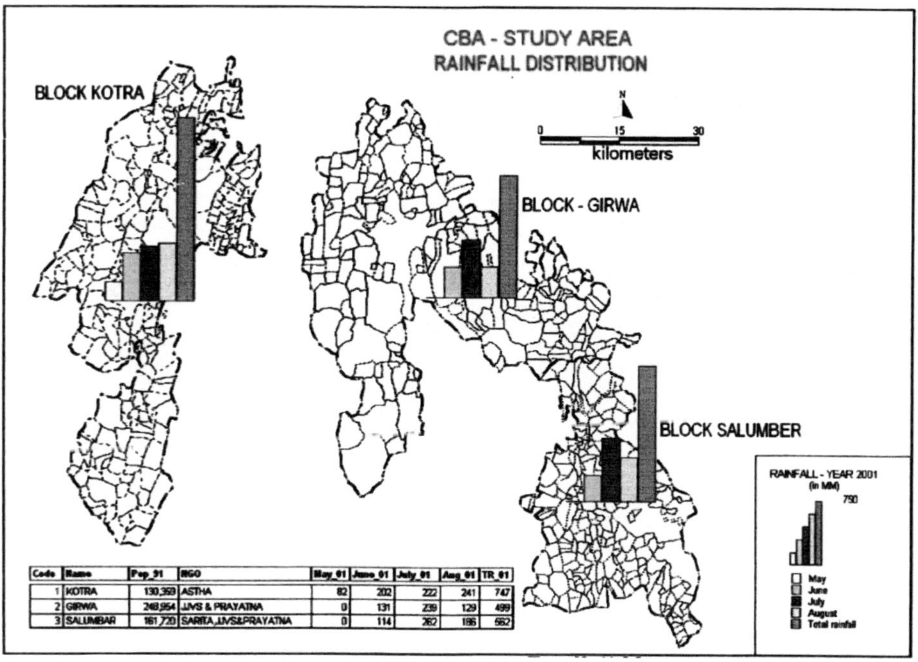

Code	Name	Pop_91	NGO	May_01	June_01	July_01	Aug_01	TR_01
1	KOTRA	130,359	ASTHA	82	202	222	241	747
2	GIRWA	248,954	JJVS & PRAYATNA	0	131	239	129	499
3	SALUMBAR	161,720	SARITA,JJVS&PRAYATNA	0	114	262	186	562

Figure 11.6.3. – Rainfall Distribution

The rainfall received during the monsoon period of year 2000 is almost close to the mean annual rainfall but due to uneven distribution scanty variation and backlog of last three consecutive year drought impact it could not produce the desired result in the fields of agricultural/ crop requirement, surface water availability and ground water recharge.

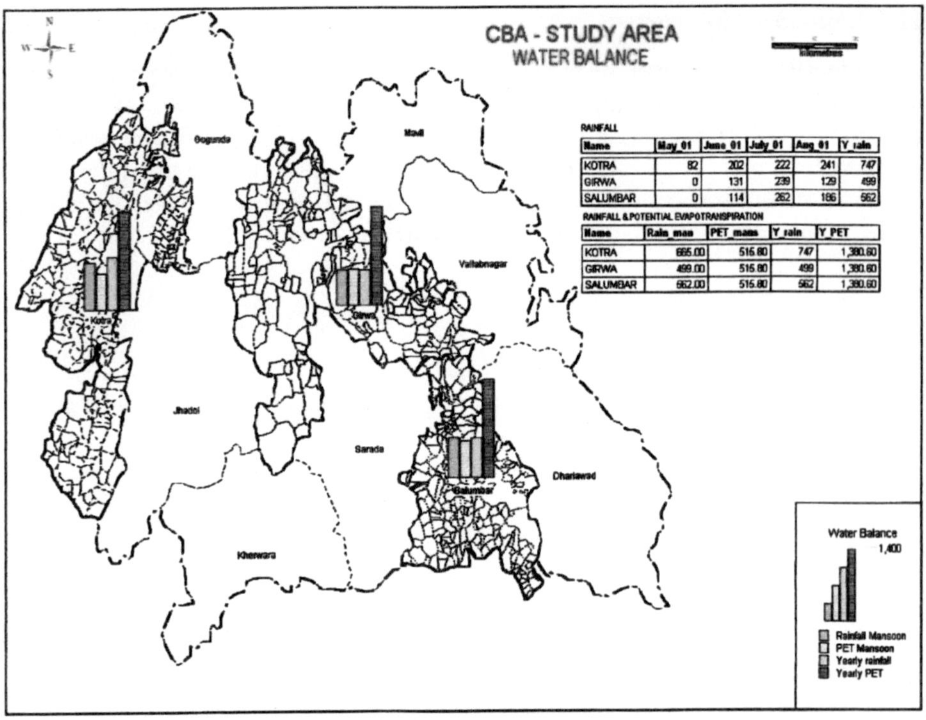

Figure 11.6.4 - Water Balance

Table 11.6.2 Rainfall Distribution (Year 2000)

S. No.	Blocks/ District	Mean Rainfall (mm)												
		Jan.	**Feb.**	**March**	**April**	**May**	**June**	**July**	**Aug.**	**Sept.**	**Oct.**	**Nov.**	**Dec.**	**Ann.**
1	Kotra	2	1	0	9	13	73	312	211	120	16	2	0	759.0
2	Girwa	3.8	4.2	0.3	7.1	12.3	86.4	198.7	150.8	98.8	12.5	1.0	2.1	578.1
3	Salumbar	3	3	0	0.5	5	86	246	213	110	12	1	2.7	683.5
	Udaipur District	8.0	3.2	2.9	1.7	9.2	68.0	228.9	204.9	107.9	10.0	4.5	2.0	650.3

Source: Irrigation Department, Udaipur

11.6.3.2 *Mean Annual Potential Evapo-Transpiration (P.E.T)*

The evaporation of water from the land and transpiration from vegetation is customarily clubbed together and is known as evapo-transpiration (ET). ET plays a dominant role in the estimate of the optimal supply of the irrigation water to farm lands. The ET estimates are used in determining seasonal crop water requirement and serve as an aid in the management of reservoirs, streams and groundwater. The estimates are also used as one of the inputs in irrigation system design and to determine season-wise consumptive use requirements of crops. It helps in matching cropping pattern and the area cropped vis-à-vis the available water resources. The

ET-values are definitely needed in the overall planning of the command development of large-scale irrigation projects. It can also be used in the hydrological water balance studies for the catchment area. Values of PET for Udaipur district is as follows:

Table 11.6.3 Mean Potential Evapo-Transpiration of Udaipur

S. No.	District	Mean Annual Potential Evapo-transpiration (cm)
1	Udaipur	130 - 140

Source: Irrigation Department, Udaipur

11.6.3.3 *Groundwater Condition*

Pre Monsoon scenario

State Ground water Department, Udaipur collects the water level depth data in regular intervals from the selected key wells scattered all over the district. For Kotra, Girwa 7 Salumbar same data have been collected from the department and analysed. The analysis reveals the fact that there are only 30 villages who are having water level depth of about 0-5m in the Salumbar block, and none in Girwa and Kotra block for the same range. 62% of the Salumbar block is having depth between 5 to 10m. 80% of the villages in Kotra lies in the same range. While the alarming conclusion is for Girwa block as about 70% of the village falls in the greater than 10m depth.

Post Monsoon scenario

The Post Monsoon analysis reveals that there are 41 villages who are having water level depth of about 0-5m in the Salumbar block (increased 11 no.'s), and nine (against none in pre monsoon) in Girwa and 29 in Kotra block for the same range. 78% of the Salumbar block is having depth between 5 to 10 m. 85% of the villages in Kotra lies in the same range. The alarming remains for Girwa block as about 36% of the village still falls in the depth category of > 10m.

Ground Water Level Trend:

The Post Monsoon season is assumed to be the recharging period to the ground water as there is no to there source of recharging. The level difference in Pre and post monsoon depicts the recharge of the particular well and thus the area. Maximum the level difference maximum is the recharge. The -ve (red) value indicate that the well and area is still depleting the ground water in spite of monsoon rains. Analysis reveals that there are 28 villages who are having water level fluctuation of > 5m in the Salumbar block, and none in Girwa and Kotra block for the same range. 60% of the Salumbar block is having depth fluctuation of 0 - 2.5m range. 89% of the villages in Kotra and 39% of Girwa villages lies in the same range. The alarming fact in the analysis is that **9 village of the Salumbar**, **12 of Kotra** and **43 villages** of the Girwa is facing the problem of water level depletion since last three year.

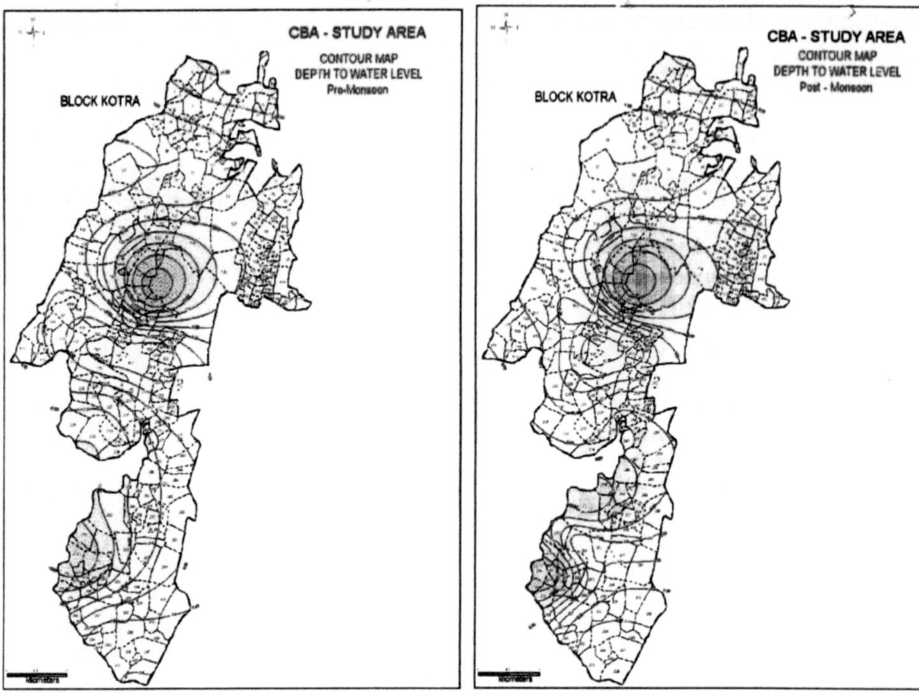

Figure 11.6.5 Pre and Post Monsoon Water Level Depth – Kotra Block

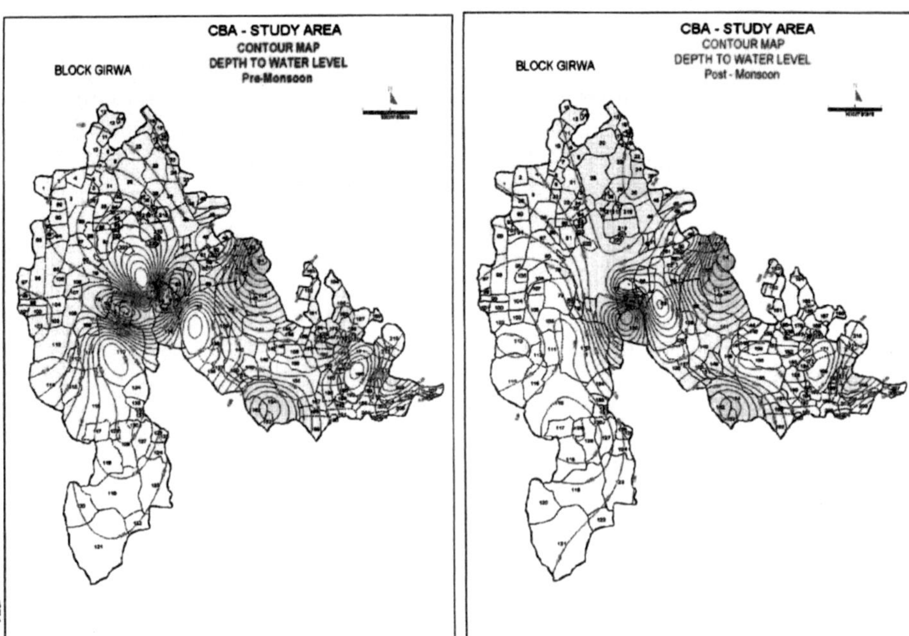

Figure 11.6.6 Pre and Post Monsoon Water Level Depth – Girwa Block

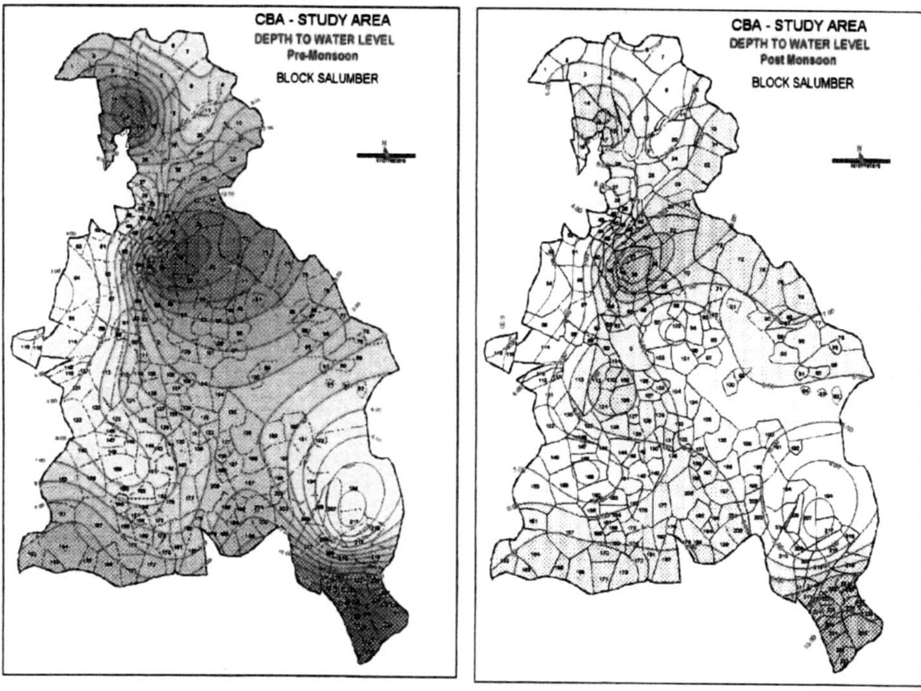

Figure 11.6.7 Pre and Post Monsoon Water Leve l Depth – Salumbar Block

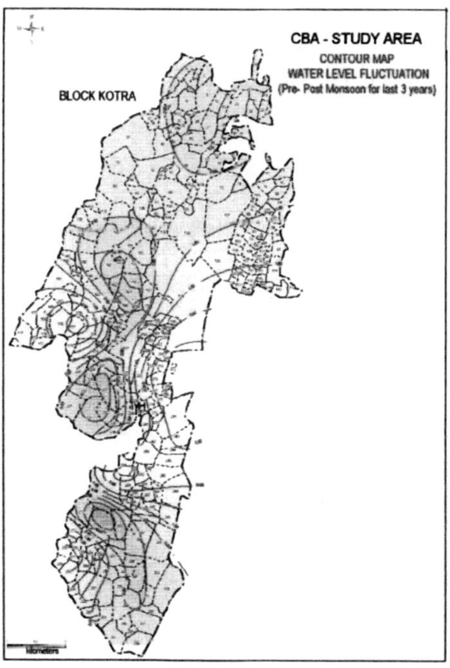

Figure 11.6.8 Water Level Fluctuation – Kotra Block

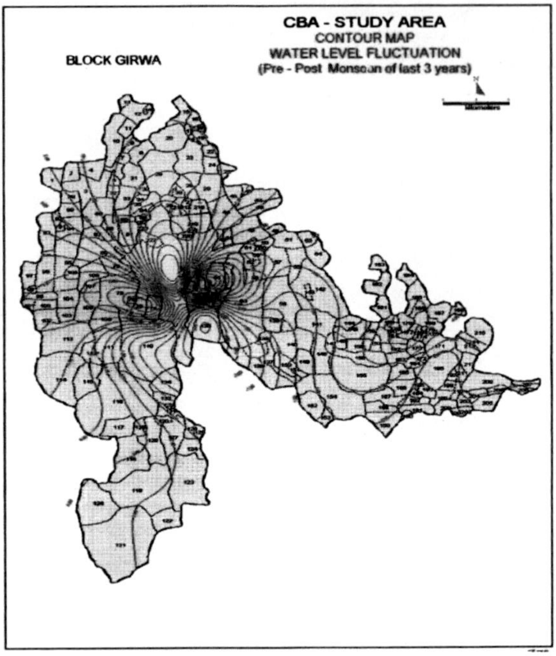

Figure 11.6.9 Water Level Fluctuation – Girwa Block

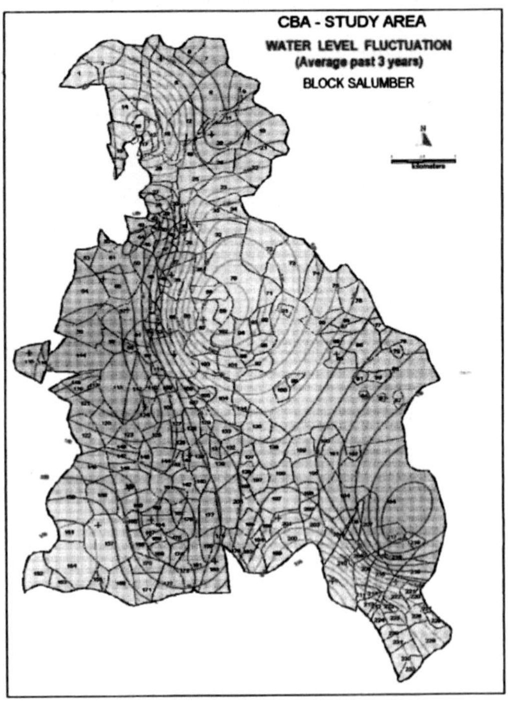

Figure 11.6.10 Water Level Fluctuation – Salumbar Block

11.6.3.4 *Groundwater Quality*

The ground water in Udaipur district is fresh to slightly saline in general. Ground water unsuitable for drinking is observed in two big patches, one in north-eastern part covering part of Mavli and Bhinder blocks, on account of salinity and another in southern part covering part of Salumber and Sarada blocks having high fluoride content. Small patches showing ground water unsuitable for drinking are also seen in block Girwa, Mavli and Bhinder blocks. The quality parameters of the key wells of the block Kotra, Girwa and Salumbar of the SGWD, Udaipur have been collected and processed. The contour maps were generated with the collected data. The map shows the distribution of the parameters over the entire block level indicating deteriorated are in red, permissible in yellow and acceptable in green colour. The standards of the parameters are mentioned in the following Table 11.6.4.

Table 11.6.4 Water Quality Standards

S. No.	Constituents/ Properties	Undesirable effects that may be produced	Highest Desirable Level	Maximum Permissible Level
1	Total Dissolved Solids TDS (mg/L)	Gastro intestinal irritation	500	1500
2	PH Range	Taste, corrosion and scale formation	7.0-8.5	6.5-9.2
3	Total Hardness as CaCO3 (mg/L)	Taste, scale formation	300	600
4	Calcium (mg/L)	Taste, scale formation	75	200
5	Magnesium (mg/L)	Taste, scale formation and gastro intestinal irritation in presence of Sulphate	Not more than 50 mg/L if there are 200 mg/L Sulphate. If there are less Sulphate, magnesium up to 100mg/L may be allowed at the rate of 1 mg/L magnesium for every 4mg/L decrease of Sulphate.	100
6	Chloride (mg/L)	Taste, corrosion in hot water system	200	1000
7	Sulphate (mg/L)	Gastro-intestinal irritation when combined with magnesium or sodium	200	400
8	Nitrate (mg/L)	Danger of infantilety methaemoglobinaemia if water is consumed by infants	20	20-100
9	Fluoride (mg/L)	Fluorosis	1	1.5

Source: Central Public Health Engineering Organisation's Manual

11.6.3.5 *Population Distribution*

As per the Census 1991 data in Udaipur district the total population is 2066580 among which 1670817 are rural and 395763 are urban population. In study blocks the population details are as follows:

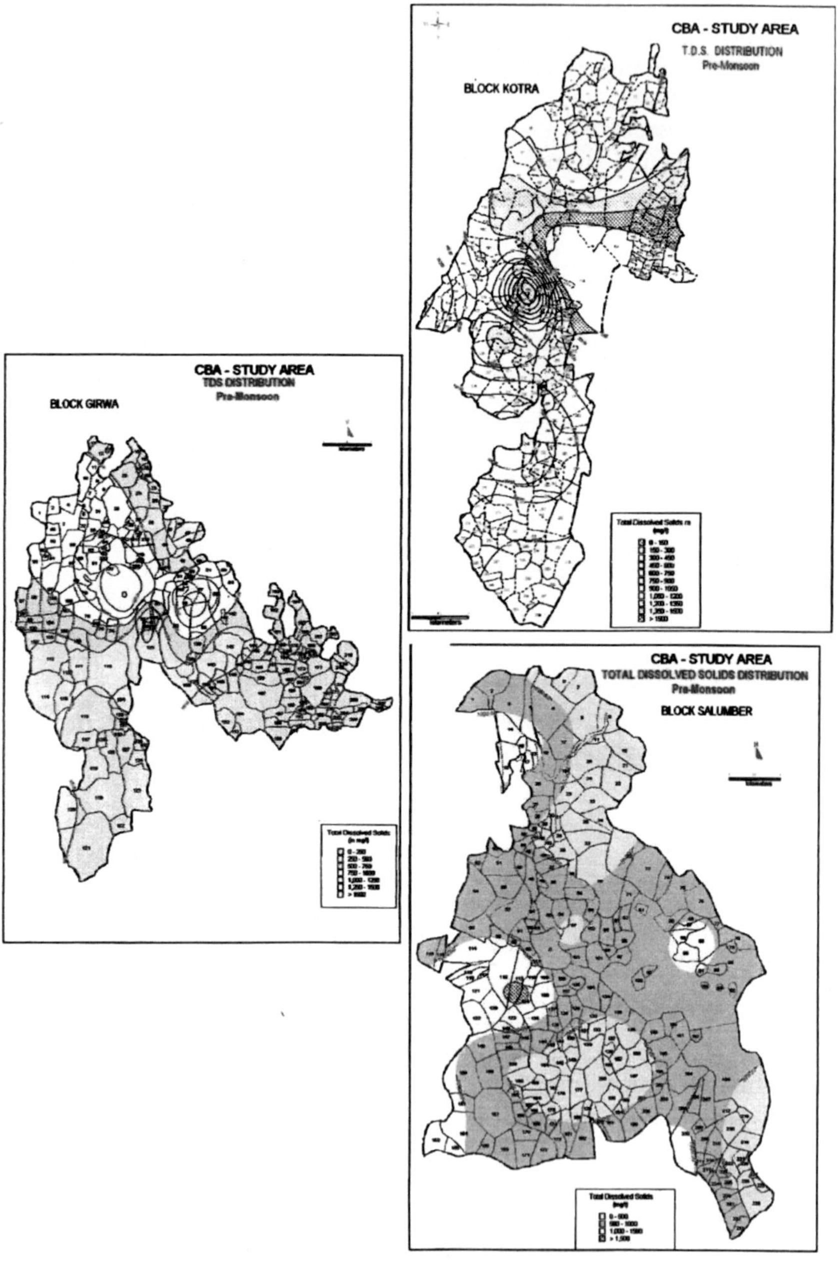

Figure 11.6.11 TDS Distribution

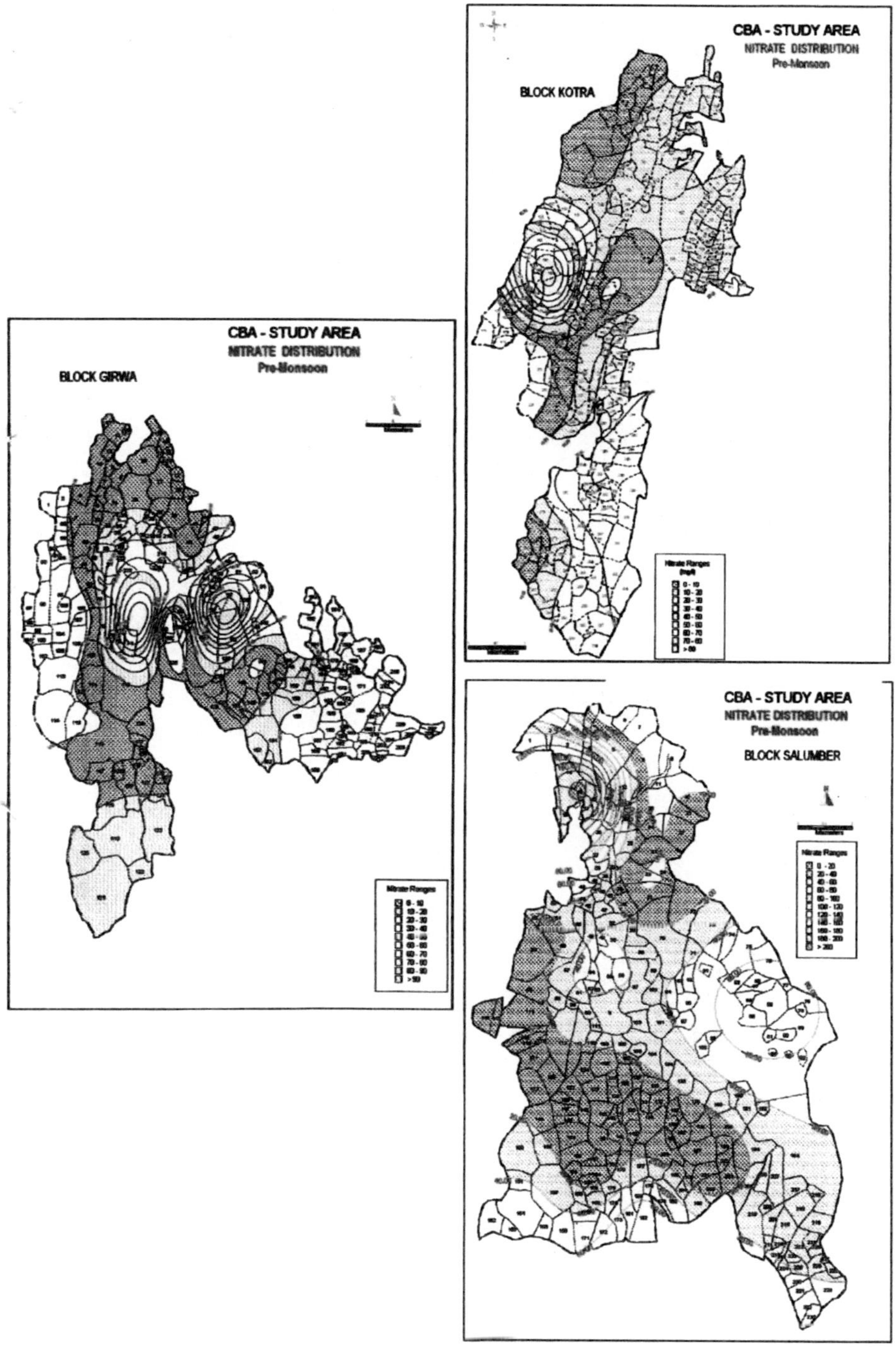

Figure 11.6.12 Nitrate Distribution

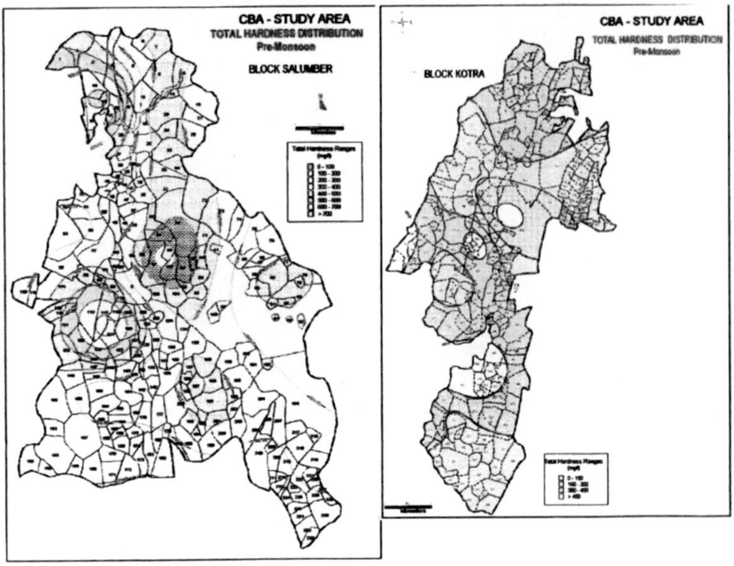

Figure 11.6.13 Total Hardness Distribution

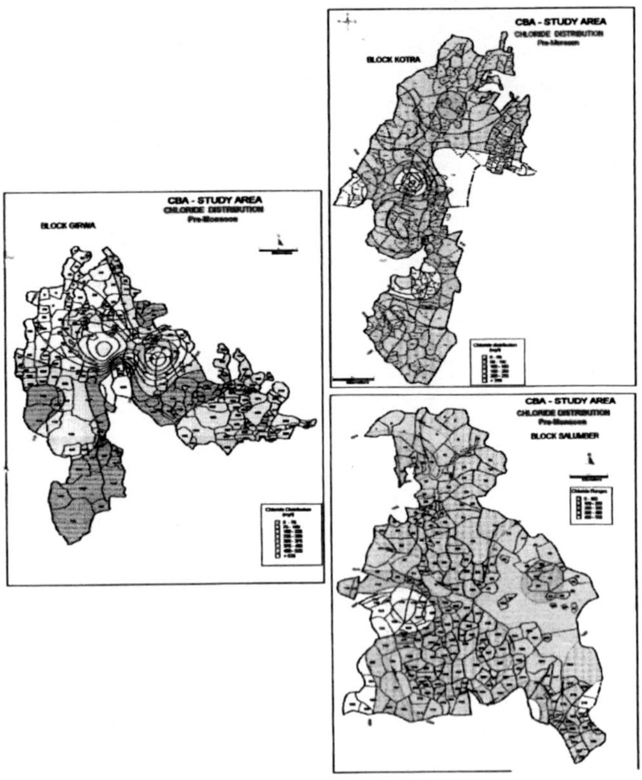

Figure 11.6.14 Chloride Distribution

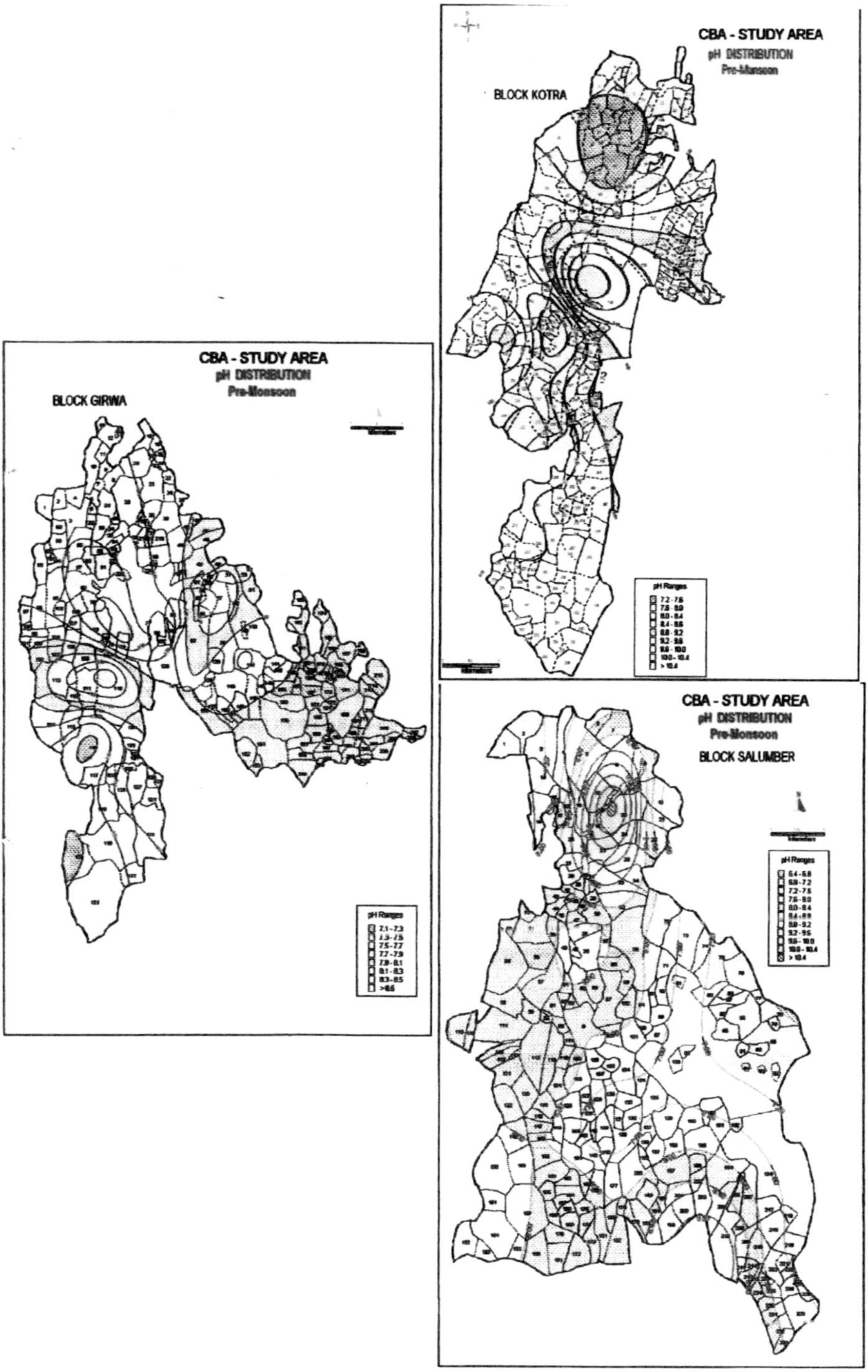

Figure 11.6.15 pH Distribution

Table 11.6.5 Population Distribution

S. No	Tehsil	Total Population - 1991	Male	Female	Schedule Tribe	Schedule Cast
1	Kotra	130449	66743	63706	116263	1724
2	Girwa	557694	293598	264096	141524	8969
3	Salumbar	175611	87933	87678	86852	8086

Source: Census of India – 1991

11.6.3.6 *Cropping Pattern*

The cropping pattern followed in the three blocks in Rabi and Kharief region is as under:

Table 11.6.6 Rabi Season

S. No.	Crop	KOTRA		GIRWA		SALUMBAR	
		Irrigated (ha)	Non-Irrigated	Irrigated. (ha).	Non-Irrigated	Irrigated. (ha)	Non-Irrigated
1	Wheat	328	5	2248	61	2027	14
2	Barley	3	-	272	38	475	28
3	Gram	1	9	25	92	145	71
4	Mustered	20	48	130	57	66	22
5	Coriander	-	-	49	1	-	-
6	Vegetables	-	-	102	-	10	-
…..	………….	….	….	….	…..	…..	…..
	Total	356	63	3138	270	2827	140

Table 11.6.7 Kharief (Rainy) Season

S. No.	Crop	KOTRA		GIRWA		SALUMBAR	
		Irrigated (ha)	Non-Irrigated	Irrigated. (ha).	Non-Irrigated	Irrigated. (ha)	Non-Irrigated
1	Maize	7	14048	162	19575	99	18765
2	Jowar	-	2	-	887	-	9
3	Bajara	-	-	-	1	-	-
4	Small Millet	-	5	-	2	-	51
5	Gram	-	96	-	2	-	2
6	Pulses	-	1889	-	294	-	1225
7	Sesamum	-	112	-	143	-	528
8	Groundnut	-	7	-	69	-	-
9	Sugarcane	14	-	202	12	5	-

10	Cotton	-	2	-		-	-
11	Chilies	-	-	12	108	6	27
12	Vegetables	-	3	253	53	12	-
…….	………..	………	………	……	….	……	……..
	Total	48	18269	739	22817	135	24418

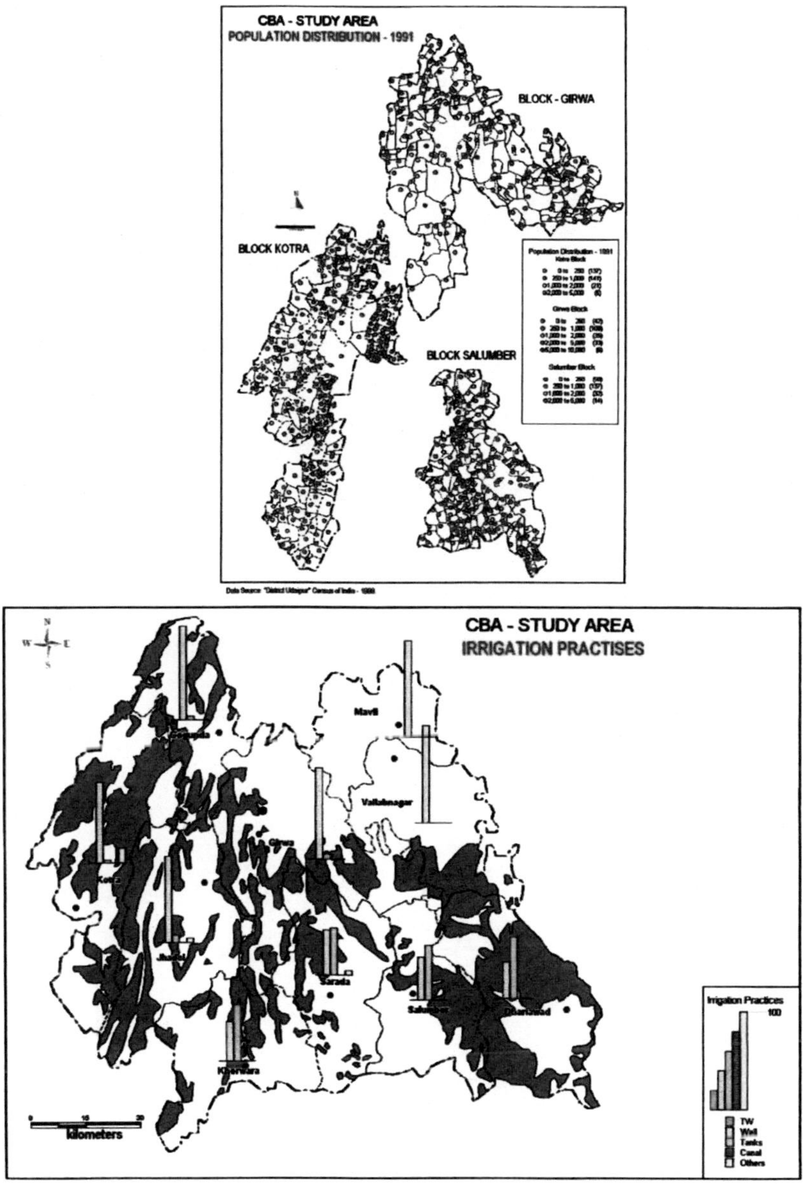

Figure 11.6.16 Population Distribution and Irrigation Methods

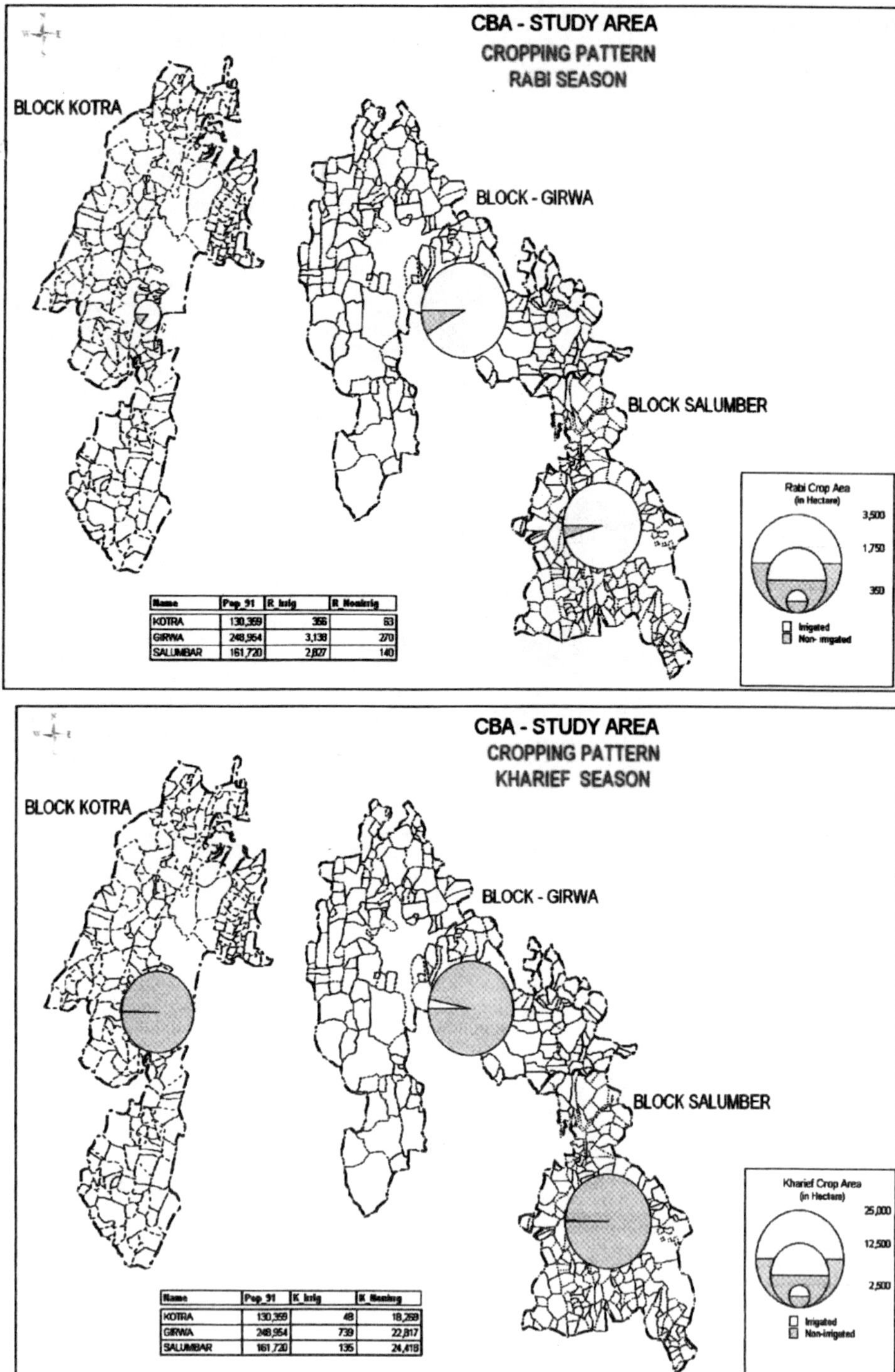

Figure 11.6.17 Cropping Pattern – Rabi and Kharief Seasons

11.6.3.7 *Cost Benefit Analysis*

Though the cost involved in the construction of the structures has been utilised to generate the employment and thus distributing money to the villagers suffering from sever drought situation since last three consecutive years, they had been an additional asset for the villages. In fact the site selection, type of construction was done on scientific and real ground conditions hence there seems no doubt about he functioning of the structure for what they were constructed. This study was mainly focused on the aspect of water availability in the form of surface or groundwater. The extent of that water availability was the basis for judging the benefit aspect of this study. The study was divided in two parts i.e., direct benefits and indirect benefits in the area after construction of the water harvesting structure. Catchment area has been calculated through the drainage map of the site and reservoir capacity of the structure constructed across the drain has been calculated through field survey. Based on these field data an analysis has been performed to work out the quantum of water for surface water and ground water before and after construction of the structure. Each site has been analysed individually and the mathematical modeling exercise has been performed to find out the quantum of water available for storage and recharge. The site data map and analysis data sheet gives a friendly user and simple computer based mathematical model to find out the quantum of water available for surface storage and groundwater recharge based on the climatological parameters and catchment characteristics.

11.6.3.7.1 *Direct Benefits*

Surface water used for irrigation for crops and drinking for human beings and cattle population of the affected cluster of the village generate the Direct Benefits from the construction of the Water Harvesting Structure. Since the water was available in small quantum as compared to the quantity required for the Rabi crops; only the impact on Kharif crop had been studied. Optimum Water use has been discussed considering the fact that drinking, milk producces and crop production, all the three factors are very important for a villager and proportionate water use in these fields generate most favorable and optimum water use of the stored surface water. Each site has been discussed in details individually and a comparison note has been prepared for all sites.

11.6.3.7.2 Indirect Benefits

Stored water will percolate in the ground and one part of it will meet to the ground water storage and second will increase the moisture contents of the soil and will be available for plants/ crops for their use. There are number open wells in the d/s of the sites, which are used as irrigation and drinking water source in the post monsoon periods. The water joined to ground water storage will be available for irrigation and drinking from these outlet structures (i.e. wells, hand pumps etc.). The water increasing the moisture contents will be utilized by the plants and grasses, which will generate the crop production and fodder production for cattle.

The overall benefit in terms on Rs comes out to be Rs. 3,35,222.00. The induced recharge has more significantly impact, which cannot be measured in term of monetary scale. One of the examples is the improvement in the quality of the ground water. Since the rainwater (or here) stored water is almost pure water in terms of chemical composition when it joins deteriorated ground water, dilutes the chemical contents. Thus it improves the quality of ground water which infect is generally deteriorating day by day as ground water table is depleting in the absence of the recharge. Second impact that is also significant is the change in the flora and fauna of the area as the available moisture contents in the soil increases. This improves overall environment and create healthy atmosphere in the nearby catchment site.

11.6.4 Study Villages

11.6.4.1 *Village - JOYRA NAKA*

Joyra Naka is situated in village Bassi-Jhunjhawat in Salumbar Block of Udaipur district. The site is situated at the northeastern part of the village. The site forms a small watershed of aerial extent of 14.52 ha. The upstream area consists of gentle sloping terrain with small hillocks and d/s area consists of agricultural area of about 20.0 ha. An old broken earthen construction exist at the construction site. The structure of Loose Stone Masonry with earthen support was constructed to store water for water harvesting.

11.6.4.1.1 *Pre-construction Status*

The Rainwater falling in the water shed was flowing without stoppage through the catchment to a pond at the 1 km downstream. The flow water was causing loss of precious rainwater and also fertile soil erosion from the agricultural fields. Since there is no structure to stop the water, there is no storage of water in the catchment, hence no surface water available during rainy, winter and summer season for drinking as well as irrigation. The ground water table was also depleting day by day and in general there is no water in the open well that exists in the catchment during winter and summer season. Quality of water in general is within the standards of the WHO so there was no problem as far as quality parameters are concerned. Since water available for irrigation is only the monsoon rainfall hence the crop production is very much limited to the Kharif crops and the production varies from 0.2 to 3.6 tons per hectare depending upon bad and good monsoon rainfall respectively. Generally Maize is sown in the season.

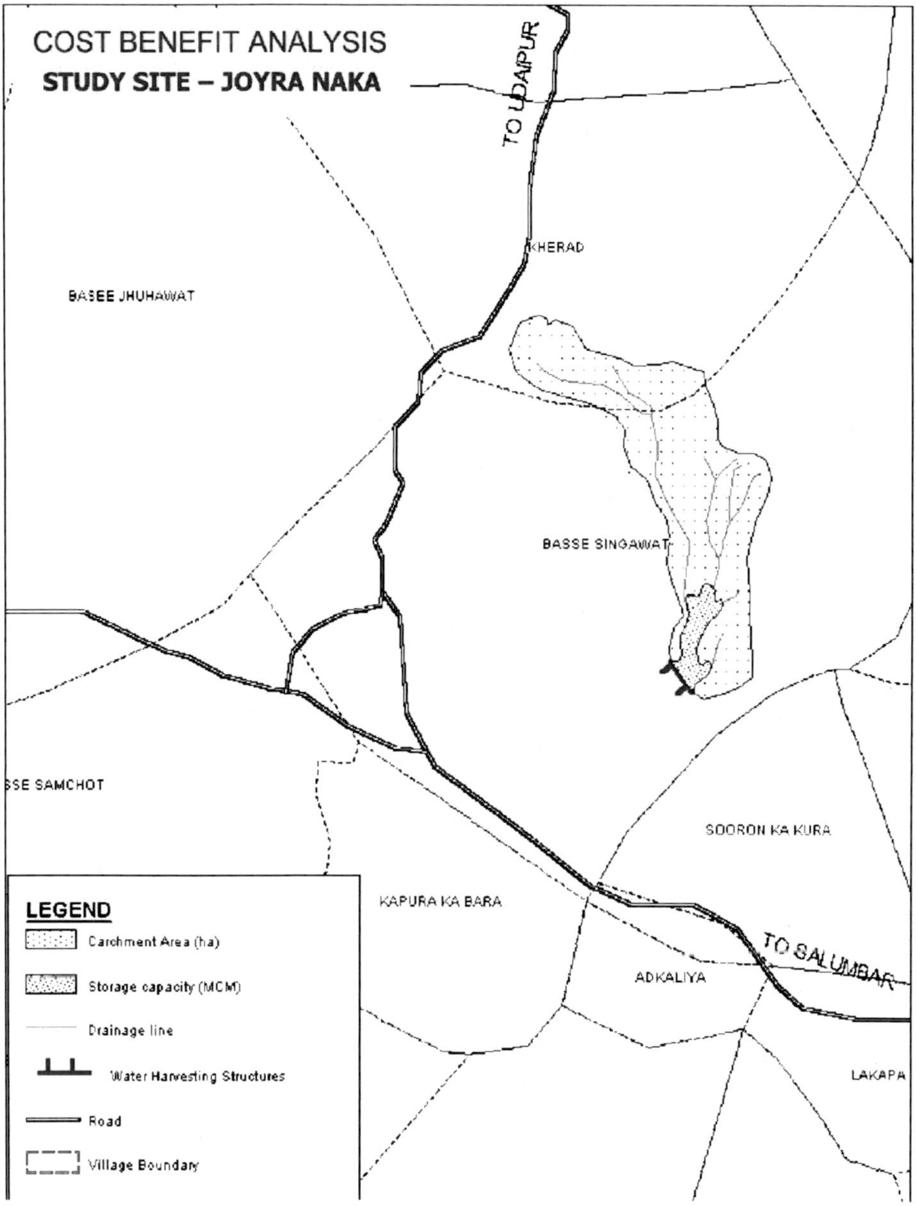

Figure 11.6.18.

11.6.4.1.2 *Post-Construction Status*

The Rain water falling in the water shed now flow through the drains to the artificial reservoir formed due to the construction of water harvesting structure, causing water to store there to a capacity of 0.5 ha-m. Surplus water passes through the structure and joins a pond in the d/s.

Since the catchment has been treated right from the up stream thus making much less the rate of flow of water. That causes more infiltration of water in the ground causing increasing in the ground water. The water flowing on the surface is stored in the reservoir the making available water to the human and cattle. The water will have some losses due to the evaporation and percolation in the ground. Rest of the water is available as surface water. Surface water availability in terms of quantity is estimated as 5000 cum. Study reveals that Surface water if utilized for drinking only would be available for 2 months to the population of the human as well as cattle living near the site. If some part of the water is utilized for irrigation than the water availability period will be 1.59 months or 47 days. Ground water will also be available drinking as well as irrigation in winter season. The recharge due to delayed flow as well as reservoir will impact favorable to the recharge of the groundwater. Hence the water will be available in the wells in the winter season also. As the water is available in reservoir and also in wells during monsoon season hence if there is no rainfall during the last watering to the crop the same could be meet out through the irrigation from the stored water sources. Hence, at least Kharif crop is assured. This will cause much increase in the crop production. Also flora and fauna will be developed due to presence of the moisture in the soil. Production of crop could be yield up to 17.60 ton per year.

11.6.4.1.3 *Cost Benefit Analysis*

1. Construction Cost:		**Rs. 84490.00**
2. Direct Benefits:		
Beneficiaries:	Human Population:	500
	Cattle Population:	100
Case a: Surface Water Irrigation use:		
Increased Agricultural Production by 17.60 Tons		**(Rs. 79.200.00)**
Case b: Surface Drinking Water use:		
Drinking water (available for 59 days)		(Rs. 15000.00)
Dairy produces		(Rs. 148980.00)
Total benefits		**(Rs. 163980.00)**
Case c: Irrigation and Drinking Water use:		
Drinking water (available for 47 days)		(Rs. 12000.00)
Surface water irrigation		(Rs. 79200.00)
Dairy produces		(Rs. 119184.00)
Total Optimum Benefits		**(Rs. 210383.00)**
3. Indirect Benefits: (Groundwater recharge)		
Increased Crop Production + Drinking water =		**Rs. 38340.00**

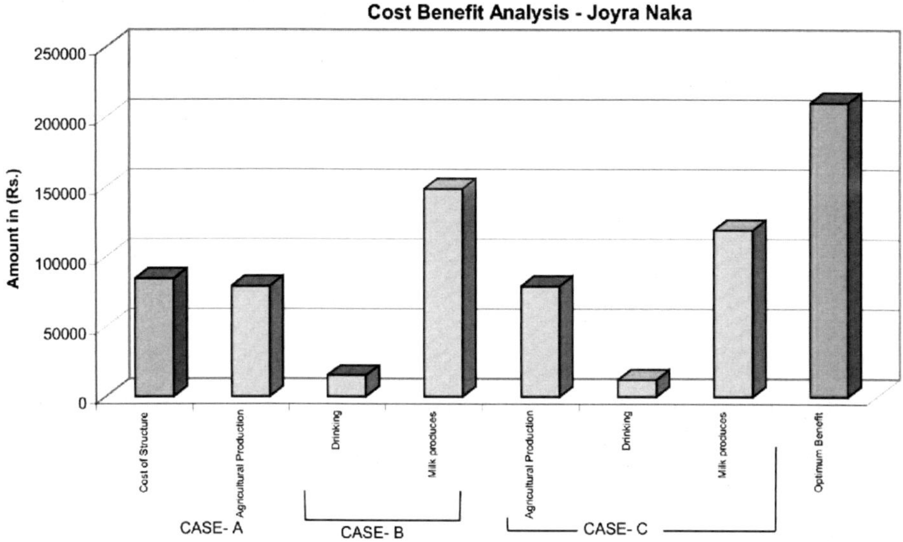

Figure 11.6.19

The Figure 11.6.19 depicts the overall benefits of the construction at site as an impact of construction of the structure. It is clear from the graph that in a single year the optimum direct benefits achieved will be about **2.5 times** the cost of construction. Spending a small amount in maintenance the villagers could derive benefits from the structure to considerable years.

11.6.4.2 *Village - HINGAWAT NAL (GAONRA PAL)*

Hingawat Nal is situated in village Ganora Pal in Salumbar Block of Udaipur district. The site is situated at the Southwestern part of the village. The site forms a medium watershed of areal extent of 70.80 ha. The upstream area consists of gentle sloping terrain with hillocks and d/s area consists of agricultural area of about 250.0 ha. The structure of small anicut of stone Masonry was constructed to store water for water harvesting.

11.6.4.2.1 *Pre-construction Status*

The Rainwater falling in the water shed was flowing without stoppage through the catchment to river at the downstream. The flow water was causing loss of precious rainwater and also fertile soil erosion from the agricultural fields. Since there is no structure to stop the water, there is no storage of water in the catchment, there is no surface water available during rainy, winter and summer season for drinking as well as irrigation. The ground water table was also depleting day by day and in general there is little water in the open well that exists in the catchment during winter and summer season. No surface water is available for drinking for human being as well as cattle population. Also the well remains with water for very short duration hence ground water is also limited.

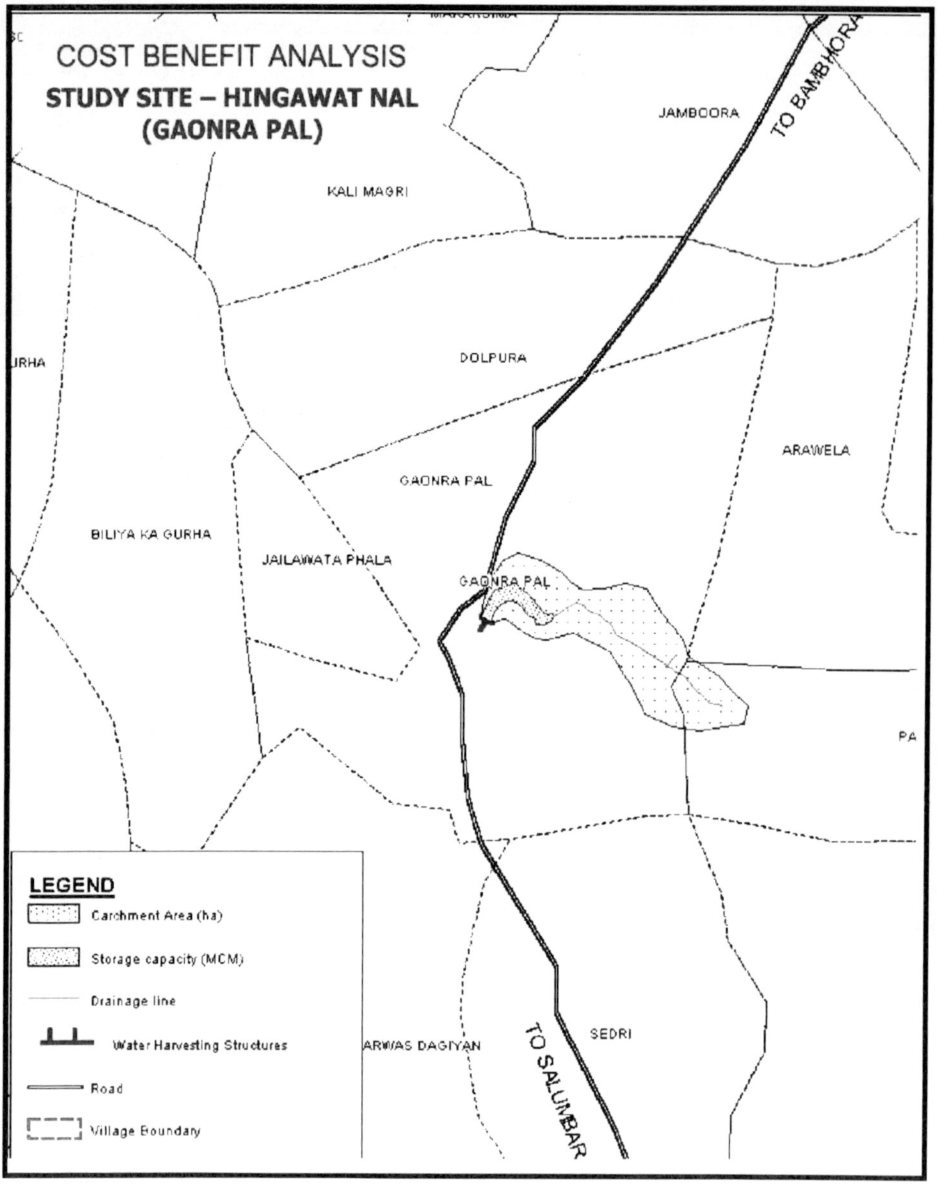

Figure 11.6.20.

Quality of water in general is within the standards of the WHO so there was no problem as far as quality parameters are concerned. Since water available for irrigation is only the monsoon rainfall hence the crop production is very much limited to the Kharif crops and the production of the crop varies from 0.2 to 3.6 tons per hectare depending upon bad and good monsoon rainfall respectively. Generally Maize is sown in the season.

11.6.4.2.2 *Post-construction Status*

The Rain water falling in the water shed is now flowing through the drains to the artificial reservoir formed due to the construction of water harvesting structure, causing water to store there to a capacity of 1.8 ha.m making available the surface water to the human and cattle. Surplus rainwater passes through the structure and joins a river in the d/s. Study reveals that Surface water if utilized for drinking only would be available for 2.5 months to the population of the human as well as cattle living near the site. If some part of the water is utilized for irrigation than the water availability period will be 1.37 months or 41 days. The recharge due to delayed flow as well as reservoir will impact favorable to the recharge of the groundwater. Hence the water will be available in the wells in the summer season also as the water is available in reservoir and also in well during monsoon season hence if there is no rainfall in last watering to the crop the same could be meet through the irrigation from the stored water sources, hence, at least Kharif crop is assured. This will cause much increase in the crop production. Also flora and fauna will be developed due to presence of the moisture in the soil. Production of crop could be yield upto 220.00 ton per year.

11.6.4.2.3 Cost Benefit Analysis

1. Construction Cost: **Rs. 136600.00**
2. Direct Benefits:

Beneficiaries:	Human Population:	700
	Cattle Population:	300

Case a: Surface Water Irrigation use:

Increased Agricultural Production by 220 Tons	(Rs. 990,000.00)

Case b: Surface Drinking Water use:

Drinking water available for 59 days	(Rs. 54000.00)
Dairy produces	(Rs. 571139.00)
Total benefits	**(Rs. 625139.00)**

Case c: Irrigation and Drinking Water use:

Drinking water available for 41 days	(Rs. 29250.00)
Surface water irrigation	(Rs. 990000.00)
Dairy produces	(Rs. 309367.00)
Total optimum benefits	**(Rs. 1328616.00)**

2. Indirect Benefits: (Groundwater recharge)

Increased Crop Production + Drinking water = **Rs. 50179.00**

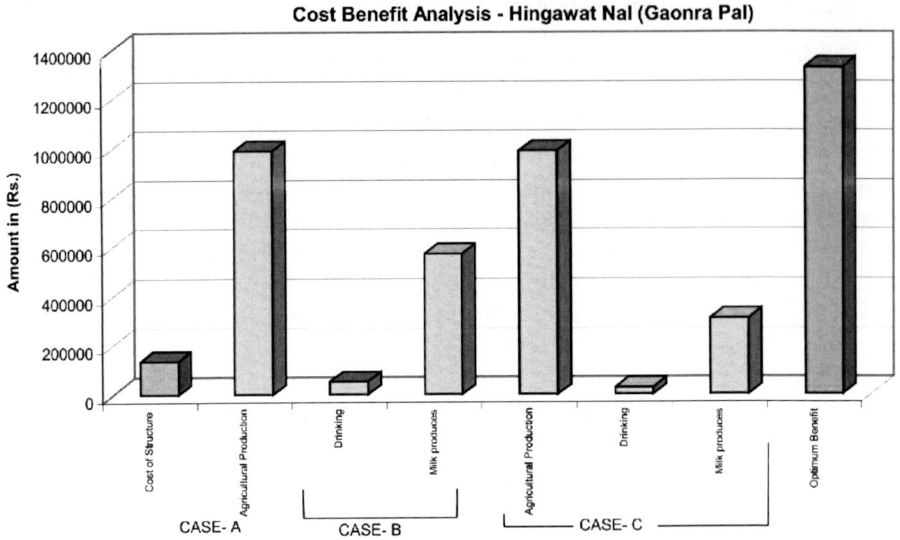

Figure 11.6.21

The Figure 11.6.21 depicts the overall benefits of the construction at site as an impact of construction of the structure. It is clear from the graph that in a single year the optimum direct benefits achieved will be about 9.72 times the cost of construction. Spending a small amount in maintenance the villagers could derive benefits from the structure to considerable years.

11.6.4.3 *Village - Hama Talav*

Hama Talav is situated in village Bassi Singawat in Salumbar Block of Udaipur district. The site is situated at the Southern part of the village. The site forms a watershed of areal extent of 42.38 ha. The upstream area consists of gentle sloping terrain with hillocks and d/s area consists of agricultural area of about 75.0 ha. The structure of Loose Stone Masonry with earthen support was constructed to store water for water harvesting.

11.6.4.3.1 *Pre-construction Status*

The Rainwater falling in the watershed flow without stoppage through the catchment to the downstream. The flow water was causing loss of precious rainwater and also fertile soil erosion from the fields. As there is no structure to stop the rainwater, no surface water is available during rainy, winter and summer season for drinking as well as irrigation. The ground water table was also depleting day by day and in general, there is no water in open wells that exists in the catchment during winter and summer season.

Quality of water in general is within the standards of the WHO so there was no problem as far as quality parameters are concerned.

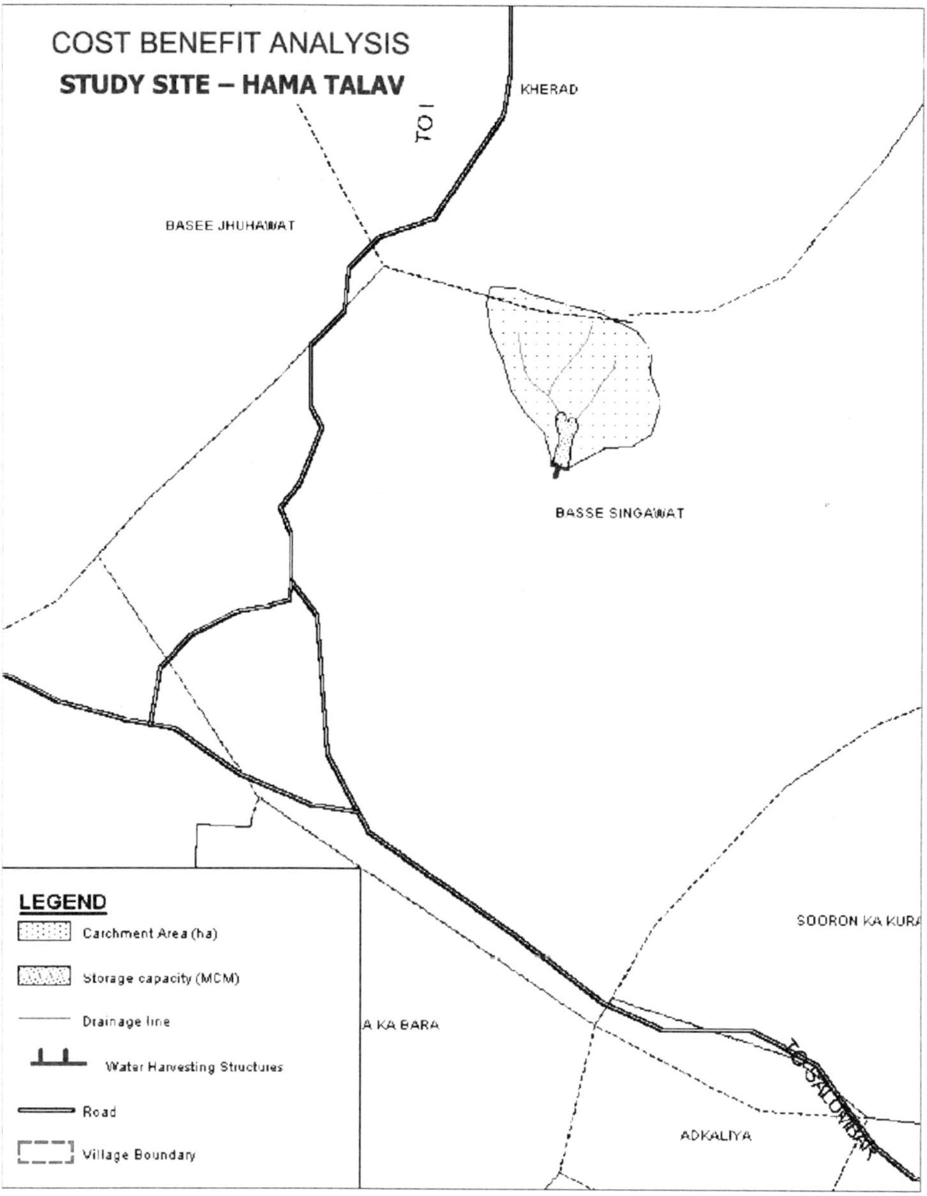

Figure 11.6.22

Since water available for irrigation is only the monsoon rainfall hence the crop production is very much limited to the Kharif crops and the production of the crop varies from 0.2 to 3.6 tons per hectare depending upon bad and good monsoon rainfall respectively. Generally Maize is sown in the season.

11.6.4.3.2 *Post-Construction Status*

The Rain water falling in the water shed now drain into the artificial reservoir formed due to the construction of water harvesting structure, causing water to store there to a capacity of 0.5 ha.m. Surplus water passes through the structure and joins a pond in the d/s. Since the catchment has been treated right from the up stream thus making much less the rate of flow of water. That causes more infiltration of water in the ground causing increasing in the ground water recharge. The water will have some losses due to the evaporation and percolation in the ground. Rest of the water is available as surface water. Surface water availability in terms of quantity is estimated as 5000 cum. Study reveals that Surface water if utilized for drinking only would be available for 2.1 months to the population of the human as well as cattle living near the site. If some part of the water is utilized for irrigation than the water availability period will be 1.32 months or 40 days. The recharge due to delayed flow as well as reservoir will impact favorable to the recharge of the groundwater. Hence ground water will also be available drinking as well as irrigation in winter season. As the water is available in reservoir and also in well during monsoon season hence if there is no rainfall in last watering to the crop the same could be meet through the irrigation from the stored water sources, hence, at least Kharif crop is assured. This will cause much increase in the crop production. Also flora and fauna will be developed due to presence of the moisture in the soil. Production of crop could be yield upto 66.00 ton per year.

11.6.4.3.3 Cost Benefit Analysis

1. Construction Cost: **Rs. 263600.00**

2. Direct Benefits:

Beneficiaries: Human Population: 000

Cattle Population: 200

(Assuming water is available for cattle only as the structure fall in the charagah land protected for cattle feeding)

Case a: Surface Water Irrigation use:

Increased Agricultural Production by 66 Tons (Rs. 297000.00)

Case b: Surface Drinking Water use:

Drinking water available for 59 days (Rs. 15000.00)

Dairy produces (Rs. 316840.00)

Total benefits (Rs. 331840.00)

Case c: Irrigation and Drinking Water use:

Drinking water available for 40 days (Rs. 9375.00)

Surface water irrigation (Rs. 198025.00)

Dairy produces (Rs. 297000.00)

Total benefits **(Rs. 504400.00)**

2. Indirect Benefits:

Increased Crop Production + Drinking water = **Rs. 20124.00**

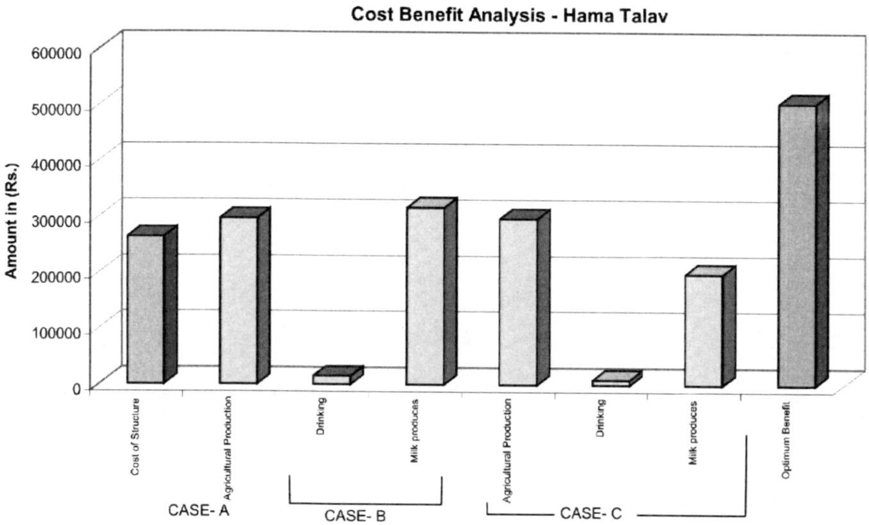

Figure 11.6.23

The Figure 11.6.23 depicts the overall benefits of the construction at site as an impact of construction of the structure. It is clear from the graph that in a single year the optimum direct benefits achieved will be about 1.91 times the cost of construction. Spending a small amount in maintenance the villagers could derive benefits from the structure to considerable years.

11.6.4.4 *Khedia (Semliya)*

Khedia is situated in village Semliya in Girwa Block of Udaipur district. The site is situated at the western part of the village. The site forms a small watershed of areal extent of 16.47 ha. The upstream area consists of gentle sloping terrain with small hillocks and d/s area consists of agricultural area of about 125.0 ha. A structure of Loose Stone Masonry with earthen support was constructed to store water for water harvesting.

11.6.4.4.1 *Pre-construction Status*

The Rainwater falling in the water shed was flowing without any stoppage through the catchment to a pond at the 2 km downstream. The flow water was causing loss of precious rainwater and also fertile soil erosion from the agricultural fields. There was no surface water storage in the catchment, hence no surface water available during rainy, winter and summer season for drinking as well as irrigation purposes. The ground water table was also depleting day by day. In general there is no water in the open wells that exists in the catchment during winter and summer season.

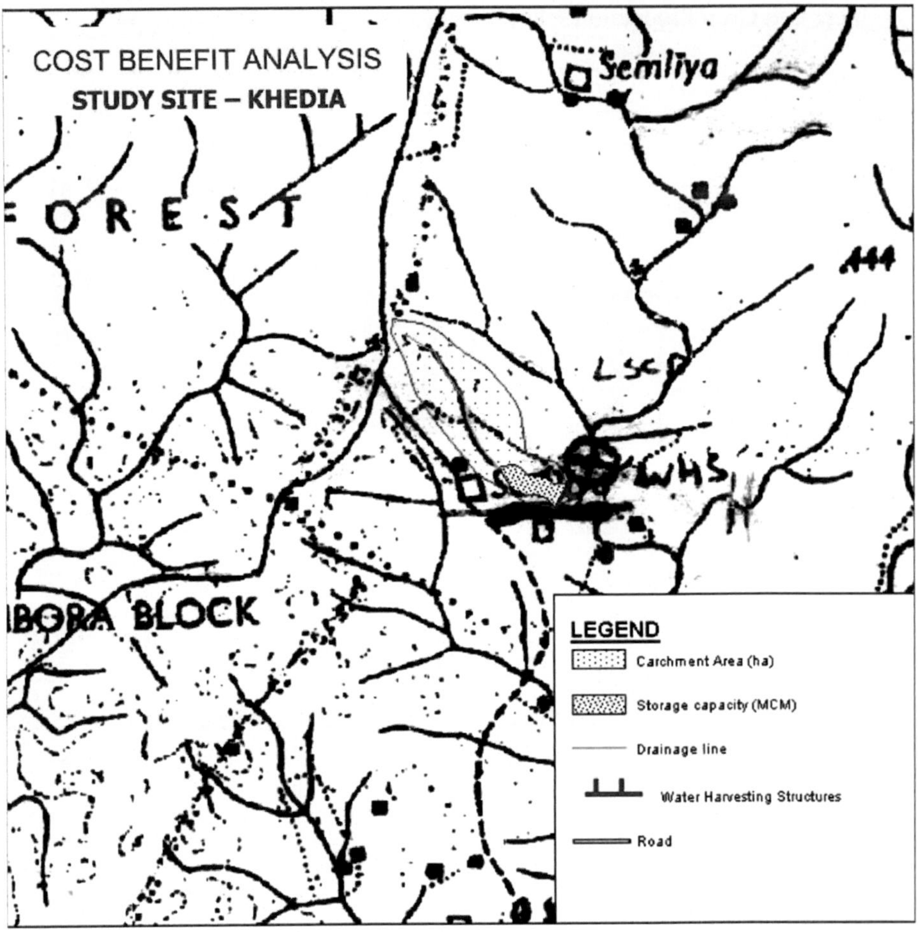

Figure 11.6.24

Quality of water in general is within the standards of the WHO so there was no problem as far as quality parameters are concerned. Since water available for irrigation is only the monsoon rainfall hence the crop production is very much limited to the Kharif crops and the production of the crop varies from 0.2 to 3.6 tons per hectare depending upon bad and good monsoon rainfall respectively. Generally Maize is sown in the season.

11.6.4.4.2 *Post-Construction Status*

The Rainwater falling in the water shed now flows through the drains to the reservoir formed due to the construction of water harvesting structure, causing water to store to a capacity of 0.6 ha.m. Surplus water passes through the structure and joins a pond in the d/s. Surface water availability in terms of quantity is estimated as 6000.00 cum. The water will have some losses due to the evaporation and percolation in the ground. Rest of the water is available as surface water The recharge due to delayed flow as well as reservoir will impact favorable to the recharge of the groundwater. Hence the water will be available in the wells in the winter season also. Study reveals that Surface water if utilized for drinking only would be available for 2 months to the population of the human as well as cattle living near the site. If some part of the water is utilized for irrigation than no water will be available for drinking. Ground water will be available drinking as well as irrigation in winter season. As the water is available in reservoir and also in well during monsoon season hence if there is no rainfall in last watering to the crop the same could be meet through the irrigation from the stored water sources, hence, at least Kharif crop is assured. This will cause much increase in the crop production. Also flora and fauna will be developed due to presence of the moisture in the soil. Production of crop could be yield upto 23.75 ton per year.

11.6.4.4.3 *Cost Benefit Analysis*

1. Construction Cost: **Rs. 44915.00**

(Total cost of works done in the village is Rs. 211914 but Water harvesting structure construction costing Rs. 44915.00 is in separate catchment from the rest of work area. Hence only the cost of WHS is being considered, as it will generate benefits without affecting other area.

2. Direct Benefits:

Beneficiaries: Human Population:	390
Cattle Population:	150
Case a: Surface Water Irrigation use:	
Increased Agricultural Production by 23.75 Tons	(Rs. 106875.00)
Case b: Surface Drinking Water use:	
Drinking water available for 61 days	(Rs. 18000.00)
Dairy produces	(Rs. 229627.00)
Total benefits	(Rs. 247627.00)
Case c: Irrigation and Drinking Water use:	
Drinking water available for 0 days	(Rs. 0.00.00)
Surface water irrigation	(Rs. 106875.00.00)
Dairy produces	(Rs. 0.00)
Total Optimum benefits	**(Rs. 106875.00)***

(*There are already wells for drinking in the village where water is available throughout the summer season hence drinking aspect is not considered in the analysis.)

2. Indirect Benefits:

Increased Crop Production + Drinking water = **Rs. 20969.00**

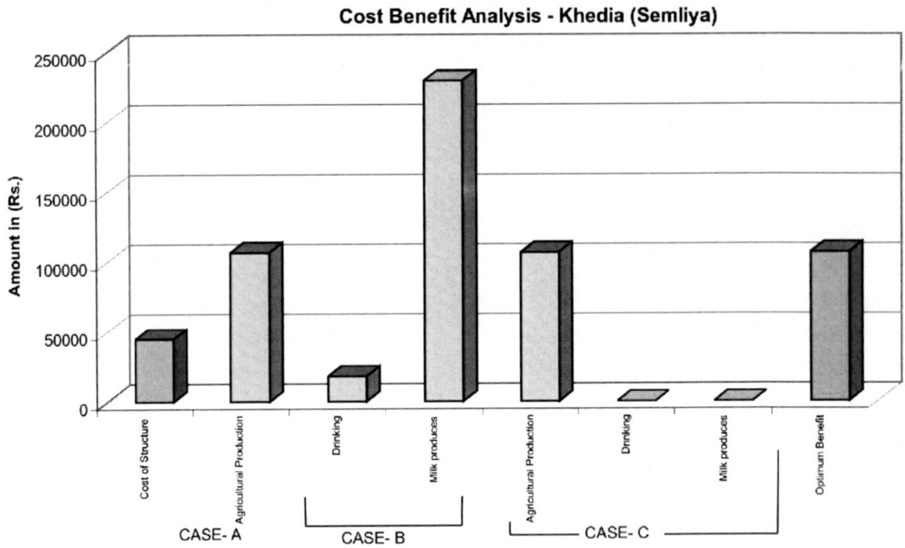

Figure 11.6.25

The Figure 11.6.25 depicts the overall benefits of the construction at site as an impact of construction of the structure. It is clear from the graph that in a single year the optimum direct benefits achieved will be about 2.38 times the cost of construction. Spending a small amount in maintenance the villagers could derive benefits from the structure to considerable years.

11.6.4.5 *Village - SIRWAL (DANG)*

Sirwal is situated in village Dang in Kotra Block of Udaipur district. The site is situated at the Northern part of the village. The site forms a small watershed of areal extent of 47.00 ha. The upstream area consists of steep sloping terrain with hillocks and d/s area consists of agricultural area of about 75.0 ha. A structure of Loose Stone Masonry with earthen support was constructed to store water for water harvesting.

11.6.4.5.1 *Pre-construction Status*

The Rainwater falling in the watershed flows without stoppage through the catchment to a drain at downstream. The flow of water was causing loss of precious rainwater and also fertile soil erosion from the agricultural fields. Since there is no structure to stop the water, there is no storage of water in the catchment, hence no surface water available during rainy, winter and summer season for drinking as well as irrigation. The ground water table was also depleting day by day and in general there is no water in the open well that exists in the catchment during winter season only.

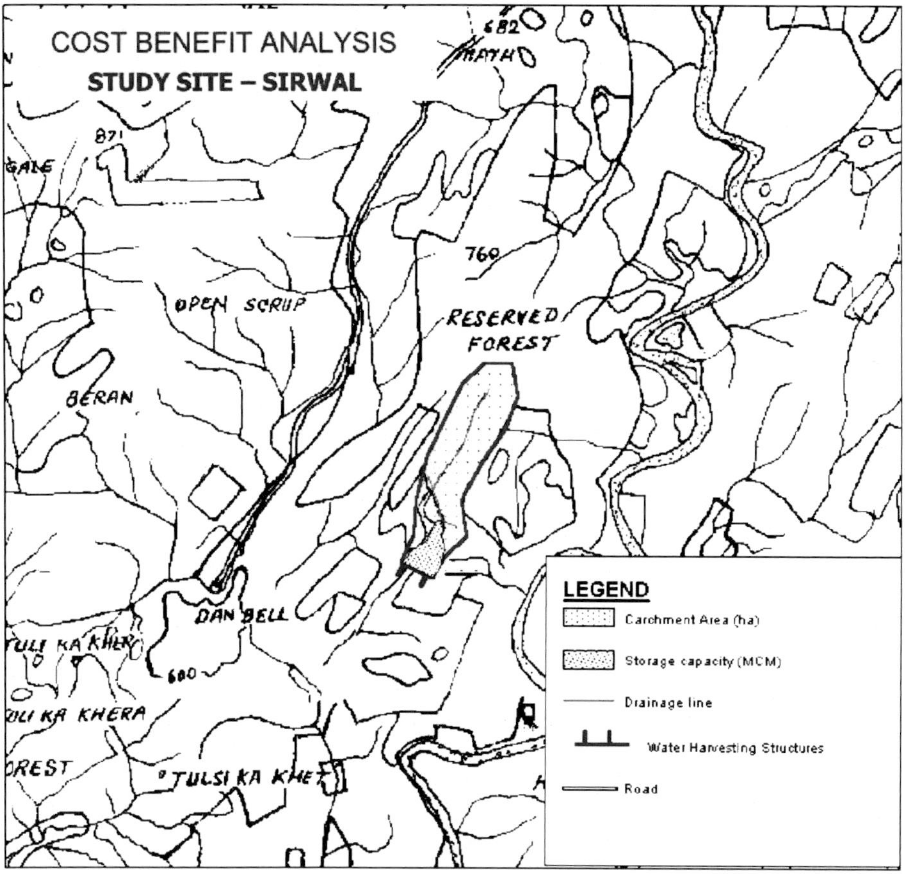

Figure 11.6.26.

Quality: Quality of water in general is within the standards of the WHO so there was no problem as far as quality parameters are concerned. Since water available for irrigation is only the monsoon rainfall hence the crop production is very much limited to the Kharif crops and the production of the crop varies from 0.2 to 3.6 tons per hectare depending upon bad and good monsoon rainfall respectively. Generally Maize is sown in the season.

11.6.4.5.2 *Post-Construction Status*

The Rainwater falling in the water shed now flows through the drains to the reservoir formed due to the construction of water harvesting structure, causing water to store there up to a capacity of 0.5 ha.m. Surplus water passes through the structure and joins a pond in the d/s. Surface water availability in terms of quantity is estimated as 5000.00 cum. Study reveals that Surface water if utilized for drinking only would be available for 2.2 months to the population of the human as well as cattle living near the site. If some part of the water is utilized for irrigation than the water availability period will be 1.35 months or 40 days.

Since the catchment has been treated right from the up stream thus making much less the rate of flow of water. That causes more infiltration of water in the ground causing increasing in the ground water recharge. The recharge due to delayed flow as well as reservoir will impact favorable to the recharge of the groundwater. Hence the water will be available in the wells in the winter season also as the water is available in reservoir and also in well during monsoon season hence if there is no rainfall in last watering to the crop the same could be meet through the irrigation from the stored water sources, hence, at least Kharif crop is assured. This will cause much increase in the crop production. Also flora and fauna will be developed due to presence of the moisture in the soil. Production of crop could be yield up to 66.00 ton per year.

11.6.4.5.3 *Cost Benefit Analysis*

1. Construction Cost: **Rs. 127170.00**
2. Direct Benefits

Beneficiaries:	Human Population:	270
	Cattle Population:	100

Case a: Surface Water Irrigation use:	
Increased Agricultural Production by 66 Tons	(Rs. 297000.00)
Case b: Surface Drinking Water use:	
Drinking water available for 61 days	(Rs. 15000.00)
Dairy produces	(Rs. 168228.00)
Total benefits	(Rs. 183228.00)
Case c: Irrigation and Drinking Water use:	
Drinking water available for 40 days	(Rs. 9000.00)
Surface water irrigation	(Rs. 297000.00.00)
Dairy produces	(Rs. 100937.00)
Total benefits	**(Rs. 406936.00)**

2. *Indirect Benefits:*

Increased Crop Production + Drinking water = **Rs. 18112.00**

The Figure 11.6.27 depicts the overall benefits of the construction at site as an impact of construction of the structure. It is clear from the graph that in a single year the optimum direct benefits achieved will be about 3.2 times of cost of construction. Spending a small amount in maintenance the villagers could derive benefits from the structure to considerable years.

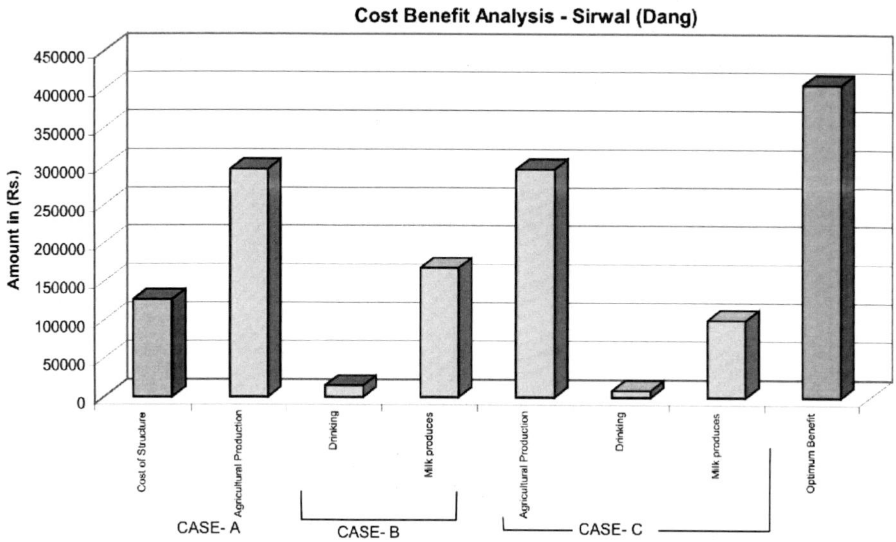

Figure 11.6.27

11.6.4.6 *Village - CHARNOT LAND (PATOO KHERA)*

Charnot Land is situated in village Patoo Khera in Girwa Block of Udaipur district. The site is situated at the eastern part of the village. The site forms a medium watershed of areal extent of 91.46 ha. The upstream area consists of gentle sloping terrain with hillocks and d/s area consists of agricultural area of about 200.0 ha. An anicut structure of stone masonry along with Loose Stone Masonry dam with earthen support was constructed to store rainwater.

11.6.4.6.1 *Pre-construction Status*

The Rainwater falling in the water shed was flowing without stoppage through the catchment to river at the downstream. The flow water was causing loss of precious rainwater and also fertile soil erosion from the agricultural fields. There was no surface water available during rainy, winter and summer season for drinking as well as for irrigation. The ground water table was also depleting day by day and in general there was a little water in the open well that exists in the catchment during winter and summer season. Last season the villagers were dependent only on a single well, which fed all the human and cattle population. There was no water for irrigation in last year crop period. Quality of water in general is within the standards of the WHO so there was no problem as far as quality parameters are concerned. Since water available for irrigation is only the monsoon rainfall hence the crop production is very much limited to the Kharif crops and the production of the crop varies from 0.2 to 3.6 tons per hectare depending upon bad and good monsoon rainfall respectively.

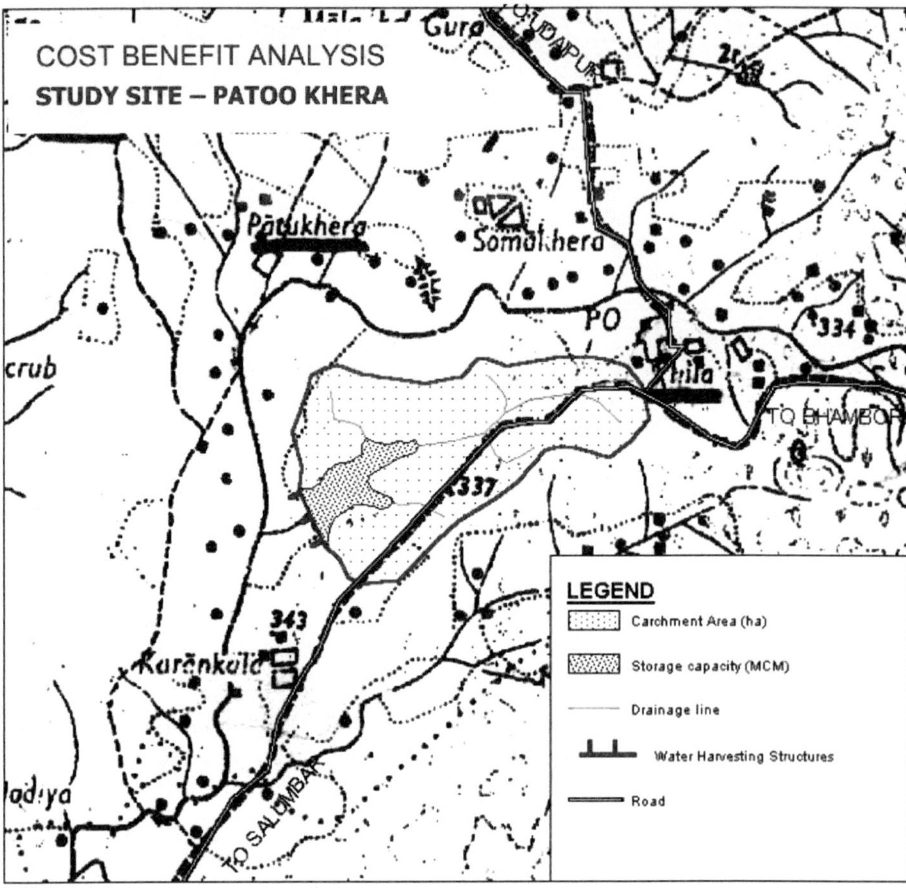

Figure 11.6.28.

11.6.4.6.2 *Post-Construction Status*

The Rainwater falling in the watershed now flows through the drains to the reservoir formed due to the construction of water harvesting structure, causing water to store upto a capacity of 2.25 ha.m. Surplus water passes through the structure and joins a river in the d/s. The stored water will have some losses due to the evaporation and percolation in the ground. Rest of the water is available as surface water. The recharge due to delayed flow as well as reservoir will impact favorable to the recharge of the groundwater. Hence the water will be available in the wells in the summer season also. Study reveals that Surface water if utilized for drinking only would be available for 2.3 months to the population of the human as well as cattle living near the site. If some part of the water is utilized for irrigation than the water availability period will be 1.30 months or 39 days. Ground water will also be available drinking as well as irrigation in winter season.

Charnot land has some good quality trees and other precious plantations, which are going to benefit by the increased moisture in the soil. As the water is available in reservoir and also in well during monsoon season hence if there is no rainfall in last watering to the crop the same could be meet through the irrigation from the stored water sources, hence, at least Kharief crop in the d/s area is assured. This will cause increase in the crop production. Also flora and fauna will be developed due to presence of the moisture in the soil. Production of crop may yield upto 176.00 ton per year.

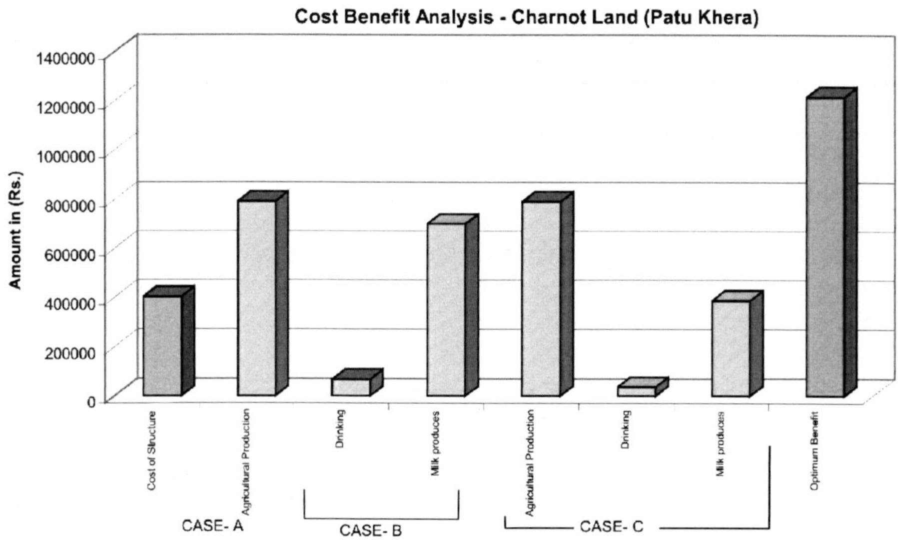

Figure 11.6.29

11.6.4.6.3 *Cost Benefit Analysis*

1. Construction Cost: **Rs. 418025.00**
2. Direct Benefits:

Beneficiaries:	Human Population:	553
	Cattle Population:	400

Case a: Surface Water Irrigation use:

Increased Agricultural Production by 176 Tons	(Rs. 792000.00)

Case b: Surface Drinking Water use:

Drinking water available	(Rs. 67500.00)
Dairy produces	(Rs. 701987.00)
Total benefits	(Rs. 769487.00)

Case c: Irrigation and Drinking Water use:

Drinking water available for 39 days	(Rs. 37500.00)
Surface water irrigation	(Rs. 389993.00)
Dairy produces	(Rs. 792000.00)

Total benefits **(Rs. 1219492.00)**

2. Indirect Benefits:

Increased Crop Production + Drinking water = **Rs. 64735.00**

The Figure 11.6.29 depicts the overall benefits of the construction at site as an impact of construction of the structure. It is clear from the graph that in a single year the optimum direct benefits achieved will be about 2.91 times of cost of construction. Spending a small amount in maintenance the villagers could derive benefits from the structure to considerable years.

11.6.4.7 *Village - KHIRWANIA KI NAL (BASSI SINGHAWAT)*

Khirwania ki Nal is situated in village Bassi Singhawat at the junction of three villages in Salumbar Block of Udaipur district. The site is situated at the northern part of the village. The site forms a medium watershed of areal extent of 158.30 ha. The upstream area consists of gentle sloping terrain with hillocks and d/s areas consist of agricultural area of about 300.0 ha. An anicut structure of stone Masonry was constructed to store water.

11.6.4.7.1 *Pre-construction Status*

The Rainwater falling in the water shed was flowing without stoppage through the catchment to river at the downstream. Causing loss of precious rainwater and also fertile soil erosion from the agricultural fields as it flows through the agricultural fields. No surface water is available for drinking for human being as well as cattle population. Also the well remains with water for very short duration hence ground water is also limited. Since water available for irrigation is only the monsoon rainfall hence the crop production is very much limited to the Kharif crops and the production of the crop varies from 0.2 to 3.6 tons per hectare depending upon bad and good monsoon rainfall respectively.

11.6.4.7.2 *Post-Construction Status*

The Rainwater falling in the watershed will flow through the drains to the reservoir formed due to the construction of structure, causing water to store there to a capacity of 4.5 ha.m. Surplus water passes through the structure and joins a river in the d/s. About 45000.00 cum water is available as surface water. The recharge due to delayed flow as well as reservoir will impact favorable to the recharge of the groundwater. Hence the water will be available in the wells in the summer season also. Study reveals that Surface water if utilized for drinking only would be available for 2.6 months to the population of the human as well as cattle living near the site. If some part of the water is utilized for irrigation than the water availability period will be 2.15 months or 64 days.

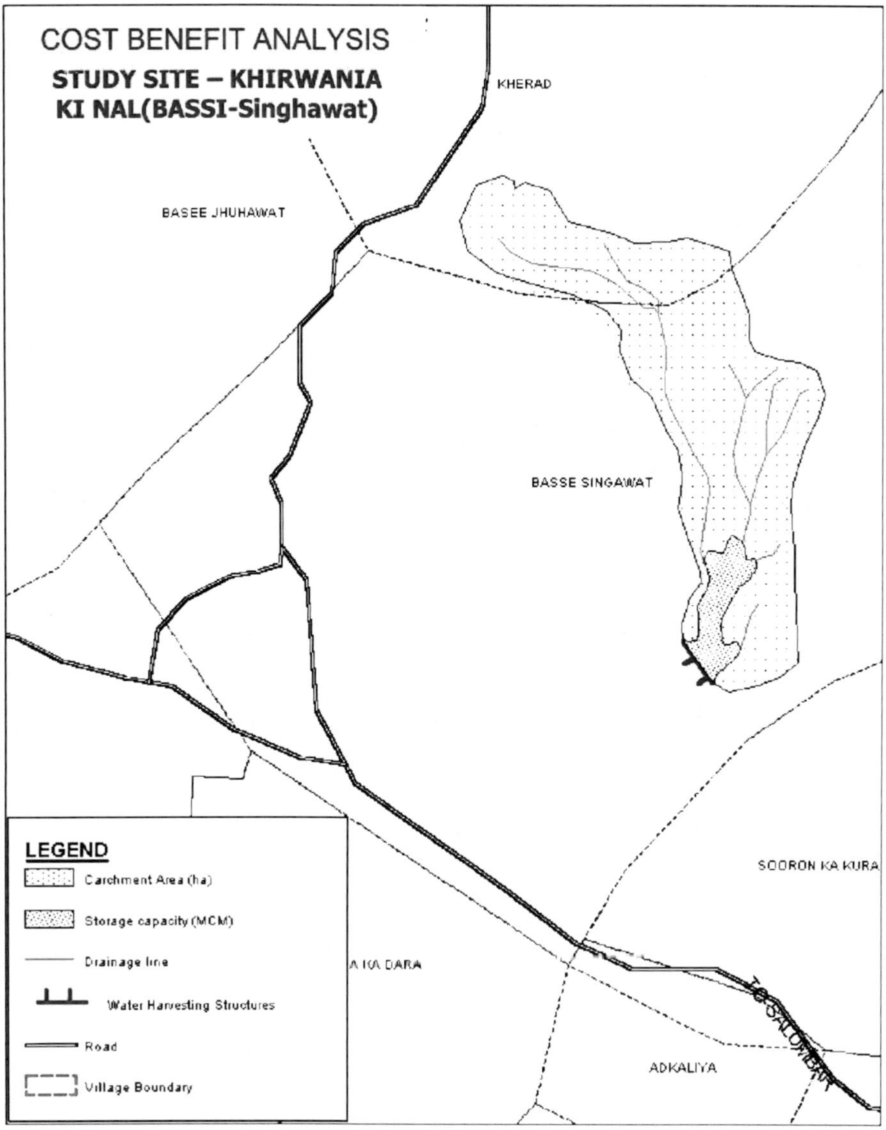

Figure 11.6.30.

Quality of water in general is within the standards of the WHO so there was no problem as far as quality parameters are concerned. As the water is available in reservoir and also in well during monsoon season hence if there is no rainfall in last watering to the crop the same could be meet through the irrigation from the stored water sources, hence, at least Kharif crop is assured. This will cause much increase in the crop production. Also flora and fauna will be developed due to presence of the moisture in the soil. Production of crop could be yield upto 264.00 ton per year.

11.6.4.7.3 *Cost Benefit Analysis*

1. Construction Cost:	**Rs. 579404.00**
2. Direct Benefits:	
Beneficiaries:	
Human Population:	1920
Cattle Population:	500
Case a: Surface Water Irrigation use:	
Increased Agricultural Production by 264 Tons	(Rs. 1188000.00)
Case b: Surface Drinking Water use:	
Drinking water available	(Rs. 135000.00)
Dairy produces	(Rs. 966786.00)
Total benefits	(Rs. 1101786.00)
Case c: Irrigation and Drinking Water use:	
Drinking water available for 39 days	(Rs. 112500.00)
Surface water irrigation	(Rs. 805655.00)
Dairy produces	(Rs. 1188000.00)
Total benefits	**(Rs. 2106155.00)**
2. Indirect Benefits	
Increased Crop Production + Drinking water =	**Rs. 122764.00**

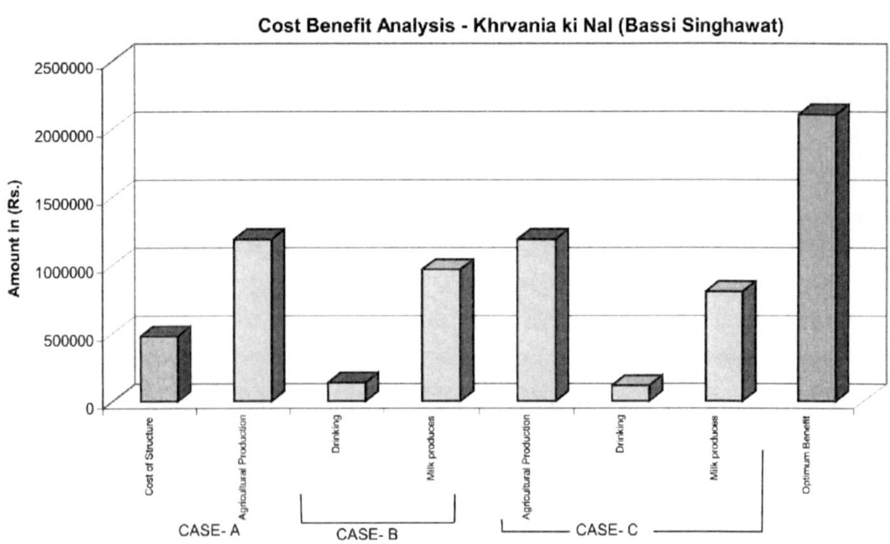

Figure 11.6.31

The Figure 11.6.31 depicts the overall benefits of the construction at site as an impact of construction of the structure. It is clear from the graph that in a single year the optimum direct benefits achieved will be about 3.60 times of cost of construction. Spending a small amount in maintenance the villagers could derive benefits from the structure to considerable years.

11.6.5. Comparative Analysis

Each site has been extensively studied and analysed. Followings are the outcome of the study that can be compared individually:

- ***Location:*** Out of 7 sites 2 sites are in Girwa Block namely, Patoo Khera and Khedia (Semliya) both near Bhambhora town; one site Sirwal (Dang) in Kotra and rest four are in Salumbar Block of Udaipur district.
- ***Catchment:*** Sites were selected in watersheds near villages at best possible location with maximum catchment area upstream. Khirwania ki nal has maximum catchment area of 158.30 hectare followed by Patoo khera 91.46, Gaonra Pal – 70.80, Sirwal – 47.0, Hama Talav – 42.38, Semliya 16.47 and Joyra naka with 14.52 hectares.
- ***Cultivable area at d/s:*** Site survey carried out at individual sites reflects that maximum land is available at site Khirvania ki nal (300 hect.) followed by Gaonra pal (250 hect) Patoo khera (200 hect), semliya (125 hect) and minimum of 20 hectare at Joyra naka.
- ***Type of Structure:*** The selection of water harvesting structure depends upon catchment characteristics and based on that Stone Masonry Anicut with cement mortar were constructed at three big catchment sites – Khirwania ki nal, Patoo khere and Gaonra Pal. At rest of the sites Loose Stone Masonry structures had been constructed as their catchments are small. Length of the structure at Patoo khera is maximum 105 m including 25m-spillway section of stone masonry stone masonry. Khirwaina ki nal has length of 50.0 m, Sirwal - 30m, Semliya – 30m, Hama talav- 30m, Gaonra pal – 15m, and joyra naka has length of 20m. Heights of the structure are limited to maximum of 3.0 m.
- ***Cost of Structure:*** Cost of structure depends upon type of structure and dimension of the structure. Cement mortar Stone Masonry structure cost more than Loose Stone Masonry structures as material and skilled labour components are involved. Khirwania ki nal is the costliest among all amounting to Rs. 579404.00 followed by Patoo Khera Rs. 418025.00, Hama talav – Rs. 263600.00, Gaonra Pal – Rs. 136600.00, Sirwal – Rs. 127170.00, Joyra naka – Rs. 84490.00 and Semliya is the cheapest – Rs. 44915.00.
- ***Storage Capacity:*** In general larger the catchment and larger the size of the structure larger will be the storage capacity of the reservoir formed by the structure. Survey carried in the fields reveal that Khirvania ki nal will have max. storage capacity of 45000.00 cum followed by Patoo khera – 22500.00,

Gaonra pal – 18000.00, Semliya – 6000.00 and rest of the sites – 5000.00 cum.

- ***Population Benefited:*** Since the sites are situated at some particular location of the village, the populations inhabited near the site are going to be directly benefited. Survey carried indicates that maximum of 1920.00 human population are going to be directly benefited at site Khirwania ki nal, 700 at Gaonra pal, 553 at Patoo khera, 500 at Joyra naka, 390 at semliya, 270 at dang. Assuming nil human population at Hama talav as that is at charagah land and only cattle are assumed to be benefited. Cattle population receives max benefits as drinking and fodder at sit Khirwania ki nal with 500 numbers followed by Patoo khera as 400, Gaonara pal as 300, Hama talav as 200, Semliya as 150, and rest with 100.

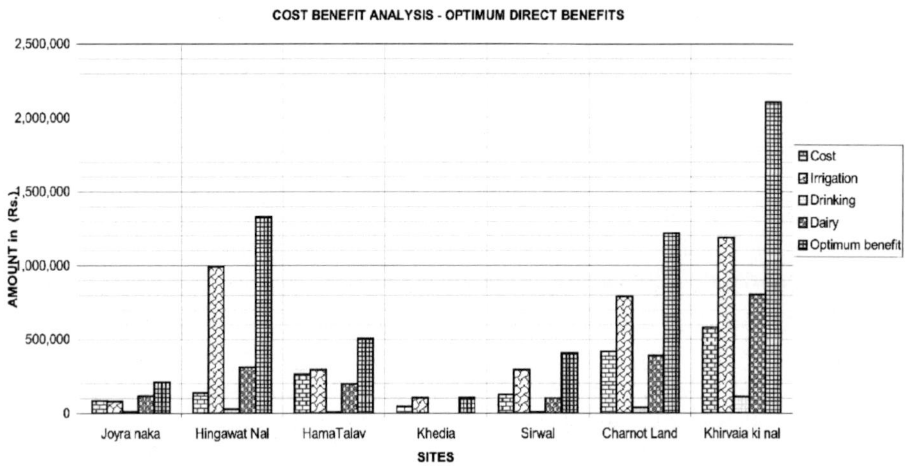

Figure 11.6.32 Comparative Graph of Optimum Direct Benefits

11.6.5.1 *Optimum Direct Benefits*

- ***Agricultural Production:*** Maximum the cultivable area and storage capacity of reservoir maximum the increase in production. Khirvania ki nal gets max. benefits by producing crop of worth Rs. 11,88,000.00, followed by Gaonra pal – 9,90,000.00, Patoo khera – 7,92,000.00, Dang and Hamatalav with 2,97,000.00 with minimum at Semliya with Rs. 1,06,875.00.
- ***Drinking Water Availability:*** Khirwania ki nal has max. surface water availability for drinking as 64 days followed by joyra naka 47 days, Gaonra pal – 41 days, Dang with 41 days… with minimum of 0 days at Semliya. In monetary terms benefit of Rs. 1,12,500.00 is achieved at Khirwania ki nal, Rs. 37,500.00 at Patoo khera, Rs. 29,250 at Gaonra Pal, Rs. 12000.00 at Joyra naka and min of Rs. 00.00 at Semliya.

Milk Produces: As drinking water and fodder is made available to the cattle population there is going to increase in the milk produces leaving behind the other produces from them. Considering only the milk produces the benefits achieved are of Rs. 8,05,655 at Khirvania ki nal, Rs. 3,89,993 at Patoo khera, Rs, 3,09,367 at Gaonra Pal, Rs. 1,98,025 at Hama Talav, Rs. 100937.00 at Dang. At semliya in optimum benefit analysis it has been assumed that all the water is used for irrigation only.

11.6.5.2. *Indirect Benefits*

The recharge due to delayed flow as well as reservoir will impact favorable to the recharge of the groundwater. The recharged ground water will generate the indirect benefits in term of making available the ground water to longer duration for drinking purpose through wells, and also for irrigation. Study with mathematical model reveals that Khirwaina ki nal is getting max. benefit from the structure of Rs. 1,22,764 followed by Patoo khera –Rs. 64,735, Gaonra Pal – Rs. 50,179, Joyra Naka – Rs. 38,340, Semliya – Rs. 20,969 with minimum at Dang- Rs. 18,112. It indicates that larger the storage capacity and cultivable area maximum the indirect benefits.

11.6.5.3 *Cost Benefit Ratio*

Results of the Cost Benefit Analysis supported by Field Survey and Computer based Mathematical model Analysis indicates that maximum C.B. is achieved at site Gaonra Pal with 9.72 followed by 3.60 at Khirvania ki nal, 3.20 at Dang, 2.91 at Patoo Khera, 2.50 at Joyra naka, 2.38 at Semliya and minimum of 1.91 at Hama Talav. These figures indicate that maximum C.B.R. will be generated at sites where agricultural land is more and cost of structure is less with larger storage capacity. Figure 11.6.32 clearly reflects this aspect. Second aspect is milk produced by cattle (small scale dairy industry). Drinking water aspect has a very limited or negligible (in comparison to other two aspect) impact on Cost Benefit Ratio.

CASE STUDY - 7

11.7 GIS BASED MAPPING OF WATER SUPPLY SYSTEM, DRAINAGE AND SOLID WASTE MANAGEMENT SYSTEM OF SUNEL TOWN OF JHALAWAR DISTRICT

The issues of drinking water scarcity have emerged as a result of inadequate provisions to conserve water, low rainfall and unorganized urban planning in Sunel town. Similarly, improper environment sanitation condition has resulted out of inadequate infrastructure and lack of awareness among the community in the town. At present, large proportion of households in Sunel town is being served with tap water supply from Public Health Engineering Department. In spite of the fact that considerable amount of runoff water is harvested at Sagaria dam (source), these households of the town in general face drinking water scarcity. This situation particularly has aggravated in last three years due to recurrent drought conditions in the area. In the wake of inadequate availability and hence supply, households are illegally using suction pumps (Tullu Pumps), leading to imbalanced drinking water distribution or low-pressure supply at the tail end. In order to improve the drinking water availability or facilitate wider population having adequate access to drinking water with in the Study, identification of specific activities was considered imperative.

Similarly, in the case of sanitation, identification of specific activities for the Study was considered necessary to lessen the problem of water logging and unhygienic conditions faced by the households in Sunel town. Sunel has open drainage lines for sewage disposal, which is managed by the Local Bodies. These open channels begin at a particular point within the town and finally drain in open fields without any adequate treatment. For proper and safe disposal it is recommended that sewage should be treated as per the measures/ technical norms before its final disposal. Due to unplanned sewage system and encroachment by the households over the natural drain, the unhygienic conditions further aggravates every year during the rainy season by way of creating waterlogged conditions in main streets. Looking into the fundamental requirement of the Study, GIS technology was applied to suggest measures through which the problem of inadequate access of piped water supply to the tail end households and deteriorating sanitation conditions could be addressed. This study analyses the gaps existing in water supply system and drainage network using GIS techniques and suggest specific measure for implementation through which the concerned problems could be addressed, leading to improved drinking water and sanitation condition in Sunel town.

11.7.1 Objective

The objective of the study was as outline below:

- Assess discharge and actual supply (ward wise) status of water, identify the gaps if any, and suggest specific technical measures/activities to bring down the gap
- Assess water demand and availability in clusters not serviced by PHED, identify the gaps if any, and suggest specific activities to fulfill the water needs of the community.
- Estimate the current and required carrying capacity of the drainage lines and suggests specific measures to streamline the sewage flow.
- Understand the composition of the sewage passed through the drainage lines and identify the treatment requirement before safe and final disposal of the sludge.
- Identify low-lying areas and locate the points where problem of choking, water logging is observed.

11.7.2 Methodology

11.7.2.1 *Data Collection*

Since the study required deeper understanding of drainage, sanitation and water supply system which includes identification of gaps, the study used both secondary as well as primary sources to generate information for further analysis. The details of the data collected from different sources based on the checklist developed is discussed as following:

11.7.2.1.1 *Secondary Sources*

The water supply and sanitation is governed by PHED and Panchayat in Sunel respectively. Thus, for the study purpose existing literature already available with these departments was collected for review and further analysis. The type of information retrieved from different sources based on the checklist is reflected in the following Table 11.7.1.

Table 11.7.1: Data Collection from secondary sources

Source	Information/ Data
Gram Panchayat – Sunel	• Base map of Sunel town with road network, wards with numbers and their boundary details – Year 2003. • Ward-wise population. • Solid Waste Disposal System – disposal point, collection system, transportation and final disposal.
Public Health Engineering Department	• Piped Water Supply System- Source, per capita supply, duration, quantity and quality. • Hand pumps- number and locations and status. • Public Stand Posts – Numbers and locations and status.

	• Open wells etc.
Irrigation Department – Jhalawar	• Rainfall data.
Local Non governmental Organisations (NGOs)	• Literature related to Sunel and Bhawani Mandi – population, household etc.

11.7.2.1.2 *Primary Sources*

As discussed earlier, collection of primary data was considered necessary for the study to understand the existing operation and maintenance system of both water supply and drainage system and enlist the problems in the system for use in further analysis. Focused group discussions with Panchayat, PHED, community members along with field survey methods were conducted to fulfill the purpose. Field survey was primarily conducted to triangulate the data collected from secondary sources. The data collected from primary sources was based on the checklist as outlined in Table 11.7.2.

Table 11.7.2 Data Collection from Primary sources

Method	**Information/ Data**
A. Focus Group Discussion (FGD)	
1.Gram Panchayat – Sunel	• Verification of Ward boundaries and its demarcation on base map. • Drainage network system and its demarcation on Base map with size and material. • Solid waste disposal points, sewage treatment and final disposal system and its demarcation on base map. • Existing condition of garbage bins being utilised by Nagar Palika. • Low lying waterlogged areas and its demarcation on base map. • Sewage discharge at the household level.
2. Public Health Engineering Department (PHED)	• Demarcation of water supply network system on Base map with pipe diameter and material of construction, valves, SR, CWR, etc. • Difficulties and constraints faced by the department during Water supply. • Serviced and un-serviced areas. • Hand pumps- demarcation on base map • Public Stand Posts – demarcation on base map. • Open wells – Demarcation on base map. • Number of households served by alternative sources. • Water demands and needs fulfillment of the households in each ward.
3. Community/ households	• Collect data regarding the availability of suction pumps, its specification and duration of its use.

B. Field Survey

1. GPS Survey	• Verification of Ward boundaries and its demarcation on base map. • Drainage network system and its demarcation on Base map with size and material. • Solid waste disposal points, sewage treatment and final disposal system and its demarcation on base map. • Low lying waterlogged areas and its demarcation on base map. • As drainage line map of Sunel is not available with Gram Panchayat, Leveling Survey using GPS was undertaken to record the levels, width of the drainage lines and sewage flow system.
2. Field observation	• To ascertain the quality of the drinking water and sewage water at key points.

The study required testing of sewage composition to ascertain its quality and identify measures for its treatment. This was however, dropped after physical examination of sewage quality. It is observed that the sewage do not get polluted by chemical substance which erstwhile makes it necessary to ascertain its quality through lab testing. The quality of sewage thus was examined through physical inspection as referred above.

11.7.2.2 *Process of Data Collection*

Before initiation of fieldwork preliminary meeting with local authorities was organised to discuss the objectives, methodology of the study. The preliminary information about the Study and related components was collected from these organisations. As required, focused group discussions were conducted with PHED, Panchayat, community members to collect primary data. Literature available with these departments on the concerned issue was also reviewed and relevant data retrieved from them. Field survey using GPS instrument was conducted to obtain ground points and their levels and identify the land use features of Sunel town.

11.7.2.3 *Data Processing and Analysis*

The collected information vis-à-vis ward wise population, households, base map with transport network and ward boundaries, water supply distribution network map, drainage map with ground levels was fed in to the computer. Base map was developed demarcating ward boundaries and road network using MAP-INFO (GIS) software. GIS based Water-CAD software was used to check the adequacy of the existing water supply network. Under this following parameters were assessed by way of simulating model in ideal conditions.

- Ward-wise present water demand.
- Availability of water in serviced and un-serviced areas of Piped water supply system.

- Present status of water availability (discharge) at source, CWR and SR and nodal points.
- Existing water supply network condition.
- Adequacy of pipe network - pipe diameter and construction material.
- Capacity of pipes network- pressure and flow at various nodal points.

The results generated were again superimposed on the base map using MAP info to compare the current system (FGD), conclude deviation in demand and supply and identify probable reasons of gaps and measures to control it. For undertaking analysis of drainage system, Surfer and Vertical Mapper (GIS modules of Map-info) were used to generate the ground level contour map basically with the purpose to understand the slope in Sunel town. This information was later used along with drainage network map in Storm CAD software to simulate the operational status of drainage system in ideal scenario. The details of the parameters assessed which were later superimposed on base map to identify gaps using storm CAD software are as outlined below:

- Ward-wise present waste water generation.
- Existing drainage network condition.
- Adequacy and capacity of V-shape and U-shape drains for domestic wastewater and storm water.
- Deviation in actual flow generation and carrying capacity of the drains.
- Measures to improve the carrying capacity of the drains.
- Ground level contour map generation and its analysis to identify low-lying area in Sunel town.

In case of solid waste, quantum of waste generated per capita per day was calculated using standard guidelines of the Manual of Solid Waste Management. Based on the calculations, the waste generated in each ward was estimated in ideal situation and superimposed on the Base Map. To draw conclusions, this data was later compared with field reality and assessed by superimposing the carrying capacity of the total points existing in each ward. The carrying capacity of each existing solid waste point was assessed on the basis of number of households disposing solid waste at that particular point. The sewage was physically inspected in order to assess its contamination status. Treatment for its safe disposal was designed according to the Manual. After assessment of water supply, drainage/sewage and solid waste, interventions were identified and suggested.

11.7.3. Findings

11.7.3.1. *The Study Area*

Sunel town is situated 45 km east of Jhalawar district headquarter spread over an area of 2.98 sq.km. As shown in figure 11.7.1 the town lies in GT sheet No. 45P/15. The town is constituted of 10 wards representing a total population of 11364 belonging to different caste communities including schedule caste/tribe and OBC represent the villages. The ward wise map of Sunel town is shown in figure 11.7.2.

STUDY AREA
LOCATION MAP

G.T. SHEET NO.: 45P/15

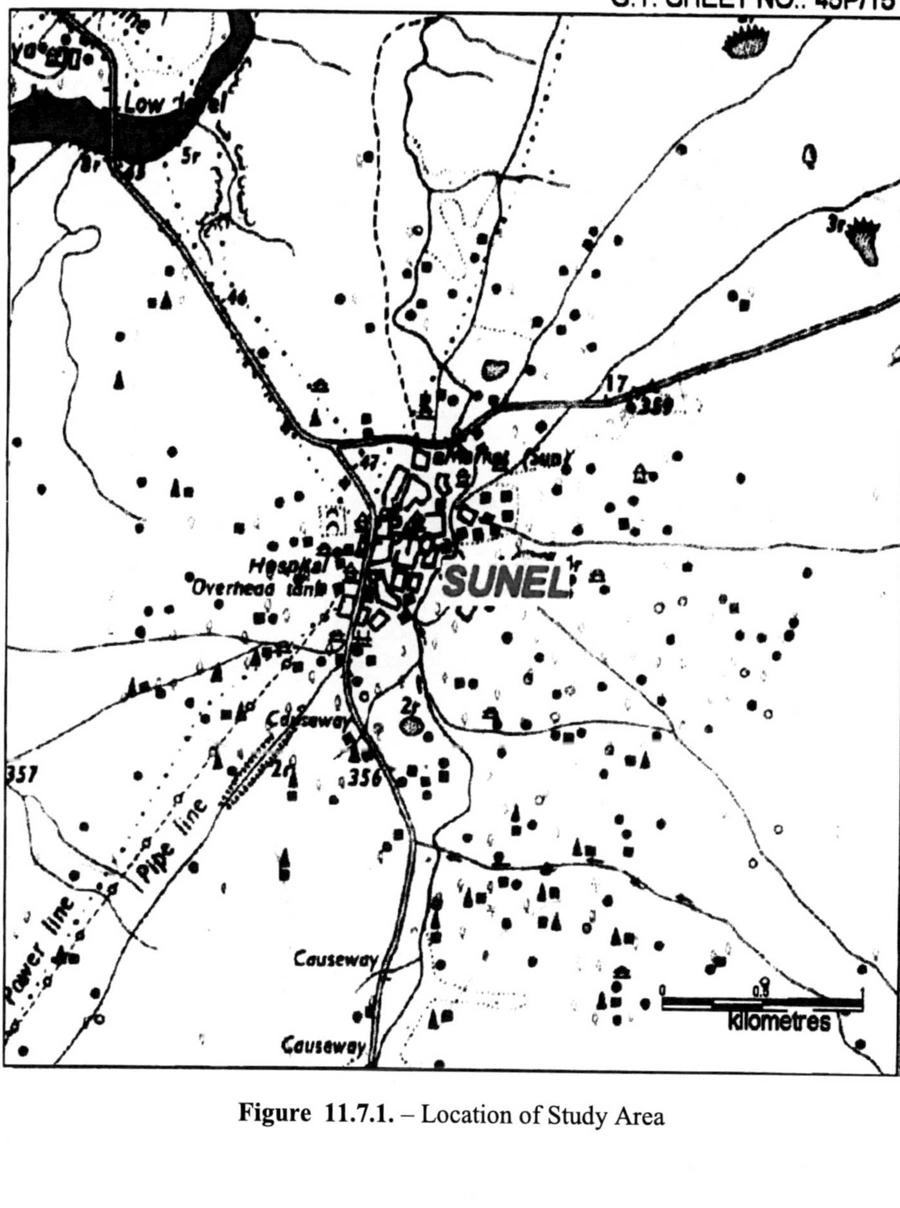

Figure 11.7.1. – Location of Study Area

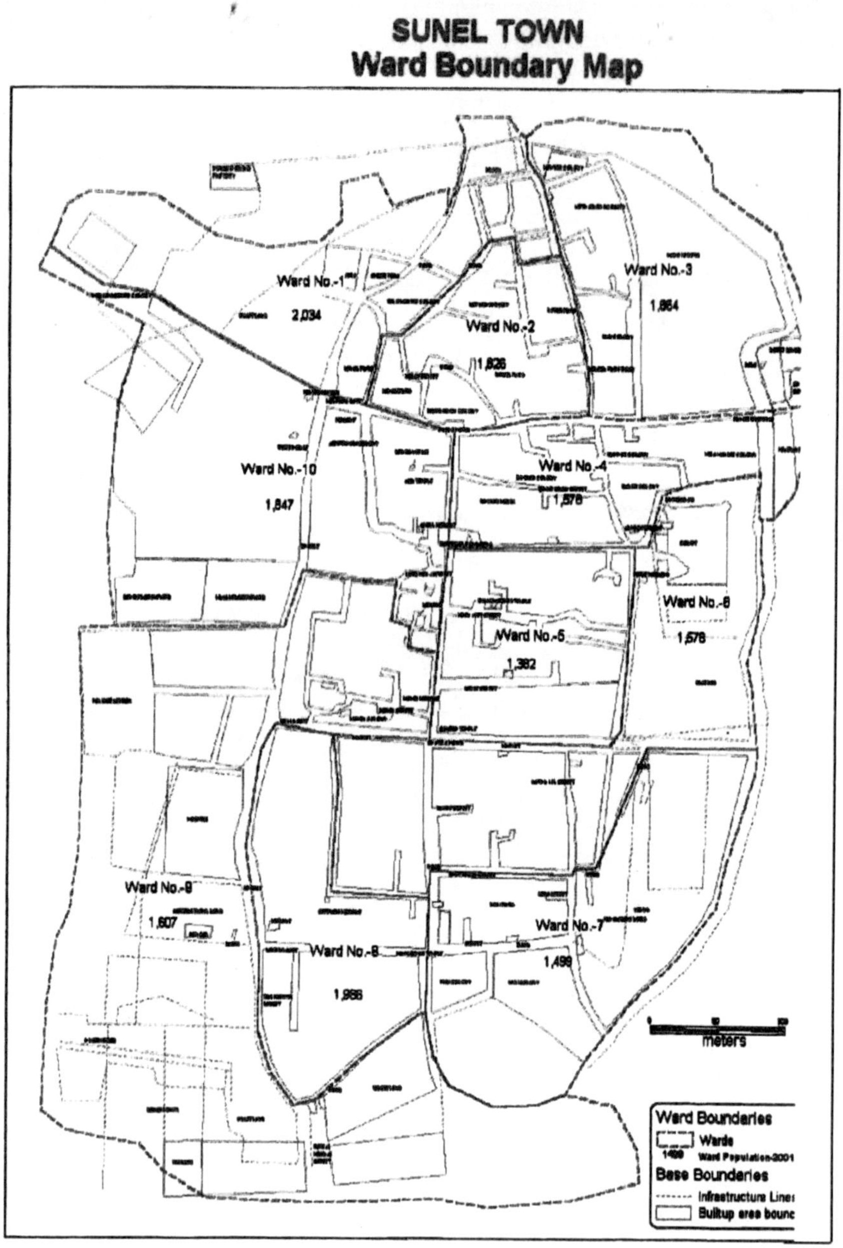

Figure 11.7.2 – Ward map of Sunel town

11.7.3.2 *Water Supply System*

The population of Sunel is largely dependent on the piped water supply system constructed and maintained by the Public Health Engineering Department of

Government of Rajasthan. In places where the households are not served with piped water supply, water requirements are met from alternative sources such as HP, open dug well and PSPs. These sources as shown in Figure 11.7.3 have been spread over the entire town covering almost each and every locality. Households served with piped water supply as discussed below also use these alternative provisions.

SUNEL TOWN
PIPED WATER SUPPLY NETWORK COVERAGE

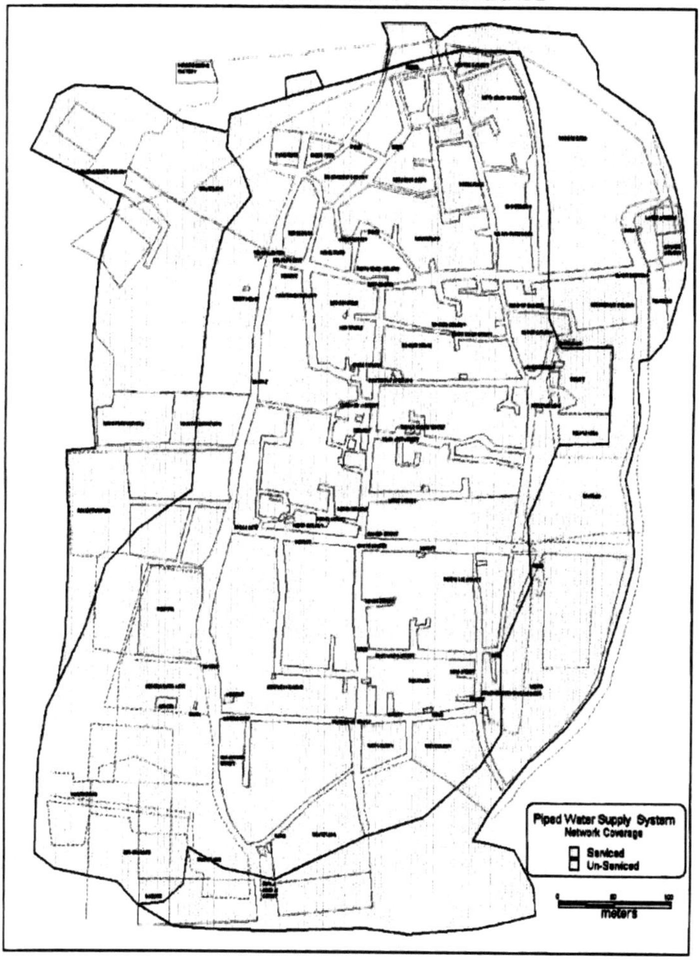

Figure 11.7.3 Water supply coverage

11.7.3.2.1 *Alternative Sources*

- ***Handpumps:*** There are 35 Hand pumps spread over Sunel town with 25 in working condition. In present scenario the Hand pumps are playing a crucial role in meeting the demand gap for drinking water. The quality of water in HP differ (saline composition) compare to the WSS but is preferred highly in the areas where there is no piped water supply system or exists problem in supply system.

- ***Public Stand Posts (PSP)***: There are 15 P.S.P's in Sunel town but most of them are not functional due to short and irregular supply of piped water.
- ***Open Dug Wells***: There are around 18 dug wells in and around Sunel. The functionality of these wells is decided by availability of water. At present, the water level has declined in almost all wells to a level that they have turned almost dysfunctional for use.

Figure 11.7.4 represents the location of above alternate sources in Sunel town.

11.7.3.2.2 *Piped Water Supply*

- ***Source:*** The main source of piped water supply is Sagaria dam on Ahu River located about 8.5 km south-west of Sunel town. The surface water stored at dam is pumped from the source by pipeline of 8" diameter to Sunel and is collected in CWR. The alternate source of the water supply is Rangbari Thala from where the water is pumped and collected in CWR through 5" dia A.C. pipe.
- ***Population Served***: The piped water supply is currently serving 10341 persons out of total population of 14773. Rest of the people depends on alternative water supply system as discussed above. Out of 1400 household's connection, 1100 has been provided with meter connection while rest is charged on a monthly flat rate basis. The charge for flat rate and meter connection has been set around Rs. 29/- per month and Rs1.25/- per 1000 liters respectively.
- ***Water Distribution Network***: As shown in Figure 11.7.5, the pipeline network covers almost entire town, covering a distance of around 5 Kms. The diameter of the pipeline of the distribution network ranges from 7 inches to 2 inches for main to service line respectively. As could be noticed from figure 5 there is one clear water reservoir and two overhead service reservoirs by which the households are being served water through gravity flow. The capacity of CWR and each SR is 1.0 Lakh Liters and 3.0 Lakh Liters respectively. The system is so designed that entire pipe network could be divided in 2 zones supported by each S.R. but at present no such zoning is in practice.
- ***Water Quality***: The quality of water falls within the permissible limits for drinking (Chlorides 45 mg/l, Nitrates traces, Total Dissolved Solids 267 mg/l, and Fluorides 0.3 mg/l). Water is disinfected before its supply to the consumers. For dis-infection with bleaching powder, chlorination method is used.
- ***Duration of Water Supply***: Looking into the condition of the source the duration of water supply is changed from time to time by PHED. From November it was changed to 1 hour supply in the morning (From 6.00AM to

7.00AM) to once in three days in the month of April. Presently the water is supplied with 45 liters per capita per day for one hour. As the source are emptying the water through tankers are used for filling the CWR and SR. When the water of sufficient quantity is stored the supply is opened for the consumers.

- **Serviced and Un-serviced Areas**: There are some area i.e Housing Board Colony (north), Shahid Park (south), western parts adjoining main drain within the town are not covered with piped water supply. As shown in Figure 11.7.3, the households of these areas thus depends on alternative sources to meet their water needs. The Table 11.7.3 depicts the statistics of such population dependent on piped water supply and alternate source in the Sunel town.

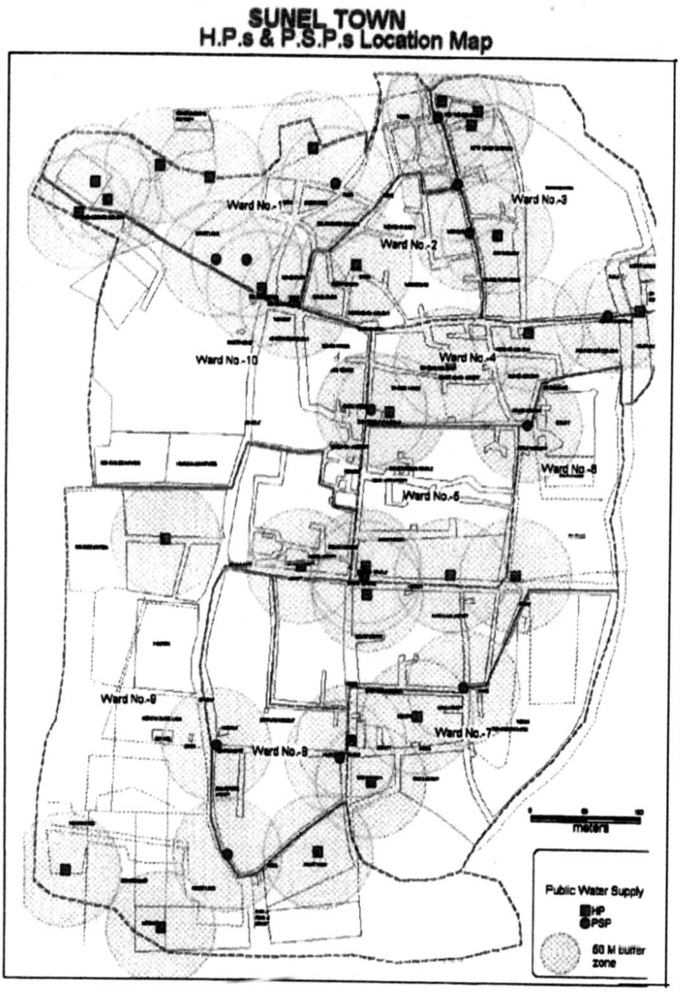

Figure 11.7.4. PSP/ WELL location map

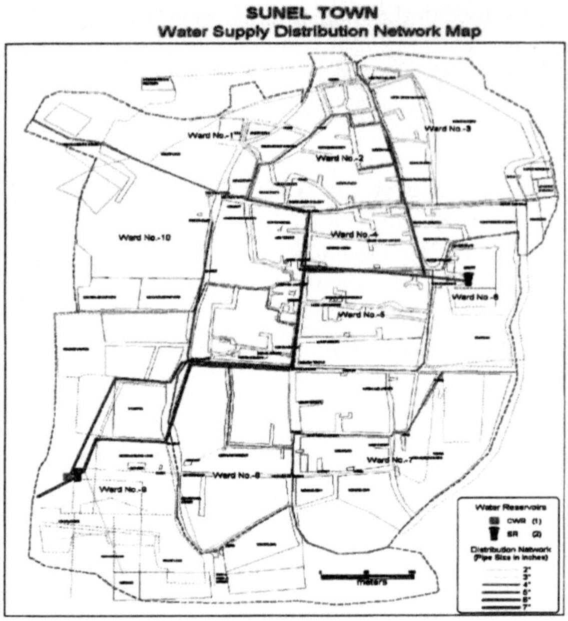

Figure 11.7.5 WS Distribution network map

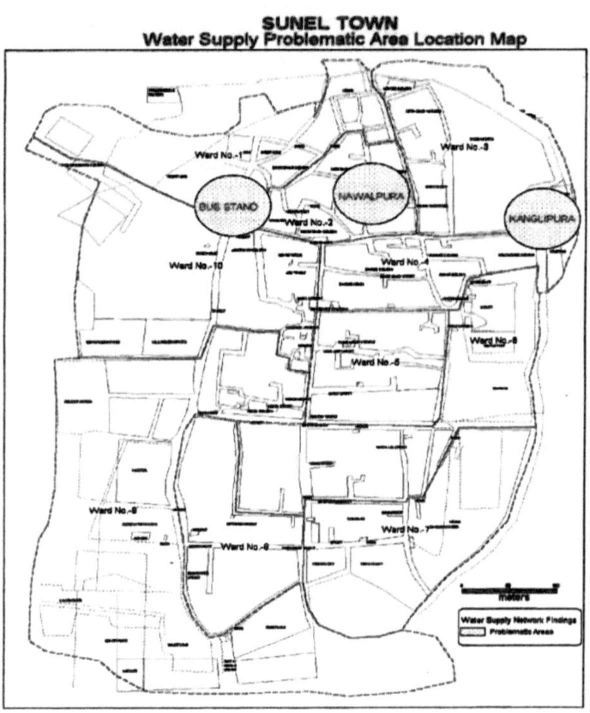

Figure 11.7.6 WS Problematic area map

Table 11.7.3. Proportion of population served with piped water supply and other alternative sources

Water Supply System	Geographical Coverage		Population Dependency		Water Demand (@40LPCD)
	(in Hectare)	In %	In numbers	In %	
Piped Water Supply System	22.746	65%	10341	70%	4.13 Lac Lit.
Alternate Water Supply (HP)	12.258	35%	4432	30%	1.77 Lac Lit.

- **Water Demand v/s Supply:** As could be noticed from Table 11.7.4 the present demand is 5.90 Lakh Liters for the entire population of Sunel town. The water supply till October 2002 was 40 lpcd once in a day, however, due to less availability of water at source, quantum of supply has been reduced in recent time i.e., once in two day till March and then once in three days after the month of March. From the table it is clear that at present 15 LPCD of water is being supplied, which merely cater the drinking water needs of the households.

As per crude network analysis, huge gap of 3.90 Lakh Liter exists between demand and supply. The data collected and the results obtained ward wise as shown in following table shows that each ward is confronting this problem with demand and supply gap ranging from 0.32 to 0.46.

Table 11.7.4: Ward-wise Gap in water demand and supply in Sunel town

Ward	Population (2001)	Drinking Water Demand (Lakh Liter Per day) @ 40 LPCD	Piped Water Supply (Per day)			Present Gap
			Oct-02	Jan - 03	March-03	
1	1749	0.70	0.71	0.36	0.24	0.46
2	1240	0.50	0.50	0.25	0.17	0.33
3	1860	0.74	0.76	0.38	0.25	0.49
4	1686	0.67	0.69	0.34	0.23	0.44
5	1377	0.55	0.56	0.28	0.19	0.36
6	1801	0.72	0.73	0.37	0.24	0.48
7	1243	0.50	0.51	0.25	0.17	0.33
8	1131	0.45	0.46	0.23	0.15	0.30
9	1468	0.59	0.60	0.30	0.20	0.39
10	1219	0.49	0.50	0.25	0.17	0.32
TOTAL	14774	5.90	6.01	3.00	2.00	3.90

To understand why such gap in demand and supply exist in different wards, theoretical assessment was undertaken using Water CAD software. Simulation analysis was carried out to judge the pressure at each node and draw conclusion theorctically whether there exist a problem of low pressure, which erstwhile studyed by PHED and members of the community in Kanglipura, Bus stand and Nawalpura.

The data as shown in Table 11.7.5 reveals that variation in the available residual pressure in the pipeline network is the major factor responsible for creating such a gap. The analysis suggests that every ward should receive water with sufficient pressure in-spite of the altitude variation or tail end location of the households. Since, the problem of low-pressure supply was practically observed in these localities, the study findings suggest that insufficient pipe sizes (over or under), choking or clogging of pipeline and leakage's due to open tap could be some of the reasons which might be blocking the flow of water. As depicted in figure 6, the problem of declining residual pressure in Nawalpura, Kanglipura and Bus stand localities could have also been contributed from use suction pumps during the supply by majority of households. The use of suction pumps empties the water in the pipeline network, which causes negative pressure (vacuum) inside the pipeline. This phenomena impact adversely during the next supply, as greater time is lost in regaining the pressure with in the pipeline network.

Table 11.7.5: Simulation of residual water pressure at tail end of the network pipeline

Ward	Population	Present Water Demand (L.L. per day)	Present Water Supply (L.L. per day)	Pressure at tail end (As per computer model analysis)
1	1749	0.7	0.24	3
2	1240	0.5	0.17	3
3	1860	0.74	0.25	4
4	1686	0.67	0.23	10
5	1377	0.55	0.19	12
6	1801	0.72	0.24	5
7	1243	0.5	0.17	5
8	1131	0.45	0.15	10
9	1468	0.59	0.20	3
10	1219	0.49	0.17	5
TOTAL	14774	5.9	2.00	

11.7.3.3 Drainage System

Sunel town is well covered with open drainage network system. Figure 11.7.7 indicates that almost all the streets are covered with side drains. These drains receive domestic wastewater from individual households in small v-shape drains and joins U-shape drains at major street junctions. The Topography of the Sunel town is such that the half of the town drains towards the southwest direction and remaining half towards the east - north direction. The nallah or natural drain carries the flows from south to west and north. The sewage water drains in this natural drain that ultimately joins Ahu River. It is observed that some parts of drains are well constructed and maintained while in few places specifically in tail ends they are not maintained. A part of Sunel town faces frequent clogging, choking of drains in low-lying localities. The field level survey results match the actual ground conditions (Figure 11.7.9). The contour map generated from the field levels is shown in Figure 11.7.10. The study findings reveals that the total length of the drainage network laid in the Sunel town is 10.941 km with maximum (1.665 km) in ward no.- 9 and minimum (0.703 km) in ward no. 3. Nearly 205.19 liter per minute sewage is generated in the town with maximum in ward number three due to its dense population.

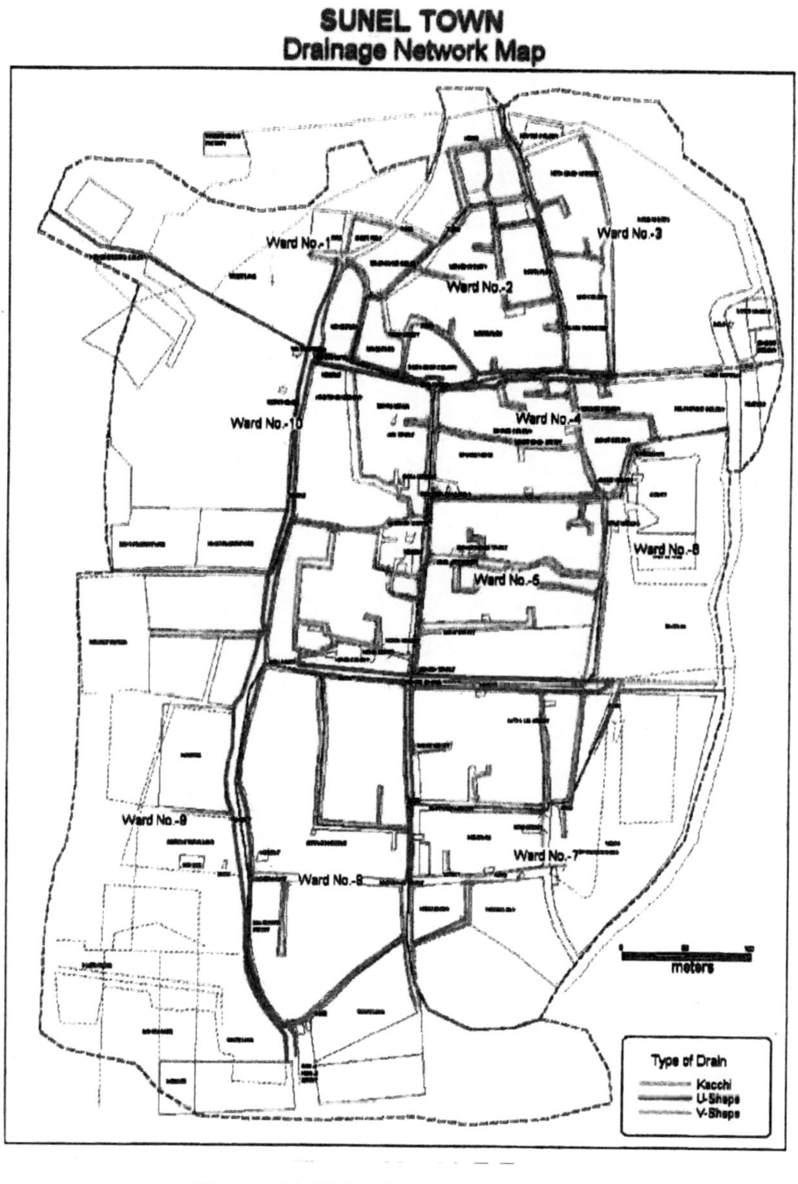

Figure 11.7.7 Drainage Network map

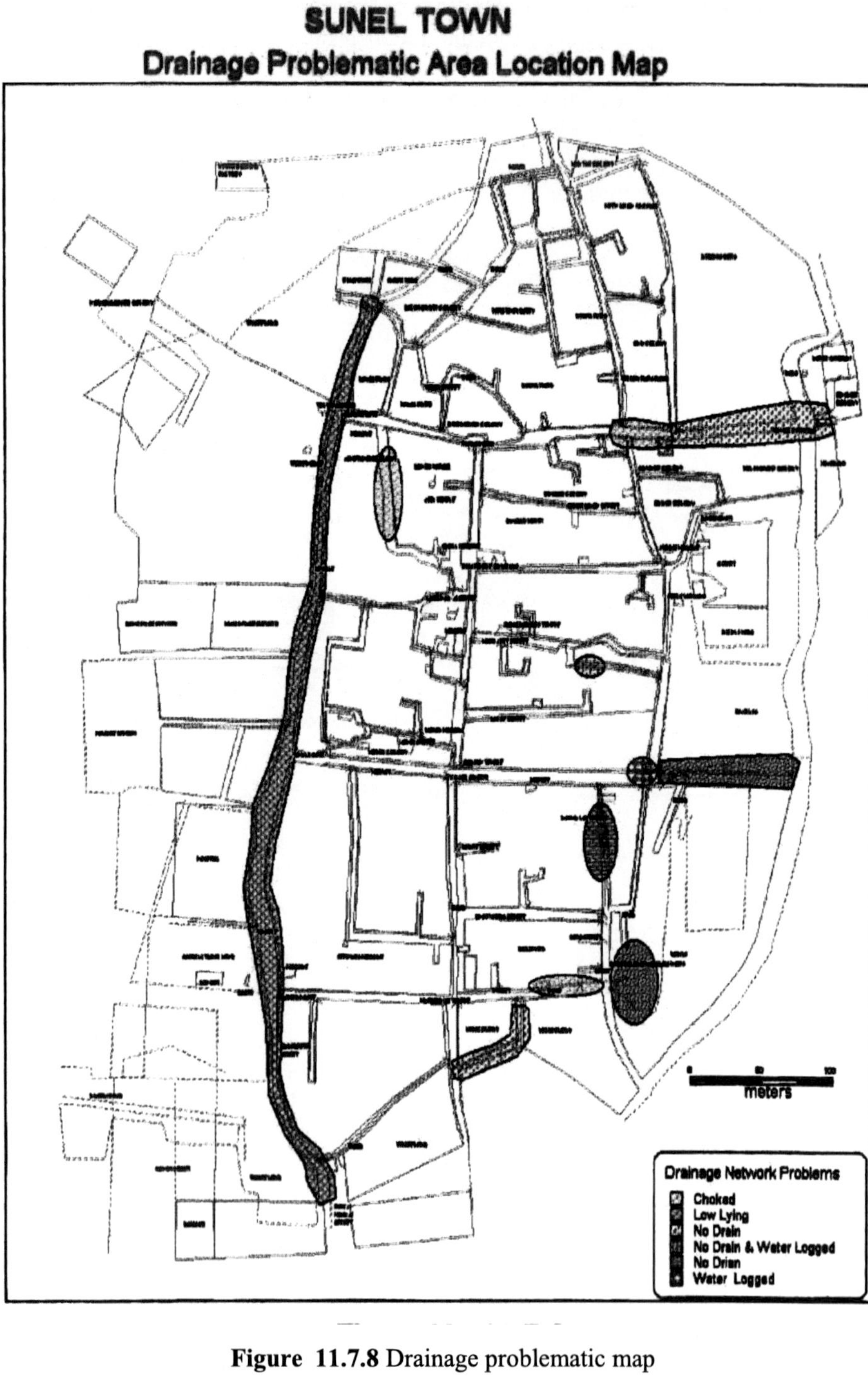

Figure 11.7.8 Drainage problematic map

SUNEL TOWN
Spot Level Map

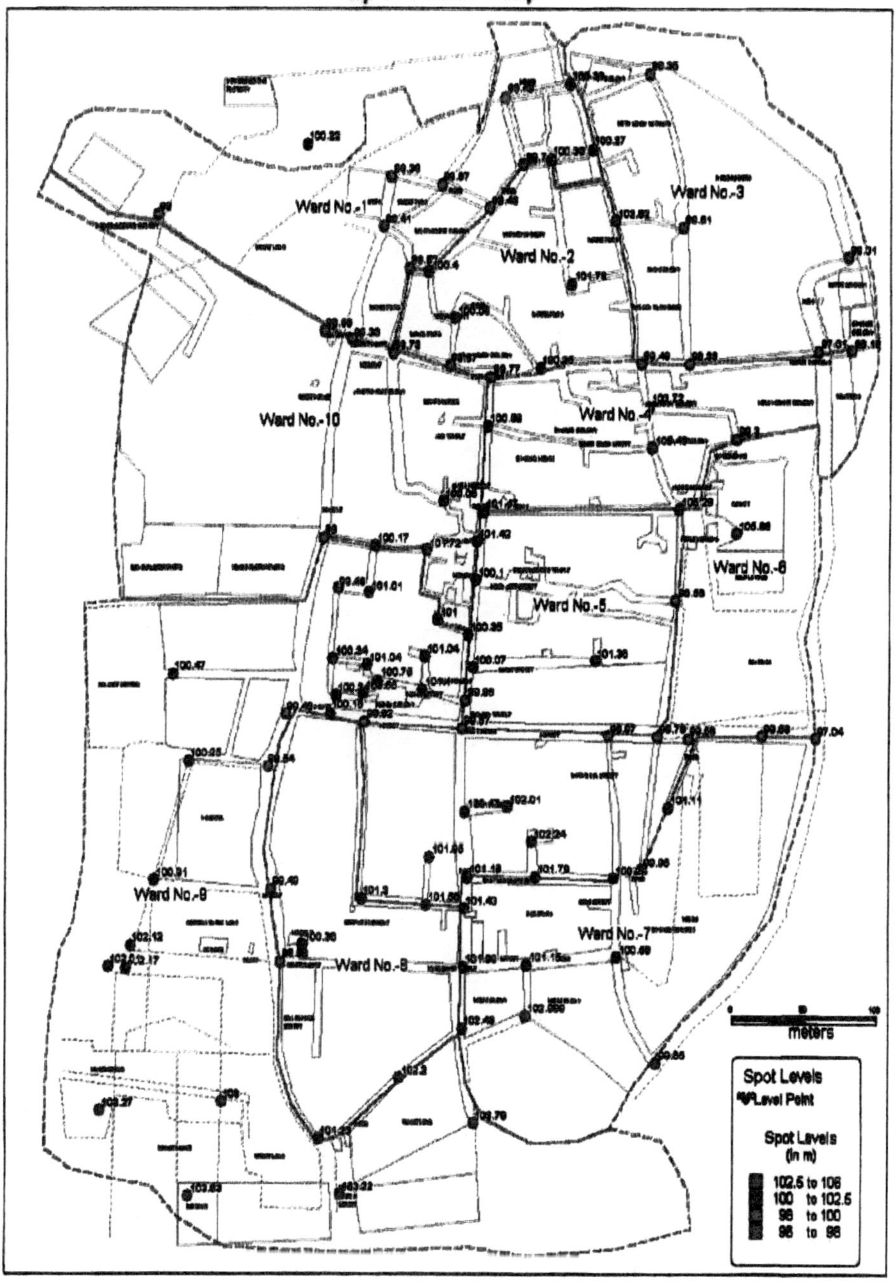

Figure 11.7.9 Spot Level Map

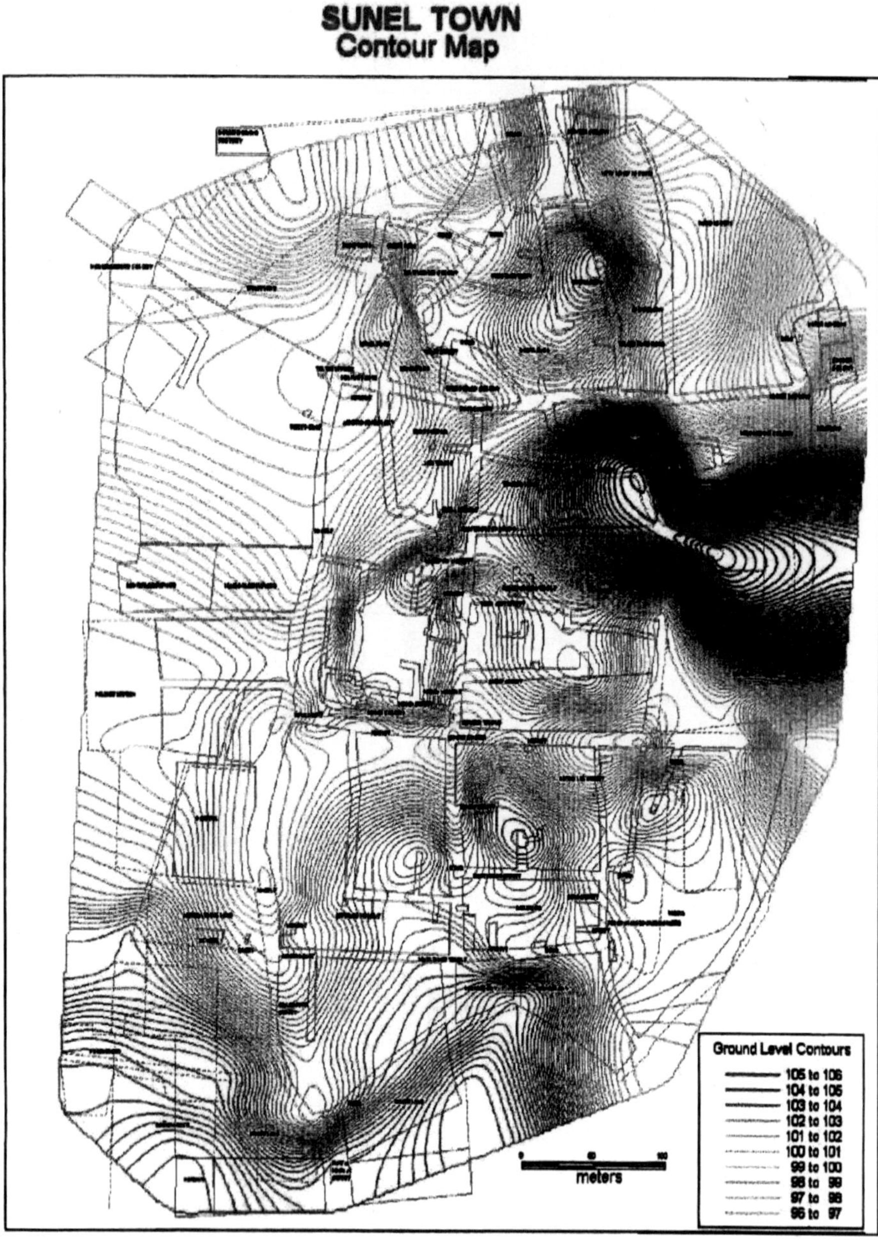

Figure 11.7.10 Contour map

The drainage network analysis as shown in Table 11.7.6 conducted to assess the carrying capacity of existing drainage lines and present discharge using Storm CAD software indicates that carrying capacity of the existing drainage network is 193.87 liter per minute, less than the quantum of sewage produced in the town. The sewage due to this fact either gets flooded on the exit ends or overflows from the choked drains. The situation becomes grave in rainy season as the extra 396.03 liter per minute water is discharged through the drainage network. Various factors are responsible for low carrying capacity of the drains, which ultimately has an implication on flooding on the street in the town. As shown in Figure 11.7.8 the problem of choking and flooding in Pirava road (approximately 1200 mtr), Baribad Street, Shiv Mandir Street and Bohra Mohalla as studyed by the community primarily occurs due to its improper design, inadequate construction (size and slope) and it's covering. The impact of slope on the sewage water flow could be validated from spot level. Kumhar Darwaza, Dhakad Mohalla, Dhakad ka nora, Kachahari Ghati, Vadi Colony (1400 mtrs) localities are either having kacchi drainage lines with no cover or has not been constructed yet, which permits choking and clogging. The sewage in these generates foul smell due to stagnation. The analysis of the modeling exercise as shown in following table indicates that out of 10 wards 5 wards are under size drains, which covers the above-mentioned localities.

Table 11.7.6: Carrying capacity of the drainage lines in each ward

Ward	Pop	HH	Waste Water Generation (Liter/min)	Storm Water flow (Lit/min) in rainy season	Length of Drainage network (m)	Capacity of drainage network (Liter/min)	Remark
1	1749	289	24.29	39.90	1003.3789	25.38	O.K.
2	1240	209	17.22	23.66	1101.6379	12.45	Under size
3	1860	326	25.83	41.70	703.6212	13.25	Under size
4	1686	302	23.42	31.40	1200.4100	9.83	Under size
5	1377	247	19.13	27.93	1402.4830	12.38	Under size
6	1801	290	25.01	45.29	1453.0562	16.35	Under size
7	1243	205	17.26	36.76	643.8601	23.30	O.K.
8	1131	191	15.71	27.33	887.4307	28.00	O.K.
9	1468	263	20.39	78.31	1665.2482	26.68	O.K.
10	1219	238	16.93	43.75	880.2630	26.25	O.K.
Total	14773	2560	205.19	396.03	10941.3892	193.87	

11.7.3.4. SOLID WASTE DISPOSAL

Secondary sources reveal that approximately 50 tons of solid waste is generated daily in Sunel however, the local institutions is able to dispose just half of the total waste twice a day i.e. in the morning and in the afternoon. Panchayat representatives informed that inadequate facilities like tractor, trolleys and inadequate provisioning of human resources are main reason limiting the appropriate lifting of waste within the town and its safe disposal outside. Field observations validates that the garbage is largely disposed in open by the households and hence this waste flushes into open drains and chokes them. Representative of local bodies informed that bins earlier kept in certain critical points have been removed due to their improper management. In some places they still rest however not used by the households. There exist inadequacy in the design of these bins, which limits easy lifting of garbage from them. The major problem of garbage accumulation as marked in Figure 11.7.11 is observed in Balai Mandir, near Panchayat, ram mandir, muslim graveyard, near hospital, kanri street, samaruya gate, kachehri ghati, peepli vraksh areas.

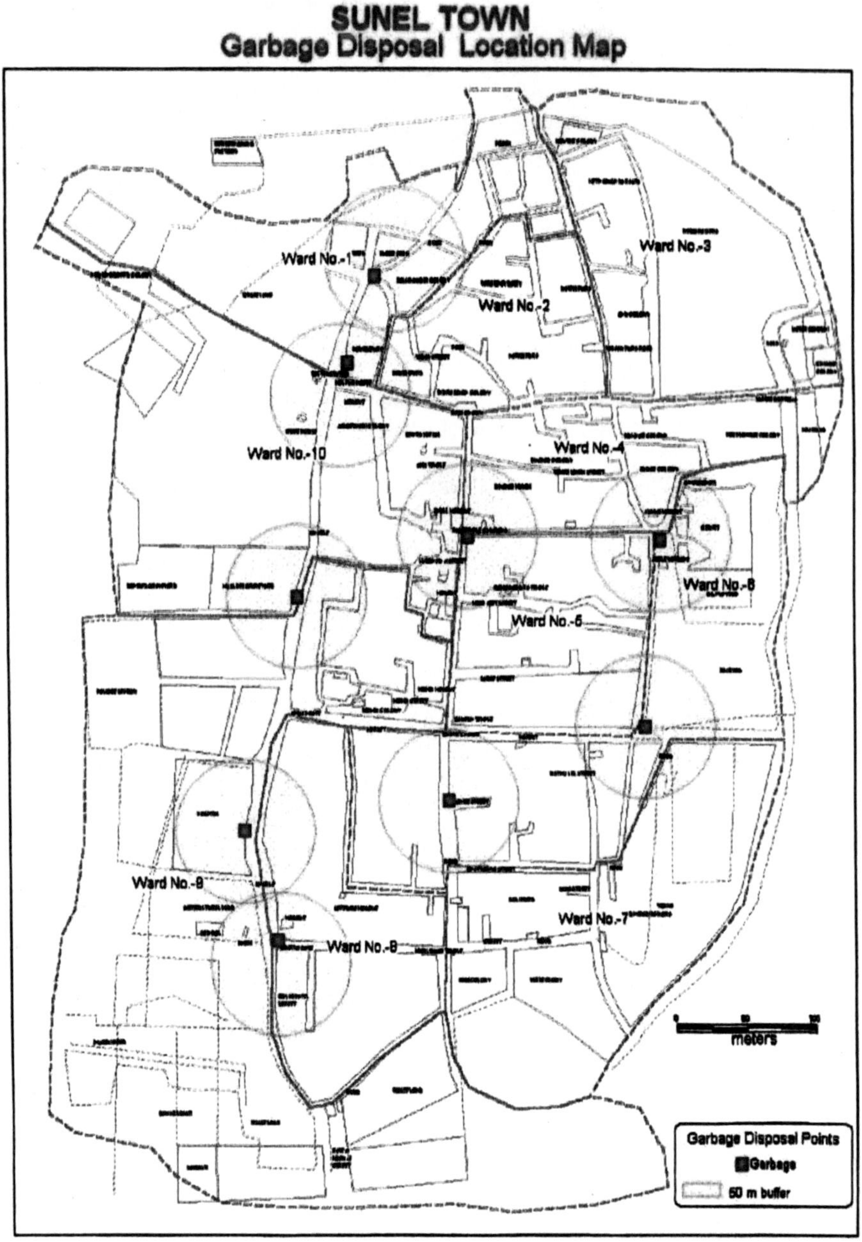

Figure 11.7.11 Garbage disposal Location map

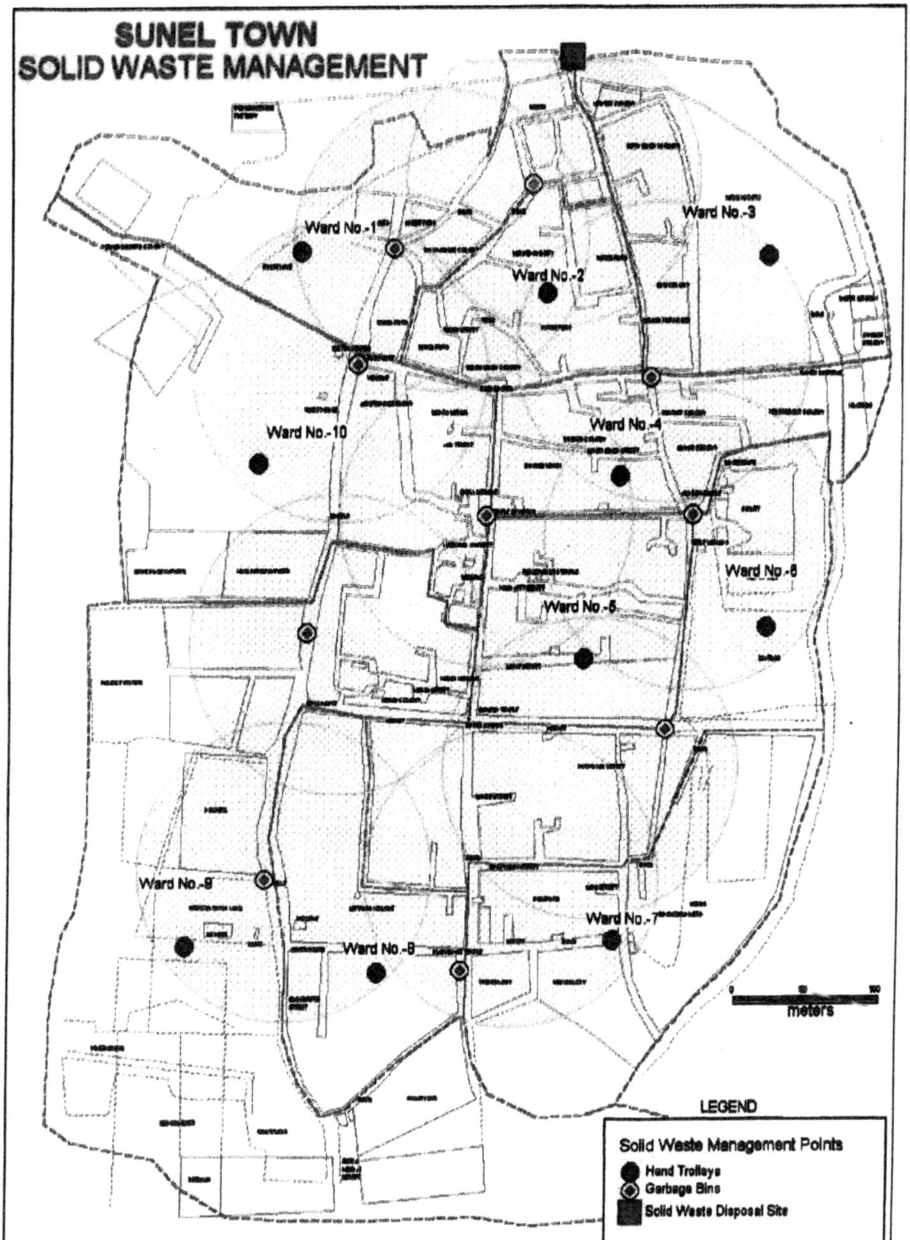

Figure 11.7.12 Solid Waste Management Points.

At present, the solid waste generated in Sunel is disposed in open outside the town. Panchayat has defined no fixed location neither the waste is properly disposed. In some localities, Panchayat is disposing the solid waste in tail ends within the town premises creating unhealthy environment condition for the households living nearby. The study focused on locating the site for safe disposal of the waste after assessing the quantum of waste generated. As per the study analysis 0.21 kg/capita/day, which amounts to total waste of 3.102 tons, is generated in the town from the households. If other waste generation from other sources is also included than the figure may mount to around 50 tons as informed by Panchayat representatives. The location for safe disposal of this quantum of waste thus defined after assessing the availability of land with Panchayat as shown in Figure 11.7.12.

Chapter - 12

SUSTAINABLE DEVELOPMENT PLAN

12.1 GENERAL

Following general conclusion are drawn from the research study:

A comprehensive digital database supported with GIS technology has been generated in the form of Rajasthan Resources Database during the thesis work. Its potential users are not limited to development planers as wide spectrum of data makes it an important reference library for anyone having interest in the state. Researches, inventor, administrators, sales persons or politicians with interest in Rajasthan could get benefit by its use. GIS developers and applications experts will find the database a variety of techniques that have been used in the analysis of data from database. GIS is a powerful tool for the management of spatial data and spatial problem solving. The enormous potential of this tool for spatial planning has, however, hardly been tapped in India. Before the fullest use of this technology as a management and planning tool can be made, a number of supportive steps are required to be taken. These include matters relating to control and distribution of maps by the government and strengthening of routine data collection. Equally important is the need for 'geometrical' definition of guidelines that often have to be used by planners. An important consideration for the future would be to create facilities for training and providing support in promoting GIS.

The SOI still is predominantly a manual map-producing organization, although it is now preparing for electronic production of maps. Clearly, it will take many years before all map features of the entire country can be digitized, but it is time to consider imminent problems that will arise with the change in the medium of map production. With the use of computers in the process of map production, there will be a great deal of flexibility in the manner in which the output will be presented. The use of the conic group of map projections preferred presently by the SOI will almost certainly be replaced by other map projections, partly because of the ease with which projections can be changed within the computer environment. The law requires that all published maps showing International boundaries of in India be certified by the SOI. However, at present the standards for maps certification is maps built on the conic or poly-conic projections and all maps using projections other than these are considered incorrect by the certification authorities. Besides the question of map projection, there is also the question of map generalization. The SOI's standard to for the purpose of certification are those made on the two million scales. In GIS, maps are usually digitized from topographic sheets of 1:250,000 or larger. The maps made from these, when reduced will never have feature that will tally identically with the features of those reduced from maps of smaller scale. Strangely enough although these are commonly known facts, they are not taken into consideration by the SOI when certifying maps for the International boundaries of India. These results in the paradoxical situation that the SOI does not accept reduction form its own large scale maps while insisting on the correctness of only the boundary reduced from a smaller scale map. It is time to resolve these discrepancies and irrationalities before the widespread digitizing of topographic maps takes place.

GIS in Rajasthan is only beginning. Much needs to be done to promote it as a tool for planning. It is important to train personnel. We need easy access to large-scale maps and most importantly, a change in the attitude to the place of data in decision-support. Even a cursory review of the present situation in the state suggests that this is very unlikely to happen in the near future. One of the most disturbing aspects of work relating to secondary data in Rajasthan is that even when data are published, the distributing authority tends to imply that it has the right to withhold them. Quite often letters of authority are demanded for departmental data that should normally be open to all. This shadow of secrecy surrounding government data and the problems encountered in getting them is one reason for data being underutilized by researchers. It also explains why the departments are often unconcerned about improving data quality. Information is quite clearly not a top priority item for most departments. The first step would be to develop skills of GIS operators and educate planners and administrators in the applications of GIS. Short workshops can be organize to show a variety of real life situations from India and other developing countries where GIS have been used to make better decisions that would otherwise have not been possible. For the GIS operators more in depth, short-training courses could be organized. The stress could be on the use of microcomputers that could be serviced locally in all states. It is not necessary to use the most sophisticated GIS software packages. Inexpensive GIS with perhaps only limited functions can be used. Once GIS skills become widespread, then better equipment and more sophisticated software can be used.

A GIS can be effective only when there is good, reliable map data. The next question then is, where do the many digitized files come from. There are several ways of solving this problem. The SOI, which has all the maps, may take on the responsibility, but that would mean delays since the SOI is presently digitizing maps of 1:250,000 scale, and this in itself will take many years to complete. The districts could do their own digitizing from the SOI maps where they can get them, or use the census maps which have very detailed records from their household surveys and village maps. This could mean that each workstation would have to be provided with digitizing facilities. As digitizing maps is laborious and requires a fair degree of skill, the other alternative would be to have one or more digitizing centers, which would supply the districts with their digital maps. This would ensure that GIS users would get the digitized maps that they need without having to wait indefinitely for the SOI to produce them. It would also save them the cost of digitizing hardware and software. There would be saving in terms of man hours needed for digitizing, and editing digitized maps. Digitizing centers are perhaps the best alternative in terms of cost saving on equipment, time saving in acquiring maps and increasing productivity in digitizing since it would be undertaken by skilled people. There are no reasons for not to suggest that a new institution be set up for the propagation of GIS. Institutions interested in planning and administration can take on these functions. But an institution devoted wholly to the study, development and application of GIS may be the answer to attracting long term commitments from GIS vendors, developers and specialists. Such center could bridge the gap between developers and users. Today there seems to be very limited communication between those who develop GIS software and those who may ultimately use them. Evidently the need to maintain secrecy in the developmental stages is a very important factor in explaining the hesitation on the part of developers and vendors in taking potential users into confidence. But it has also resulted in the creation of software that is poorly designed and utterly inadequate. Such a center would also be ideal for rapidly digitizing maps for other users.

There is need for an advisory body to inform potential administrators and operators of GIS on the sources of data, hardware and software in the state. Initially information on capabilities of different software packages for meeting specific requirements will also be needed. Such a body could advice on the probable sources of map data, budgeting and time scheduling for different kinds of work. This body can be formed out of a resources group on GIS. Although this body may not have any regulatory functions, it can still play a very constructive role in promoting GIS in the state.

Need of an Information Policy on GIS

At present there seems to be an urgent need for all those who are interested in the future of the development of GIS in Rajasthan to define clearly the parameters within which one can work. It is crucial time a clear-cut policy be enunciated. Such a policy should consider the fact that regarding spatial information, the most outstanding feature so far has been the great degree of suppression and control exercised by the government. Many of these controls are unnecessary today with the major changes that have taken place in the world on information technology. At present the high cost of hardware and software is a deterrent to widespread use of GIS in Rajasthan. It also seems to be the reason for its gaining ground only in specialized government institutions such as the Indian Space Research Organization and the Space Applications Centers. It is important for the policy makers to appreciate that while this is commendable, there is rear scope for both macro and micro-level physical planning in the GIS environment using map and government departmental data that is already available. Stress should now be laid on developing inexpensive user-friendly software, even perhaps with limited capabilities. This would ensure a larger user community than is presently possible. The development of GRAMS by the Department of Science and Technology is a step towards this end.

Any information policy must recognize the importance of spatial information as an important component. It should address itself to the ease of access and distribution of such information. Besides supporting the adaptation of appropriate technology by GIS users and stress in the importance of training in the field, the government can play an active role in removing discriminating financial burdens suffered by information users. An information policy supporting the use of map data and remotely sensed data can only work to the advantage of manufactures, developers of software and those providing GIS services. It also means that when the planner makes informed decisions, the chances of his making incorrect decisions lessen. The demand for skilled people will also increase if the application of GIS is more widespread. Lastly, and perhaps most important, GIS is the only contemporary tool that is suited to supporting space-related decisions of a large and populous country like ours.

Imminent Impact of Modern Technology

Rapid changes in the technology of data capture from the Earth's surface and the processing of these data in the computer environment may ultimately be the reasons for a change in the national policy towards spatial information. It is widely known that now a days satellites (and not spy satellites only) record data all over the earth (including those areas on map that are not available to the general public in India) at resolutions as high as one meters. These data and images are commercially available (IKONOS data). The unfortunate truth regarding the SOI maps today is that Indians living in India cannot find map necessary for their use, while those outside the country who have any need for maps of the same areas can easily buy them. The encouraging thing is that privately many of those enforcing these restrictions of access to the maps agree that they should no longer be doing so. If there is sufficient demand for digitized map data, it may be expected that there would be companies and individuals who will be in a position to supply them. Then there would be the need to ensure a certain degree of accuracy of the electronic spatial data. Who should be responsible for ensuring that strict standards are observed in digitizing maps? At present the SOI only insists that the international boundaries of India on a paper map should look identical to its standard reference map used for certification.

If the SOI takes upon itself the right to certify for standards then it should be tempered with the responsibility of providing high standard digital cartog-raphic data to the public. Alternatively, digital cartographic data can be generated by anyone with the facilities and the expertise. But these maps need not necessarily have the degree of accuracy that is essential for the purpose of producing topographic maps. Its important to make a distinction between very accurate maps such as those useful for military purpose and those which are less, but equally useful for many other purpose, such as in macro level regional planning. It is quite possible that many agencies such as urban development authorities and environmental planning groups would in the future use Global Positioning System (GPS) to generate spatial data and update maps for their use. There is no record of the use of GPS in India yet, but it appears to be potentially a very suitable tool for use in areas where there is a dearth of maps.

E Governance: To overcome the hurdles as well as to establish sustainable delivery mechanism it is necessary to

Provide adequate funds for applications like SETU / Citizen Facilitation Centre at district level and Taluka level.

It is also necessary to provide funds to establish data centre at district level.

Policy decisions regarding IT systems should not be shrouded by financial dependency.

It is necessary to have a single government portal so citizen / employee / other departments can avail of existing e-governance applications. It is also necessary to build a mechanism for database integration, application integration, process automation and ultimately all G to G integration. This is possible only if Vertical silos of information are not allowed and strong discipline is laid down to other to common architecture, platform, database prescribed by central agency like Bureau of Economics and Statistics. In entire IT system implementation and its success, software / hardware doesn't contribute more than 25% of total effort but what lacks is administrative will and support. Department has to exert substantially for necessary reengineering of the government procedures and systems. This change management has to be methodical, preplanned and not by fluke or coincident.

Availability of clean, useful, up-to-date data obtained from root level is the core necessity of any IT system of government. Therefore it is necessary to have data centers where such huge data is generated. These data centers can be established in most economic manner.

Geoinformatics: Potential of Geo-informatics Technology has been demonstrated in all above areas. However, there is need to spread awareness along with offering customized and turnkey solution in order to ensure successful exploit-ation of technology to make e-Governance sufficiently responsive and accountable. C-DAC along with many national institutions/organizations has taken initiative towards offering solutions of such class.

12.2 CASE STUDY

Conclusions have also been drawn after appropriate analysis in the case studies carried during the research in various parts of the Rajasthan State using different application of the Geographical Information System. Study wise conclusions are presented in the following paragraphs:

Case Study 1: GIS Based Economic Development Levels Assessment Study of Osian – Mathania Tract in Jodhpur District

Apart from climatic conditions prevalent in a region, it is the geomorphic setting that ultimately decides the levels of economic development. Thus Mathania situated in the middle of a trough like basin in rocky desert receives all the runoff water from whatever little precipitation that occurs from the surrounding hills and virtually is perched on a high ground water reservoir. Exploitation of this groundwater for irrigation and consequent raising of double crops particularly cash crops like chilies, cumin and also wheat has ushered in relative economic prosperity to the village Mathania compared with Khudiyala or Osian which are situated in transitional sandy zone and sand dune desert respectively. However, there appears to be over exploitation of ground water in Mathania and this should be checked by optimization of use of water, otherwise the groundwater resources will be depleted in the not distant future. The present pumpage of water from this zone is estimated at 25 million cubic meters which appears to be higher than the anticipated ground water recharge.

The over drafting can be avoided by use of sprinkler and drip irrigation systems for economic utilization of groundwater and yet maximizing crop production. The ground water potential in Khudiyala is meager and hence no wells can be further considered for irrigation. In south and west of Osian there is good ground water storage available. This is recommended for exploitation by way of sinking tube wells, installation of pump sets and their early electrification. The wells should be spaced at least 500 meters apart in order to avoid mutual interference. If this is done, than Osian can also hope to develop on the lines of Mathania but for sustained use the modern methods of sprinkler and drip irrigation should be employed from the beginning itself. Being dune infested area much of land north of Osian, on the basis of landuse capability, should not be put to crop production as annual tillage practices for latter would accentuate wind erosion of soil. These are, therefore, best kept under permanent vegetation cover as pasture land and forestry or what is popularly termed the silvipastoral system. Adoption of such practices would help sustain livestock population very efficiently which is already quite high in Osian.

It is surprising that despite such good population and bread of livestock available in Osian, almost nil population in Osian have taken to animal husbandry as occupation. Studies have clearly shown that the per hectare returns from animal husbandry cum silvi-pasture is much greater than from grain crop production which when pricing the labour cost becomes begets negative returns. Therefore, there is great scope for diversification from predominantly agricultural to animal husbandry occupation for improving the rural economy. Above all rational utilization of any natural resource be it ground water, pasture forestry or mining, is the golden thumb rule for its sustained yield/production over an indefinite period.

Case Study 2: Impact Assessment Study of Rainwater Harvesting Structures in Bikaner District

WHRS activities significantly contribute in improving quality of life of the local community. It also influence the consumption pattern of the individual member in the family and positively impact the nutritional status. The study has estimated the benefits, which matches the cost of construction/ renovation of water harvesting/ recharging structures. As per analysis, the local farmers through increased cop production would yield direct benefit amounting to Rs. 45,93,750.00. The same would apply in case of drinking water and milk production. The study findings suggest that monetary value of increased drinking water availability would amount to Rs 39,76,929.75. Also, the local community would gain in terms of milk production. The monetary value of increased milk production is estimated to be around Rs. 1,58,26,692.00. The overall benefit thus amounts to Rs. 2,43,97,371.75, which is 98.88% of the cost, invested in construction/renovation of WHRS would be generated by the study. As far as indirect benefits are concerned, the increase in storage capacity of the structures would enhance the groundwater recharge. Further it will improve the quality and quantity of water in the adjoining sources. This contribution of the study is significant as 90% of the study villages are having un-potable groundwater.

Case Study 3: Assessment of Potential Recharge Areas in Four Districts of Rajasthan

The Rain Water Harvesting and Groundwater Availability area selection approach should include some practical aspect by field visits and discussion with various government departments already working in the area. As the prime objective of the study was to find out the potential areas with respect to the Groundwater and to carry out labour oriented Water Harvesting works so as to provide the employment thus wages to the drought affected people, other aspects such as agriculture production, domestic water consumption etc. has to be given some sort of priority.

Case Study 4: Mapping of High Risk Behaviour Population Groups in Rajasthan

This exercise was an attempt to assess the vulnerability to HIV/AIDS of the districts in Rajasthan on the basis of attributes of high-risk behaviours and associated environmental conditions. A series of secondary data drawn from different sources was used for analysis. Following conclusions are drawn from the study:

The selection of variables was governed by the assumption that the factors depicting high-risk behaviours are present in every society. It is their intensity and combinations, which makes population groups particularly vulnerable. UNAIDS states that the factors in vulnerability can best be understood in the context of poverty, ignorance, discrimination and principal of human rights. The attributes of risk behaviour and environmental factors considered in this analysis included; demographic, socio-economic, cultural, health, infrastructure and other development, health factors etc. The vulnerability to HIV/AIDS was assessed with the help of a vulnerability index constructed by using several alternative methods. The vulnerability index is used to rank the districts on the basis of their vulnerability to HIV/AIDS.

The secondary data used for analysis were subjected to a lengthy process of sorting, refining and adjustment. The number of variables was short listed on the basis of their correlation with the prevalence of STI among men and inter-correlations among themselves. The number of variables used for final analysis was 25. The alternative methods used for developing vulnerability index included; method of difference, method of standard deviation and GIS mapping. The vulnerability indices were also constructed specific to environmental conditions and priority areas for targeted interventions.

The results obtained showed similarities in the pattern of vulnerability. Different methods produced similar ranking of the districts on the basis of their vulnerability to HIV/AIDS. However, the specific vulnerability index due to environmental conditions and priority areas for targeted interventions showed a mixed pattern of ranking, which was obvious also. On the basis of composite index of vulnerability, districts of Jaipur, Sawai Madhopur, Bharatpur, Udaipur and Tonk were considered as more vulnerable to HIV/AIDS. By the same token, districts of Jhunjhunu, Churu, Jalore, Jaisalmer and Sikar had emerged as less vulnerable to HIV/AIDS. The districts of Banswara, Alwar and Ajmer were also among the higher vulnerability districts.

Vulnerability Index specific to demographic characteristics had placed Banswara, Ganga Nagar, Kota and Sawai Madhopur as more Vulnerable to HIV/AIDS. The economic development parameters had identified Barmer, Jalore, Rajsamand and Sirohi as more vulnerable districts. The health parameters have put Bharatpur, Dungarpur, Karauli and Sawai Madhopur at the top of vulnerability list. The social development factors have indicated Kota, Bikaner, Jodhpur and Hanumangarh as more vulnerable districts. On the basis of socio-cultural and other variables, Jaipur, Jodhpur, Alwar and Udaipur have emerged as more vulnerable districts. An interesting and important feature of this analysis was situational variation in the contribution of different variables in determining the vulnerability to HIV/AIDS. For example, in one situation economic development contributes to impede the vulnerability to HIV/AIDS, in an alternative situation it contributes to the vulnerability.

Vulnerability Indices specific to components of targeted interventions were calculated for rural traditional sex workers, length of the national highways passing through a district, number of reported cases of AIDS and TB, and number of inmates in jail. The analysis shows that with respects to present of traditional sex workers, districts of Alwar, Tonk, Dholpur and Sawai Madhopur are more vulnerable to HIV/AIDS. With respects to the length of the national highways passing through a district, Jaipur, Bikaner, Churu and Jaisalmer have been identified as more vulnerable. In relation to reported cases of HIV/AIDS, Jaipur, Sikar, Udaipur and Ajmer occupies the top positions. The reported cases of TB have placed Jaipur, Udaipur, Bhilwara and Alwar at the top of vulnerability list. Data on number of inmates in jail indicate that Jaipur, Kota, Jodhpur and Udaipur are more vulnerable districts. Vulnerability Indices specific to environmental conditions and components of targeted intervention particularly gave useful insight in the pattern of vulnerability. Firstly, it indicates that the central and southern belts of the state have a higher vulnerability to HIV/AIDS on most of the time. Secondly, it clearly identifies districts for planning targeted intervention specific to different components.

Case Study –5. GIS Based Mapping of Mines and Mineworkers in Rajasthan

From the study following conclusions were drawn:

There are 9103 registered leased mine scattered over the entire geographical area of the state of Rajasthan. Rajsamand district leads the state with highest number of the total registered mines followed by Bundi. Sri Ganganagar and Hanumangarh are having the lowest number of the leased mines. Nagaur district geographically covers the largest leased mining area followed by Udaipur and Baran contributing the lowest mining area. The total registered Minor mineral mines are more that 6 times of the Major mineral mines in the state of Rajasthan.

There are about 43 minerals in Major category and 22 minerals in Minor mineral category presently mined in the state of Rajasthan. Soapstone leads the tally followed by Quartz in terms of largest lease numbers in Major mineral category. Soapstone mineral also leads the state in term of largest geographical lease (mine) area in Major mineral category. In terms of Revenue, Limestone is the largest revenue-generating major mineral followed by Lead Zinc in major minerals in the year 2002-2003.

Ajmer district is having the maximum number of leased mines in major mineral category. Bhilwara district leads in revenue generation from major mineral category. Major mineral category constitutes 14% of total registered leased mine generating 56.5% of the total revenue from all type of mines. Masonry Stone mineral has the largest number of mines followed by Marble in minor mineral category. Sandstone mineral leads the state in term of largest geographical lease area in minor mineral category, Marble is the largest revenue-generating minor mineral in the state of Rajasthan

Bhilwara district leads in total number of leased minor mineral mines. Rajsamand leads in revenue generation from minor mineral category. Minor mineral category comprises 86% of the total leased mines generating 43.5% of the total revenue from all type of mines. Largest numbers of mineworkers are employed in Masonry stone mines followed by Sandstone mines. Highest number of mine workers are employed in Jaisalmer in major leased mines while about 16% mine workers are employed Jodhpur district in Minor leased mines category. Collectively, Jodhpur leads the tally with highest mine workers population.

Population of RTFSW is Maximum in Alwar followed by Tonk district. Population of FSW is maximum in Ajmer followed by Alwar district. Population of Out migrants is maximum in Dungarpur followed by Nagaur district. In-Migrants population is maximum in Sikar followed by Churu district. Truckers population is maximum in Ajmer followed by Jaipur. Street children's population is maximum in Jodhpur followed by Udaipur. Eunuchs are maximum in Ajmer followed by Jaipur. MSM are maximum in Jaipur followed by Ajmer. IVDUs are maximum in Ajmer followed by Bharatpur districts. The Priority rating indicates that Rajsamand is most potential study district for future study followed by Udaipur, Bhilwara, Jodhpur, Nagaur, Jaipur, Bundi, Bharatpur, Ajmer and Jaisalmer.

Case Study –6: GIS Based Water Resource Mapping and Benefit Assessment of Rain Water Harvesting Structures in Three Blocks of Udaipur District

As per the analysis total cost of the structures in all the focus study sites is Rs.16,54,204.00 and the benefited crop production from these structure in one single cycle is Rs. 37,50.075.00 which is more than two times in the single yearly only. The impact of surface water by making them available for drinking to cattle population of the affected cluster of village indicates increase in dairy business. The cattle like buffalo, cow and goats are the main component of the cattle in these selected villages. These cattle's will help in the over all betterment of their owners by producing the milk or by- products like ghee, butter etc. Sine all the construction site are inside the village common land or charnot land hence there will be no difficulties and objection for cattle grazing and drinking. The overall benefits from drinking and dairy business are estimated as Rs. 34,23,066.00, which is about 2 times of the cost input in construction works.

The impact of surface water by making them available for drinking to Human beings and cattle population and increased cropping production of the affected cluster of village gives the maximum direct (estimated) benefit incurred from the surface water use. The optimum benefit figures out to be Rs. 58,82,860.00 which is about 3.5 times the cost of the construction. Total Indirect Benefits from the structures is of Rs. 3,35,222.00 which is about 0.2 time of the cost of construction. Results of the Cost Benefit Analysis indicates that maximum CB ratio is achieved at site Gaonra Pal followed by Khirvania ki nal, Dang, Patoo Khera, Joyra naka, Semliya with minimum of at Hama Talav. The study results indicate that maximum C.B.R. will be generated at sites where agricultural land is more and cost of structure is less with larger storage capacity followed by milk produced by cattle. Increased drinking water availability and quantity results a limited or negligible impact in terms of monetary value on Cost Benefit Ratio.

Case Study –7: GIS Based Mapping of Water Supply System, Drainage and Solid Waste Management System of Sunel Town of Jhalawar District.

As study was focused on mapping of existing basic infrastructure facilities and gaps in availability and demand, following conclusions are drawn:

Piped Water Supply System: In Sunel there exist two categories of households, one which has been provided with piped water by PHED and the other depending alter-native sources like HP, open dug wells etc. Since both of these categories of households have specific problems in meeting their drinking water requirement, the Study needs to address those applying appropriate strategies and implementing specific activities. The focus of the study was on piped water supply, thus the conclusions has been drawn in the same context. As referred in the study, the households depending on piped water supply are facing drinking water shortages due to variety of reasons. This includes low pressure caused by improper sizes of the network pipes (either under or over size pipes), choking/ clogging, less water available for supply at source, leakage in pipeline due to open taps at the consumers end and illegal operation of Tullu pumps/Khechu pumps. The low pressure exists in different localities, however, its severity is primarily noticed in Bus Stand (Tail end), Nawalpura (High ground level) and Kangalipura (Tail end and poor maintenance).

These households in the case of less availability of water are meeting their requirements from alternative sources, which also are fast drying up on account of drought conditions. On the basis of the assessment, the study therefore suggests following interventions from the Study end to address the problem of low pressure thereby improve the supply status.

- Prevent exploitative use of Tullu and Khenchu pumps by way of awareness, enforcing legal provisions and introducing social sanction.
- Develop zonal water supply system for Sunel to address the problem of low-pressure supply. Unrder this entire water supply network should be divided

into smaller zones in which water supply is regulated through sluice valves in rotation basis. As no zoning system exist at present in Sunel town the study suggest Sluice should be broadly placed in school campus, Luhar Darwaja, Chatar Chowk, Court Campus near S.R., Raja Chowk and Kumhar Colony. Since identification of exact numbers and locations of sluice valve requires detailed network analysis, pressure measurement activity at probable low pressure, leakage, clogging should be examined. The data collected would facilitate in locating the specific points for placement of sluice valves for zoning system.

Drainage Network: Study findings indicates that the town lacks proper sewerage system due to which water logging, flooding in choked drains frequently occurs. The Study should therefore, suggest following specific intervention to improve the drainage system.

- Construction/ renovation of V shape drains (1300.00 m)
- Construction of U shape drains (1200.00 m)
- Construction of drain junction chambers including screens, draining bed and sludge removal system (15.00 nos.)
- Covering of u shape drain (1200.00 m)
- Construction of Waste weir channel (100.00 m)
- Construction of Diversion structure (30.00 m)
- Construction of diversion channel (50.00 m)

Solid Waste Management: Any improvement in drainage system should be complimented by system for safe disposal of solid waste. As observed during the study Panchayat lacks adequate bins and trolleys to lift solid waste from the town effectively. Similarly they also not fully capacitated in safe disposal of solid waste. Both of these facts combined with lack of awareness among the masses towards practice of appropriate behaviour deteriorating the sanitation condition of Sunel town. It becomes imperative therefore, those adequate systems/provisions should be developed enabling Panchayat to effectively manage the solid waste that is generated every day in Sunel. The study suggests that Study should provision Panchayat one hand trolley per ward of designed capacity for effective operation of garbage collection from narrow streets. They should be also facilitated in appropriate placement of garbage bins. Capacity building measures should be undertaken for Panchayat representatives for effective management of garbage bins and use of trolleys. Solid waste disposal site located in north near the nallah should also be constructed scientifically enabling Panchayat to dispose off the waste safely and at appropriate location. The efforts will not pay, unless awareness campaign is supplemented with technical interventions. Awareness campaigns should be organised as an ongoing process to propagate messages among the community members on adoption of appropriate sanitation practices.

BIBLIOGRAPHY

Addison, H (1955) Land, Water and Food. Chapman and Hall Ltd., London.

Agrawal, G. C. (1972) Base Maps for Exploitation of Natural Resources. Presented at "Phone-Interpretation Application Seminar" held at Indian Photo-Interpretation Institute, Dehradun.

Alien, S. W. and Leonard, J.W. (1966) Conserving Natural Resources. McGraw-Hill, New York.

All India Soil and Land-use Survey (1990) Watershed Atlas of India, Department of Agriculture and Co-operation. IARI Campus. New Delhi

An Operational frame work for Arid, Semi-Arid and Mountain Areas. Center for Environment Education, Environment and Development Series. C.E.E. Ahmedabad.

Bahuguna, Shiva, Vandana; Buch (1992) Environment Crisis and sustainable development, Dehradun, Natraj Publishers.

Bakewell, D.R. (1971) Maximising our Resources in Resource Management. The Forest Chronicle, Vol. 47(3), U.S.A.

Balakrishnan, P. (1987) Remote Sensing Application in water Resource in India, Proceedings, 8th ACRS, Jakarta, Indonesia.

Barbier, Edward B. (1989) Economics, Natural Resources. Scarcity and Development. Conventional and Alternative Views. Earthscan Publishers, London

Barlowe, R. (1958) Land Resources Economic. Princeton Hall, London.

Binswanger, Hans; Hazell Peter etc. LaToushe, T.D. (1902) Geology of Western Rajputana. Memories of the Geological Survey of India, Vol. 35 (Part- 1), Dehradun.

Biswas, A.K. (1976) Systems Approach in Water Management. Tata McGraw-Hill, Tata Publishing Company, New Delhi.

Biswas, Asit Jellali, Mohammad; Stout, Glenn ed. (1993) Water sustainable development in the 21st Century Water Resources Management Series New Oxford.

Brich, J.W. (1972) Geography and Resource Management. Journal of Environmental Management, U.K.

Bryant, Dirk and Woodward, Richard. Performance in the context of sustainable development World Resources Institute.

Carley, Michael Christie, Ian (1992) Managing sustainable development. London, Earthscan.

Carpenter, R.A. and Harper D.E (1989) "Towards a science of sustainable upland development in developing countries" Environmental Management, 13(1): 43-54.

CAZRI (1977) Summer Institute on Resource Inventory and Land-use Planning, Central Arid Zone Research Institute, Jodhpur.

Census of India (1981) Rajasthan Districts Census (1991) Handbooks.

Center for Development Studies (1999) Population sustainable Development programme Handbook - 1998-99, London.

Central Research Institute for Dryland Agriculture (1990) Field Manual on Watershed Management, CRIDA, Hyderabad.

Chambers, Robert (1987) Sustainable rural livelihoods: a strategy for people, environment and development. Institute of Development Studies, Sussex.

Chattopadhyay, S.K. and H.S. Sharma (1998) Sustainable development: Issues and case studies, concept publishing Company, Delhi.

Chaturvedi, M.C. (1976) Water in Second India-Second India Studies, Ford Foundation Macmilan.

Chouhan, T.S. (1985) Resources for Planning in the Arid Areas of Rajasthan: A case study of Malani Region. Dry-land Resources and Technology, Scientific Publishers (India), Jodhpur.

Chouhan, T.S. (1987) Agricultural Geography: A Study of Rajasthan State, Academic Publishers, Jaipur (India).

Chouhan, T.S. (1988) Integrated area Development of Indian desert Geo-Environ Academia, Jodhpur.

Chouhan, T. S. (1993) Rajasthan Atlas, (Hindi) Vigyan Prakashan, Jodhpur

Chouhan, T.S. (1994) Geography of Rajasthan, Vol. 1 and 2, Vigyan Prakashan, Jodhpur.

Chouhan, T.S. (1994) Natural and Human Resources of Rajasthan, Scientific Publishers Jodhpur.

Conway, Gordon R; Barbier, Edward B. (1990) After the green revolution Sustainable agriculture for development. London, Earthscan.

Daly, H.E. (1983) "Sustainable Development: From concept and theory towards operational principles" Population and Development Review.

Davis, Shelton H; Ebbe, Katrinka ed (1995) Traditional Knowledge and sustainable development Proceedings of a conference held at the Word Bank (Sept. 27-25, 1993). World Bank.

Dhabriya, S.S (1988) Eco.-Crisis in the Aravalli Hills Region, Environmentalist, 1/210, H.I.G. Malviya Nagar, Jaipur

Dhabriya,S.S. (1987) Rajasthan Aravalli Hill Area Development programme, project profile Vol.-I and II, Birla Institute of Scientific Research, Jaipur.

Dixen, J.A. and L.A. Fallon (1989) The concept of sustainability-: Origins, extensions and useful-ness for policy, W.P. No. 1.

Elliott, J.A. (1995) Introduction sustainable development, Order to UBS on 17-1-95.

Fiery, W. (1960) Man Mind and Land- A Theory of Resource Use. Glinecee Illinois, 14-38.

Ford Foundation. (1977) Forestry for sustainable rural development

Geological Survey of India (1976) Atlas of Rajasthan, Geology and Mineral, Jaipur.

Gilbert, A.J. (1991) Modelling for population and sustainable development. Routledge, London,

Ginsburg, N. (1957) Natural Resources and Economic Development. Annals of the Association of American Geographers, Vol. 47, U.S.A.

Golan, Elise H. (1992) Soil conservation and sustainable development in the Sahel International Labour Organisation.

Gopal Iyer, K. (1996) Sustainable development Ecological socio-cultural and dimensions, Vikas, New Delhi.

Gosal, G.S (1987) Water management in arid and semi arid tracts of India. Perspective aspects, In: Regional Imperatives in Utilisation and - Management of Resource India and U.S.S.R. Concept Publishing Company. New Delhi.

Gupta, N.L. et al. (Eds.) (1993) Sustainable Development (2 Vols.) Rawat Publication, Jaipur.

Gupta, N.L. et al. (Eds.) (1994) Resource Management, Rawat Publication, Jaipur.

ICSSR (1984) A Survey of Research in Geography. 1972-75. Concept Publishing Company, New Delhi.

Jodha . N.S. (1995) Sustainable Development in Fragile Environment.

Johnsson, Per-Olov; Lofgres, Kari Gustaf (1985) Economics of Forestry Kari Natural Resources. Basil Blackwell, London.

Joshi. Y.G. Verma D.K. (Ed.) (1998) Social environment sustainable development. Rawat Publication, Jaipur

Karale. R.L. (Ed.) (1993) Natural Resources Management A New Perspective National Natural Resources Management System, ISRO, Bangalore.

Kohli. Anju Shah, Farida Chowdhary, A P (Ed.) (1997) Sustainable development in tribal and backward areas. Indus Pub. Co., New Delhi

Lele, S.M. (1989) Sustainable Development A critical review, "World Development"

Lutz, Ernst (1998) Agriculture and the environment

Mandaal, R.B. (1982) Statistics for Geographers and Social Scientists, *Concept* Publishing Company, New Delhi.

Martin-Brown, Joan (Eds.) Investing in cultural and natural endowments. Pp. 195 World Bank.

Mathur, U.B. (1971) Administrative Atlas Series 18, Rajasthan, Part IX-A, Census of India.

Mathur. C.M (1960) Forest Types of Rajasthan, Forest Department Government of Rajasthan, Bulletin No. 15.

Narayan, L,R,A., D.P. Rao, and N.C. Gautam (1989) Wasteland identification in India using satellite remote sensing data, Int. J. remote sensing 10(1): 93.309.

NATMO (1983) National Atlas of India, Resource Atlas Series, Vol. V National Atlas and Thematic Mapping Organisation, Calcutta.

Negi, B.S. (1980) Geography of Resources, Kedar Nath, Ram Nath, Merrut.

Newson, Malcolm (1993) Land, Water and development River basin systems and their sustainable management. New York, Routledge.

NRSA (1986) Manual of Procedure for Watersheds Mapping using Remote sensing Techniques. National Remote Sensing Agency, Department of Space, Hyderabad.

NRSA (1989) Manual of Nationwide Land Use/Land Cover Mapping using Sensing Agency. National Remote Sensing Agency, Department of Space, Hyderabad.

NRSA (1995) A technical guidelines for Intergrated Mission for Sustainable Development, NRSA Hyderabad.

Osmoston, F.C. (1984) The Management of Forest; International Books Distributioin, Dehradun.

Owen, O.S. (1971) Natural Resource Conservation: An Ecological Approach, McMillan, New York.

Padhauri, M.K. (1983) Conceptualization, Classification and Management of Resources: A Discussion, Environmental Management. The Allahabad Geographical Society, Allahabad.

Pillai, K.M. (1987) Water Management and Planing, Himalayan Publishing House, Bombay.

Ramakrishnan P.S. (1993) Shifting Agriculture and Sustainable Development. Oxford University- Press, New Delhi.

Roonwal, M.L. (Ed.) (1977) The Natural Resources of Rajasthan, Vol. I and II, University of Jodhpur, Jodhpur.

Roy, B.C. (1959) The Economic Geology and Mineral Resources of Rajasthan and Ajmer. Geol. Surv. India.

Serageldin, Ismail; Steer, Andrew (Eds.) (1994) Making Development sustainable: From concepts to action. World Bank Environmental - Sustainable Development Occasional Paper Series No. 2.

Serageldinl Ismail; Barrett, Richard (1996) Ethics and spiritual values promoting environmental -sustainable development World Bank.

Singh, Katar (1993) Free trade, foreign aid and sustainable development. Institute of rural Management Anand.

Singh, Katar and Viswa Bullabh (Ed.) (1996) Cooperative Management of Natural Resources. Sage, New Delhi.

Singh, Naresh Titi, Vangile (Eds.) (1995) Empowerment towards sustainable development. Uk, Zed Books.

Singh, V.P. and Archana Gupta (1996) Integrated evaluation of Natural resources for sustainable development, Subimitted to the NRSA,DOS, Hyaderabad.

Society for International Development (1989) Towards sustainable development Emerging economic scenario and planning in Rajasthan.

Space Application Center (1992) GIS based Information System for Regional Planning- A Case Study of Bharatpur District. Report No. SAC/RSA/NRIS- URI S /P R- 20/- July 1992, ISRO, Ahmedabad.

Space Application Center (1992) Remote Sensing in Sustainable Development- A Case Study in Akole Takuka, Ahmednagar District, SAC/RSA/NRIS-DLP/SNn 10/1992, ISRO, Ahmedabad.

Srivastava, P.R. and Shukla, (Ed.) (1992) Water Resources Planning and Development, Common Wealth, New Delhi.

UNDP (1998) Sustainable balanced development in the Asia Pacific, Korea Development Institute; EDAP Joint Policy Studies.

United Nations Development Programme (1994) Sustainable human development and agriculture.

Verstappen, H. Th. and Van Zuidam (1975) ITC System of Geomorphological Survey- ITC Text Book VII-2, 3rd ed. 52 p.

World Bank (1994) World Bank Group Environment. and the Making sustainable Dvelopment Sustainable.

World Bank (1996) Environmentally Sustainable Development. Participation Sourcebook.

World Bank (1997) Expanding the measure of wealth. Indicators of environmentally. Sustainable development.

Zobler, L. (1962) An Economic-Historical view Natural Resources: Use and Conservation, Economic Geography Vol. 38 USA.

■■■■